Professor Sopchak
Memorial Collection

Fundamentals of HORTICULTURE

Fundamentals of HORTICULTURE

J. B. EDMOND
Professor Emeritus of Horticulture
Mississippi State University

T. L. SENN
Professor of Horticulture
Clemson University

F. S. ANDREWS
Associate Professor Emeritus
of Horticulture
Virginia Polytechnic Institute
and State University

R. G. HALFACRE
Associate Professor of Horticulture
Clemson University

Fourth Edition

McGraw-Hill Book Company
New York St. Louis San Francisco Auckland
Düsseldorf Johannesburg Kuala Lumpur London
Mexico Montreal New Delhi Panama Paris
São Paulo Singapore Sydney Tokyo Toronto

Library of Congress Cataloging in Publication Data
Main entry under title:

Fundamentals of horticulture.

Previous eds. entered under J.B. Edmond
1.Horticulture. I. Edmond, Joseph Bailey,
date Fundamentals of horticulture.
SB91.E3 1975 635 74-20881
ISBN 0-07-018985-4

Fundamentals of HORTICULTURE

67890 DODO 8765

This book was set in Times Roman by Creative Book Services, division of McGregor & Werner, Inc. The editor was William J. Willey; the production supervisor was Judi Frey.

Contents

Preface

In 1936, the first courses in fruit production and vegetable gardening at Clemson Agricultural College were combined into one course entitled "General Horticulture," and a mimeographed text was prepared for the use of the students. A second and revised edition of this text was brought out in 1941, and, following the suggestions and encouragement of several colleagues, the material was prepared as a published text in 1951 under the title *Fundamentals of Horticulture*. Since 1951, *Fundamentals of Horticulture* has been adopted by several junior and senior colleges and universities throughout the country as a basis for the work in the first, or basic, course in horticulture. Thus, the text seems to be meeting a real need and suggests possibilities of going through several more editions.

The purpose of the text is threefold in providing an opportunity for the student (1) to acquire a working knowledge of the fundamentals of crop-plant growth, (2) to develop the ability to apply the fundamentals in solving practical problems, and (3) to obtain an appreciation of the significance of horticulture in human affairs. To meet these objectives, the text discusses the fundamentals, particularly the processes of photosynthesis, respiration, and water absorption and transpiration, and applies the fundamental processes to the growth and development of plants and to the solution of plant-production problems. Thus, the book is primarily a "why" book. As Gardner, Bradford, and Hooker,[1] Gardner,[2] Thompson,[3] and others have pointed out, the primary purpose of a college education is to teach students to think for themselves. For this reason, particular emphasis is placed on teaching students to understand the fundamentals underlying a given practice rather than to acquire proficiency in any given operation.

In common with other life sciences, horticulture is dynamic. As a result of research, new knowledge is leading to a more complete understanding of crop-plant behavior. Accordingly, the fourth edition retains the emphasis on the manufacture and disposition of carbohydrates which was characteristic of previous editions and, in addition, discusses the significance of phytohormones and growth regulators in crop-plant growth and development. Further, the text presents the subject matter from the point of view that the crop plant is essentially a biochemical factory. From their classroom experiences, the authors believe that the presentation of the subject matter from this point of view will lead to a more accurate understanding of crop-plant behavior and crop-plant production problems.

As in previous editions, the text is divided into three parts: Part I, a study of fundamental processes; Part II, the application of fundamental processes to horticulture crop practices; and Part III, a discussion of horticulture crops. The fourth edition, however, differs from previous editions in that the discussions in Part III on disposition of carbohydrates, water, temperature, and light requirements have been transferred to and consolidated with appropriate chapters in Part I; and in like manner, the discussions on propagation, soils, commercial fertilizers, pest control, harvesting, and marketing have been transferred to and consolidated with appropriate chapters in Part II. In this way, considerable duplication has been eliminated, and space has been made available for a discussion on the use of plant-growth regulators—an important feature of modern horticultural crop production, an extension of the chapter on marketing, and the inclusion of two new chapters—Chapter 8, on the development of superior varieties, and Chapter 26, on the scope of and careers in horticulture. Further, three chapters in Part III discuss each of the three main groups: fruits, vegetables, and ornamentals. In other words, *Fundamentals of Horticulture* transcends departmentalization of knowledge within the field of horticulture. Thus, it should appeal to and be equally valuable for students who are interested primarily in the growing of fruits, or vegetables, or ornamentals, or in other phases of crop-plant production.

In conclusion, the authors express appreciation to all colleagues and coworkers who have assisted in any way in the preparation of this edition. In particular, they are indebted to Dr. L. D. Moore of Virginia Polytechnic Institute and State University for making suggestions for the improvement of Chapter 1; to Dr. E. W. McElwee of the University of Florida and Professor F. S. Batson, formerly of Mississippi State University, for assistance in the development of Chapter 23;

[1] V. R. Gardner, F. C. Bradford, and H. D. Hooker, *Orcharding,* New York: McGraw-Hill, 1927.
[2] V. R. Gardner, *Basic horticulture,* New York: Macmillan, 1942.
[3] H. C. Thompson, *Vegetable crops,* 2d ed., New York: McGraw-Hill, 1931.

to the late Dr. J. B. Hester, Elkton, Maryland, for the use of cuts for Figures 10-4 and 13-1; to the several colleagues who provided photographs; and to Professor Alex Laurie, formerly of Ohio State University, Dr. William L. Giles, Dr. Clyde C. Singletary, and Dr. Walter J. Drapala of Mississippi State University for their encouragement. The authors also extend thanks to Mrs. Kathy Anderson, Mrs. Betty Darden, and Mrs. Janice Johnson for assistance in typing the manuscript.

J. B. Edmond
T. L. Senn
F. S. Andrews
R. G. Halfacre

Fundamentals of HORTICULTURE

Part One

Plant Growth and Development

Chapter I

Fundamental Crop-Plant Reactions and Processes

And ye shall know the truth and the truth shall set you free.
John 8:32

ENERGY AND LIVING THINGS

Although man[1] is continously exploring the regions above and below the earth, he is primarily interested in the environment in which nature has placed him, the environment on the surface of the earth. This environment is living and dynamic, and it is crowded with living things—countless lower forms, the bacteria, fungi, nematodes, and protozoa, and billions of higher forms, the many kinds of green plants and animals.

Life cannot exist without a source of energy. Without energy plants could not grow and man could not work. For example, crop plants need the energy of light to make foods and man needs the energy in foods in order to perform his various activities.

[1]The term as used here includes men and women.

Energy may be classified as potential or kinetic. *Potential energy is energy of position or energy in storage*. An example is the potential energy stored in the valency bonds which bind the atoms of a chemical compound together. In sharp contrast, *kinetic energy is energy in motion*. Examples are the kinetic energy of water falling to the bottom of a dam, the flow of molecules of water in an osmotic system, and the flow of electrons in an electric current. In general, potential energy may be transformed into kinetic energy, and kinetic energy may be transformed into potential energy. The amount of energy required for each of these transformations is the same; for example, the amount of energy required to drive the turbines at the bottom of a dam is equal to the amount of energy required to return the water to the top of the dam, or the amount of energy required to combine six molecules of carbon dioxide and six molecules of water in the formation of one molecule of glucose is the same as the amount of energy released in the decomposition of one molecule of glucose. These transformations follow the first law of thermodynamics which states: Energy cannot be created or destroyed; it can only be transformed.[2] In other words, the total amount of energy in the universe is constant. Man has always been primarily interested in transforming this energy from one form to another in order to promote his happiness and welfare.

Energy can also be classified as free, or useful, energy and as useless, or degraded, energy, or entropy. In general, free energy, or useful energy, is available for the use of all living things, whereas useless, or degraded, energy is not available. The relation of free energy to degraded energy, or entropy, is shown in Figure 1-1. The block at the top of the plane possesses a certain amount of potential energy. As it moves down the plane, the block gains kinetic energy at the expense of potential energy. However, when it reaches the bottom of the plane, the block still contains some potential energy. It will lose all of its potential energy only when the plane is infinitely long. Since this does not happen under natural conditions, a certain amount of energy will always be

[2]In the vernacular, this law may be stated: One cannot get something for nothing.

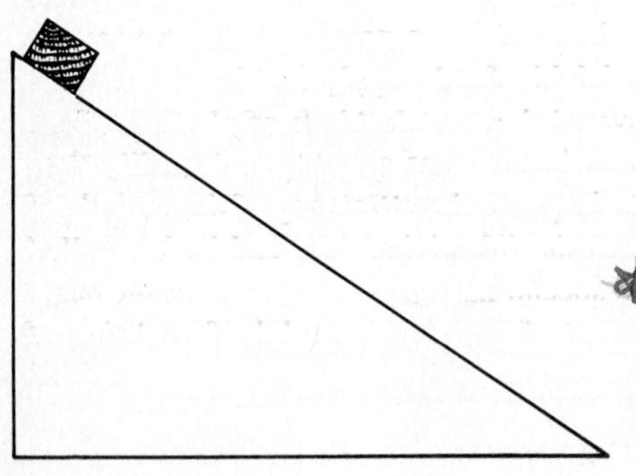

Figure 1-1 The relation of potential energy to kinetic energy. As the block moves down the slope, it loses potential energy and gains kinetic energy. At the bottom of the slope, the block still contains some potential but useless energy.

unavailable to living things and this energy is called *entropy*. This situation illustrates the second law of thermodynamics which states that there are two forms of energy: free, or useful, and useless, or degraded.[3]

Free Energy and Living Things

In general, living things contain two types of cells: living and dead. The living cells must have a constant supply of free energy in order to do their work, in much the same way a steam engine must have a continuous supply of steam or an internal combustion engine must have a continuous supply of gasoline. If steam for the engine or gasoline for the motor was not available, the engine or motor would not be able to operate. In like manner, if a supply of free energy for living cells is not available, the highly organized system within the living cells would break down, the cells would no longer be able to do their work and would die. In other words, free energy and living things and their ability to function go together. This illustrates the third law of thermodynamics which may be stated as follows: The maintenance of any given highly organized system requires a constant supply of free energy.

Autotrophs and Heterotrophs

All living things may be placed in one of two groups: the autotrophs and the heterotrophs. In general, autotrophs have the ability to capture their energy directly as they manufacture substances. These, in turn, comprise the chemomorphs and the photomorphs. The chemomorphs consist of a few species of bacteria and have the ability to transform potential energy in certain inorganic compounds containing sulfur, iron, and nitrogen into the chemical energy in foods, whereas the photomorphs comprise all living things which contain chlorophyll and have the ability to transform the kinetic energy of light into the potential chemical energy in foods and other manufactured compounds.

In sharp contrast to the autotrophs, the heterotrophs do not have the ability to obtain free energy directly. They obtain their free energy from the food which the photomorphs make. Man considers himself as the dominant or most important heterotroph. From time immemorial his food supply has been a serious problem. Over the years, he has discovered that certain kinds of plants produce more food per unit plant or per unit area in terms of quantity and quality than other kinds. Continually, he has tried to improve the efficiency of these kinds of plants by transforming light energy from the life-giving sun, and he has called them *agricultural crops* and given them special names. These crops are the basic elements of agriculture, the source of most of our wealth, and the foundation of our present day civilization.

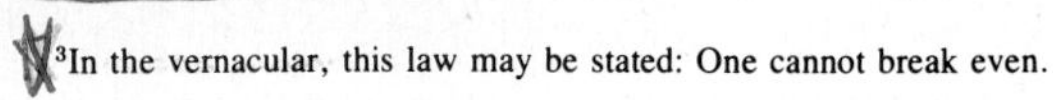

[3]In the vernacular, this law may be stated: One cannot break even.

BASIC BIOCHEMICAL REACTIONS IN CROP PLANTS

Photosynthesis

Photosynthesis is essentially an energy-fixing or light-trapping reaction. As is well known, this energy comes from the sun in the kinetic form and is changed to the potential form found in foods and other manufactured compounds. Thus, although man is dependent on the sun for his existence, he cannot use this energy directly. He can only use the potential energy stored in foods. As pointed out previously, these foods are made by a special group of plants called crop plants. For this reason, crop plants, the only connecting link between ourselves and the life-giving sun, assume great practical significance.

Photosynthesis and the High Energy Phosphate Bond Scientists are gradually unraveling the mysteries of photosynthesis. They have discovered that two remarkable compounds are involved. These are adenosine-diphosphate, abbreviated ADP, and adenosine-triphosphate, abbreviated ATP. There are three important features regarding these compounds: (1) large quantities of free energy are released when the high energy phosphate bond of ATP is broken, (2) this free energy is available for the making of sugars, fats, proteins, and other manufactured compounds, and (3) crop plants have the ability to make ATP from ADP. Note the equation:

$$\text{AD-P} + \text{phosphoric acid} \leftrightarrows \text{AT-P} \sim \text{P}$$[4]

Note that the reactions are reversible with the fixation of free energy in the formation of AT-P ~ P and the release of free energy in the formation of AD-P.

Photosynthesis and Its Light and Dark Reactions Scientists have also discovered that in photosynthesis there are two rather distinct but directly related reactions: the photochemical, or light, reaction, and the thermochemical, or dark, reaction. In the photochemical, chlorophyll absorbs the kinetic energy of light, and this energy is used in breaking the stable bonds between the hydrogen and hydroxyl ions of water. In this way, the lively and unstable hydrogen ions are released, and they combine with an appropriate carrier, usually nicotinamide-adeninedinucleotidephosphate (abbreviated NADP) in the formation of NADPH, the reduced form of NADP, and the release of molecular oxygen. Thus, water is an important compound in the photochemical reaction since it is a primary source of oxygen for all living things. The reaction is set forth as follows:

$$\text{Light} + \text{chlorophyll} + H_2O \rightarrow \overset{+}{H} + \overset{-}{OH}$$

$$\overset{+}{H} + \text{NADP} \rightarrow \text{NADPH} + O_2$$

In the thermochemical reaction, carbon dioxide and the hydrogen of the

[4]The wavy line indicates the presence of a large amount of free energy.

hydrogen carrier react in the formation of three or four simple carbon groups. These, in turn, react with the ADP-ATP cycle to form the initial manufactured compounds, and from these the finished manufactured products are made—the carbohydrates, fats, proteins, and related substances. The thermochemical reactions are set forth as follows:

$$CO_2 + \text{NADPH} + \text{ribulose diphosphate} \rightarrow \text{Triosphosphates}$$

$$\text{Triosphosphates} + \left(\begin{array}{c}\text{ADP}\\ P_1\\ \text{ATP}\end{array}\right) \rightarrow \text{Sugars and other manufactured compounds}$$

Respiration

In sharp contrast to photosynthesis, respiration is essentially an energy-releasing reaction. The potential chemical energy of foods is transformed into various kinds of kinetic energy. In this way, the light-energy reserve built up by crop plants becomes available not only for the crops themselves, but also for all mankind. Thus, along with photosynthesis, respiration serves as the connecting link between mankind and the life-giving sun. Photosynthesis fixes, or stores, the free energy from the sun, whereas respiration releases it.

Respiration and the High Energy Phosphate Bond As with photosynthesis, scientists are constantly working to increase our knowledge of respiration. As with photosynthesis, they have found that the unique and remarkable compound AT-P with the high-energy phosphate bond is necessary, AT-P ~ P. This energy is apparently needed to raise the free-energy level of glucose and other substances decomposed in respiration.

Respiration and Its Two Stages As with photosynthesis, scientists have discovered two rather distinct but closely related reactions in respiration. These are the fermentative and the oxidative. In the fermentative stage, no free oxygen is used, and intermediate compounds are formed, such as two carbon alcohols, three carbon acetones, and three carbon phosphoglycerides, with the liberation of hydrogen and the release of relatively small quantities of free energy for the changing of AD-P to AT-P ~ P. In the oxidative stage, free oxygen is necessary. The intermediate compounds are decomposed to carbon dioxide and water, and large quantities of free energy are liberated for the changing of AD-P to AT-P ~ P. The fate of one molecule of glucose for each of these stages is illustrated as follows:

$$\text{Glucose} + \text{no oxygen} \rightarrow \text{intermediate compounds} + \overset{+}{H}$$

$$\overset{+}{H} + \text{NAD}$$
(nicotinamideadeninedinucleotide)

or FAD ⟶ NADH or FADH
(flavinadeninedinucleotide)

$$\text{Intermediate compounds} + \begin{matrix}\text{NADH}\\ \text{or}\\ \text{FADH}\end{matrix} + O_2 \rightarrow H_2O + CO_2 + \begin{pmatrix}\text{ATP}\\ P_1\\ \text{ADP}\end{pmatrix} + \text{Useful energy}$$

Photosynthesis and Respiration Contrasted

The student will note that, as far as energy relations are concerned, respiration is exactly the reverse of photosynthesis. A contrast between these reactions will help clarify the differences between them. (See Table 1-1)

Total, or Gross, Photosynthesis versus Apparent, or Net, Photosynthesis

The rate of photosynthesis may be determined by using either of these two measurements: the rate of gross, or total, photosynthesis (abbreviated P_g) and the rate of apparent, or net, photosynthesis (abbreviated P_n). If a measure of total, or gross, photosynthesis is desirable or necessary, the rate of respiration is determined during a dark period at the same temperature level as during the light period, and the results are added to the rate of photosynthesis during the day. Thus, $P_g = P_n + R$; transposing $P_n = P_g - R$, since $P_g - R = Y$, it follows that

Table 1-1 Crop Plant Photosynthesis and Crop Plant Respiration Contrasted

Photosynthesis	Respiration
The kinetic energy of light is transformed into the potential chemical energy of foods.	The potential chemical energy of foods is transformed into various forms of kinetic energy.
The hydrogen of water is transferred to carbon dioxide, and the oxygen of water is given off.	The hydrogen of organic compounds is transferred to oxygen, and water is formed.
Carbon dioxide is reduced and water is oxidized.	Carbon compounds are oxidized and oxygen is reduced.
It takes place in chloroplasts only.	It takes place in all living tissues of both plants and animals.
It takes place in light only.	It takes place in light and darkness.
It is an endogonic, energy-absorbing process.	It is an exogonic, energy-releasing process.
It always increases the dry weight of crop plants.	It always decreases the dry weight of crop plants.

$Y = P_n$. In other words, the rate of P_n may be considered as a measure of the marketable yield.

C_3 and C_4 Plants

The terms C_3 and C_4 refer to the number of carbon atoms in the initial, or first, product of photosynthesis. With C_3 plants the first product is C_3-phosphoglycerate, whereas with C_4 plants it is C_4-dicarboxylic acid. Recent investigations have shown marked and significant differences in the anatomy, physiology, and biochemistry between these two groups. Pertinent differences are set forth in Table 1-2. Note that C_4 plants, as compared to C_3 plants, have (1) chloroplasts in both the mesophyll and bundle sheaths, (2) low rates of transpiration, (3) a low P-R compensation range, (4) a low rate of photorespiration, (5) a high light-saturation range, and (6) two pathways for the production of the first product of photosynthesis. In addition, all of these factors combine to promote high rates of P_n, particularly under conditions of high temperature and high light intensity.

The presence of two pathways in C_4 plants is particularly significant. One pathway results in the formation of C_4-dicarboxylic acid and the other, consisting of the decarboxylation of the acid, results in the formation of the reductive pentose-phosphate cycle. In other words, there are two sinks in C_4 plants and only one sink in C_3 plants. According to the law of mass action, two sinks are more efficient than one. Examples of C_3 crops are alfalfa, barley, beet, cabbage, carrot, crotalaria, cucumber, lettuce, mustard, oak, onion, potato, protepea, rice, spinach, snap bean, and wheat. Examples of C_4 crops are coastal bermuda, corn, sorghum, and sugar cane.

Table 1-2 Differential Characteristics of C_3 and C_4 Plants

Major Characteristic	C_3 Plants	C_4 Plants
Tissue with chloroplasts	Mesophyll	Mesophyll and bundle sheaths
Rate of transpiration	High	Low
Optimum temperature range for photosynthesis	15 – 25°C	30 – 40°C
Light intensity saturation foot-candles	3000	6000+
P-R compensation range ppm CO_2	30 – 70	0 – 10
Rate of photorespiration	Measurable	Not measurable
Pathways for formation of initial compound	One	Two
Rate of P_n under conditions of high T and high light intensity	Low	High

Source: Adapted from Table 1. *Adv. in Ecol. Res.* 7:100, 1971.

BASIC BIOPHYSICAL PROCESSES

Diffusion with a Gradient

Diffusion with a gradient, or simple diffusion, involves the movement of ions or molecules of a given substance from regions of high concentration of ions or molecules to regions of low concentration of the same substance. This type of diffusion is analogous to the movement of a ball down a slope. The ball moves down the slope by virtue of its own kinetic energy and, as a result, no external source of free energy is necessary. With simple diffusion, ions and molecules always move from regions of high to regions of low concentration by virtue of their own kinetic energy. Thus, with this type of diffusion, no external source of free energy is necessary. An example of simple diffusion is the movement of molecules of carbon dioxide from the ambient air to the surface of the chloroplasts inside the leaves or green stems. When photosynthesis is taking place, carbon dioxide is constantly being taken out of solution in order to form triose phosphates. As a result, the concentration of carbon dioxide at the surface of the chloroplasts becomes lower than that of the ambient air. Another example is the movement of the molecules of water vapor from the moist cells of the stomatal chambers to the outside air. Usually, the vapor pressure of the water films is relatively high, while that of the ambient air is relatively low. As a result, the molecules of water vapor move from regions of high pressure (high concentration) to regions of low pressure (low concentration). In general, simple diffusion may be defined as *the movement of ions or molecules by virtue of their own kinetic energy from regions of high concentration to regions of low concentration*.

Diffusion Against a Gradient

Diffusion against a gradient involves the movement of ions and molecules of any given substance from regions of low concentration to regions of high concentration of the same substance. In sharp contrast to simple diffusion, this type of diffusion is analogous to the movement of a ball up a slope. In order for the ball to move up the slope, an external source of free energy is necessary. Crop plants have the ability to move certain ions against a diffusion gradient; for example, the potato plant has the ability to move both potassium and magnesium ions against a concentration gradient of these ions. For this movement a source of free energy is necessary, and this free energy comes from compounds made by photosynthesis and decomposed by respiration. Therefore, both photosynthesis and respiration are needed for the uptake of ions or molecules against a diffusion gradient. In general, diffusion against a gradient may be defined as *the movement of ions or molecules from regions of low to regions of high concentration, and for this movement, the free energy of foods is necessary.* (See Fig. 1-2)

Osmosis

Osmosis is a special type of simple diffusion. Suppose we have two adjacent cells, A and B, each of which is equal in volume and is separated by a membrane

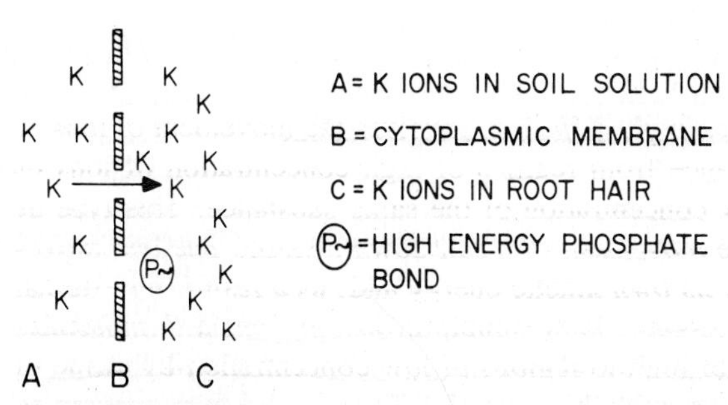

Figure 1-2 Diffusion against a gradient. Note that free energy is required to move the potassium ions from regions of low concentration to regions of high concentration.

which allows only the molecules of water to pass through. Suppose cell A contains water only, and cell B contains both sugar and water. Because of the greater concentration of water in cell A than in cell B the molecules of water will pass from A to B at a greater rate than from B to A, until the rate of diffusion of water molecules is the same in both directions. *This movement of water from regions of higher concentration of water to regions of lower concentration or from regions of lower concentration of solution to regions of higher concentration is called osmosis*. (See Fig. 1-3)

Water enters the plant by means of osmosis. The plasma membranes of the root hairs behave as semipermeable membranes. They allow water molecules to diffuse through them more readily than sugar molecules. Since the concentration of water is less within the cell than in the soil, by virtue of the kinetic energy of its molecules, water passes from soil to cell and from cell to cell until it reaches the conducting tubes—the xylem. Thus, osmosis is simply the passage of water through a semipermeable membrane and the means by which water gets into the plant.

Water Absorption and Transpiration

Water absorption and transpiration are essentially biophysical processes. Water is absorbed in the liquid state in the regions of cell elongation and the root-hair zone, and water is lost in the vapor state from the tissues of plants. This loss, or

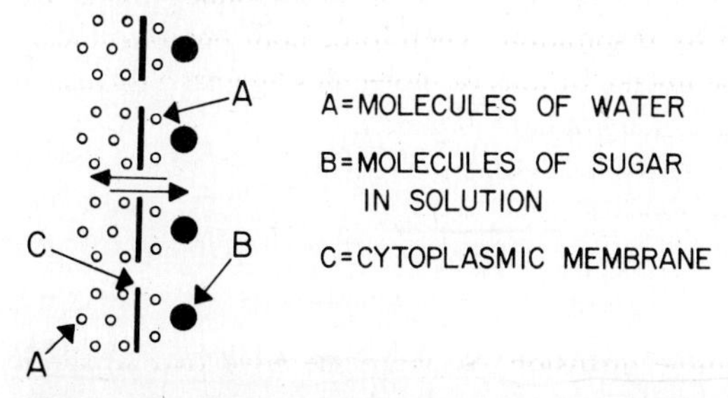

Figure 1-3 Osmosis. The movement of water from regions of high concentration of water to regions of low concentration.

outgo, of water in the form of vapor is called *transpiration*. Although any part of a plant exposed to the air may give off water vapor, the leaves, because of their structure, shape, position, and function, give off the most.

In relation to photosynthesis, a prime function of water within the plant is to maintain turgor within the guard cells of the leaves. The guard cells are in a state of turgor when they are fully stretched. To maintain turgor the amount of water absorbed must, in general, equal the amount of water transpired. When the rate of absorption is much less than the rate of transpiration, the guard cells lose turgor, the stomates partially or completely close, carbon dioxide cannot rapidly diffuse into the leaves, and photosynthesis slows down or entirely stops while respiration continues. As a result, very few carbohydrates are available for growth and development, and growth and yields are low. Thus, when plants are growing in soils with insufficient water, as many of them are for occasional periods in regions of poorly distributed rainfall and lack of irrigation, water absorption and transpiration assume great practical significance.

MANUFACTURED COMPOUNDS

Significance of the Element Carbon

In general, of the 16 elements crop plants need in making compounds, the element carbon seems to be the most important. There are several reasons for this. (1) Carbon has the remarkable ability to combine with itself forming either single or double bonds between adjacent atoms in the creation of compounds. All of these carbon atoms may be in straight chains only or in the form of rings, or some of the carbon atoms may be in straight chains and the remainder in the form of rings. (2) Carbon also distributes each of its four bonds in space so that when four different elements or radicals are attached to each atom, two isomers are formed.[5] One isomer is a mirror image of the other, as the right hand is to the left hand. One isomer turns the plane of polarized light to the right, and the other to the left. Of greater significance in the agricultural industry is the biological activity of these isomers. For example, the glucose known as D glucose, and made by crop plants is dextrorotary, and all the amino acids known as L amino acids and made by crop plants are levorotary, simply because crop plants have the enzymatic machinery to make only the D forms of glucose and the L forms of amino acids.

Compounds Essential to the Life and Well-Being of Crop Plants

Carbohydrates Carbohydrates contain the elements carbon, hydrogen, and oxygen, usually with the carbon atoms combined with hydrogen and oxygen in the same proportion that hydrogen and oxygen exist in water. Examples are

[5] The two isomers are compounds with the same chemical composition, but they have different physical and chemical properties.

$C_6H_{12}O_6$ and $C_{12}H_{22}O_{11}$. As stated previously, of the two isomeric forms, crop plants in general produce the dextroform. Thus, the D forms of glucose assume great biological importance. There are differences in arrangement in the hydrogen and hydroxyl groups in carbon in alpha D glucose and in beta D glucose; the alpha form is the mirror image of the beta form. This slight difference in spatial arrangement determines the fate of each of these forms. Investigators have shown that the alpha form constitutes the building block needed to make the storage forms of carbohydrates—sucrose, dextrin, starch, inulin, and hemicellulose—substances which contain free energy in the potential form. These storage forms are not only necessary for the life and well-being of crop plants, but they also provide mankind with available supplies of free energy in the potential form. For example, the grape stores large quantities of glucose in its fruit; sugar cane and sugar beets store large quantities of sucrose in their stems and roots respectively; the potato stores large quantities of starch in its tubers; and the sweet potato stores large quantities of starch in its roots.

Investigators have also shown that the beta form constitutes the building blocks which make up the structural forms of carbohydrates: the celluloses and the lignocelluloses of the cell walls. These structural forms are not only necessary in building the plant body, but they also provide mankind with the material for making textiles. In fact, certain crops are grown primarily for their cellulose fibers. Examples are cotton, flax, and ramie.

Pigments Pigments are chemical compounds which stimulate the retina of the eye and give the sensation of color. In other words, they provide for distinctive color. In general, there are two kinds: plastid and sap. Plastid pigments occur in the surface of plastids—small blobs of protoplasm within certain cells. Principal plastid pigments are (1) chlorophyll, (2) carotene, (3) xanthophyll, and (4) lycopene. Chlorophyll is the green pigment of higher plants. Actually it consists of two pigments, chlorophyll a and chlorophyll b. In both pigments magnesium is the center of the molecule, and pyrrole rings containing nitrogen are on the outside. Thus, both magnesium and nitrogen are essential for chlorophyll formation. *The function of chlorophyll is the absorption of light for the combination of carbon dioxide and water in the photosynthetic reaction.* In this way, a source of energy is made available for practically all living things. For this reason, chlorophyll is considered to be the most important organic compound known to man.

Carotene and xanthophyll are yellow pigments. Carotene is associated with chlorophyll and thus occurs in all green tissues. In addition, it is stored in the fleshy roots of carrots, in yellow-fleshed varieties of sweet potatoes, turnips, and rutabagas and occurs in tomato fruits and yellow-fleshed varieties of peaches. Since carotene is the precursor, or "mother," substance of vitamin A, the compound necessary for normal vision, it is particularly important in human nutrition. Lycopene is one of the pigments of red varieties of tomatoes and peppers and the pigment of red-fleshed varieties of watermelons and pink-fleshed varieties of grapefruit.

The principal sap pigments are the anthocyanins and the anthoxanthins. The anthocyanins are responsible for the reds, blues, and purples of many flowers, fruits, and vegetables; the anthoxanthins are responsible for the yellows and ivories. These compounds are rather complex in chemical constitution and contain, with other groups, one or two molecules of sugar in their molecular structure; hence sugars are necessary for their formation.

Lipids Lipids comprise oils, fats, waxes, and phospholipins. Oils and fats contain the elements carbon, hydrogen, and oxygen and the structural unit is alpha D glucose. In general, some of these units are changed to unsaturated fatty acids and others are changed to glycerol, a trihydric alcohol. These combine in the formation of fat or oil. Since oils and fats contain relatively large quantities of hydrogen in proportion to oxygen, they contain large quantities of potential free energy. Thus, fats and oils from the standpoint of the crop plant and human welfare are excellent energy-reserve materials. Many crops are grown for their ability to make and store large quantities of oils or fats. Examples are soybean, cotton, and flax for their seed; corn for the young embryos of the seed; and coconut palm for its fruit.

Waxes are also derived from D glucose, but they differ from oils or fats in that (1) they are made by unsaturated fatty acids combining with long chain 24-36 carbon monohydric alcohols, and (2) they are not foods but structural materials for the plant body. In particular, waxes are part of the cuticle on the epidermis of plant leaves, the outer layer of tissues of fleshy roots, and the outer layer of tissues of fleshy structures or fruits, such as the apple and the grape, and storage organs, such as the fleshy roots of the sweet potato.

The phospholipins are similar in chemical makeup to the oils or fats, but they differ from the oils or fats in that (1) at least one of the fatty acids has been replaced by phosphoric acid, and (2) they are not reserve foods as in the case of oils or fats, but like waxes, they are part of the structure of plants. In general, the phospholipins seem to be necessary to maintain the stability of membranes, and at the junction of two substances with unlike properties, oil and water for example, the phosphorus group is hydrophilic and attracted to water, whereas the fatty acid group is hydrophobic and attracted to the fat interface.

Nucleic Acids Nucleic acids consist of two closely related kinds: deoxyribonucleic acid, called DNA and ribonucleic acid, called RNA. The building blocks of each of these two acids contain a molecule of the five-carbon sugar pentose, a molecule of orthophosphoric acid, and a molecule of one of seven nitrogen bases. In general, these components are arranged in the form of a double helix, somewhat like a winding staircase, with the pentose sugar-phosphoric acid group making up the edge, or side, of the stairs and paired nitrogen bases representing the steps. The differences between DNA and RNA pertain to slight differences in degree of oxidation of the pentose sugar, and to the paired nitrogen bases and their respective location and function. The pentose in DNA has one less hydroxyl group than that of RNA, hence the name deoxyribonucleic acid.

The paired nitrogen bases in DNA are adenine-guanine and cytosine-thymine, or methylcytosine, whereas those in RNA are adenine-guanine and cytosine-uracil. The nucleus is the home of DNA, whereas the cytoplasm is the place of operation of RNA. Finally, the molecules of DNA are considered to be the genes of heredity and as such assume the function of genes in crop plant behavior, whereas RNA carries out the instructions from DNA in the formation of enzyme systems and other proteins characteristic of each crop plant species.

Proteins Crop plants make their proteins from amino acids. Amino acids have two active groups: the amino (NH_2) basic group and the carboxylic (COOH) acid group. When any two amino acids react, it is through the amino group of the one and the carboxylic group of the other bringing about the elimination of water. Thus, the formation of proteins in plants is a condensation process.

Most crop plant proteins are made from alpha amino acids; that is, the NH_2 group is always attached to the alpha carbon atom which is next to the carbon atom of the carboxyl group. There are about 22 alpha amino acids in crop plants. Some of the more common are lycine, alanine, valine, leucine, aspartic acid, glutamic acid, methionine, and tryptophan. With 22 amino acids, the formation of a large number of combinations is possible. In general, crop plants make two types of proteins: the dynamic and the static. The dynamic proteins enter into the formation of enzymes and protein compounds of living protoplasm, whereas the static proteins enter into the formation of storage forms, compounds which contain free energy in the potential form.

Nonfoods Nonfoods are compounds which do not store or liberate free energy, as in the case of foods, nor do they furnish substances for the making or repair of living tissues. They do, however, facilitate the fixation or release of free energy in the numerous chemical reactions taking place in crop plants. These are known as enzymes, coenzymes, and phytohormones.

Enzymes In crop plants, thousands of chemical reactions are taking place at the same time. These reactions occur at a relatively low temperature level, and, in a sense, certain substances are necessary to take the place of high temperatures. These substances are the enzymes. In general, enzymes are in the colloidal state, they consist of proteins with or without some other compound, they are effective in minute concentration, and they are specific in action. For example, the carbohydrases work on carbohydrates only, the lipases react with the oils and fats only, and the proteases react with the proteins only. This specificity is related to the configuration of both the enzyme itself and the substance upon which the enzyme works, in much the same way as the indentations of a specific key fit into the lock for which the key is made. In this way, the numerous chemical reactions within crop plants are more or less controlled.

Since enzymes are specific in action, they are classified according to two criteria: (1) the type of compounds upon which they work, and (2) the type of reaction they mediate. Important systems in crop plants are presented in Table 1-3.

Coenzymes Coenzymes are organic compounds which are necessary for the activity of certain enzymes. In other words, the enzyme itself will not operate in the absence of its corresponding coenzyme. Both are necessary. For example, thiamin, called vitamin B_1, is a coenzyme for the enzyme system responsible for the respiration of all living tissues. Vitamin B_1 is made in young, physiologically active leaves and is translocated to all living tissues in both the root system and the stem system. Other vitamins associated with respiration are riboflavin, called vitamin B_2, niacin, and pyridoxine, called vitamin B_6. All the vitamins or their immediate precursors needed for human nutrition are made by plants.

Phytohormones Phytohormones are similar to enzymes in that they are effective in very low concentration, though they differ from the enzymes in that they are less numerous than enzymes and are made in one part of the crop plant and translocated to another. In accordance with our present knowledge, they may be placed in one of two groups: (1) hormones which promote or stimulate crop growth, and (2) hormones which suppress or retard crop growth.

Hormones which promote or stimulate growth and development are the cytokinins, the auxins, and the gibberellins. In general, the cytokinins are made in the region of cell division of the root and are translocated to the region of cell elongation of the stem, where they seem to be necessary for the making of new cells. The auxins are made in the regions of cell division of both root and stem and are translocated basipetally to the region of cell elongation, where they seem to give the cell walls the ability to stretch. Finally, the gibberellins are made in physiologically active leaves and are translocated in the xylem to the regions of cell elongation, where, with the auxins, they facilitate cell elongation.

The interaction of the cytokinins, auxins, and gibberellins—three kinds of phytohormones—is illustrated by the behavior of the potato. This crop plant is grown for its tubers—the fleshy underground stems which store large quantities

Table 1-3 Examples of Enzyme Systems in Crop Plants

Enzyme system	Subsystem	Type of reaction
Hydrolases	Carbohydrases	Starch + water ⟶ glucose
	Lipases	Fats + water ⟶ unsaturated fatty acids and glycerol
	Proteases	Proteins ⟶ polypeptides amino acids
Transferases	Hexokinase	Glucose + AT-P ⟶ P glucose 6 phosphate + AD-P
	Transaminases	Glutamic acid ⟶ alanine
Dehydrogenases		DPN + H ⟶ DPNH
		4DPNH + 0_2 ⟶ 2DPN + $2H_2O$
Oxidases	Catalase	Hydrogen perozide ⟶ water and oxygen
	Tyrosinase	Phenols ⟶ quinols
	Ascorbic acid Oxidase	Ascorbic acid ⟶ dehydroascorbic acid

of starch. If the apices of the stem of a growing plant are removed, the stolons become erect and develop leaves instead of tubers. However if indoleacetic acid, an auxin-type hormone, and gibberellic acid are applied to the stems from which the apices were removed, the stolons develop normally and form tubers. Finally, if a plant is decapitated and grown without roots, stolons develop normally but do not develop into tubers. Thus, the cytokinins made in the roots and the auxins and gibberellins made in the tops interact, or work together, in the development of crops of tubers. Further research may show that these three types of phytohormones are needed for the profitable production of other crops.

Hormones which suppress or retard crop growth are abscisic acid, or dormin, courmarin, phenolic acid, naringenin, and ethylene. Of these compounds, abscisic acid has been investigated the most. In general, abscisic acid is made in physiologically active leaves and translocated in the phloem to the vegetative buds, where it causes the young leaves to develop scalelike leaves for the protection of the growing point during the winter. Further, abscisic acid is an isoprenoid and as such it is closely related to the gibberellins, the compounds which promote cell elongation. In fact, many plant scientists believe that the onset of dormancy is regulated by the balance between gibberellins on the one hand and growth inhibitors on the other.

Growth Regulators Growth regulators are man-made phytohormones. Some have the same chemical structure as natural phytohormones, whereas others are closely related chemically to those natural substances. As with the phytohormones, they are placed in one of two groups: (1) compounds which promote plant development and (2) compounds which retard or inhibit growth and development.

Compounds which promote growth and development are the naturally occurring but synthetically made gibberellins, indoleacetic acid (IAC), indolebutyric acid (IBA), and naphthalene acetic acid (NAA). The effect of gibberellic acid 3 (GA_3) has undergone the most investigation. In general, GA_3 promotes cell elongation by increasing the plasticity of the cell walls and by promoting the growth of the radical in the germination of seeds.

Compounds which suppress or retard growth and development are maleic hydrozide (MH), 2-4 dichlorobenzyltributylphosphonium chloride (Phosphon), 2 chloroethyltrimethylammoniumchloride (Cycocel), N dimethylaminosuccinamic acid (B-nine), and N^6-denzyladenine. In general, both the stimulatory and the suppressive types of synthetic compounds are widely used in many horticultural industries throughout the world. Their nature and application are discussed in appropriate chapters in Parts 1 and 3.

Compounds Which Have No Known Function in the Development of Crop Plants but are Important in Human Affairs

In general, these compounds may be placed in one of four groups: (1) alkaloids, (2) tannins, (3) essential oils, balsam, and resins, and (4) latex.

Alkaloids may be categorized as one of two groups: (1) mildly physiologic stimulants, or (2) drugs. Examples of the first group are caffeine in the leaves of tea and the seed of coffee, theobromine in the seed of cacao, and nicotine in the leaves of tobacco. Examples of the second group are morphine from the opium poppy, quinine from the bark of cinchona, colchicine from meadow saffron, and reserpine from snakeroot. Tannins are used in small quantities to improve the flavor of tea, coffee, cider, and beer and in large quantities to change animal hides to leather. Important sources are leaves of Sicilian sumac, leaf bolls of Chinese sumac, bark of red mangrove and oak, and the hardwood of quebrachos. Essential oils are used in the making of incense, perfume, and flavoring substances; balsam and resins are used in the making of turpentine, varnish, lacquers, and paints. Important sources of the former are the petals of certain flowering crops, for example, rose and violet, the leaves of certain herbaceous crops, for example, spearmint and peppermint, and the rind of the fruit of the citrus family, for example, the rind of a lemon. Important sources of balsam and resins are the exudates of the loblolly and the slash pine. Latex often contains rubber; rubber, in turn, consists of a large number of isoprene units (500 to 5,000) linked together in long unbranched chains. A principal source of supply is the South American rubber tree, *Hevea braziliensis L.*

The Crop Plant—A Biochemical Factory

As previously stated, the initial food substances are made in the green tissues, usually the leaves. From these substances a wide variety of other compounds are made. These compounds may be made in the leaves or in many other parts of the plant. Two examples should suffice. (1) When the meristem of the roots or stems is dividing, it is making protoplasm. Protoplasm requires the formation of proteins, and proteins, in turn, require the combination of certain sugars and certain compounds containing nitrogen. The sugars are made in the leaves and translocated to the meristem of the roots or stems, where they combine with compounds containing nitrogen in the formation of proteins. (2) When the tubers of the potato are developing, large quantities of starch are stored in the tissues. This starch is made in the tubers from an initial food substance called *glucose*. Here again, the glucose is made in the leaves, where it is changed to sucrose, and sucrose, in turn, is translocated to the storage tissue of the tubers, where it is changed to starch. The same situation exists in the formation of many other compounds, such as pigments, fats, cellulose, and compound celluloses. Thus, two facts should be kept in mind: (1) the initial food substances are made in the leaves; and (2) from these substances, a wide variety of other compounds are made not only in the leaves, but also in other parts of the plant. For these reasons the entire crop plant should be considered as a highly organized, well-integrated biochemical factory.

The Diurnal Cycle and the Carbohydrates

The diurnal cycle refers to the rate of photosynthesis, the rate of respiration, and the amount of carbohydrates available for growth and development during any given 24-hour period. The effect of three possible cases is presented in Figure 1-4. Note that in all three cases the rate of photosynthesis is lower than the rate of respiration just before sunrise, that the rate of photosynthesis equals the rate of respiration at sunrise or immediately thereafter. This is called the *compensation point,* that is that point at which the amount of carbohydrates made equals the amount used and that there are high rates of photosynthesis in A, low rates in B, and very low rates in C. Thus, abundant carbohydrates are available for growth and development in condition A, small quantities in B, and very few, if any, are available in C. Further, with abnormally high rates of respiration, in each of these three cases, there will be correspondingly lower amounts of carbohydrates available for growth. Thus, the amount of carbohydrates available for growth and development assume great practical significance. In general, when the rate of photosynthesis equals the rate of respiration and when the crop plant uses all of the carbohydrates it has previously stored, a condition eventually takes place known as *carbohydrate exhaustion*, or *starvation*. Gardner states, "Carbohydrate starvation is probably 10 times as common as potash starvation, 100 times as common as calcium starvation, and 1,000 times as common as boron starvation. Common symptoms are slow growth of roots, stems, leaves, flowers, and fruits, and such storage organs as tubers, bulbs, corms, and fleshy roots."[6] In general, this explains why seedlings growing under crowded conditions are weak and spindly, why young trees growing in the forest floor frequently lack vigor, why

[6] V. R. Gardner, *Principles of horticultural production,* East Lansing, Mich.: Michigan State University Press, chap. XII, 1966.

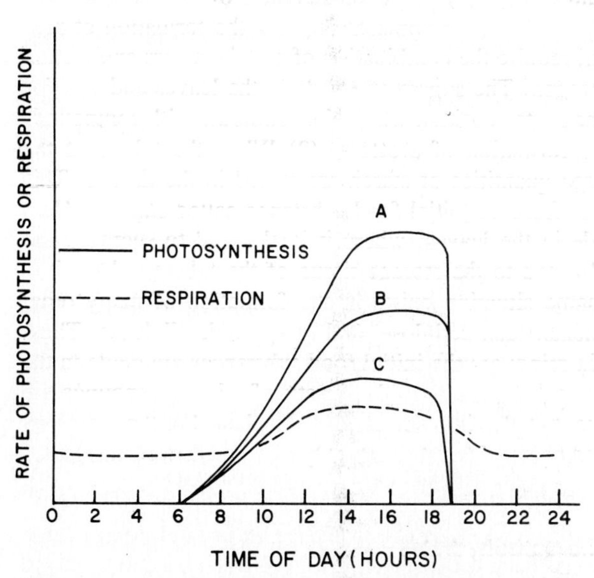

Figure 1-4 Diurnal changes in photosynthesis and respiration with high, moderately high, and low rates of photosynthesis.

overplucking the leaves of the tea plant eventually kills the plant, and why overtapping the gum arabic tree shortens the life of the tree. Many other examples could be cited.

QUESTIONS

1 Life cannot exist without a source of energy. Explain.
2 Differentiate between potential energy and kinetic energy.
3 How does the statement "You cannot get something for nothing" conform to the first law of thermodynamics?
4 Differentiate between useful energy and useless or degraded energy. What is entropy?
5 How does the statement "You cannot break even" conform to the second law of thermodynamics?
6 The crop plant binds all mankind and all animals to the life-giving sun. Explain.
7 What is meant by the statement "We are all a little bit of the sun"? Explain.
8 Describe the environmental conditions under which the light reaction is the limiting factor in photosynthesis? The dark reaction?
9 What is phosphorilation? Discuss briefly the place of the energy-rich phosphate bond in human affairs.
10 Set up the overall equation for photosynthesis.
11 What is the function of chlorophyll?
12 In general, a plant with a large quantity of chlorophyll (large leaf area) makes more food in a given time than a plant with a small quantity of chlorophyll (small leaf area). Explain.
13 Disease infection and insect infestation of the leaves reduce photosynthesis, growth, and yield. Explain.
14 What is meant by the slogans: "A blanket of green for South Carolina," "Every acre continuously at work in Mississippi," "Keep North Carolina clean and green"?
15 Set up the equation for respiration.
16 Contrast photosynthesis and respiration from the point of view of (1) seat of operation, (2) time of operation, and (3) weight and energy relations.
17 The dry weight of all green plants decreases at night. The dry weight of deciduous perennials decreases during the winter. Explain.
18 What is wrong with this statement: "Plants give off oxygen only in the light and carbon dioxide only in the dark"?
19 Show the relation of the rate of photosynthesis and the rate of respiration to the yield of crop plants.
20 When the stomates are closed during the day, the green plant is not making the initial food substances. Explain.
21 Differentiate: diffusion with a gradient and diffusion against a gradient.
22 What is osmosis?
23 Name six groups of compounds which are essential to the life and well-being of crop plants and three groups which are not essential.

SELECTED REFERENCES FOR FURTHER STUDY

Black, C. C. 1971. Ecological implication of dividing plants into groups with distinct photosynthetic production capacities. *Adv. Ecol. Res.* 7:87–114. A discussion on the discovery and characteristics of high and low, photosynthetic-capacity crop and weed plants, including 96 citations.

Gardner, V. R. 1966. *Principles of horticultural production*. East Lansing, Mich.: Michigan State University Press, Chap. XI. A discussion of the manufactured compounds made by crop plants by an outstanding teacher of horticulture.

Lehninger, A. L. 1961. How cells transform energy. *Sci. Amer.* 205:63–73. An application of the first and second laws of thermodynamics and the role of the AD-P ↔ AT-P cycle in plant and animal life.

Van Overbeck, J. 1966. Plant hormones and regulators. *Sci.* 152:721–731. A review of research in plant hormones and growth regulators during the past 40 years—1926 to 1966—by an eminent authority in the field, which includes 136 citations.

Chapter 2

Crop-Plant Anatomy and Morphology

The crop plant is the foundation of agriculture–the basic and only essential industry.

CELL TYPES

Crop plants are composed of cells. These cells may be divided into two great groups: living and dead. Living cells, in turn, may be divided into two significant groups: cells which contain chlorophyll and cells which do not contain this important pigment.

Cells with Chlorophyll

The pigment chlorophyll exists within the chloroplasts. Studies with the light microscope show that the chloroplasts arrange themselves in a definite order with respect to location and alignment. In general, they are embedded in the cytoplasm next to the inner walls and spaced from each other at definite intervals, presumably to increase their efficiency in trapping the light energy from the sun.

Note the position and regularity of spacing of the chloroplasts in Figure 2-1. Studies with the electron microscope show that within an individual chloroplast there are alternating molecular thick layers of chlorophyll and proteinaceous materials. Within these layers, photosynthesis takes place. Chlorophyll absorbs the kinetic energy of light. This absorbed energy breaks the bonds between the hydrogen and hydroxyl ions of water, enzymes facilitate the breaking of the bonds, and coenzymes combine temporarily with the unstable hydrogen which is incorporated with carbon dioxide in the formation of simple sugars, the initial manufactured substances. As explained in Chapter 1, these substances are the precursors of all manufactured compounds made by crop plants. Thus, the tissues containing the chloroplasts are the connecting link between mankind and their crop plants and animals and the life-giving sun. For this reason, the green tissues of crop plants should be considered the most important tissues on the surface of the earth.

Cells With or Without Chlorophyll

Studies with the light microscope show that most living cells contain a nucleus, and some of them contain nongreen plastids, the chromoplasts and the leucoplasts. The nucleus contains the chromosomes which, in turn, contain genes. The chromoplasts contain carotene or xanthophyll, and the colorless leucoplasts store starch. Further, studies with the electron microscope show that all living cells, both with and without chlorophyll, contain definite entities within the cytoplasm called *organelles*. These are the mitochondria, the endoplasmic reticulum, the ribosomes, the Golgi bodies, and the vacuoles. In general, an individual mitochondrion is the site of the oxidative phase of respiration. Electrons or hydrogen ions are removed from partially decomposed carbohydrates, fats, or proteins on an assembly line of coenzymes, and finally the hydrogen combines with oxygen in the formation of water. The biologic, or useful, energy is used to change ADP to ATP, the compound with two high energy phosphate bonds. The endoplasmic reticulum consists of a series of parallel cytoplasmic

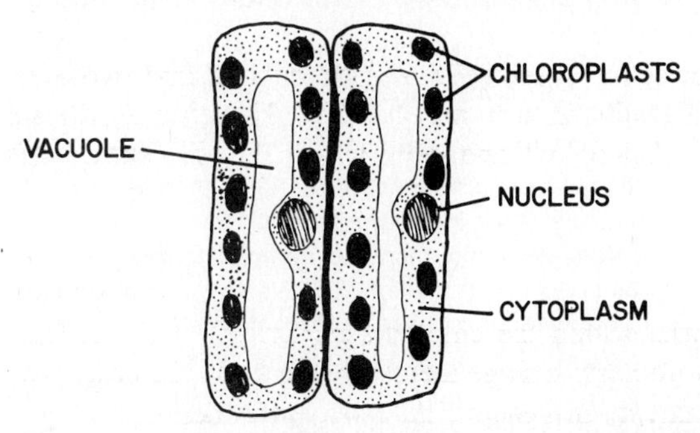

Figure 2-1 The arrangement of chloroplasts in a typical manufacturing cell.

strands which contain in a definite linear fashion the ribosomes, the organelles in which enzymes and other proteins are formed. The Golgi bodies are parallel strands of dense cytoplasm consisting largely of emulsified fat which seem to be necessary for the formation of membranes. Finally, the vacuoles are sites for the income and outgo of substances in solution with the plants sap; examples are the ions of essential elements and the molecules of simple sugars and amino acids. (See Table 2-1) Thus, the living cell with its various organelles is a highly organized and coordinated system for the transformation of free energy. The cells with chlorophyll capture the kinetic energy of light and transform it into the chemical energy of foods, and the cells with or without chlorophyll transform the potential energy of foods into various forms of kinetic energy. Note the relative size and location of the organelles shown in Figure 2-2.

Dead Cells

Dead cells develop from living cells, usually from parenchyma. In general, the walls of the living parenchyma become thick, hard, and rigid, and the nucleus and protoplasm finally disappear. There are four types: (1) stone cells, (2) fiber-like cells, (3) xylem vessels or tracheids and (4) cork or bark.

Stone cells are irregular in shape and thick-walled, with exceedingly small cell cavities. They occur mostly in fruits and seeds, e.g., the grit cells in the fruit of certain varieties of pear and the shells of certain nut fruits—pecan, walnut, almond, and coconut. *Fiberlike cells* are extremely long, thick-walled, and tapering. The tapered ends dovetail with each other so that the plant can withstand stresses and strains, as, for example, the forces of the wind. These fibers may occur, as necessity requires, in the cortex, pericycle, phloem, or xylem of stems and roots. *Xylem vessels* are long and pipelike in appearance, and the walls are strengthened according to various patterns and designs. Figure 2-3 shows the three types of sclerenchyma.

Table 2-1 The Organelles of Living Cells of Crop Plants and the Function of Each

Organelle	Function
Nucleus	Direction of biochemical activities
Chloroplasts	Manufacture of the initial compounds
Leucoplasts	Temporary storage of manufactured compounds
Mitochondria	Source of energy or seat of oxidative phase of respiration
Lysosome	Seat of fermentative phase of respiration
Endoplasmic reticulum with ribosomes	Manufacture of enzymes and other proteins
Vacuole	Storage of soluble foods and essential ions

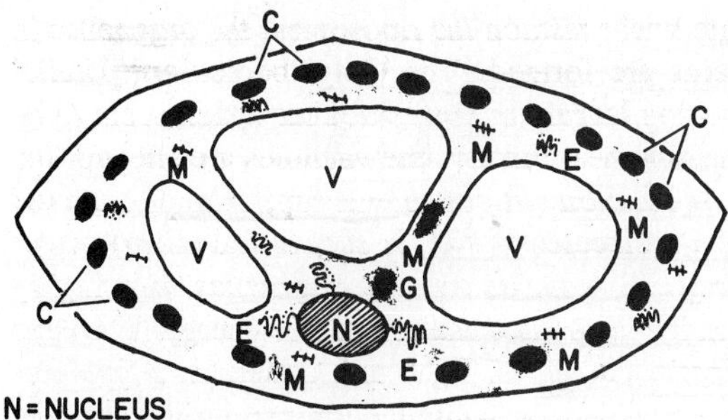

Figure 2-2 The organelles of a typical manufacturing cell.

Woody plants, the old stems of herbaceous plants, and certain storage organs are covered with a tissue called *cork* or *bark*. This issue takes the place of the epidermis and is formed by the division of the cork cambium. Individual cells are rectangular, with no intercellular spaces. The walls are relatively thin and are impregnated with a fatty-like substance called suberin and with tannin and other substances. The suberin makes the tissue impermeable to water and gases, and the tannin renders the tissue resistant to the attacks of rot-producing organisms. Thus, the function of the cork layer is to protect the stem or root from excessive loss of water and against the attack of rot-producing organisms. Figure 2-4 shows the replacement of the epidermis by a layer of cork.

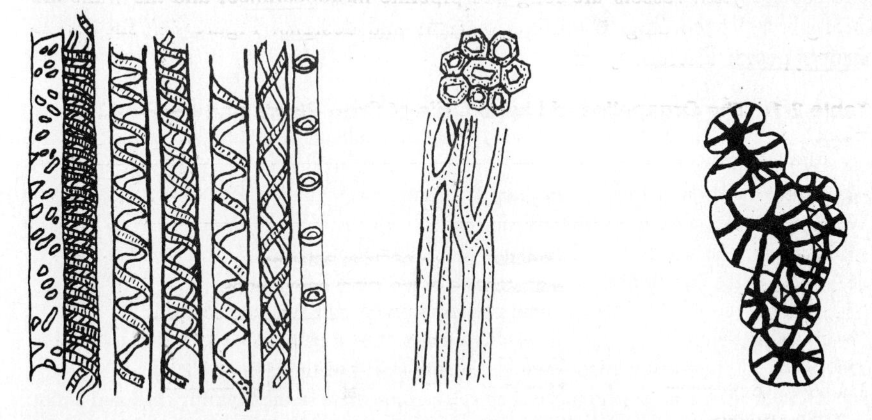

Figure 2-3 Types of sclerenchyma. Left: xylem vessels. Center: fiber-like cells. Right: stone cells.

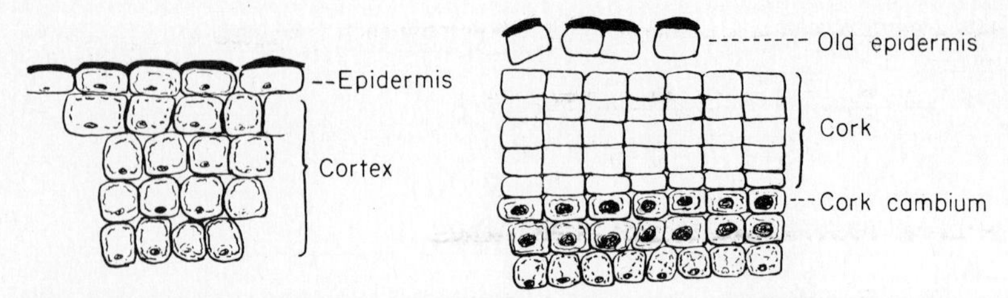

Figure 2-4 Cross section of primary stem and secondary stem showing replacement of epidermis by layer of cork cells.

Cells and Tissue System

Cells do not work alone; they work together in groups. For example, the group of cells at the growing points of the roots and stems has the common function of cell division, and the group of cells which contain chlorophyll has the common function of the manufacture of initial food substances. Groups of cells which have a common function are called *tissues*. Tissues, in turn, join with adjacent tissues to promote the work of any given structure or organ. For example, the tissues of the root combine to promote the functions of the root. In like manner, the tissues of the stem, leaf, flower, or fruit combine to promote the functions of the stem, leaf, flower, or fruit respectively. Thus, the crop plant consists of tissue systems which combine to promote the work of the plant as a whole. In other words, the various tissue systems are interdependent. For example, the chlorenchyma, the food-manufacturing tissue, depend on the water-absorbing tissue and the xylem for their water supply. The water-absorbing tissue, however, depend on the chlorenchyma and the phloem for their sugar supply. Without the sugars the water-absorbing tissue would not be able to absorb water. Can you think of any other examples of interdependence?

QUESTIONS

1. Do cells work alone? How do they work? Explain fully, giving an example.
2. What is a tissue?
3. Name the principal tissues in plants.
4. Draw a meristematic cell. Label all parts.
5. Why are the walls of meristematic cells thin and pliable?
6. What is the function of the meristem?
7. What is the fundamental process concerned with meristem?
8. How do parenchyma cells differ from meristematic cells?
9. Name the types of parenchyma tissues and state the function of each.
10. Draw from memory an epidermal cell, a carbohydrate-manufacturing cell, and a root hair. Label all parts. Show how the size, shape, and structure of each cell is adapted to its particular function.

11 Draw a pipelike cell and a fiberlike cell. Show how each is adapted to its particular function.
12 What is the function of the phloem? The xylem?
13 Photosynthesis is necessary for the absorption of water. Explain.

CROP-PLANT PARTS AND THEIR FUNCTIONS

Most horticultural plants are spermatophytes; that is, they produce pollen, which eventually contains sperms, and ovules, which eventually contain, with other cells, an egg. When a sperm and an egg unite, a new individual is formed which, together with nourishing and protecting tissues, constitutes the seed. In other words, spermatophytes produce seed.

Spermatophytes are divided into two groups: gymnosperms and angiosperms. In general, gymnosperms produce neither flowers nor fruit and hence nonenclosed or naked seed, e.g., pine, spruce, fir, yew, hemlock, juniper, and sequoia. On the other hand, angiosperms, with a few exceptions, produce flowers and fruit and hence enclosed seed. They comprise a large group and are divided into two subgroups: monocots and dicots. In general, monocots develop one seed leaf, parallel-veined leaves, closed vascular bundles, and flower parts in groups of three or in multiples of three; whereas dicots develop two seed leaves, net-veined leaves, open vascular bundles, and flower parts in groups of four or five or in multiples of four or five.

The body of spermatophytes is divided into two parts: primary and secondary. The primary body develops from the meristem at the root and stem tips, and the secondary body develops from the cambia. In general, the body of monocots consists of primary tissues only throughout the life cycle of the plant. However, the body of gymnosperms and dicots consists of primary tissues during the first stages of their development and secondary tissues during the later stages.

The Primary Body

From the standpoint of structure and function, the primary body is divided into two parts: primary root and primary stem.

The Primary Root System The primary root system provides for growth in length or extension and consists of four distinct but overlapping regions: (1) cell division, (2) cell elongation, (3) water and essential-element absorption and initial differentiation, and (4) lateral root formation and further differentiation.

The region of cell division consists of a growing point and, in most crops, a root cap.[1] The growing point is a group of meristematic cells. Individual cells are small and boxlike. They possess thin, pliable walls and dense cytoplasm, very

[1]The roots of aquatic plants have no root caps or root hairs.

small vacuoles, and a large centrally located nucleus. When the root is growing, these cells are dividing and using large quantities of sugars and available nitrogen for the making of the protoplasm. The root cap is a thimble-shaped mass of cells which covers the growing point. Its function is to surround and protect the delicate meristematic cells of the growing point. Obviously, the primary function of the region of cell division is the making of new cells.

The region of cell elongation is located just back of the growing point. It consists of meristematic cells which have elongated; that is, the longitudinal axis has increased more than the transverse axis. These cells absorb large quantities of water and require auxinic (or auxin-like) hormones to give the cells walls the ability to stretch. This abundant absorption increases the turgor pressure within the cells and the walls become permanently stretched. Thus, *the region of cell elongation is responsible for most of the growth in length*.

The region of water and essential-element absorption and of the initial stages of differentiation is located just back of the region of elongation and is often referred to as the root-hair zone. The cells are differentiated into distinct tissues: (1) epidermis, (2) cortex, (3) endodermis, (4) pericycle, and (5) vascular bundles.

The epidermis consists of a single layer of cells, and its function is the absorption of water and the essential raw materials dissolved in the water. To absorb water and these essential materials most effectively, the outer wall of many cells elongates, and these cells are called root hairs. In general, their walls are very thin and are made of cellulose and pectin or of pectinaceous material only, and they are also short-lived (a few hours to a few weeks). When the meristematic cells of the growing points are dividing, root hairs are forming continuously just back of the region of elongation. In this way, new stores of water and essential raw materials become available to the plant.

The cortex consists of relatively large, nearly rounded, thin-walled parenchyma cells. In the root-hair zone, this tissue serves as a system for the diffusion of water and solutes in the water from the root hairs to the endodermis and the diffusion of sugars and other manufactured compounds from the endodermis to the root hairs.

The endodermis consists of a single layer of specified parenchyma cells. In the root-hair zone, its radial and transverse walls are impregnated with a layer of oxidized fat called *suberin*. Suberin, in common with all fatty-like substances, is impervious to water and substances in solution in water, so that the sugars and other foods which pass to the root hairs are under protoplasmic control.

The pericycle consists of one to several layers of cells, and the vascular bundles consist of radially arranged primary phloem and primary xylem. Note that this radial arrangement allows the water and solutes to pass directly to the primary xylem. The phloem provides a channel for the passage or translocation of foods, hormones, and vitamins, and the xylem provides a channel for the transportation of water and essential raw materials. Figure 2-5 is a cross section

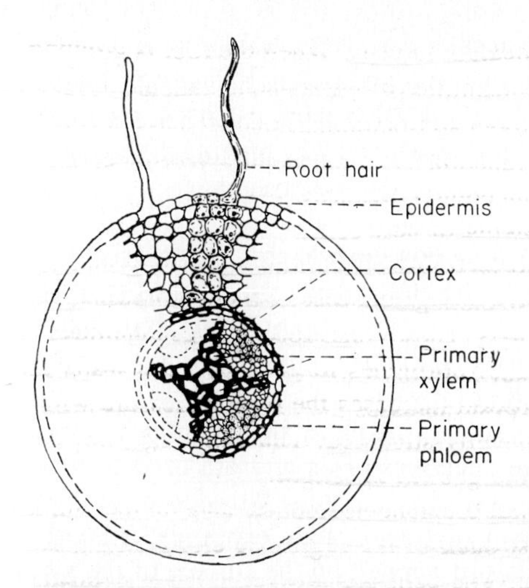

Figure 2-5 Cross section of a root in the root-hair zone. *(Redrawn by permission from M. C. Coulter,* The Story of the Plant Kingdom, *Chicago: University of Chicago Press.)*

of the root-hair zone. Note the position, thickness, and structure of the cells of each tissue.

The region of lateral root formation and further differentiation occurs just back of the root-hair zone. In this region the walls of the epidermis, cortex, and endodermis increase in thickness and become impervious to water, but the pericycle remains meristematic and forms growing points. These growing points pass through the endodermis, cortex, and epidermis and into the soil. Thus, *the function of the pericycle in this region is the formation of lateral roots.*

The main functions of the primary root system are (1) providing for growth in extension and (2) absorbing water and essential raw materials in the water. Other functions are holding the plant and soil in place, conducting water and solutes in the water through the xylem, conducting the manufactured compounds through the phloem, and storing carbohydrates.

As previously stated, the root system does not make the initial food substances—the simple carbohydrates. These substances are made in the leaves. Since the initial food substances are the precursors of all other manufactured compounds, the growth and development of the root system of any given plant depend largely on the health and abundance of the leaves. Why are the health and abundance of the leaves so important? Plants with healthy leaves have more chlorophyll per unit area than plants with diseased leaves. Thus, they absorb more light and, with other factors favorable, make more carbohydrates within any given period. As a result, more carbohydrates are available for the growth and development of the root system. Similarly, plants of a given kind with a large leaf area (a large number of leaves) have more chlorophyll than plants with a small leaf area (a small number of leaves). Thus, they absorb more light and, with other factors favorable, make more carbohydrates per unit time. As a result, more carbohydrates are available for growth and development of the root system.

This explains why plants with healthy leaves or dark green leaves develop more active and extensive root systems than plants with diseased or light green leaves, why plants with large tops (a large leaf area) develop more extensive root systems than plants with small tops (a small leaf area), and why pruning of the top is likely to reduce the growth and extensiveness of the root system.

Since the root system contains living tissues, it is always respiring. Therefore, the root system is always giving off carbon dioxide and always taking in oxygen. This oxygen supply is present in the atmosphere and diffuses into the pore spaces of the soil. In general, the root systems of most of our crop plants require abundant quantities of oxygen, particularly when the growing points are making new cells and when large quantities of water and essential raw materials are being absorbed. When these processes are taking place, the rate of respiration is usually high and, consequently, large quantities of oxygen are needed. This explains why most crop plants require well-drained soils; why loose, friable, highly colored subsoils provide for deeper root penetration than tight or poorly drained subsoils; and why a layer of clay, if applied on the surface of the soil under a large tree, is likely to retard the growth and development of the tree.

The Primary Stem System The primary stem system provides for growth in height or extension and consists of three distinct but overlapping regions: (1) cell division, (2) cell elongation, and (3) cell differentiation, or tissue formation.

The region of cell division exists within the buds of stems. These buds are really undeveloped stems. Groups of meristematic cells occur at the terminal portions and at the nodes of these young stems. The mass of cells at the terminal portion or apex is dome-shaped and is surrounded by young developing leaves. When buds are expanding and growing, the meristematic cells are dividing. Thus, *the primary function of this region is the making of new cells*.

The region of cell elongation exists from the base of the buds downward through several nodes and internodes. Like the region of elongation in the root system, individual cells develop large vacuoles for the absorption of abundant quantities of water. This absorption of water, together with the cell-stretching hormones, elongates the cells.

When the buds are expanding, the regions of cell division and cell elongation are centers of intense metabolic activity. In particular, there are high rates of respiration, sugar utilization, and water absorption. Large quantities of sugars are needed for the formation of the walls and protoplasm of the new cells and for the liberation of free energy in respiration.

The region of cell differentiation or maturation consists of the remainder of the stem which grows in length. In this region, the cells differentiate into distinct tissues: (1) epidermis, (2) cortex and pericycle, (3) vascular bundles, and (4) pith.

The epidermis consists mainly of cells with thick radial walls and toothed and flanged tangential walls. Usually, the outer wall contains a layer of waxlike material called the cuticle. The thickening and dovetailing of the walls and the cuticle combine to greatly reduce the rate of transpiration of stems. In this way, most of the water absorbed by the roots can get to the leaves. *Thus, an important function of the epidermis in stems is to keep the absorbed water within the plant.*

The cortex usually contains both living and dead tissues. In general, the living tissue stores carbohydrates, usually as grains of starch; and the dead tissue, the thick-walled, elongated fibers, gives strength and rigidity to stems. In this way, the stems have the ability to hold the leaves in the light and support large crops of flowers and fruit.

The vascular bundles consist of (1) primary phloem and (2) primary xylem. As explained previously, the function of the phloem is the transportation of the many manufactured compounds—the many types of foods, hormones, and vitamins. The function of the xylem is the transportation of water and the essential raw materials in the water. In addition, the xylem tubes or vessels, because of their structure, give strength and rigidity to stems.

The pith is found in the central portion of stems. This tissue is made up of relatively large, thin-walled cells, frequently with numerous intercellular spaces. The function of the pith is the storage of food, particularly starch—the most important reserve carbohydrate.

What is the relative metabolic activity of the region of differentiation? As previously stated, the regions of cell division and elongation are centers of intense metabolic activity, since all the cells in these regions are alive and require a continuous oxygen supply for their respiration. However, the region of cell differentiation contains both living and dead tissues. Since dead tissues do not respire and use foods and since the living tissues do not divide and make new cells, but only keep the protoplasm in good repair, the metabolic activity of the region of differentiation is much lower than that of the regions of division and elongation.

The Stems of Monocots and Dicots As previously pointed out, the plant body of monocots is derived from the meristem at the tips, and that of dicots is derived first from the meristem of the tips and finally from the cambia—the vascular cambium and the cork cambium. In other words, the vascular cambium in monocots exists for a short time only, whereas that in dicots exists from the time it is formed until the plant dies. Thus, mature vascular bundles of monocots lack a cambium and are called *closed bundles*, whereas mature vascular bundles of dicots contain a cambium and are called *open bundles*. Since monocots contain a vascular cambium for a limited period only, the stems grow in diameter during the development of the primary tissues only; and since dicots contain a vascular cambium throughout the life of the plant, they grow in diameter throughout the life of the plant. Further, in monocots the primary vascular bundles serve as the transportation system throughout the life of the plant, regardless of whether the plant lives for one growing season or for several seasons. However, in dicots the primary bundles serve as the transportation system until the secondary bundles are developed. This period varies with the life of the plant; for annuals it is a relatively long time, and for perennials a relatively short time. Note the lack of a cambium and the arrangement of the bundles in corn, an important monocot, in Figure 2-6 and the cambium and the arrangement of the bundles in a typical dicotyledonous plant in Figure 2-7.

The Functions of the Primary Stem The main functions of the primary stem are (1) providing for growth in length, or extension, and (2) serving as the connecting link between the leaves and the roots. Other functions are holding the leaves in the light; supporting flowers, fruit, and seed; and storing manufactured foods, particularly reserve carbohydrates.

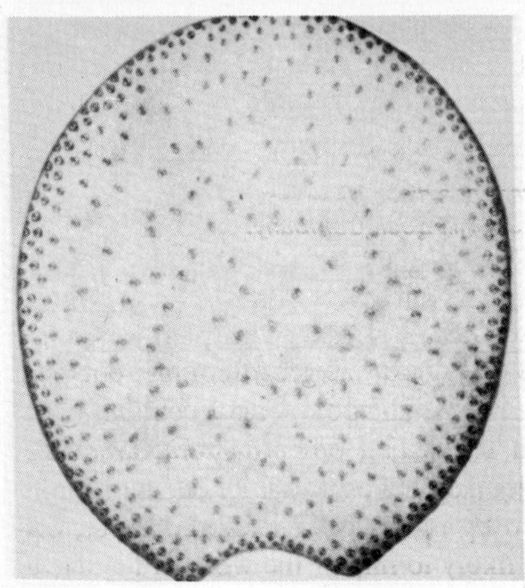

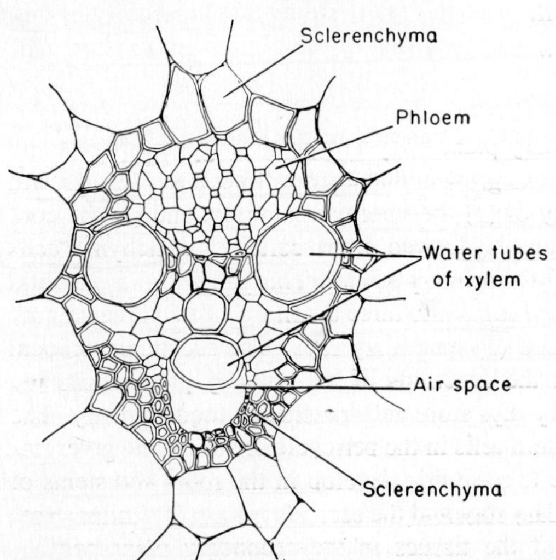

Figure 2-6 Cross section of stem and vascular bundle of corn—an important monocotyledon. *(Courtesy, F. W. Emerson,* Basic Botany, *New York: McGraw-Hill Book Company, Inc.)*

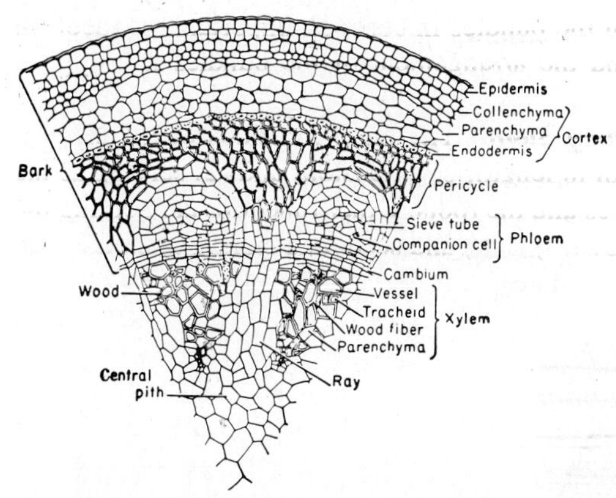

Figure 2-7 Cross section of the herbaceous stem of a dicot. *(Courtesy, F. W. Emerson,* Basic Botany, *New York: McGraw-Hill Book Company, Inc.)*

As previously stated, the initial food substances and certain hormones and vitamins are made in the leaves. These are needed for the growth of the roots. On the other hand, the leaves require abundant water and compounds containing the essential elements for making the initial foods and other manufactured substances. The water and all the compounds containing essential elements, with the exception of carbon dioxide, are absorbed by the roots. Therefore, insect, disease, or mechanical injury to the stems is likely to impair the work of the leaves and the work of the roots as well.

The Secondary Body

As stated previously, the secondary plant body is limited to gymnosperms and dicots only. In these plants the stems grow in diameter, or thickness. This growth in thickness is due to the activity of (1) the vascular cambium and (2) the cork cambium. The vascular cambium is derived from certain parenchyma cells between the primary phloem and the primary xylem. This cambium divides and forms new cells. These cells differentiate into three distinct tissues: (1) secondary phloem, (2) secondary xylem, and (3) medullary rays. The secondary phloem and secondary xylem take over the functions of the primary phloem and the primary xylem; and the medullary rays store and translocate foods radially. The cork cambium develops from certain cells in the pericycle or cortex and gives rise to the layer of cork. Since these two cambia develop in the roots and stems of dicots, the anatomy of the secondary root and the secondary stem is similar. Note the arrangement and structure of the tissues in the secondary plant body in Figure 2-8.

The main functions of the secondary plant body are (1) providing for growth in thickness, or diameter, and (2) serving as the connecting link between the

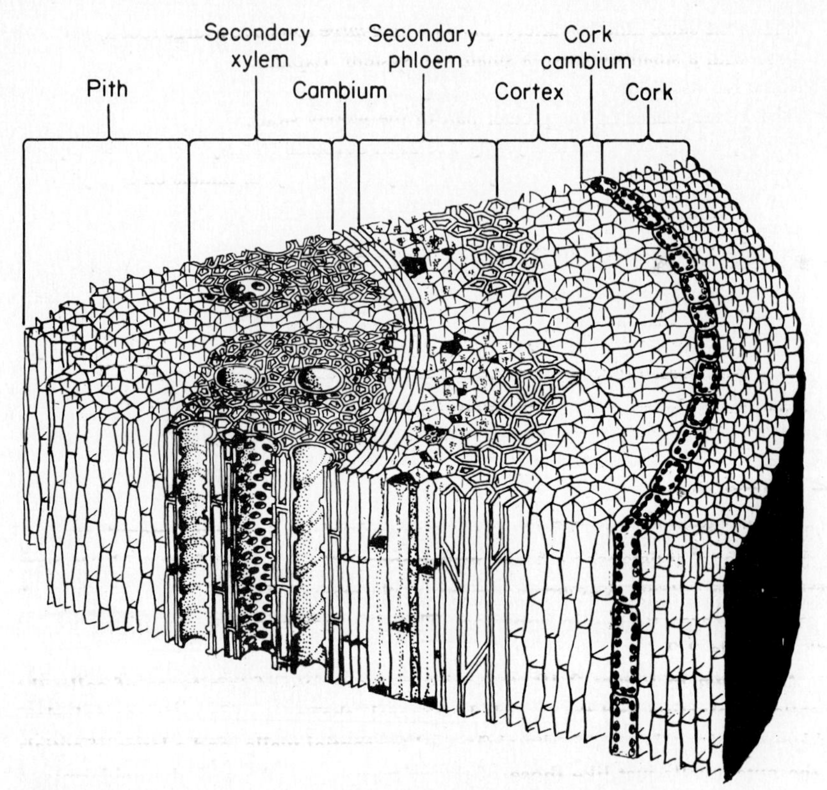

Figure 2-8 Cross section of a sector cut from a dicot stem. This stem is diagrammatic for a woody plant, showing well-defined cork cambium and vascular cambium.

primary root system and the primary stem system. Other functions are support and the storage of reserve carbohydrates, particularly starch and related substances.

QUESTIONS

1. What are spermatophytes?
2. Distinguish between gymnosperms and angiosperms.
3. Compare monocots and dicots from the standpoint of (1) number of seed leaves, (2) venation of the leaves, (3) type of vascular bundles, and (4) number of flower parts.
4. Name the regions of growth in length of the root system.
5. What is a root hair? What is its function?
6. What is the primary function of the root system?
7. To what extent is the root dependent on the top? The top dependent on the root? Explain.

8 Within the same kind or variety, plants with a large top have a large root system and those with a small top have a small root system. Explain.
9 Name the tissues from which the woody stem arises.
10 What tissue makes up the greater part of the woody stem?
11 A nail partially driven into a young tree soon becomes entirely embedded. Explain.
12 Suppose that when you were 15 years old, you carved your initials on a 15-year-old tree 5 feet from its base. Would your initials be at the same or at a greater height when you and the tree are 30 years old? Explain.
13 Suppose you nail a wire fence to a row of five-year-old trees. Would the fence gradually move upward as the trees increased in height or remain at the original height? Give reasons.
14 The trunks of palm trees and bamboo are as large in diameter when the plants are short (3 to 4 feet high) as when they are tall (50 to 100 feet high). Explain.
15 What is the function of the medullary rays?

The Leaves

In general, leaves are modified stems designed primarily for the manufacture of the initial food substances. As is well known, the leaves of crop plants vary greatly in size, shape, and arrangement on the stems. Despite this variation, all leaves have three tissues in common. These tissues and their functions are set forth in Table 2-2.

As shown in Figure 2-10, the epidermis consists of two types of cells: the so-called protective and the guard. In general, the protective cells are colorless and have thick radial, toothed and flanged tangential walls, and a layer of cuticle on the outer walls just like those of the primary stem. As with the epidermis of the primary stem, thickening and dovetailing of these walls, combined with the layer of cuticle, greatly reduces the rate of transpiration through the protective cells. In fact, under conditions of high transpiration, only 1 to 2 percent of the enormous quantities of water transpired from the leaves is lost through the protective tissue.

The guard cells differ from the protective cells in that they exist in pairs and contain chlorophyll; thus they manufacture food. The walls of these cells vary in

Table 2-2 Principal Tissues of Leaves and Their Primary Functions

Tissues	Primary functions
Epidermis:	
Protective cells	To keep water within the leaf
Guard cells and stomates	To allow CO_2 and O_2 diffusion in photosynthesis and respiration
Chlorophyll-containing cells	To manufacture the initial food substances
Veins	To translocate water and raw materials to, and manufactured compounds away from, the manufacturing cells

thickness, and the inner walls are next to an opening, the stomate, through which gases, carbon dioxide, oxygen, and water vapor diffuse. Because of the varying thickness of the walls and because the walls are pliable, changes in shape of these guard cells alter the size of the pore. Thus, since the primary function of the leaf is the manufacture of the initial food substances and since the stomates must be open for the diffusion of carbon dioxide to the manufacturing cells, the behavior of the guard cells and stomates assumes great practical significance. Important environmental factors which influence the shape of these guard cells, which, in turn, influence the opening and closing of the stomates, are (1) light and (2) the water supply within the plant.

Light With most crop plants light directly influences the shape of the guard cells and the opening of the stomates. In other words, when light is available and the water supply is favorable, the guard cells are fully stretched and the stomates are usually open. Conversely, when light is absent the guard cells are flaccid and the stomates are closed. How does light influence the shape and turgor of the guard cells? Two explanations are presented herewith. The first and older is based on the manufacture of sugars in the guard cells. In the morning just before sunrise the supply of sugars in the guard cells is low and the osmotic pressure is correspondingly low. With sunrise the guard cells begin the manufacture of sugars. This increases the osmotic pressure of these cells, and water is absorbed from the adjacent cells. In this way, the guard cells become turgid and the stomates open. With sunset, because of the decreasing light, sugar manufacture declines, but respiration continues. This decreases the sugar supply, and the osmotic pressure within the cells correspondingly decreases. As a result, water is withdrawn from the guard cells and the stomates close. The second and most recent explanation is based on the ability of the leaf to transform the kinetic energy of light directly into useful chemical energy in the formation of ATP. With the onset of light, photophosphorization takes place with the formation of the ADP-ATP cycle and the formation of glucose. This, in turn, increases the flow of water into the guard cells and the stomates open. With the onset of dark, photophosphorization stops, but sufficient ATP is available to provide the free energy to change the glucose to starch. This, in turn, decreases the flow of water into the guard cells and the stomates close.

Water Supply within the Plant When a water deficit occurs within the plant, even in the presence of light, the guard cells lose turgor and the stomates close. In general, this deficit occurs when the rate of absorption of water is less than the rate of transpiration. As a result, the water supply within the plant decreases and the guard cells, because of their readily pliable, easily stretched walls, respond to this decrease in the water supply. Thus, with a decrease in the supply of water within the plant, the guard cells change in shape, which, in turn, decreases the size of the stomates. This, in turn, decreases the amount of water going out of the plant. However, it also decreases the rate of diffusion of carbon dioxide into the plant. With this decrease in rate of diffusion of carbon dioxide, a decrease in the rate of photosynthesis takes place with a corresponding decrease in the rate of sugar manufacture and a corresponding decrease in growth and yield. Figure 2-9 shows an open and closed stomate of apple. Note the differences in shape of the guard cells.

In general, the structure of the epidermis, the position of the guard cells and stomates, and the size, shape, and arrangement of chlorophyll-containing tissues vary with crop plants. Note the differences in these tissues in the leaves of the apple, Kentucky bluegrass, and white pine presented in Figures 2-10 to 2-12. Note that the apple's guard cells and stomates are on the lower epidermis only, and the chlorophyll-containing cells are of two types: the elongated cells at right angles to the upper epidermis, called palisade cells; and the round cells on the lower part, called spongy cells. In Kentucky bluegrass the guard cells and stomates occur on the upper surface only, and all the manufacturing cells are round. In white pine each needle, or leaf, is triangular, and the needles occur in groups of five. The epidermis is very thick, and the guard cells and stomates are situated on the sides facing the adjacent leaves. Can you think of any reason for this?

The veins of leaves are relatively small vascular bundles. They are in close contact with the manufacturing cells and conduct the foods, hormones, and vitamins from, and water and essential raw materials to, these cells. In large veins, in addition to the vascular tissues, there are sheaths of thick-walled cells which give strength and support to the leaves.

The leaves of many crop plants have petioles. In general, the petiole, or leafstalk, consists mainly of vascular bundles and mechanical tissue. Thus, the petiole serves as a transportation system between the manufacturing cells and the stem and holds the leaf blade in the light. The petioles of some crops are fleshy and form the edible portion, e.g., rhubarb and celery. See Fig. 20-2.

Rate of Translocation of Manufactured Sugars As previously stated, the initial food substances are soluble in the sap of plants. During the day these

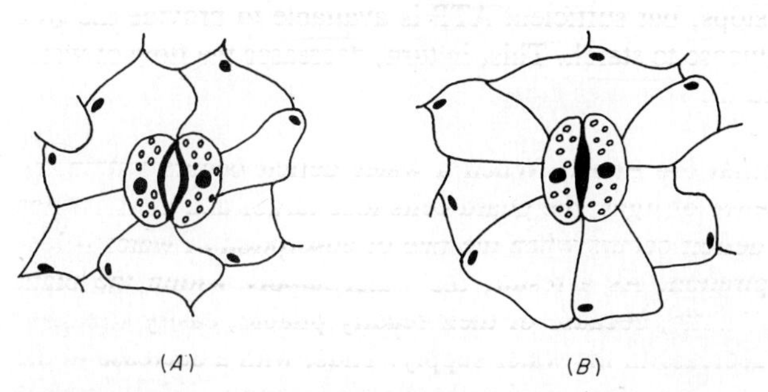

Figure 2-9 Open (A) and closed (B) stomate of apple (*Redrawn from O. F. Curtis and D. G. Clark,* An Introduction to Plant Physiology, *New York: McGraw-Hill Book Company, Inc.*)

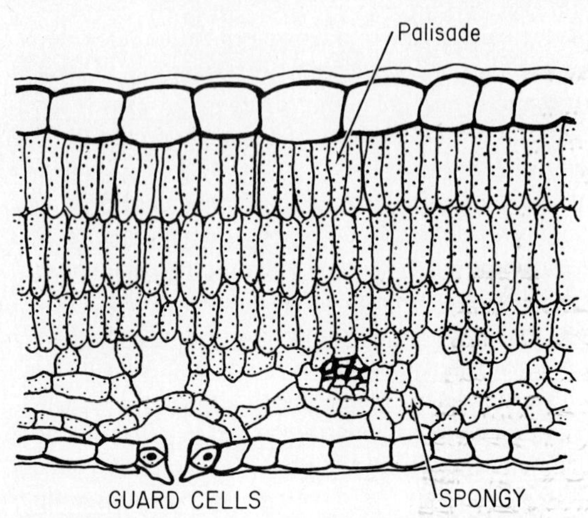

Figure 2-10 Cross section of leaf of apple. (*Courtesy, A. J. Eames and L. H. MacDaniels,* An Introduction to Plant Anatomy, *New York: McGraw-Hill Book Company, Inc.*)

soluble carbohydrates are changed into insoluble forms and are stored temporarily in the chloroplasts. During the night these compounds are changed to soluble forms, usually sucrose, and are translocated to other parts of the plant. If, for any reason, not all the sugars are translocated from the leaves, the storage capacity of the chloroplasts is decreased and the manufacture of sugar is accordingly decreased. Thus, a slowing down in the growth and development of roots, stems,

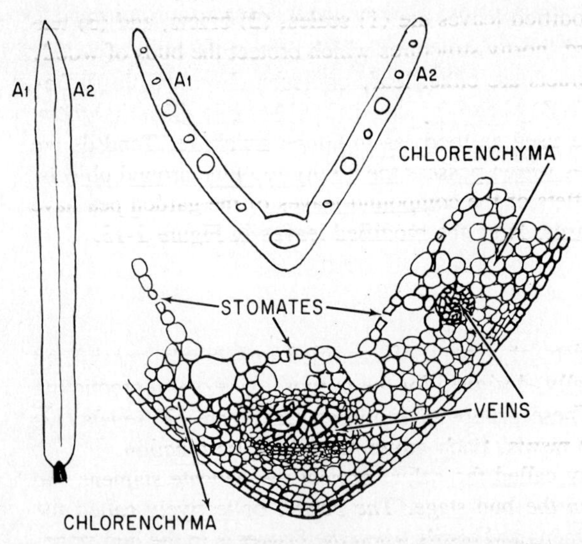

Figure 2-11 Sections of leaf of Kentucky bluegrass. (*Courtesy, F. W. Emerson,* Basic Botany, *New York: McGraw-Hill Book Company, Inc.*)

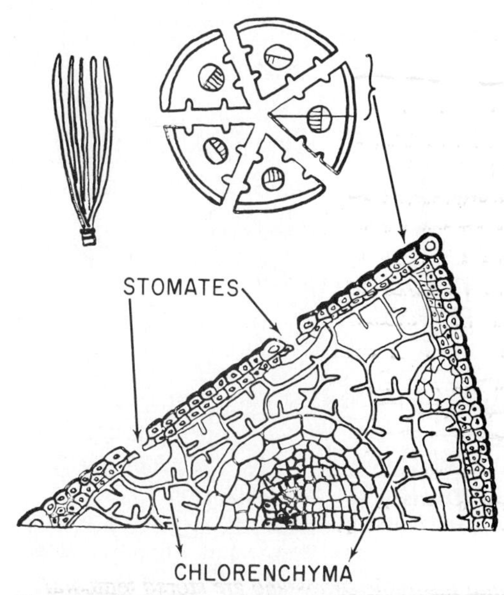

Figure 2-12 Sections of leaf of white pine. (*Courtesy, F. W. Emerson,* Basic Botany, *New York: McGraw-Hill Book Company, Inc.*)

flowers, or fruit may correspondingly slow down the rate of sugar manufacture. This, you will recall, is a practical application of the law of mass action.

Modified Leaves Modified leaves are (1) scales, (2) bracts, and (3) tendrils. Scales are usually hard, horny structures which protect the buds of woody plants during the winter. Bracts are either leafy or fleshy. In some plants they take the place of petals, as in dogwood, poinsettia, and bougainvillea; in others, they become fleshy and are used as food, as in Globe artichoke. Tendrils are slender, threadlike structures which possess the ability to twine around objects. For example, the upper leaflets of the compound leaves of the garden pea have been modified to form tendrils. Note the modified leaves in Figure 2-13.

The Flowers

An individual flower consists of groups of modified and highly specialized leaves, arranged concentrically, designed for the purpose of sexual reproduction or reproduction by seed. These groups of modified leaves are (1) sepals, (2) petals, (3) stamens, and (4) pistils. Each group has a specific function.

The sepals, collectively called the *calyx,* protect the delicate stamens and pistils when the flower is in the bud stage. The petals, collectively called the *corolla,* also protect the stamens and pistils when the flower is in the bud stage, and large, highly colored petals attract pollinating insects.

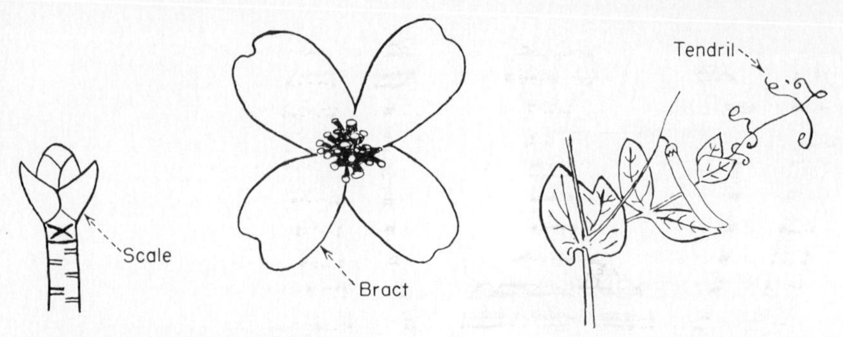

Figure 2-13 Types of modified leaves.

An individual stamen consists of a stalk called the *filament* and a pollen sac called an *anther*. *The function of the stamen, particularly the anther, is to produce pollen which in due time contains sperms*. For this reason the stamens are called the male organs of the plant.

An individual pistil consists of an ovary, style, and stigma. The ovary is the enlarged portion at the base of the pistil, the stigma is the flattened portion at the apex, and the style is the connecting tissue between the two. *The function of the pistil, particularly the ovary, is to produce one or more ovules, each of which in due time contains, with other cells, an egg*. For this reason the pistils are called the female organs of the plant.[2]

Figure 2-14 shows a longitudinal view of a complete flower. It illustrates the transfer of pollen within the flower, the germination of pollen grains, and the growth of the pollen tube down the style.

Complete and Incomplete Flowers The flowers of some crops contain all four of the main parts: sepals, petals, stamens, and pistils. This type of flower is called a complete flower. However, the flowers of other crops contain no petals and are called apetalous flowers; others contain functional stamens and nonfunctional pistils and are called staminate flowers; and others contain nonfunctional stamens and functional pistils and are called pistillate flowers. All these types of flowers—apetalous, staminate, and pistillate—are called incomplete flowers.

Sex Expression The sex expression of crop plants is based on whether one or both of the sex organs are in the same flower. In general, there are three main types: plants with functional stamens and functional pistils in the same flower—the perfect flowering crops; plants with functional stamens and functional pistils in separate flowers on the same plant—the monoecious crops; plants

[2]The authors are aware that morphologists regard the n generation as "sexual" and the $2n$ generation as "nonsexual." In this text both generations are considered one entire life cycle, and the stamen is considered the male organ, the pistil the female organ.

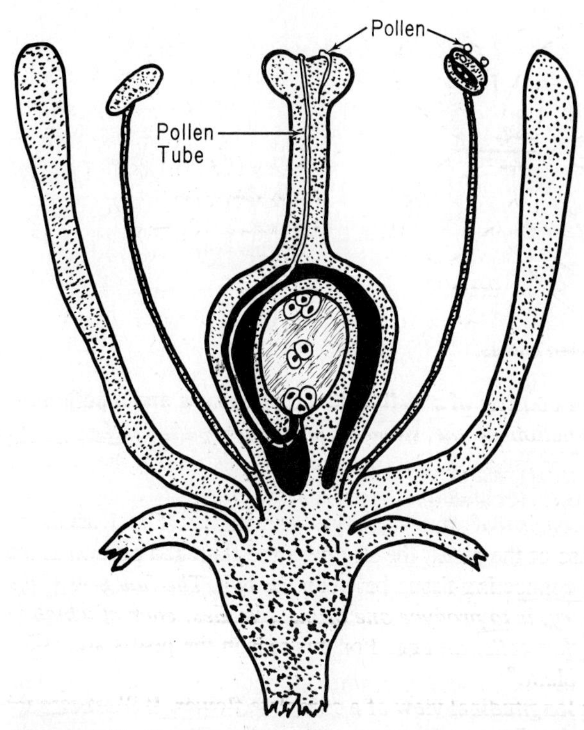

Figure 2-14 Longitudinal section of a complete flower. (*Courtesy, F. W. Emerson*, Basic Botany, *New York: McGraw-Hill Book Company, Inc.*)

with functional stamens and functional pistils in separate flowers on different plants—the dioecious crops. Examples of each type follow:

Crops with Perfect Flowers Apple, pear, peach, plum, vinifera and labrusca grape (most varieties), raspberry, blackberry, strawberry, gooseberry, cranberry, lemon, orange, grapefruit, avocado, almond, cabbage, radish, carrot, celery, sweet potato, tomato, capsicum pepper, eggplant, bean, pea, okra, rose, chrysanthemum, carnation, violet, sweet pea, and snapdragon.

Monoecious Crops Banana, coconut, pecan, walnut, filbert, chestnut, tung, cucumber (most varieties), cantaloupe or muskmelon (European varieties), pumpkin, squash, watermelon and sweet corn. Modifications of the monoecious type are andromonoecious and gynoecious. In the andromonoecious, perfect and staminate flowers occur on the same plant, and this type of expression takes place in American varieties of cantaloupe or muskmelon. In the gynoecious, all of the flowers are pistillate, and this type of expression occurs in certain varieties of

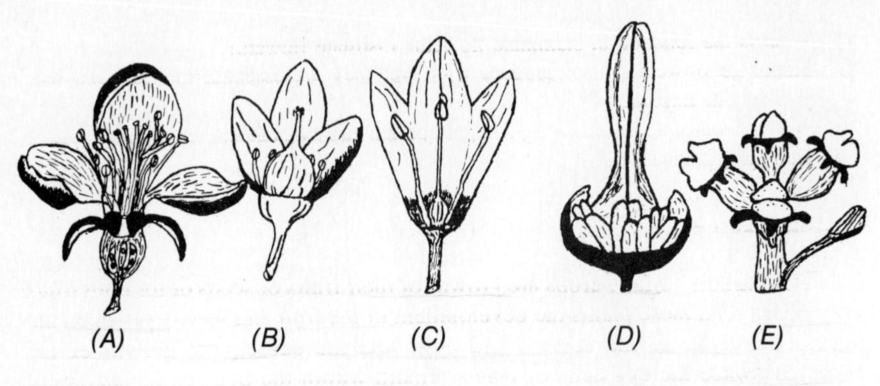

Figure 2-15 Examples of perfect, staminate, and pistillate flowers. A—perfect flower of apple. B, C—pistillate and staminate flowers of asparagus. D, E—staminate and pistillate flowers of pecan.

cucumbers grown for pickling. A monoecious variety is interplanted with the gynoecious variety to provide pollen.

Dioecious Crops Date palm, papaya, persimmon, muscadine grape (certain varieties), spinach, asparagus, and holly. (Figure 2-15 shows the perfect flower of apple and the functional pistillate and functional staminate flowers of asparagus and pecan.)

QUESTIONS

1. What is the primary function of the leaves?
2. Draw from memory a diagram of the tissues of a leaf of a major crop, showing the upper and lower epidermis, the palisade and spongy tissue, the veins, the guard cells, the stomates.
3. Show how decreasing supplies of sugar and water in the guard cells induce closing of the stomates.
4. In general, the longer the stomates are open during the day, the longer photosynthesis takes place and the greater the yield. Explain.
5. When the stomates are closed, the plant is not making carbohydrates. Explain.
6. In the leaf petiole the corners of certain cells are greatly thickened. Explain.
7. What is the principal function of the flower?
8. Name the four parts of a complete flower and give the function of each.
9. What part of the flower produces pollen? Ovules?
10. Name the different ways in which the sex organs of plants are arranged.
11. What is a perfect flower, a staminate flower, a pistillate flower?

12 What is the function of staminate flowers? Pistillate flowers?
13 Staminate flowers do not produce fruit, yet they are essential for fruit and seed production. Explain.
14 Monoecious and dioecious crops require pollen-carrying agents for the production of fruit and seed. Explain.

The Fruit and Seed

Pollination Many crops are grown for their fruits or seeds or for both fruits and seeds. With these plants the development of the fruit and seed depends on the successful union of the sperms and eggs and the subsequent growth of the embryos. Since the egg cells of plants remain within the ovules, the sperms of plants must go to the eggs. With many plants the journey of an individual sperm to an individual egg is not an easy one. Instances are frequent in which the sperm fails to reach the egg. Sometimes the pollen which carries the sperm is not transferred to the stigma of the pistil; sometimes the weather retards or prevents the germination of the pollen; sometimes the rate of growth of the pollen tube down the style is so slow that the egg dies before the sperm arrives. To more fully understand some of the difficulties met by the sperm, its journey to the egg is considered in three stages: (1) the transfer of pollen, (2) time of pollen shedding and pistil receptivity, and (3) growth of the pollen tube down the style to the embryo sac.

Transfer of Pollen *The transfer of pollen from the anther of the stamen to the stigma of the pistil is called pollination.* Four general methods of pollen transfer are used: (1) by force of gravity, (2) by contact, (3) by wind, and (4) by insects. Pollen transfer *by the force of gravity* or *by contact* usually takes place when the pistil(s) and stamens are in the same flower. The force of gravity is effective when the flower is in the pendent position and the stigma extends beyond the anthers. Transfer by contact is effected when the stamens, as they elongate, shed their pollen as the anthers come in contact with the receptive stigma. Pollen transferred *by wind* or *by insects* is necessary when the pistil and the stamens are in different flowers or when, for some reason, the stamens fail to pollinate the pistil in the same flower.

Wind- and insect-pollinated crops present somewhat different problems with respect to adequate pollination. Pollen transferred by wind is subject to chance currents of air. Thus, it lacks directness; large quantities of pollen are necessary; and the grower should know how far such pollen is carried. On the other hand, pollen transferred by insects is more direct, and lesser amounts of pollen are necessary. Thus, wind- and insect-pollinated plants have marked differences in adaption to pollination. In general, wind-pollinated plants have inconspicuous flowers, produce small, dry pollen in large quantities, and possess long, branched, or feathery styles in order to catch the pollen grains. On the other hand, insect-pollinated plants have large, highly colored petals or bracts, produce large, sometimes sticky, pollen, and possess well-developed nectaries. These nectaries secrete sugars and other substances for the attraction of insects. They are so situated that, when insects visit the flowers, their bodies which carry the pollen come in contact with the stigma. The principal insects are certain species of bees, most important of which is the honeybee.

The needs and the physical equipment of honeybees are well adapted for the work of pollination. Their bodies are covered with numerous hairs to which the pollen clings, and their action on the flower is gentle. Thus, the delicate tissues of the pistil are not injured. Important environmental factors affecting the activity of honeybees are (1) the temperature of the air and (2) the food (nectar and pollen) supply. Investigations have shown that honeybees are comparatively inactive at temperatures lower than 50°F (10°C) and cannot use their wings at temperatures lower than 40°F (4°C). Consequently, honeybees do not fly very far in cool weather. To help the bees perform the task of pollination of many crops, growers distribute colonies of bees throughout the orchard, greenhouse, or plantation. Figure 2-16 shows a honeybee visiting a cucumber flower.

The production of honeybees for pollination purposes in many parts of the world is an important industry. In fact, some authorities have stated that honeybees are about fifty times more valuable for their work of pollination in orchards, gardens, and fields than they are for the honey they make. In the United States and other parts of the world, the year-round activity of an individual hive or colony may be divided into two distinct phases: the passive, or static, and the active, or dynamic.

The passive period takes place during the late fall, winter, and early spring; an individual colony consists of a queen and workers only; and they live on the honey they have made during the preceding season. Since honeybees are very sensitive to low temperatures, they are stored in heated places in severe climates, and in outdoor sheltered places in mild climates.

In sharp contrast, the active period takes place during the late spring, summer, and early fall. An individual colony consists of a queen, the young brood in various stages of development, workers (undeveloped females), and drones (males). During this period the demand for fresh supplies of food is quite high. Fortunately, the temperatures are sufficiently high to permit the workers to range over a wide territory in their search for pollen and nectar. To provide adequate pollination consistent with low cost of production, hives are overwintered in regions characterized by mild winters, and bees are shipped in small lots at the beginning of the blossoming season to regions characterized by severe winters. These small lots are called *package bees*. In general, information on the distribution of hives for any given orchard or planting can be obtained from the local experiment station.

Self- and Cross-pollination In horticulture there are two distinct concepts of self- and cross-pollination. These different concepts are based on differences in the behavior of asexually and sexually propagated crops. In general, the plants of any given variety of an asexually propagated crop, for example, the apple, are descended from a single plant or a single bud, and unless mutations take place,

Figure 2-16 A honeybee visiting a cucumber flower. (*Courtesy, W. S. Anderson, Mississippi State, Mississippi.*)

the plants of any particular variety are identical. For example, all the apple trees of the Delicious variety are descendants from one original bud. Since the apple is propagated by using a vegetative structure, the many thousands of Delicious trees in the country are identical to the original tree. Thus, the pollen of all Delicious trees is identical, and in like manner the pollen of any given asexually propagated variety is identical. In other words, in asexually propagated crops the *variety* is considered the unit. Accordingly, *with asexually propagated crops self-pollination is the transfer of pollen from a flower of one variety to a flower of the same variety.* The two flowers may be on the same tree or plant or on different trees or plants of the same variety. *Cross-pollination is the transfer of pollen from a flower of one variety to a flower of another variety.*

On the other hand, the plants of any given variety of a sexually propagated crop, for example, the tomato, are produced from seed, and the plants from the seed may or may not be identical. Thus, the pollen may or may not be identical. In other words, in sexually propagated crops the *plant* is considered the unit. Accordingly, *with sexually propagated plants self-pollination is the transfer of pollen from the stamens to the pistil within the same flower or between flowers of the same plant,* and *cross-pollination is the transfer of pollen from the stamens of a flower of one plant to the pistil of a flower of another plant.* This plant may belong to the same strain or to another variety or even to a related species.

Table 2-3 Examples of Wind- and Insect-Pollinated Crops

Agent	Crops
Wind	Gymnosperms, date palm, corn, sweet corn, chestnut, filbert, pecan, walnut, beets, spinach
Insects	Avocado, cacao, vanilla, apple, pear, peach (a few varieties), plum, cherry, vinifera and labrusca grapes, brambleberries, asparagus, cabbage, lettuce, onion, carrot, the vine crops, aster, azalea, calceolaria, pansy, zinnia
Both wind and insects	Muscadine grape, citrus, coffee, tung

Each type of pollination has its place in horticultural crop production. With asexually propagated plants, many varieties of certain crops produce more fruit when they are cross-pollinated. For example, many varieties of the apple and all varieties of the sweet cherry fail to set complete crops of fruit unless pollen of other varieties is used. Those varieties which are necessary for the pollination of other varieties are called pollinators. Consequently, the choice of a variety for pollination purposes is particularly important. With sexually propagated crops, self-pollination maintains uniformity and reduces variability of any particular strain or variety.[3] On the other hand, cross-pollination introduces new characteristics and increases diversity.

Time of Pollen Shedding and Time of Pistil Receptivity When the egg is ready to receive the sperm, the pistil becomes receptive. Pistil receptivity is indicated by the secretion of sugars and other food substances and hormones on the surface of the stigma. These materials are necessary for the germination of the pollen tube. Obviously, pollen shedding and pistil receptivity should take place at the same time. With some crops these processes do not always take place simultaneously.

When the pollen of any given variety is shed before the pistils of that same variety are receptive, the variety is said to be *protandrous*. Conversely, when the pollen of any given variety is shed after the pistils of that same variety are receptive, the variety is said to be *protogynous*. In general and for any given variety, protandry or protogyny may be complete or partially complete. In other words, no overlapping or some overlapping may take place with each of these two processes. This seems to vary with the variety and with the season. For example, some varieties of corn shed practically all their pollen before the pistils are receptive; others shed most of their pollen after the pistils are receptive; and the same variety may shed pollen too soon in some seasons, about the right time in others, and too late in others. Certain varieties of pecan and walnut shed most of their pollen before their pistils are receptive, and other varieties shed their pollen after the pistils are receptive. Certain varieties of avocado open their flowers in the morning, at which time the pistils are receptive; close them again

[3]Some cross-pollination usually takes place in sexually propagated, self-pollinated plants.

in midday; and open them again the following afternoon, at which time the anthers shed pollen. Other varieties open their flowers in the afternoon, at which time the pistils are receptive; close them again in the late afternoon; and open them again the following morning or morning of the second day, at which time the anthers dehisce. Within any variety, marked changes in the weather may greatly change the normal schedule. Other crops in which protandry or protogyny is a problem are chestnut, lyechee, and mango. Information on the variety which should be interplanted for any given crop or for any given location can be obtained from the local agricultural experiment station or agricultural extension service.

Growth of the Pollen Tube Down the Style On the surface of the stigma the pollen grain germinates; that is, it produces a slender tube which grows down the style. The temperature of the air, the food supplied by the stylar tissue, and the compatibility of the pollen tube and stylar tissues are important to the growth of the tube. Thus, when the weather is cold in the orchard or garden during the pollinating season, the set of fruit is likely to be low. Naturally, if the food supply of the tissues of the style is low, the growth of the tube will be correspondingly slow. Of more importance is the compatibility of the pollen. Compatible pollen possesses the ability to grow down the style in time for the sperm to unite with the egg. Incompatible pollen does not possess this ability; that is, the pollen tube may grow slowly or not at all, even though the temperature and food supply are favorable.

There are two types of incompatibility: self-incompatibility and cross-incompatibility. When the pollen tube of any given variety fails to grow down the style in time to release the sperms for the fertilization of the egg and the endosperm nuclei of the same variety, the variety is said to be *self-incompatible*. Similarly, when the pollen tube of any given variety fails to grow down the style in time to release the sperms for the fertilization of the egg and endosperm nuclei of another variety, the two varieties are said to be *cross-incompatible*. Investigations have shown that self- and cross-incompatibility exist in many valuable agronomic and horticultural crops. Examples of each type follow:

Self-incompatibility Apple (most varieties), sweet cherry (practically all varieties), Duke cherry (some varieties), Japanese plum (most varieties), European plum (some varieties), filbert (most varieties), and many coffee seedlings and clones.

Cross-incompatibility Sweet cherry (some varieties), Japanese plum (some varieties), European plum (some varieties).

To overcome the effect of self-incompatibility and cross-incompatibility and to assure the annual production of high yields of high-quality fruit, a suitable cross-compatible variety is interplanted with the incompatible sort. For the most part, the requirements of a satisfactory interplant for any given industry have been developed by the local agricultural experiment station.

Fertilization When the pollen grain germinates, the tube in some plants pushes its way down the style from cell to cell; in others it passes through a channel between the cells; and in either case it finally passes through the micropyle of the ovule and arrives at the embryo sac. Within the embryo sac the tube liberates two sperms. One sperm unites with the egg in the formation of the zygote, or new plant; the other unites with the endosperm nuclei in the formation of the endosperm. This union of the one sperm with the egg and the other sperm with the two endosperm nuclei is called double fertilization.

Immediately after fertilization, the embryo makes new cells and increases in size. At the same time, the ovary increases in size to make room for the rapidly developing embryo(s).

As with the development of other crop plant organs, there are definite stages in the development of fruit. In general, these are (1) the making of new cells, (2) the enlargement of the cells, (3) the accumulation of sugars and starch and compounds associated with quality, flavor, and palatability, and (4) the reduction of the accumulated compounds within the cells. For example, an individual fruit of the apple makes from ½ to 1 million cells during the first stage; enlarges these cells with the nucleus and cytoplasm occupying the space next to the walls and the large vacuoles occupying the space within the inner part of the cell during the second stage; accumulates sugars, starch, and compounds associated with aroma, flavor, and quality during the third stage; and decreases the amount of accumulated compounds during the fourth stage. The fundamental process concerned is respiration. A typical respiration curve during the development and ripening of a fruit is shown in Figure 2-17. Note that the rate of respiration is moderately high during the first stage, that it declines slightly during the second, increases rapidly during or right after the third, and decreases slowly during the fourth stage. This sharp rise in the rate of respiration is associated with a physiologic condition known as the *climateric,* and this, in turn, is positively associated with high eating quality, flavor, and palatability. Examples in which the climateric occurs are fruit of apple, pear, and avocado.

The fruit of many valuable crops develops without fertilization, e.g., banana, navel orange, seedless grape, pineapple, and certain varieties of cucumber. These fruits are called parthenocarpic fruits irrespective of whether pollination has occured. Recently, scientists have secured parthenocarpic fruits of crops which normally require pollination and fertilization by applying certain hormonelike chemicals called *growth regulators* to the stigma of the pistils. Examples of the chemicals used are naphthaleneacetic acid and indolebutyric acid.

Fruits of horticultural crops vary greatly in size, shape, color, and chemical composition. In general, they may be classified as fleshy or dry. Fleshy fruits may be further divided into simple and accessory. Simple fruits have developed from ovarian tissue and usually from flowers with superior ovaries, whereas accessory fruits have developed from ovarian and adjacent tissue and from flowers with superior and inferior ovaries. Dry fruits may be further divided as dehiscent and indehiscent. Dehiscent fruits shed their seed when dry; indehiscent fruits do not. Examples of fleshy and dry fruits are presented in Table 2-4 and are illustrated in Figure 2-18.

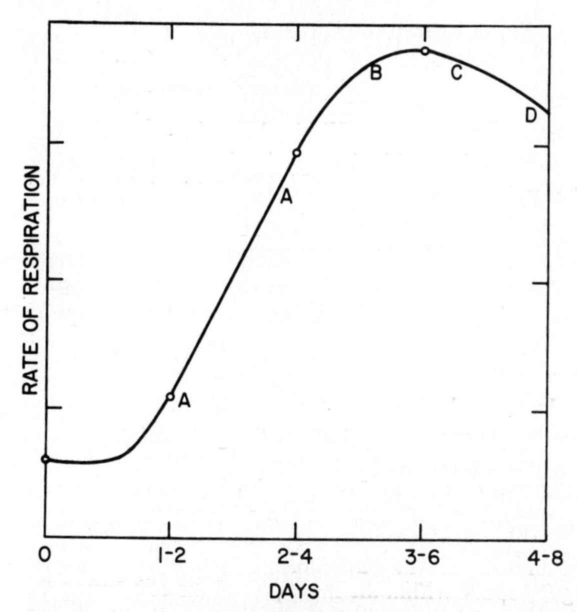

Figure 2-17 The rate of respiration of fruit with the climateric. A–B fruit becoming ripe and approaching the climateric. B–C fruit ripe and at the climateric. C–D fruit overripe and in the stage of senescence.

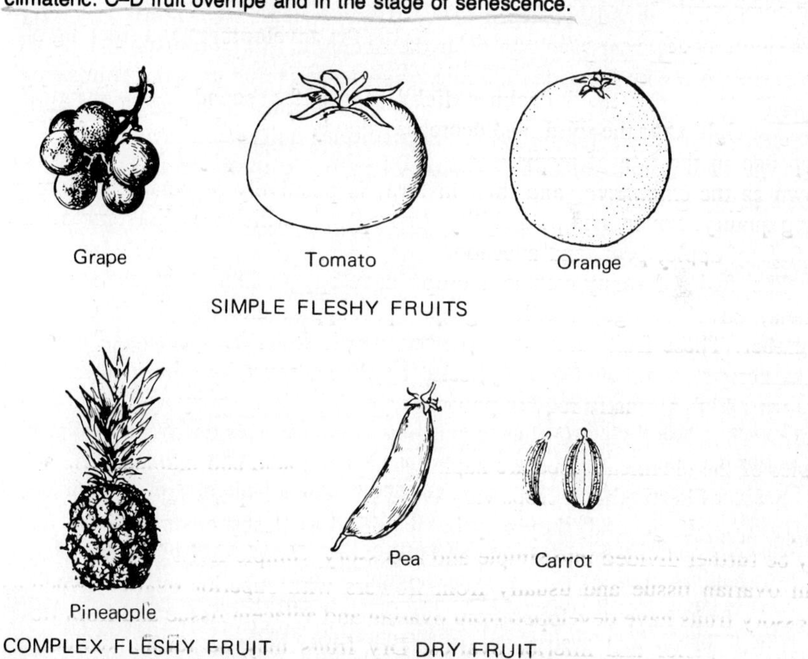

Figure 2-18 Fleshy and dry fruits. (*Redrawn by permission from M. C. Coulter,* The Story of the Plant Kingdom, *Chicago: University of Chicago Press.*)

Table 2-4 Types of Fruits of Some Horticultural Crops

Fleshy, high moisture content when mature		Dry, low moisture content when mature	
Ovary tissue only	**Ovary(s) and adjacent tissue**	**Shedding when mature**	**Not shedding when mature**
Peach, plum, grape, tomato, pepper, eggplant, currant, blueberry, date, gooseberry, olive, orange, avocado, persimmon	Apple, pear, strawberry, fig,* blackberry, pineapple, the cucurbits, dewberry	Peas, beans, okra, larkspur, lily, peony	Celery (seed), carrot (seed), sweet corn

*Most of the flesh consists of peduncle.

The Storage Organs

Storage organs are modified stems or roots designed primarily for the storage of manufactured food, usually reserve carbohydrates and related compounds. In general, they have the same anatomy as ordinary stems and ordinary roots except that certain tissues, usually the storage parenchyma, have developed many cells and have become greatly enlarged. Thus, storage stems have nodes and internodes and develop buds and shoots at the nodes just like ordinary stems; and storage roots are without nodes and internodes and develop growing points from the pericycle just like ordinary roots. Examples of storage stems are tubers of the potato and Jerusalem artichoke, corms of gladiolus, and rhizomes of canna, iris, lily of the valley, and rhubarb. Examples of storage roots are fleshy roots of dahlia and sweet potato.

QUESTIONS

1 Define pollination.
2 In general, how do flowers pollinated by wind differ from those pollinated by insects?
3 Growers place bouquets of apple blossoms under the trees during the pollinating season. Explain.
4 Observations have shown that frequently the windward side of apple trees possess less fruit than the leeward side. Can you think of any reason for this?
5 What is meant by self- and cross-pollination of asexually progagated crops? Self- and cross-pollination of sexually propagated crops?
6 What is protandry? Protogyny?
7 What is meant by compatible pollen? Incompatible pollen?
8 What is the main function of a pollinator variety?

9 Pollinator varieties are necessary in apple and sweet cherry orchards. Explain.
10 Large blocks of a single variety of most varieties of peaches can be planted, and large crops can be produced. Explain.
11 A peach orchard may consist of a single variety, but an apple orchard should contain at least two varieties. Explain.
12 In general, a single and isolated pecan tree fails to produce fruit consistently from one year to the next. Explain.
13 What is double fertilization?
14 Can you think of any advantages of parthenocarpy? Any disadvantages?

SELECTED REFERENCES FOR FURTHER STUDY

Esau, K. 1960. *Anatomy of seed plants.* New York: Wiley. An excellent text for students who wish to acquire a detailed knowledge of the anatomy of crop plants.

Siekeritz, P. 1957. Powerhouse of the cell. *Sci. Amer.* 197 (1):131–140. A discussion of the location, structure, and function of mitochondria in physiologically active cells.

Troughton, J., and L. A. Donaldson. 1972. *Probing plant structure.* New York: McGraw-Hill. A scanning electron-microscope study of the epidermis of leaves, stomates, mesophyll, chloroplasts, xylem, flowers, and seeds and their relation to certain physiologic and biochemical processes in crop plants.

Chapter 3

Phases of Growth, the Carbohydrates and the Hormones

First the blade, then the ear and the full corn on the ear.

THE VEGETATIVE PHASE AND THE REPRODUCTIVE PHASE

The growth and development of crop plants consist of two distinct, though overlapping, phases: the vegetative and the reproductive.

The Vegetative Phase and Carbohydrate Utilization

The vegetative phase concerns essentially the development of the stems, leaves, and absorbing roots. This phase is associated with three important processes: (1) cell division, (2) cell enlargement, and (3) the initial stages of cell differentiation.

Cell division involves the making of new cells. These new cells require large quantities of carbohydrates since the walls are made of cellulopectinaceous materials and the protoplasm is made mostly from sugars. Thus, with other factors in favorable supply, the rate of cell division is dependent upon an adequate supply of carbohydrates. As previously stated, cell division takes place within the meristematic tissues, the growing points at the stem and root tips, and the cambia. Therefore, these tissues must be provided with the manufactured foods, hormones, and vitamins in order to form new cells.

Cell elongation concerns the enlargement of the new cells. This process requires (1) abundant supplies of water, (2) the presence of certain hormones which give the cell walls the ability to stretch, and (3) the presence of sugars. As discussed in Chapter 2, the region of cell enlargement occurs just back of the growing points. When the cells in this region begin to enlarge, they develop large vacuoles. These vacuoles absorb relatively large quantities of water. As a result of this absorption of water and the presence of the cell-stretching hormones, the cells elongate. In addition to this increase in cell size, the walls become thicker, owing to the laying down of additional cellulose made from the sugars.

The initial stages of cell differentiation, or tissue formation, involve the development of the primary tissues. Their development requires carbohydrates, e.g., the thickening of the walls of the protective cells of the epidermis of the stem and the development of the water-conducting tubes of both the root and the stem. Thus, when a plant is making new cells, elongating these cells, and laying down its tissues, it is actually developing its stem, leaf, and root systems. If the rate of cell division and elongation and tissue formation is rapid, growth of the stems, leaves, and roots will be rapid also. Conversely, if the rate of cell division is slow, growth of the stems, leaves, and roots is accordingly slow. Since division, enlargement, and tissue formation require a supply of carbohydrates and since carbohydrates are used in these processes, the development of stems, leaves, and roots requires the utilization of carbohydrates. Thus, *in the vegetative phase of plant development, the crop plant is using most of the carbohydrates it is making.*

The Reproductive Phase and Carbohydrate Accumulation

The reproductive phase concerns the formation and development of flower buds, flowers, fruit, and seed or the enlargement and maturation of storage organs—fleshy stems and fleshy roots. This phase is associated with several important processes: (1) the making of relatively few cells; (2) the maturation of the tissues; (3) the thickening of the fibers; (4) the formation of hormones necessary for the development of flower-bud primordia; (5) the development of flower buds, flowers, fruit, and seed; (6) the development of storage organs; and (7) the formation of the water-retaining substances—the hydrophilic colloids. The student will note that all these manifestations of the reproductive phase require a supply of carbohydrates. In most cases these carbohydrates are the sugars and the starches. In other words, when a plant is developing its flowers, fruit, seeds, or storage organs, not all the carbohydrates are being used for the development of the stems, leaves, and absorbing roots; some are left for the development of the flowers, fruit, or seed, or for storage organs. For example, the flesh of the apple, the tubers of the potato, and the fleshy roots of the sweet potato all contain relatively large quantities of starch and sugars. Thus, *in the reproductive phase of plant development, the crop plant is storing most of the carbohydrates it is making*. Figure 3-1 shows the vegetative and reproductive primordia of cabbage.

THE VEGETATIVE-REPRODUCTIVE BALANCE

The vegetative and reproductive phases of plant development may be likened to a balance. One side of the balance may be considered the vegetative phase—the development of stems, leaves, and absorbing roots. The other side may be considered the reproductive phase—the development of flowers, fruit, seed, or storage organs. This concept presents three possible cases: (1) the vegetative phase may be dominant over the reproductive with the balance tipped on the vegetative side; (2) the reproductive phase may be dominant over the vegetative with the balance tipped on the reproductive side or (3) neither the vegetative nor the reproductive phase may be dominant with both sides of the balance practically equal. However, the student should guard against forming the idea that the vegetative phase takes place without the reproductive or that the reproductive phase takes place without the vegetative. If the vegetative phase is dominant with any crop, there is always some reproduction. On the other hand, if the reproductive phase is dominant, there is always some vegetation. For example, cell division is necessary for the development of reproductive and storage organs. However, the number of cells necessary for the development of these structures is small compared with the number necessary for the complete development of the stems, leaves, and absorbing root system of any given plant. Thus, the term *balance* refers to a matter of emphasis rather than to the presence or absence of either phase of plant development.

When the vegetative phase of plant development is dominat over the reproductive, carbohydrate utilization is dominant over accumulation. More carbohydrates are used than are stored. When the reproductive phase is dominant

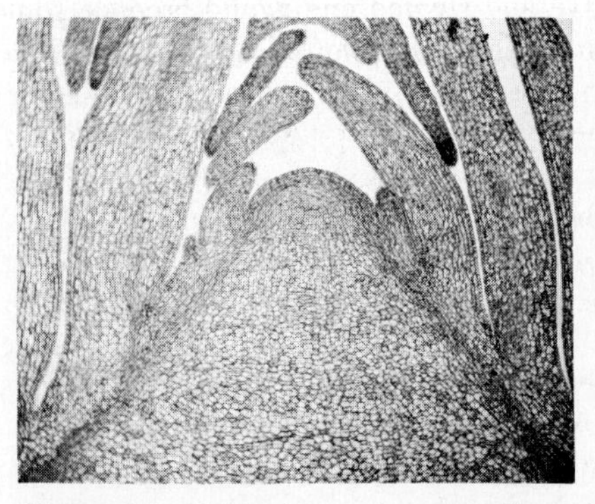

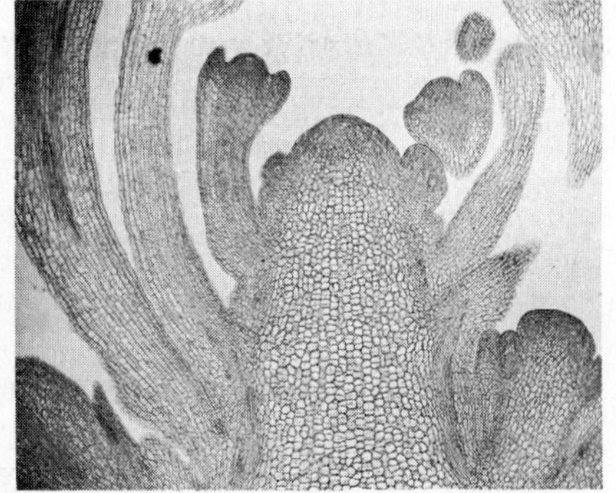

Figure 3-1 Vegetative and reproductive primordia of cabbage. Note the vegetative growing point of the plant on the left, and the flower primordia of the plant on the right. (*Courtesy, E. L. Moore, Mississippi State University.*)

over the vegetative, carbohydrate accumulation is dominant over utilization. More carbohydrates are stored than are used. When the vegetative and reproductive phases are in balance, utilization and accumulation are also in balance. Practically equal amounts of carbohydrates are used and stored.

As an example, take three tomato plants—A, B, and C—each of which has grown for a period of 150 days. In plant A the vegetative phase has been dominant over the reproductive; in plant B the reproductive phase has been dominant over the vegetative; and in plant C both phases have proceeded in practically equal magnitudes.

How would these plants appear? Plant A would be extremely vegetative; that is, there would be an abundant development of stems, leaves, and absorbing roots. The stems would be succulent, the leaves would be large with little development of cuticle—the waxlike substance on the surface. Flowering and fruiting would not occur or would be suppressed, the cell walls would be thin, and the strengthening tissue would be poorly developed. In other words, most of the carbohydrates would be used for the development of the root system, the stem system, and the leaves. As a result, very little of the carbohydrates would be left for the development of flower buds, flowers, fruit, and seed. In this case the vegetative phase was dominant over the reproductive and carbohydrate utilization was dominant over carbohydrate accumulation.

Extreme vigor of the top, combined with a lack of, or reduction in, the growth of flowers, fruit, and seed, generally develops under the following conditions: the plants are in the early stages of growth; they have a rapid rate of photosynthesis; the temperature favors a *rapid rate of cell division*; and water and essential raw materials are abundant; and the large quantity of carbohydrates made combines with the nitrogen compounds to form the protoplasm made in the growing points of the stems and roots. As a result, vegetative processes are dominant over reproductive processes. Note the plant on the extreme right in Figure 3-2.

Plant B would be poorly vegetative and stunted and would produce some fruit. There would be little development of leaves and stems. The stems would be woody, the internodes would be short, the leaves would be moderately small with the development of a thick cuticle. Flower and fruiting would be evident, the cells walls would be very thick, conducting tissues would be well developed, and storage tissues would be packed with starch. Since stems are necessary to support flowers and fruit and since in plant B there was a relatively low development of stems and leaves, yields would be accordingly low. In this case the reproductive phase was dominant over the vegetative and carbohydrate accumulation was dominant over carbohydrate utilization.

Weak, stunted plants generally result under the following conditions: the plants have a low or moderately low rate of photosynthesis; and the temperature, the water supply, or the essential-element supply, or some other factor *is unfavorable for even a moderate rate of cell division*. As a result the carbohydrates accumulate and are used for reproductive processes to a greater extent than for vegetative processes. Note the plant on the extreme left in Figure 3-2.

Figure 3-2 The vegetative-reproductive balance in tomatoes. Note that plant 197 is weakly vigorous and nonfruitful, plants 199 and 202 are moderately vigorous and fruitful, and plant 208 is extremely vigorous and nonfruitful. (*Courtesy, the late Jackson B. Hester, Elkton, Maryland.*)

Plant C would be moderately vegetative and fruitful. The stems would be moderately succulent, the internodes moderately long, and the leaves moderately large with normal development of cuticle. Flowering and fruiting would proceed simultaneously with the development of stems and leaves and absorbing roots. The cell walls would be fairly thick, and there would be normal development of the conducting tissues. In this case moderate amounts of carbohydrates were used for the development of stems, leaves, and absorbing roots and the remainder was used for the development of flowers and fruit or storage structures. Since both the vegetative and reproductive phases lacked dominance, carbohydrate utilization and accumulation lacked dominance also and each phase proceeded in practically equal magnitudes. Note the two plants in the middle of Figure 3-2.

Moderate vigor of the top, combined with the development of flowers, fruit, or seed, generally takes place under the following conditions: the plants have a high rate of photosynthesis, and the temperature and other environmental conditions favor a *moderately rapid rate of cell division*. As a result, not all the carbohydrates are used for the development of stems and leaves; some are left for the development of flowers and fruit. Vegetative and reproductive processes lack dominance and the plants are moderately vegetative and fruitful. Thus, the growth of stems and leaves is associated with carbohydrate utilization, and the development of flowers, fruit, seed, or fleshy structures is associated with carbohydrate accumulation. The student should also keep in mind (1) that associated with carbohydrate utilization are extreme succulence, crispness, and juiciness, properties which are highly desirable in some plants, and (2) that associated with carbohydrate accumulation are nonsucculence, woodiness, and resistance to cold and heat, properties which are highly desirable in other plants.

The Vegetative-Reproductive Balance and the Type of Growth and Disposition of Carbohydrates

In general, all crop plants require a dominance of the vegetative phase during the germination and seedling stages. The length of this phase varies with the type of crop. For example, with herbaceous annuals it is relatively short, a period of one, two, or more months; with woody perennials it is relatively long, a period of one, two, or more years. In either case, most of the carbohydrates are used for the development of the absorbing root system, the stems, and the leaves. Beyond this stage, that is, during the productive period, crop plants may be placed in one of five groups. These groups and their disposition of the carbohydrates are presented in Table 3-1. Note that with Group 1, the citrus trees, there are two cycles of shoot growth each year. This is in sharp contrast to the plants of Group 2, the deciduous trees, grapes, and small fruits, which have one cycle of shoot growth each year. In both groups, the period of shoot growth is followed by a period in which there is a cessation of new growth and a rapid enlargement of the fruit, and finally this period is followed by maturization of the fruit and new wood. Thus, for any given year, carbohydrate utilization is dominant during early spring; utilization becomes less dominant and accumulation becomes more evident during late spring and early summer; and carbohydrate accumulation becomes more dominant during the late summer and fall. Accordingly, large quantities of the cytokinins, auxins, and gibberellins are needed during spring and early summer, and relatively large quantities of the florigens, or growth inhibitors, are needed during late summer and fall.

With Group 3, the herbaceous crops grown for their storage organs, tubers, bulbs, and fleshy roots, the vegetative and the reproductive phases are more or less distinct. Since the storage organs are the places where carbohydrates accumulate, the yield of the storage organ is an expression of carbohydrate accumulation. Before storage organ formation can take place, the stem, leaf, and absorbing root systems must be formed. Hence, when the plants of this group are developing their tops and absorbing roots, they are making numerous new cells, and large quantities of carbohydrates are necessary for their formation. Consequently, when the plants are developing their stems, leaves, and absorbing roots, the utilization of carbohydrates is dominant over accumulation, and when the plants are developing their storage organs, accumulation is dominant over utilization. Accordingly, relatively large quantities of the cytokinins, auxins, and possibly gibberellins are needed during the vegetative phase, and large quantities of florigens, or growth inhibitors, are needed during the reproductive phase.

Table 3-1 Type of Growth and Disposition of the Carbohydrates during the Productive Period of Crop Plants

Group	Disposition of carbohydrates		
	Utilization dominant	**Utilization and accumulation in balance**	**Accumulation dominant**
1 Evergreen tree fruits	First cycle of shoot growth and growth of young fruit	Second cycle of shoot growth and continued growth of young fruit	Maturation of young wood and fruit
2 Deciduous tree and small fruits	Growth of shoots and young fruit	Cessation of shoot growth and continued growth of fruit	Maturation of young wood and fruit
3 Herbaceous crops grown for storage organ or with determinate types of growth	Growth of young plants	Cessation in growth of tops and initial formation of storage organ or flowers and fruit	Maturation of storage organ or flowers and fruit
4 Herbaceous crops with indeterminate type of growth	Growth of young plants	Growth of stems and leaves with simultaneous development of flowers and fruit	
	Depletion	**Utilization dominant**	**Accumulation dominant**
5 Herbaceous perennial crops	Spear, petiole, or flower head removal	Growth of stems or leaves	Maturation of stems or leaves

With Group 4, the herbaceous crops with the indeterminate type of growth, such as tomatoes (certain varieties), capsicum peppers, eggplant, and the vine crops, there are two distinct phases of growth and development: prefruiting and fruiting. During the prefruiting stage, the plants develop their stems, leaves, and absorbing roots, and during the fruiting stage, they develop stems, leaves, and absorbing roots simultaneously with flowers and fruit. Consequently, utilization of carbohydrates is dominant during the prefruiting stage, and a balance between utilization and accumulation is necessary during the fruiting stage. Accordingly, large quantities of the cytokinins, auxins, and possibly gibberellins are needed during the prefruiting stage, and a balance between these compounds and the florigenic, or, growth inhibitor, hormones are needed during the fruiting. Compare the vegetative vigor of the vines to the fruitfulness of the tomato plants presented in Figure 3-2.

Finally, with Group 5, the herbaceous perennials, for any given growing season there are three distinct periods: (1) carbohydrate depletion; (2) carbohydrate utilization, and (3) carbohydrate accumulation. Thus, these plants are unique in that carbohydrates are depleted but not exhausted during the first part of the growing season, utilization is dominant during the second part, and accumulation is dominant during the third.

The Vegetative-Reproductive Balance and the Environment

As previously pointed out, all crop plants require a dominance of the vegetative phase during the germination and seedling stages. Beyond this stage they may be divided into three distinct groups: (1) woody plants which require a dominance of the vegetative phase during the first part of the growing season and a dominance of the reproductive phase during the latter part, (2) herbaceous crops which require a dominance of the vegetative phase during their first period of growth and a dominance of the reproductive phase during their latter, and (3) herbaceous crops which require both phases proceeding in practically equal magnitude. Examples of each group are presented in Table 3-2. The question arises: Can the vegetative and the reproductive phases be controlled by the environment and environmental practices? As is well known among crop-plant growers, certain factors of the environment and certain cultural factors markedly affect the two phases of crop plant growth and development. Principal environmental factors are (1) the water supply, (2) the temperature, (3) the light supply, and (4) the

Table 3-2 Examples of Crops According to Disposition of Carbohydrates as Shown in Table 3-1

Group	Crops
1	Orange, lemon, grapefruit, lime
2	Apple, pear, peach, cherry, plum, pecan, walnut, filbert, grapes, small fruits
3	Tomato,* snap bean,* potato, onion, root crops, sweet potato, rose chrysanthemum, carnation
4	Tomato,† snap bean,† cucumber, cantaloupe, watermelon, capsicum pepper
5	Asparagus, rhubarb, globe artichoke

*Determinate varieties.
†Indeterminate varieties.

essential-element supply. Since these factors usually limit the growth and development of crop plants, they are called the *limiting factors* in crop-plant production.

THE MEANING AND IMPORTANCE OF LIMITING FACTORS

In 1905, F. F. Blackman, a plant biochemist, studied the effects of two variables, the concentration of carbon dioxide and the intensity of light on the rate of photosynthesis of excised leaves of cherrylaurel. He used varying concentrations of carbon dioxide and two levels of light intensity: a low intensity which enabled the leaves to assimilate 5 cubic centimeters of carbon dioxide per leaf per hour, and a high light intensity which enabled the leaves to assimilate 10 cubic centimeters of carbon dioxide per leaf per hour. This investigator presented his results in the form of a graph as shown in Figure 3-3. Note that with the low light intensity, the rate of photosynthesis increased from zero to an assimilation rate of 5 cubic centimeters of carbon dioxide per leaf per hour. Thus from A to B, the concentration of carbon dioxide was the limiting factor because only through an increase in the supply of carbon dioxide would there be an increase in the rate of photosynthesis. However, at point B the light intensity and not the carbon dioxide supply became the limiting factor, and only through an increase in light intensity would there be an increase in the rate of photosynthesis. Note also that

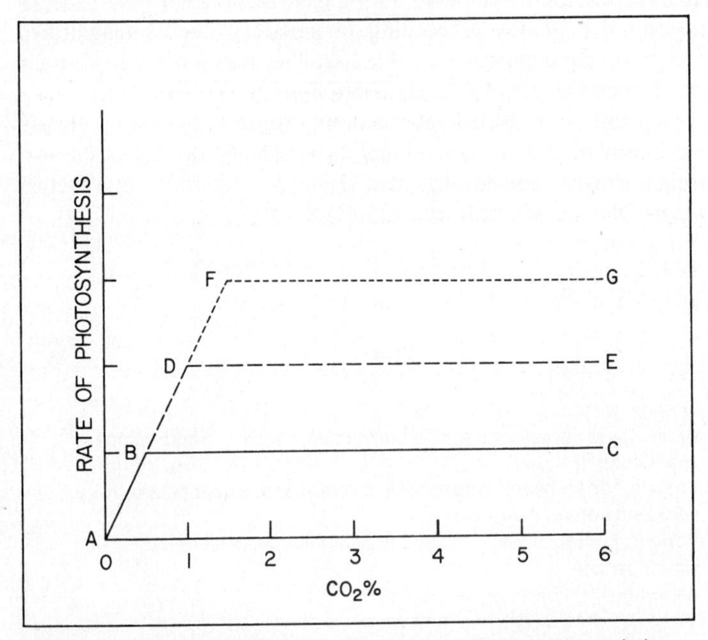

Figure 3-3 The principle of the limiting factor as illustrated by variation in supplies of carbon dioxide and light. (*Adapted from Blackman, F. F.,* Ann. Bot., *19:281–295, 1905.*)

at the higher light intensity the rate of photosynthesis increased in proportion to the concentration of carbon dioxide, and at point B the intensity of light became the limiting factor and not the concentration of carbon dioxide. From this work Blackman developed the principle of the limiting factor. He states, "When a process is conditioned as to its rapidity by a number of separate factors, the rate or speed of the process is governed by the pace of the slowest factor."

In general, the growing of crops is conditioned by a large number of factors. Thus, the principle, or law, of the limiting factor has wide application. Example 1, the rate of development of stems and leaves, is dependent largely on the amount of sugars made and the amount of nitrate or ammonium-nitrogen absorbed and assimilated with the sugars in the formation of active proteins. If the supply of available nitrogen in the soil is low, the production of new protoplasm is accordingly low and the development of stems and leaves is correspondingly low. In this case, the available nitrogen supply is the limiting factor because only through an increase in the available nitrogen supply can an increase be obtained in the growth of the stems and leaves. It would be useless to modify the water supply, the light supply, the phosphorus supply, or any other factor except available nitrogen. Example 2, crops grown in green and plastic houses, produce greater marketable yields in summer than they do in winter. In winter, light intensity and duration are relatively low—lower than the heat, water, and essential-element supply—and are usually limiting factors in greenhouse culture. In winter it would be useless for the grower to modify the temperature, the water supply, or the fertility of the soil. It is only through an increase in the light supply that an increase in growth can be obtained. Example 3, yields of crop plants, are low after they have been subjected to a period of drought. Usually under these conditions the temperature, the light supply, and the essential-element supply are quite favorable to growth and development. It is only through an increase in the water supply that an increase in yields can be obtained.

In other words, the growth and ultimate yield of any crop are regulated largely by a limiting factor. This does not mean that any one factor remains the limiting factor throughout the entire life of the plant. Sometimes it may be the carbohydrate supply, sometimes the available nitrogen supply, sometimes the water supply, or sometimes the heat supply. Since the limiting factor largely determines the rate of plant growth and development, an appreciation on the part of the student of the importance of the limiting factor is necessary. In many cases the solving of the grower's problems consists in discovering and dealing with the limiting factor. In the discussion of the principal factors limiting plant growth and development in Chapters 4 through 7, it is assumed that *all environmental factors except the one under discussion are in plentiful supply*.

QUESTIONS

1 Carbohydrates and proteins are used in the making of cells. Explain.
2 Abundant water within the plant is necessary for cell enlargement. Explain.

3 Liberal supplies of free energy are necessary for the making and enlargement of cells. Explain.
4 When a plant is developing its stems and leaves rapidly, it is using sugars and proteins rapidly. Explain.
5 Thickening of the cell walls and the storage of sucrose, starch, gums, and mucilages are manifestations of carbohydrate accumulation. Explain.
6 Mild nitrogen deficiency and carbohydrate accumulation frequently go together. Explain.
7 The behavior of crop plants is centered around the manufacture of carbohydrates and their utilization and accumulation. Explain.
8 In general, crops in the vegetative phase utilize more available nitrogen in sunny weather than in cloudy weather. Explain.
9 In the growing of spinach and leaf lettuce, what phase of growth should be dominant? Give your reasons.
10 What balance would you maintain during the flowering and fruiting of tomatoes? Give your reasons.
11 Study the tomato plants in Figure 3-2. Which plants are moderately vigorous? In your opinion, what is the limiting factor in the growth of the plant at the extreme left? Explain.
12 Excessive vegetative growth of young apple trees is likely to delay the time they begin to bear fruit. Explain.
13 Sweet peas growing in the greenhouse in rich soil are likely to produce very long and vigorous vines and long-stemmed flowers late in the season. Explain.
14 The quality of grape juice and wine varies with the nature of the growing season. Explain.
15 With other factors favorable, stem cuttings with large quantities of carbohydrates stored in their tissues produce roots at a greater rate than those with small quantities of carbohydrates. Explain.
16 What is meant by the limiting factor in growth and development of crop plants?
17 If air, water, temperature, light, or essential elements become deficient or unfavorable, high yields of high-quality crops will not materialize. Explain.
18 What environmental factor is likely to be the limiting factor in the growth of a plant during the summer at sunrise? At noon? At sunset? Explain.

SELECTED REFERENCES FOR FURTHER STUDY

Blackman, F. F. 1905. Optima and limiting factors, *Ann. Bot.* 19:251–295. The elucidation of a useful concept in crop-plant production.

Buckovac, M. J. et al., 1965. Effects of excessive fruiting on the vegetative development of young apple trees and a proposed chemical means of defruiting. *Mich. Agr. Exp. Sta. Quar. Bul.* 47:364–372. An illustrated research report on the effect of semi-dwarfing rootstocks on the vegetative and reproductive development of apple trees of the Golden Delicious and Jonathan varieties.

Kraus, F. J., and H. R. Kraybill. 1918. Vegetation and reproduction with special reference to the tomato. *Oregon Agr. Exp. Sta. Bul.* 149. One of the first classic investigations on the mineral nutrition of crop plants. The results reported in this publication laid the foundation for an understanding of commercial fertilizer practices in orchards, gardens, and fields.

Loomis, W. C. 1932. Growth-differentiation balance versus carbohydrate-nitrogen ratio. *Proc. Amer. Soc. Hort. Sci.* 29:240–245. A broadening of the concept developed by Kraus and Kraybill elucidating a definite role of the water supply within the plant.

Chapter 4

The Water Supply

The water supply is a basic and valuable resource. It should never be wasted or polluted.

LIFE-GIVING PROPERTIES AND FUNCTIONS OF WATER

Within a narrow range of temperature, water may exist as a solid (ice), as a liquid, or as a gas (vapor). As a liquid, (1) water has high specific heat; that is, it has the capacity to absorb large quantities of heat with relatively low changes in its temperature. (2) Water has high latent heat of vaporization; that is, large quantities of heat are required to change water from the liquid state to the vapor state. Thus, when water evaporates from any given surface of a crop plant, the cooling effect on that surface is quite pronounced. (3) Water has high heat of fusion of ice; that is, large quantities of heat are given off when water changes from the liquid state to the solid state. (4) Further, water has its maximum density as a liquid at 4°C (40°F). This explains why water in the solid state—ice—floats on water in the liquid state. (5) In addition, water has a high dielectric constant. This means that numerous compounds go into solution with water and some of their molecules dissociate into ions which, in turn, promote chemical reactions.

(6) Of great significance is the arrangement of each of the two hydrogen atoms to the oxygen atom. Note in Figure 4-1 that the two hydrogen atoms diverge from the oxygen atoms at about 105°. As a result, the oxygen of any given molecule of water shares the hydrogen of adjacent molecules by a phenomenon known as *hydrogen bonding*. This bonding, or sharing, of hydrogen with adjacent molecules gives water its great cohesive properties—a phenomenon important in the movement of water in the water-conducting vessels, the xylem. (7) Moreover, this unique compound combines with CO_2 in the formation of the initial manufactured substances, the compounds which contain potential energy, and (8) combines in hydrolytic reactions in the decomposition of these manufactured substances and the liberation of free energy.

THE WATER ABSORPTION—TRANSPIRATION STREAM

The water absorption—transpiration stream refers to the path through which the water flows. In general, it enters the plant in the finer portions of the root system and, with most crops, in the root hair zone and flows laterally in the walls of the cortex, through the protoplasm of the endodermis, and into the vessels of the xylem. (See Fig. 4-2). From the xylem in the root system, the water flows vertically into the xylem of the stem system and finally into the xylem of the leaves. From the xylem of the leaves, it flows laterally through the chlorenchyma

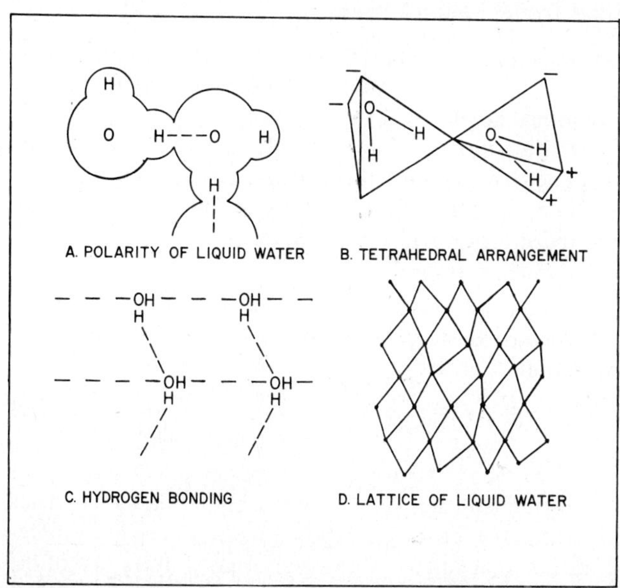

Figure 4-1 The arrangement of hydrogen and oxygen atoms, and the positive negative charges in the molecules of water. *(Courtesy, A. C. Leopold, Purdue University and McGraw-Hill Book Company, Inc.)*

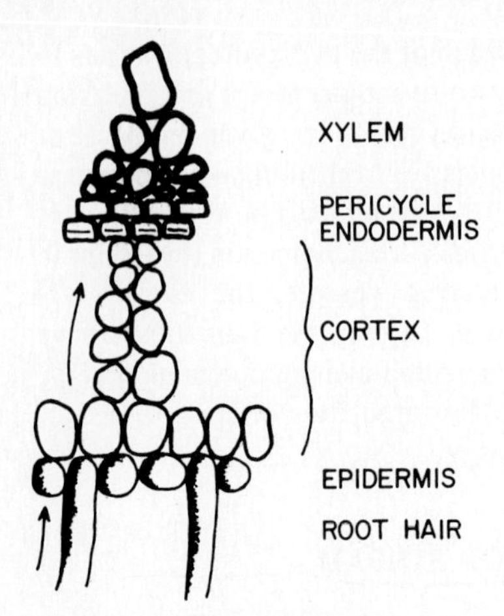

Figure 4-2 The path of water from the root hairs to the endodermis. *(Adapted from Figure 5, How crops absorb water,* Conn. Agr. Exp. Sta. Bul. *708, 1969.)*

and finally to the stomatal chamber. In the stomatal chamber, the liquid water changes to the vapor form and, if the stomates are open, the molecules of vapor escape to the outside air. Since large quantities of heat are required to change liquid water to the vapor form, transpiration helps to keep the leaves cool. This seems to be particularly the case during periods of intense sunshine.

The Main Features of the Transpiration Stream

The transpiration stream is continous throughout the plant body from the region of absorption to the region of vaporization, and its rate of movement through the plant body seems to be governed largely by the rate of transpiration. This is made possible in both cases by the enormous force of cohesion between adjacent molecules of water. As pointed out previously, this enormous force is due to the sharing of hydrogen atoms by adjacent oxygen atoms of the water molecule. Thus, despite the stresses and strains to which the column of water is subjected, it remains unbroken, at least under average conditions.

The Need for Balance Between the Rate of Water Absorption and Transpiration

In crop plants most of the water is lost through the leaves. There are several reasons for this: (1) the leaves of most crop plants are flat and broad, and as such they present a large external surface to the forces of evaporation; (2) in order for photosynthesis to take place the stomates must be open; (3) when the stomates are open, the moist cells within the leaves are exposed to the forces of evaporation; and (4) the area of the internal surface of leaves is several times that of the external surface. (See Fig. 4-3) Thus, if photosynthesis is to take place under conditions of high rates of evaporation, high rates of transpiration are inevitable;

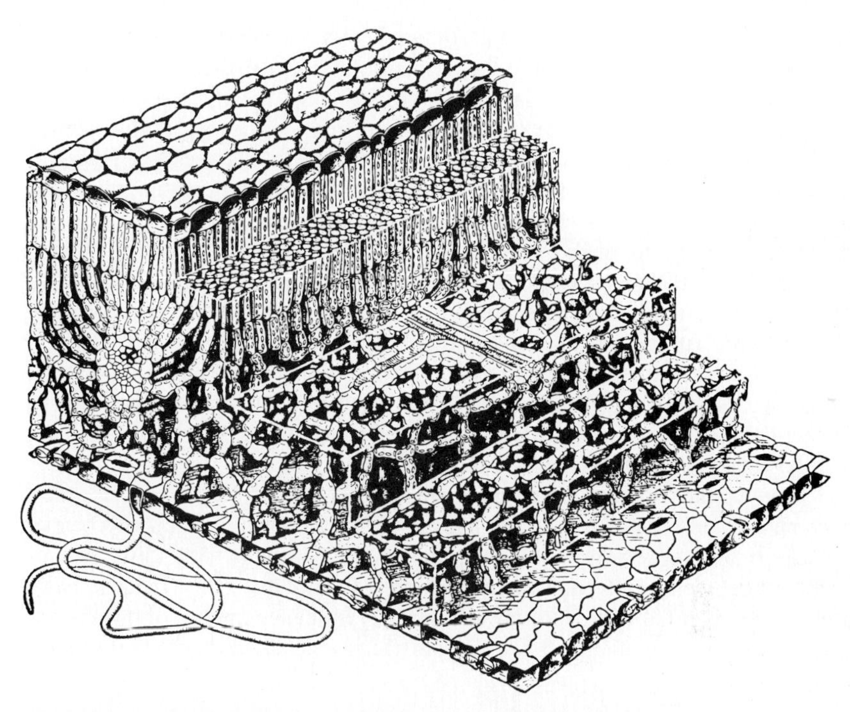

Figure 4-3 Internal structure of leaf of apple. Note the intercellular spaces below the palisade cells. *(Courtesy, A. J. Eames and L. H. MacDaniels,* An Introduction to Plant Anatomy, *New York: McGraw-Hill Book Company, Inc.)*

and if high rates of transpiration take place, high rates of water absorption should take place also. In other words, the amount of water going into a plant per unit of time should, in general, equal the amount going out. When the rate of absorption is less than the rate of transpiration, the guard cells lose turgor and thus become flaccid, and the stomates partially or completely close. This results in decreases in the rate of diffusion of carbon dioxide into the chlorophyll-containing cells with resultant decreases in the manufacture of the initial food substances and corresponding decreases in growth and yield. To more fully understand how deficits occur within plants and how this results in unsatisfactory growth and usually low yields, the student should have a working knowledge of the factors which influence the rate of absorption, or the income of water, and of the factors which influence the rate of transpiration, or the outgo of water.

The Absorption, or Uptake, of Water

The absorption, or uptake, of water involves the movement, or diffusion, of water from the soil in the immediate vicinity of the plant's roots to the xylem of the root system. In general, and in accordance with our present knowledge, the diffusion of water with a pressure gradient and the diffusion of water against a pressure gradient are involved. In either case, work is done and a source of free energy is necessary. This free energy comes from the decomposable simple sugars, probably glucose. In the diffusion of water with a pressure gradient, sugars are needed to maintain a relatively high concentration of water in the soil, and in the diffusion of water against a pressure gradient, sugars are needed to provide the energy to push the molecules of water against its own diffusion pressure. From the standpoint of thermodynamics and differences in water potential,[1] this relation may be expressed as follows. The free energy of water in soil is greater than the free energy of water in the root hair, cortex, endodermis, and pericycle; this, in turn, is greater than the free energy of water in the xylem. In other words, the water potential in the soil is greater than the water potential in the root-hair zone, cortex, endodermis, and pericycle; and this, in turn, is greater than the water potential in the xylem.

The student will note that sugars are needed for the absorption of water. Since photosynthesis makes the sugars, the compounds which contain potential free energy and since respiration decomposes these sugars, with the release of kinetic energy, environmental factors which influence the rates of these basic reactions are likely to either directly or indirectly influence the rate of the absorption of water by crop plants. The principal factors are divided into two groups: soil and plant.

Soil Factors

Amount and Kinds of Water in Soils When well-drained soils are saturated with water, as for example after a heavy rain, a certain amount of water percolates through the pore spaces and drains away because of the force of gravity. The water which remains is held against the force of gravity, and when measured it is called the field capacity. Thus, *the field capacity of soils may be defined as the amount of water held against the force of gravity.*[2] The field capacity varies greatly with the type of soil. Study the field capacity of the soils presented in Table 4-1. Note that sand and sandy loams have relatively low field capacities and that silt and clay loams have high field capacities. Since these data are more or less representative of soils as a whole, we can say that, in general, coarse-textured soils have low field capacities and fine-textured soils have relatively high field capacities.

Not all the water held by soils against the force of gravity is available for the use of the plant. Even when the guard cells lose turgor and the leaves droop or wilt, soils still contain some water; thus plants are needed in order to determine

[1]*Water potential* is the difference between the energy of water in any given system and the energy of pure water at the same temperature.

[2]The field capacity, as defined, is a rough measure at best since some soils lose water by drainage over a relatively long period.

Table 4-1 The Water Content of Certain Soils

Soil type	Location	Field capacity,* %	Permanent wilting, %	Available water, %	Available water, in./ft
Plainfield fine sand	Ohio	2.4	1.4	1.0	0.2
Yuma sand	Arizona	4.8	3.2	1.6	0.3
Delano sandy loam	California	9.1	4.2	4.9	0.8
Redman sandy loam	Oregon	18.8	6.6	12.2	2.0
Yolo fine sandy loam	California	16.8	8.9	7.9	1.3
Wooster silt loam	Ohio	23.4	6.1	17.3	2.9
Dunkirk silty clay	New York	21.7	5.0	16.7	2.7
Aiken clay loam	California	31.1	25.7	5.4	0.7

*Determined as moisture equivalent.
Source: Adapted from *Proc. Am. Soc. Hort. Sci.* 40:485, 1940.

the amount of water in the soil which they cannot absorb. In general, this is done by placing samples of soil in standard galvanized cans, each with a hole in the top for the stem of a plant; by planting a seed of a herbaceous plant, usually that of sunflower, in the soil of each can; by bringing the soil to and keeping it at or near the field capacity until the plant has developed three or four leaves and then allowing the leaves to wilt; and finally by placing the plant in a dark chamber with a high relative humidity. When the leaves do not recover from wilting, the amount of water in the soil is measured to determine the permanent wilting percentage. Note that the plants are placed in a dark, moist chamber, a condition which promotes a low rate of transpiration. Thus, *the permanent wilting percentage may be defined as the amount of water in soils when rapidly growing plants fail to recover from wilting under conditions of low transpiration.*

As with the field capacity, the permanent wilting percentage is not the same for all soils. Study the permanent wilting percentage of the soil types presented in Table 4-1. Note the low permanent wilting percentage of the fine sand and the lack of a definite relation of texture to the permanent wilting percentage of the other types. For example, the Yolo fine sandy loam has a somewhat higher permanent wilting percentage than the Wooster silt loam and the Dunkirk silty clay loam. Thus, the data indicate that very coarse-textured soils have a low permanent wilting percentage, and no definite or well-defined relation of texture to permanent wilting percentage exists with the other types.

The difference between the field capacity and the permanent wilting percentage is usually called the available water, that is, the water plants can absorb. However, as previously stated, the permanent wilting percentage is determined by placing plants in a dark, moist chamber, a condition which is accompanied by low rates of transpiration. Since the rate of transpiration is relatively high in the light and in dry air, the permanent wilting percentage, as determined by this method, would seem to be too low for crops growing under conditions of high transpiration. Nevertheless, the present method can be used to show the marked variation in the amount of the so-called available water in soil. Here again the student is referred to the data in Table 4-1. Note the low quantities of available water in the Plainfield, Yuma, and Aiken soils and the high quantities in the Redman, Wooster, and Dunkirk soils. Thus, soils with low field capacities necessarily have low quantities of available water, but soils with high field capacities may or may not have high quantities of available water. In general, from the standpoint of the water supply, ideal soils have high field capacities and low permanent wilting percentages, e.g., the Redman sandy loam in Oregon, the Wooster silt loam in Ohio, and the Dunkirk silty clay loam in New York.

Tension on the Capillary Films In general, the capillary water moves under the forces of capillarity, and it is always under a state of tension. However, the degree of tension varies greatly between that at field capacity (FC) and that at the permanent wilting percentage (PWP). In general, the tension varies from 0.3 to 0.4 atmospheres at FC to 15 atmospheres at the PWP and increases rapidly at one atmosphere for light sandy loams and at three or more atmospheres for silt and clay loams. The free energy of the capillary films decreases accordingly. In general, this explains why the amount of available water is low near the PWP, why in irrigation practices the zone of soil occupied by plant roots should be wetted to the soil's field capacity, and why irrigations should be more frequent under conditions of high transpiration than under conditions of low transpiration.

Rate of Movement of Available Water Important factors are soil temperature and concentration of the soil solution. Temperature affects the movement of available water in three ways. It influences the kinetic energy and viscosity of the molecules and the tension on the surface of the capillary films. In general, rising temperatures increase the kinetic energy and decrease the viscosity and surface tension, and these effects of temperature hold usually between 32 and 95°F (0 and 35°C). Thus, water, moves less rapidly in cold soils than in warm soils. (See Fig. 4-5) This effect of temperature explains, partially at least, why growers use bottom-heated propagation beds in the greenhouse, why they use warm water in the irrigation of warm-season crops in the greenhouse in winter, why rapidly growing herbaceous crops in the greenhouse may wilt if irrigated with cold water during sunny weather, and why broad-leaved evergreen trees may wilt in winter, especially during periods of intense sunshine.

Concentration of the soil solution refers to the number of solute particles per unit volume of solution. Water has great solvent properties. As a result, the water in the soil contains many substances in solution. The substances in solution, the solutes, naturally get in the way of water molecules and retard their movement. Thus, the greater number of solute particles per unit volume of solution, the greater is the retardation of movement of the water molecules. Usually, the water of soils contains a relatively low concentration of solutes and the available water

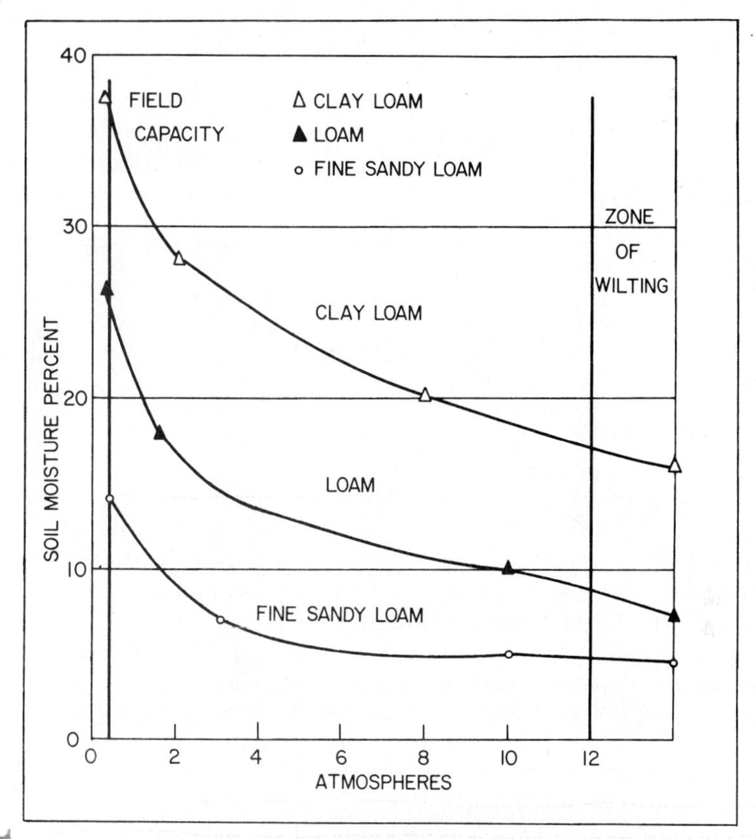

Figure 4-4 The tension of capillary films of water in three types of soil. Note that as the soil dries the amount of available water decreases. (Adapted from Figure 3, *U.S. Dept. Agr. Yearbook,* 1957, p. 64.)

moves freely to the water-absorbing region. Sometimes, however, the concentration of the solutes becomes so high that the available water moves slowly into the plant and the direction of movement may be reversed. Instead of moving into the plant, the water moves out of the plant and into the soil. Sometimes the grower inadvertantly causes this reversal of water movement when he applies large quantities of soluble fertilizer close to the plant's roots or to the seed.

Depth of Water Table The water table refers to the surface of the zone in soils which is saturated with water. In this zone, the pore space is entirely filled with water, and as a result insufficient quantities of oxygen are available for the growth and respiration of the root system. In fact, because of capillary movement, insufficient quantities of oxygen are present for a distance of 12 to 18 inches (30 to 46 centimeters) from the water table. Thus, with other factors

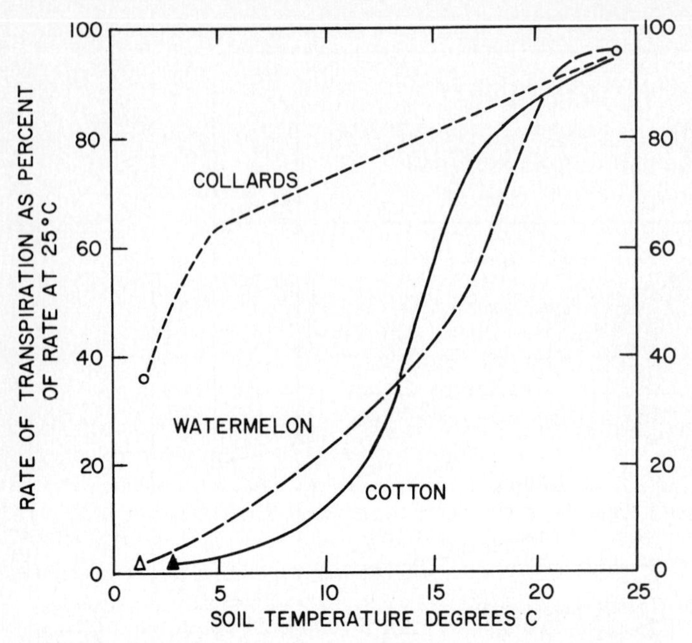

Figure 4-5 The relation of soil temperature to the rate of water absorption. Note that low soil temperature retarded the rate of absorption in all three crops. *(Adapted from Figure 7, How crops absorb water,* Conn. Agr. Exp. Sta. Bul. *708, 1969.)*

favorable, the depth of root penetration is limited to a distance which extends from the upper surface of the soil to about 12 inches (30 centimeters) above the water table. For example, if a water table is 6 feet (1.8 meters) below the surface, the depth of root penetration would be limited to 5 feet (1.5 meters). The question arises: What is the optimum depth of the water table for crop production? In general, soils with a water table close to the surface throughout the period of growth of a crop are unsatisfactory for crop production. For example, studies of the relation of height of the water table to productivity of raspberry plants in a Michigan plantation showed that plants with their root systems confined to the upper 7 inches (18 centimeters) of the soil, because of a high water table, were small, weakly vigorous, and unproductive. On the other hand, plants with root systems which penetrated to a depth of 14 inches (35.6 centimeters) were large, moderately vigorous, and productive. Similar studies have shown that the highbush blueberry requires a water table from 12 to 24 inches (30 to 61 centimeters) from the surface; and onions grown on muck require a water level of 24 to 36 inches (61 to 91 centimeters) from the surface. In fact, many growers of onions and celery on muck regulate the height of the water table by means of dams in the drainage ditches. In this way a uniform supply of moisture to the roots is assured, and excessive oxidation of the muck does not take place. In general, soils with water tables close to the surface, usually 1 to 4 feet (0.3 to 1.2

meters), depending on the transpiring surface of the plant and the type of soil, are likely to seriously limit the growth of the root system and the growth and productivity of the plants.

Plant Factors

Rate of Photosynthesis As previously stated, the water-absorbing region exists just back of the growing points of the root system. Its function is the absorption of water and solutes in the water. In general, the water-absorbing power of this region is conditioned largely by its osmotic pressure. This, in turn, is determined by differences in the concentration of water on each side of the cytoplasmic membranes of the absorbing cells. These living membranes are semipermeable; that is, they permit some substances to pass through and do not permit others. Usually, these membranes are permeable to mineral solutes and water and impermeable to organic substances, such as sugars and proteins in solution. These sugars and proteins are in solution with water within the aborbing region and are in greater concentration than the minerals in solution with water in the soil. Because of the lower concentration of water in the absorbing cells, water diffuses from the soil into the roots. The relatively low concentration of water in the absorbing region is due largely to the sugars in the cells of the absorbing region. Photosynthesis makes the sugars. These sugars pass down the phloem of the stem into the root system. Consequently, with other factors favorable, plants with high rates of photosynthesis can absorb more water per unit time than can plants with low rates. Thus, plants with dark green leaves or with healthy leaves have the ability to absorb more water per unit time than plants with light green or diseased leaves.

Rate of Respiration As stated previously, the region of water absorption contains living cells. These cells are constantly respiring. Thus, they are continually giving off carbon dioxide and taking in oxygen. Experiments have shown that if the oxygen of the soil air is displaced by nitrogen or carbon dioxide, water absorption is reduced or entirely stopped. This need of oxygen for the absorption of water emphasizes the importance of adequate drainage. If the pore space of the soil is saturated with water, oxygen for the respiration of the living cells is limited. As a result, the protoplasm dies and the absorption of water ceases. In order to get excess water out of the soil and air with its oxygen into the soil, drainage is necessary. This explains why growers drain relatively low places in crop-growing fields; why they drain soils which would otherwise remain saturated for considerable periods; and why water standing on fields of rapidly growing crops prevents the absorption of water and causes injury, if not death, to the plants. Most fruit, vegetable, and flower crops require well-drained soils.

Depth and Density of the Absorbing Surface The depth of the absorbing surface refers to the depth of soil penetrated by the roots. In general, depth of penetration varies with the stage of growth, the kind of plant, and the type of soil.

Table 4-2 Extent and Depth of Root Systems of Certain Crops When Grown in Well-Drained Soils

Nonextensive and shallow	Extensive and moderately deep	Extensive and deep
Celery, lettuce, onion	Beet, cabbage, carrot, cucumber, lima bean, potato, snap bean, sweet corn, summer squash	Asparagus, cantaloupe, sweet potato, tomato, watermelon, winter squash

In well-drained soils some plants develop rather shallow root systems, others develop moderately deep root systems, and others develop deep root systems. Naturally, plants with deep root systems can obtain more water than plants with shallow root systems. This is particularly true during conditions of high transpiration. Note the differences in depth of root penetration of the crops presented in Table 4-2.

The density of the absorbing surface refers to the number of root hairs and fine roots which occupy each unit volume of soil. Take two plants, A and B. The root system of plant A has 1 million root hairs for every cubic foot of soil for a depth, width, and length of 10 feet (3 meters), and the roots of plant B have only 10,000 root hairs for every cubic foot of soil occupied by the roots. Since the capillary water moves for very short distances, plant A, because of its greater root density, will obtain greater quantities of water than plant B. Thus, both depth of root penetration and degree of branching or root ramification are important, particularly during periods of high transpiration. A prime characteristic of drought-resistant crop plants is that they develop deep, extensive, and much-branched root systems.

QUESTIONS

1 Enumerate the properties and functions of water in crop-plant growth.
2 Define transpiration.
3 Most of the water is lost through the leaves. Explain.
4 Under conditions of high transpiration, a high rate of photosynthesis requires a high rate of water absorption. Explain.
5 Distinguish between field capacity, permanent wilting percentage, and available water in soils.
6 What is the relation of texture to the field capacity of soils?
7 From the standpoint of the water supply, ideal soils have high field capacities and low wilting percentages. Explain.
8 Under conditions of high transpiration, plants in full leaf and in cold soils are likely to wilt. Explain.
9 In general, the amount of water available to plants varies inversely with the concentration of soluble salts in the soil solution. Explain.

10 Growers sometimes withdraw water from their growing crops. Explain how they do this.
11 It is largely a matter of the plant roots growing to the water supply rather than the water supply moving to the roots. Explain.
12 A crop plant with a deep root system withstands drought to a better extent than a crop plant with a shallow root system. Explain.
13 Plants with dark green leaves have a greater water-absorbing capacity than plants with light green leaves. Explain.
14 Plants with healthy leaves have a greater water-absorbing capacity than plants with diseased leaves. Explain.
15 Plants growing in well-drained soils have a greater water-absorbing capacity than those growing in saturated soils. Explain.
16 Rapidly growing crops frequently wilt immediately after a heavy rain during the summer. Explain.
17 A heavy sod may limit and eventually prevent the development of root systems of orchard trees. Explain.
18 The presence of a hardpan in the upper level of soils is injurious to fruit trees. Explain.
19 Why does a large deciduous tree decline in growth and finally die after 2 or 3 feet of soil is added to the surface of the original soil?
20 Show how the breaking up of tight subsoils may increase crop yields.
21 In general, farmers should consider the water content, aeration, and fertility of the subsoil. Explain.

TRANSPIRATION, OR OUTGO, OF WATER

Area of Transpiring Surface

In general, with other factors equal, the amount of water transpired by any crop plant is more or less directly proportional to its area of transpiring surface. Thus, plants without leaves transpire much less than plants with leaves. In fact, the loss of water from leafless deciduous plants in late fall, winter, and early spring, is very low when compared to the loss when they are in full leaf. An important practical application is the transplanting of deciduous trees and shrubs. The removal of the trees or shrubs destroys at least some of the absorbing roots. A new absorbing system must be formed before absorption can supply the demands of transpiration. This explains why young deciduous trees and shrubs are transplanted during the nongrowing season. During this period the deciduous trees and shrubs are not in full leaf and the amount of water transpired is relatively small.

Internal Structures of the Leaves

In crop plants the internal structure of the leaves may vary greatly. Note the differences in the leaves of the apple, Kentucky bluegrass, and white pine presented in Figures 2-10, 2-11, 2-12, in the amount of intercellular space and compactness of the chlorenchyma, the tissue containing chloroplasts, in the number of layers of epidermis, and in the arrangement and position of the guard cells. In general, the leaf of the apple has an outer surface only, its chlorenchyma has large and frequent intercellular spaces, and its epidermis consists of one layer of cells only; whereas the leaves of Kentucky bluegrass and white pine have both an outer and an inner surface, their chlorenchyma is compact with very few if any intercellular spaces, and their epidermises consist of two or more layers of tightly compact cells. More particularly, under conditions of high transpiration the inner surfaces of adjacent leaves come close to each other so that there is a reduction of transpiration with little or no interference with the diffusion of carbon dioxide into the leaves and the rate of photosynthesis.

Environmental Factors

As previously stated, transpiration is the diffusion of water vapor from any plant surface to the surrounding or ambient air. With crop plants, most of this water is lost through the leaves or the photosynthetic stems. The water vapor comes from films of liquid water on the surface of cells containing chloroplasts. As with other gases these vapor molecules, as they escape from the plant surface, exert a pressure. The ambient air also contains molecules of water vapor, and these, in turn, exert a pressure. Thus, the rate of transpiration is directly proportional to the difference between the vapor pressure of the transpiring surface and that of the ambient air. This difference is called the *vapor pressure deficit,* abbreviated VPD. The relation of the rate of transpiration (Tr) to the vapor pressure deficit (VPD) expressed mathematically is as follows:

1. Rate of Tr = VP (Tr surface) − VP (Ambient air)

2. VP (Tr surface) − VP (Ambient air) = VPD

3. Rate of Tr = VPD

Thus, with other factors not limiting, the rate of transpiration is directly proportional to the vapor pressure deficit.

Light Intensity and Temperature of the Transpiring Surface and the Ambient Air Of the total amount of light energy which impinges on chlorophyll-containing tissues, about 10 percent is reflected, about 10 percent is transmitted, and about 80 percent is absorbed. Of this absorbed energy, only about 1 to 2 percent is fixed in the potential form in the manufactured compounds, and the remainder is changed to heat. This conversion of light energy to heat energy increases the temperature of the leaves above that of the ambient air, and it increases the vapor pressure of the transpiring surface above that of the ambient air accordingly. Study the data presented in Table 4-3. In this instance the

Table 4-3 Effect of Absorbed Light and Relative Humidity on Vapor Pressure

Factor	Effect of absorption of light			Effect of relative humidity			Effect of light and relative humidity		
	°C*	RH†	VP‡ mm Hg	°C	RH	VP mm Hg	°C	RH	VP mm Hg
Transpiring surface	30	100	31.82	20	100	17.54	30	100	31.82
Ambient air	20	100	17.54	20	50	8.77	20	50	8.77
Vapor pressure deficit			14.28			8.77			23.05

*Degrees C.
†Relative humidity percentage.
‡Vapor pressure deficit.
Source: From O. F. Curtis and D. G. Clark, *An introduction to plant physiology,* New York: McGraw-Hill, 1950.

absorbed light increased the temperature of the leaves from 20 to 30°C (68 to 86°F), and this, in turn, increased the vapor pressure from 17.54 millimeters of mercury to 31.82 millimeters. Thus with increases in light intensity, there will be corresponding increases in leaf temperature and corresponding increases in the rate of transpiration.

Relative Humidity of the Transpiring Surface and the Ambient Air As is well known, the atmosphere has the ability to hold water vapor. In general, moist air contains relatively large quantities of water vapor per unit volume, whereas dry air contains small quantities. This ability of the air to hold water in vapor form varies directly with the temperature. Thus, warm air holds more water as vapor than does cold air.

The amount of water vapor in the air compared with the amount when the air is saturated for any particular temperature is known as *relative humidity*. Thus, when the relative humidity is high, the number of vapor molecules per unit volume of air is high; and when the relative humidity is low, the number of vapor molecules per unit volume of air is also low. Since the outer walls of the manufacturing cells are usually surrounded by films of water, a high relative humidity exists within the chamber of each stomate, and since the air outside the stomates usually has a lower humidity, the molecules of water vapor diffuse from the stomatal chamber to the outside air. Thus, the rate of diffusion depends on the difference in relative humidity of the stomatal chambers and the outside air. Note the data in Table 4-3. In this instance the temperature of the leaves and that of the ambient air are the same, but the relative humidity of the transpiring surface is saturated and that of the ambient air is one half saturated. Thus, with the temperature of the leaf and that of the surrounding air the same, the rate of transpiration is directly proportional to the differences in vapor pressure of the transpiring surface and that of the ambient air, and inversely proportional to the relative humidity of the ambient air.

Air Movement

When air movement is rapid, the molecules of water vapor immediately above a free water surface are rapidly carried away and the rate of diffusion is accordingly increased. Conversely, when air movement is slow, the molecules are displaced less rapidly and the rate of evaporation is relatively low. Thus, with other factors constant, the rate of evaporation of water is governed by the wind velocity. However, the student should remember that transpiration is biophysical in nature and that the rate of absorption of water and the turgor of the guard cells should be considered. When the rate of absorption becomes less than the rate of transpiration, the guard cells would be expected to lose turgor and change in shape. This would result in partial or entire closing of the stomates and corresponding decreases in the rate of transpiration. In fact, experiments have shown that a wind velocity as low as 6 miles (or 9.7 kilometers) per hour induces at least partial closing of the stomates with a resultant decrease in the diffusion of carbon dioxide into the leaves and the rate of photosynthesis, growth, and yield. This explains why unshaded and sheltered cabbage and tomato plants grow faster than nonsheltered plants and why growers of flower and vegetable plants use various devices to protect plants from the deleterious effects of high winds.

All Factors Operating Together

The student should keep in mind that the factors which influence the absorption of water operate together. A warm, moist soil which has low osmotic pressure and which is thoroughly ramified by the plant's roots promotes the absorption of large quantities of water. The warm temperature promotes rapid movement of the capillary water, the moist condition provides abundant quantities of available water and adequate aeration for carbon dioxide and oxygen exchange in the respiration and growth of the roots, the soil water with a low concentration of solutes promotes rapid diffusion of water into the plant. Conversely, a cold or dry soil or a soil saturated with water or a high concentration of soluble salts limits water absorption.

As with the factors that influence the intake of water, those which influence the outgo of water operate together. Thus, high light intensities and high temperatures combined with low relative humidities induce high rates of transpiration; whereas low light intensities and low temperatures combined with high relative humidities promote low rates of transpiration. In like manner, light winds induce higher rates of transpiration than calm or still air. This explains why transpiration is greater during the day as compared with the night, on sunny days as compared with cloudy days, on cloudy days as compared with rainy days, on windy days as compared with calm days, and during the summer months as compared with the spring, fall, and winter months.

Evapotranspiration

Evapotranspiration considers both transpiration of the crop plant and evaporation of water from the soil in which the crop plant is growing. Since both processes are basically the same in that they require free energy to change liquid water to the vapor form, they may be considered collectively. In general, experiments on evapotranspiration supply basic information on the water requirement of crop plants, such as the amount of water necessary to mature a given crop, the proportion of water applied as rain or as irrigation, the proportion of water that runs off the surface, the proportion which drains away, the amount of water needed for different stages of growth, and the effect of soil covers on the water supply. To obtain data from which to answer these and other questions, the use of a special type of plant and soil container is necessary. This type of container is called a *lysimeter*. In general, lysimeters contain large quantities of soil in a tanklike box with concrete sides, an open top, and a perforated bottom through which drainage water runs into a collecting vessel, and rain gauges to measure the amount of water applied either as rain or as irrigation water. In addition, some lysimeters have devices to maintain the water supply within the tank at a definite level. Thus, with the knowledge of the amount of water applied, the amount of water which runs off the surface, and the amount which drains away, a measure of the efficiency of the rainfall and/or irrigation practices can be determined and evaluated.

WATER SUPPLY AND GROWTH AND DEVELOPMENT

The student should bear in mind in this discussion that the water supply is the limiting factor. In other words, all other factors are assumed to be favorable for growth and development: the day and night temperatures and light intensities are within the optimum range, the relative length of the light and dark periods is favorable for the type of growth the plant should be making, and all the essential elements are in favorable supply. With the water supply as the limiting factor, its effect on plant growth and development is discussed from three standpoints: (1) favorable supplies, (2) deficits, and (3) excesses.

Favorable Supplies

The effect of favorable supplies under conditions of high transpiration is shown in Figure 4-6. In this experiment, young sunflower plants were grown in a greenhouse during a period of bright sunshine and in cans containing a fertile sandy loam, continuously kept near the field capacity. The amount of water absorbed was determined by using porous cups embedded in the soil and by filling the cups at intervals of two hours, and the amount of water lost was determined by weighing both the soil and the plants also at intervals of two

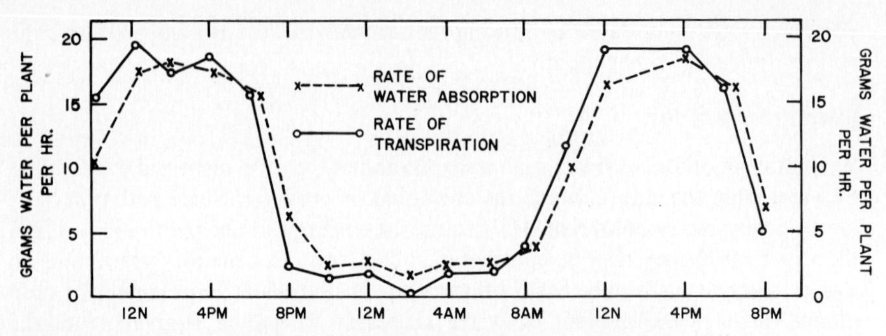

Figure 4-6 Diurnal changes in the rate of water absorption and the rate of transpiration of the sunflower. *(Adapted from Figure 3, How crops absorb water,* Conn. Agr. Exp. Sta. Bul. *708, 1969.)*

hours. Note the lines showing the rate of absorption and the rate of transpiration. In general, the rate of transpiration exceeded the rate of absorption from early morning into the middle of the day, the rate of absorption caught up with the rate of transpiration during late afternoon, and the rate of absorption exceeded the rate of transpiration during the night. Experiments with other plants, loblolly pine, willow, coleus, pelargonium, and broad bean have shown similar diurnal changes in the rate of absorption and the rate of transpiration. Thus, with crop plants growing in soil maintained at field capacity and in bright sunshine, the rate of absorption may be expected to be lower than the rate of transpiration during the morning, the rate of absorption may be expected to catch up with the rate of transpiration during the late afternoon, and the rate of absorption may be expected to exceed the rate of transpiration during the night. What would be the turgor of the living cells under these conditions? In general, in early morning, the guard cells would be in a state of turgor, the stomates would be wide open, and photosynthesis would proceed at a rapid rate; in the afternoon, the guard cells would lose some of their turgor, and the stomates would begin to close with corresponding reductions in the rate of photosynthesis; during the night, however, with the stomates closed and no light available for photosynthesis, the guard cells or, more particularly, the other living cells of the growing plant would regain turgor—a condition necessary for the elongation of the growing cells. The effect of favorable supplies is diagrammatically represented by the line AB as follows:

A **The rate of absorption = the rate of transpiration** **B**

The guard cells are turgid.
The stomates are open.
Carbon dioxide diffuses rapidly into the leaves.
The rate of photosynthesis is high.
The rate of respiration is normal.
Abundant carbohydrates are available for growth.

Deficits

Plants do not always receive favorable supplies of water; unfortunately, deficits occur. What are the immediate, subsequent, and extreme effects of water deficits within crop plants? In general, the immedite effect is a reduction in the size of the cells in the region of cell elongation. Thus, the cells which are made are small. This explains why plants grown under mild deficiencies produce stems with short internodes and why their leaves, flowers, and fruits are small.

The subsequent effect is a reduction in the rate of net photosynthesis. The rate of absorption is much lower than the rate of transpiration, the guard cells lose turgor and become flaccid, and the stomates either partially or completely close. Consequently, the rate of diffusion of carbon dioxide into the manufacturing cells is low, and the rate of manufacture of the initial food substances is accordingly low. Very few carbohydrates, pigments, fats, proteins, and other substances are made; growth is slow, and the marketable yields are low. Investigations have shown that the guard cells of many crops—apples, pears, peaches, plums, pecans, and lima beans—are markedly sensitive to water deficits. As soon as moderately severe water deficits occur within the plants, the guard cells lose turgor and the stomates begin to close. When the rate of transpiration is high and the rate of absorption is low, the stomates begin to close in the early afternoon. Under extreme conditions, that is, very low rates of absorption combined with high rates of transpiration, they close in the morning and remain closed for the remainder of the day. The effect of deficits explains why marketable yields are low after a dry spell, why orchard trees and other woody plants become susceptible to winter injury, why low water supplies during the summer in an apple orchard usually mean a crop of small apples with a dull finish, and why water deficits during the summer in a pecan orchard are likely to produce crops of poorly filled nuts. Figure 4-7 shows the effects of varied quantities of

Figure 4-7 Relation of the amount of available water to the growth of young apple trees. *(Courtesy, A. L. Kenworthy, Michigan State University.)*

water on the growth of young apple trees. Each tree is representative of an experimental group. These trees, from left to right, were given abundant, moderately abundant, and low quantities of available water.

The extreme effect results in wilting. When plants are wilted, the guard cells are fully flaccid and the stomates are closed, and, as a result, photosynthesis practically ceases. Since respiration continues, the plants begin to decrease in dry weight. With continued wilting, plants continue to starve, and finally they die. Note the appearance of the leaves of the wilted corn plant in Figure 4-8. The effect of water deficits is diagrammatically represented by line XA as follows:

Wilting	**Decreasing supplies of water**
X	**A**
No food manufacture Plants living on reserve substances only	Decreasing turgor of guard cells Decreasing size of stomates Decreasing rate of food manufacture Decreasing growth and yield With overwintered plants, decreasing resistance to winter injury

Since water deficits within plants retard growth and development, a knowledge of the symptoms of water deficiency should be helpful in plant-production practices. In general, the first symptom for plants in the vegetative phase is a reduction in the rate of extension of stems and twigs accompanied with the development of relatively small, healthy, dark green leaves. This is followed by the development of slender stems or twigs, small flowers, and small, poorly colored fruit. With certain crops—lemons, peaches, and tomatoes—the leaves, under conditions of low rates of absorption and high rates of transpiration, actually draw water from the fruit. This causes a marked decrease in volume of the fruit of lemons and a pathologic condition in tomatoes known as blossom end rot.

How can water deficits be prevented or reduced? Growers avail themselves of various practices depending on the kind of crop and enterprise. For example, florists reduce the light intensity and florists, gardeners, and orchardists transplant and cultivate carefully and apply irrigation water. These practices—reducing light intensity, transplanting, cultivating, and irrigating—are discussed more fully in Chapters 6 and 13.

Excesses

Under certain conditions with certain crops, excessive supplies within the plant produce unfavorable effects. In general, these effects include the development of leggy seedlings and the occurrence of growth cracks. Leggy seedlings usually develop under the following conditions: when the plants are set close together,

Figure 4-8 A wilted corn plant. Note the rolling of the edges of leaves. Photograph taken during period of drought. *(Courtesy, F. W. Emerson,* Basic Botany, *New York: McGraw-Hill Book Company, Inc.)*

when the soil is kept warm and moist, when the air temperature is within the optimum range, and when light intensity is relatively low. A warm, moist soil and a well-developed root system ensure abundant absorption; but close planting, favorable temperature, reduced light intensity and low wind velocity combine to bring about relatively low rates of transpiration. Thus, with high absorption on the one hand and relatively low transpiration on the other, turgor pressure in the region of cell elongation is high and the cells are unduly stretched. This often happens in greenhouses and hotbeds in early spring.

Growth cracks occur under similar conditions of water absorption and transpiration, e.g., the bursting of cabbage heads and the cracking of tomato fruits and carrot and sweet potato roots. The wet weather provides abundant supplies of available water, which, for plants with an extensive root system, promote a high rate of absorption. The wet weather also, with its comparatively low temperatures, low light intensities, and high relative humidities, causes a low rate of transpiration. Thus, the high absorption of water on the one hand and the low transpiration on the other are associated with the growth cracks. Figure 4-9

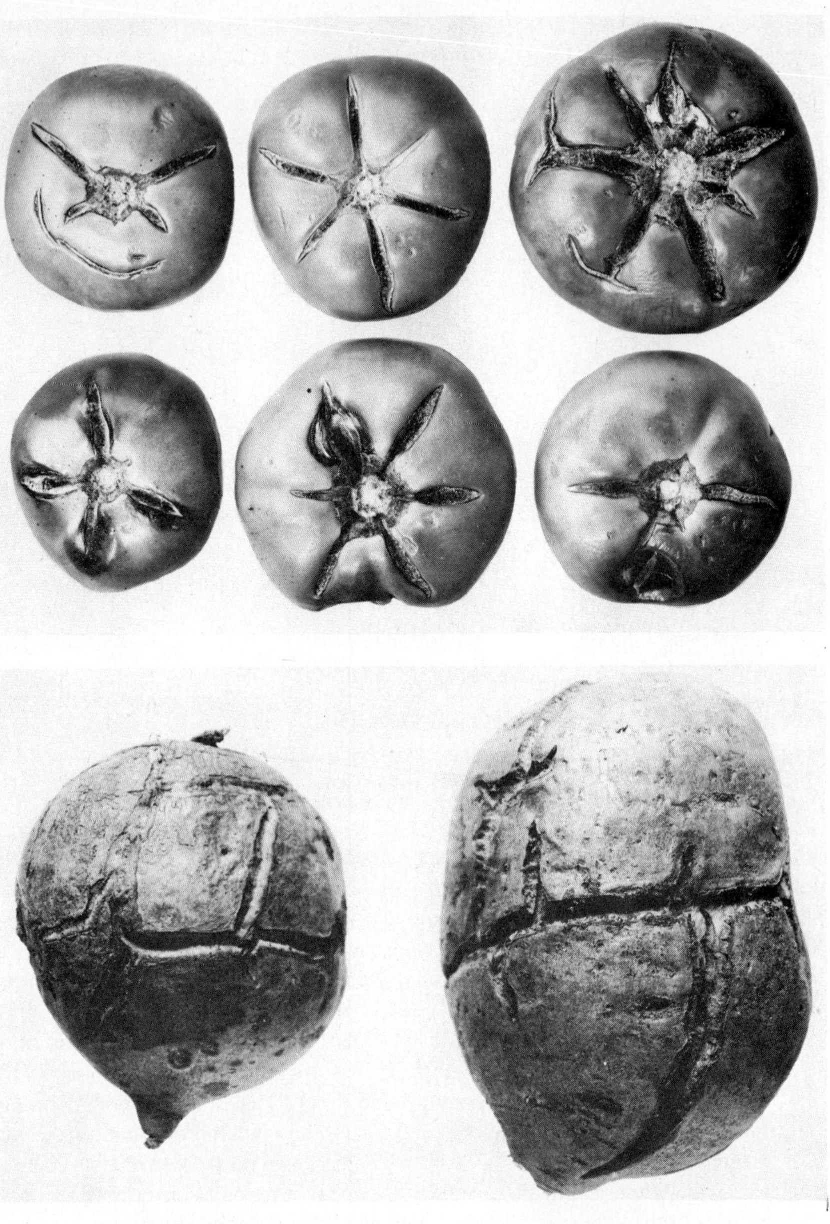

Figure 4-9 Cracked tomato fruits and sweet potato roots. *(Courtesy, V. R. Boswell, U.S. Department of Agriculture.)*

shows cracked fruits of the tomato and cracked roots of the sweet potato. The effect of excess water within the plant is diagrammatically represented by line BC as follows:

The rate of absorption is greater than the rate of transpiration.

B ———————————————————————— **C**

Increasing cell size
Long internodes and leggy growth
Bursting cells
Growth cracks

QUESTIONS

1 Define transpiration.
2 In general, the rate of transpiration is greater during the summer as compared with the winter, during sunny days as compared with cloudy days within the same season, and during the light period as compared with the dark period of any given 24-hour day. Explain.
3 Sweet potato and cotton plants frequently wilt during the middle of the day in July, August, and September. They become turgid at night. Explain.
4 In general, when the rates of water absorption and transpiration are practically the same, high rates of photosynthesis take place. Explain.
5 During the summer months in warm, temperate climates, a high rate of photosynthesis requires a high rate of water absorption. Explain.
6 What is the immediate effect of a water deficit within growing plants?
7 A continual mild deficiency of water within growing plants results in the development of short internodes and small leaves. Explain.
8 After a drought, yields are low when the plants are grown in nonirrigated soils. Explain.
9 Investigations have shown that temporary wilting of pecan leaves is accompanied by marked reduction in photosynthesis and transpiration. Explain.
10 Show how a severe drought during the latter part of the growing season makes woody plants susceptible to winter injury.
11 A low water supply during the summer in an apple orchard usually means small apples with a dull finish. Explain.
12 A low water supply during the summer in a pecan orchard is likely to produce a crop of small, poorly filled nuts. Explain.
13 When a plant is wilted, it is not making carbohydrates. Explain.
14 In general, with continuous wilting, plants finally die. Explain.
15 Do plants growing in shade (on the floor of a forest or under crowded conditions in a greenhouse) reach for the light? Explain.

16 Tomato seedlings in the center of flats in a greenhouse frequently become more "leggy" than those on the outside. Explain.

17 Cabbage heads burst open and carrot and sweet potato roots crack during wet weather, particularly when the plants are grown in moderately heavy soils. Explain.

18 The cracking of tomato fruits is severe during a rainy period preceded by drought. Explain.

SELECTED REFERENCES FOR FURTHER STUDY

Kramer, P. J. 1969. How crops absorb water. Cited from S. W. Johnson, How crops grow, a century later. *Conn. Agr. Exp. Sta. Bul.* 708. An account of current ideas by an authority in the field on the rate of water absorption and the rate of transpiration.

Magness, J. R. 1934. Status of soil moisture research. Presidential address. *Proc. Amer. Soc. Hort. Sci.* 32:651–661. One of the first accounts showing how the behavior of stomates relates to photosynthesis, growth, and the production of deciduous tree fruits.

Slatyer, R. O. 1967. *Plant Water Relationships.* New York: Academic Press. The application of the basic laws of physics and chemistry to problems pertaining to the water supply of crop plants.

Chapter 5

Heat and Temperature

To everything there is a season and a time to every purpose.

Ecclesiastes 3:1

NATURE AND SOURCE OF HEAT

According to the second law of thermodynamics, the molecules of any given substance are motionless at absolute zero (−273°C). However, at temperatures above absolute zero the molecules are in motion, that is, they bounce around each other in random fashion, and the speed of this bouncing is directly proportional to the amount of heat which they receive. This indirect, or random, motion of the molecules is due to their kinetic energy. Thus, heat is a form of kinetic energy.

The primary source of heat is the sun. As explained in Chapter 6, the short rays—the cosmic, the gamma, and the short ultraviolet—are absorbed by layers of ozone in the atmosphere, some of the moderately long rays—the red and the blue—are absorbed by chlorophyll and other pigments, and the remaining rays pass through the atmosphere to the surface of the earth. In general, when these rays impinge on the earth's surface, they lose some of their kinetic energy and change to the infrared or heat rays. These infrared, or heat, rays are absorbed or reradiated in practically all directions. Some are absorbed by the soil, some are absorbed or reflected by the clouds, and others are absorbed or reflected by plants and animals. This explains, partially at least, why soils warm up more rapidly on sunny days than on cloudy days and why the air is relatively cooler at high elevations than at low elevations, even though a higher elevation receives as many of the sun's rays as a lower.

Heat and Temperature

As previously stated, heat is due to random, or unordered, vibration of molecules, and temperature is a measure of these vibrations. Thus, with high rates of molecular vibration, the intensity of heat is high and the temperature is accordingly high, whereas with low rates, the intensity of heat is low and the temperature is accordingly low. As is well known, many types of measuring instruments have been developed to measure the temperature, or the degree, of heat. Two of the most common are the Centigrade and the Fahrenheit thermometers. Both are based on the freezing and boiling points of water under standard conditions of temperature and pressure with the Centigrade scale divided into 100 parts and the Fahrenheit into 212 parts. The relation of the one to the other is as follows:

$$\frac{C}{100} = \frac{F - 32}{180} \quad \text{or} \quad C = F - 32 \times 0.5555$$

Temperature Range for Crop Plants

In general, the biochemical reactions within crop plants, in common with those of other living things, take place within a very narrow range of temperature—from temperatures somewhat below the freezing point of liquid water to temperatures at which enzymes and other proteins are degraded, or denatured. At temperatures below the freezing point of water, liquid water changes to the solid state and the catalytic action and other remarkable properties of liquid water are lost, and at temperatures of 30°C (86°F) or above, the molecular structure of enzymes and other proteins is irreversibly altered. In other words, the integrity of the numerous enzyme and protein systems in crop plants is maintained only within a narrow range of temperature. Accordingly, the numerous chemical reactions within crop plants can take place only within this range. This does not mean that a constant temperature within this range produces the highest marketable yield. For example, a constant temperature on the lower part of the range for any given crop would be associated with low rates of gross photosynthesis, low rates of growth; and a constant temperature on the upper part would be associated not only with high rates of gross photosynthesis, but also with excessively high rates of respiration. In either case low marketable yields would be produced. Figure 5-1 shows how the frost-free growing season and the temperature level in July relates to crop production. These factors frequently determine the crops which are grown for profitable production.

To obtain a working knowledge of the influence of temperature on crop-plant growth, temperatures within crop-producing areas are classified as either favorable or unfavorable for growth and development.

FAVORABLE EFFECTS OF TEMPERATURE

Diurnal Temperatures

Diurnal temperatures refer to the alternating day and night temperatures within any given 24-hour period. In general, because of the flow of light energy from the sun, temperatures are higher during the day than they are during the night. During the night period the absorption of water is relatively high, the rate of transpiration is relatively low, and, as a result, turgor pressure for the elongation of the new cells is high. Thus, these alternating diurnal temperatures coincide with two important crop-plant activities: the making of abundant quantities of initial foods and related substances during the day, and the making of new cells and the elongating of these new cells during the night. Note the optimum

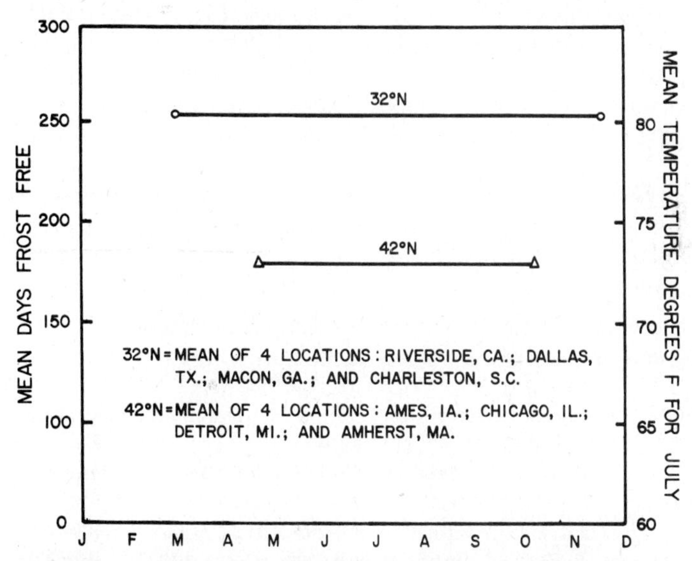

Figure 5-1 Relation of length of the frost-free growing season and temperature level in July to crop-plant production. These factors frequently determine the kind of crop that can be produced profitably.

day and the optimum night temperature ranges of the plants presented in Table 5-1. In all cases, the optimum day temperature range is higher than the optimum night temperature range. This indicates that the optimum temperature range for gross photosynthesis is higher than the optimum temperature range for cell division and cell elongation.

Favorable or Optimum Temperature Range

The favorable temperature range for the growth and development of any particular plant is known as the *optimum temperature range*. Within this range the two fundamental processes, photosynthesis and respiration,[1] are proceeding in such a way throughout the life cycle of the plant that the highest marketable yields are produced. In other words, from the time the crop is established to the time the product is ready for the harvest, the rate of photosynthesis is high and the rate of respiration is normal. As a result, large quantities of carbohydrates are available for growth and development. Thus, if plants of any given crop are handled properly, particularly with respect to the vegetative and reproductive phases, high marketable yields are obtained. Therefore, *the optimum temperature range may be defined as the range within which maximum photosynthesis and normal respiration take place throughout the life cycle of the plant*[2] and within which the highest marketable yields are obtained.

Not all crops have high rates of photosynthesis combined with normal rates of respiration within the same temperature range. In general, some crops have high rates of photosynthesis combined with normal rates of respiration within a relatively low range, and other crops have high rates of photosynthesis combined with normal rates of respiration at a relatively high range. On this basis, horticultural plants are classified as follows: (1) crops which produce their highest yields at a low temperature range, (2) crops which produce their highest yields at a

[1]In this chapter, it is assumed that the rate of water absorption and the rate of transpiration are practically the same.

[2]As is well known, the time factor should always be considered in any discussion of temperature. In this discussion, the time factor is the period of growth of any given crop.

Table 5-1 Optimum Day and Night Temperature Ranges as Determined by the Earhart Plant Research Laboratory, Pasadena, California, 1957.

Kind of plant	Optimum temperature range °F		Kind of plant	Optimum temperature range °F	
	Day	Night		Day	Night
English Daisy	58–60	46–52	Zinnia	77–83	58–68
Stocks	59–63	52–56	Petunia	82–85	58–63
Ageratum	64–67	53–56	Tomato	75–80	59–64
China Aster		56–64	Coleus	73–78	64–68

Source: F. W. Went, Climate and Agriculture, *Sci. Amer.*, 196:83–94, 1957.

moderately high temperature range, and (3) crops which produce their highest yield at a high temperature range. Note the optimum day and night temperatures for the kinds of plants listed in Table 5-1, and the heat and temperature requirements of the crops listed in Table 5-2.

Optimum Night Temperature Range and Phases of Growth In general, crop plants for the most part make new cells and the protoplasm for these cells during the night. As previously stated, protoplasm is made by sugars, usually glucose, combining with certain compounds containing nitrogen. Thus, the making of new cells is essentially a biochemical reaction; and, as in other biochemical reactions, with other factors favorable, temperature directly influences the rate of the process. This effect of temperature on the vegetative and reproductive phases may be illustrated by dividing the optimum night temperature range into two parts: the upper half and the lower half. Since, within the optimum range, the rate of cell division is more or less directly proportional to the temperature,

Table 5-2 Heat and Temperature Requirements of Horticulture Crops

Crops which require a comparatively low temperature range, 45–55°F (7–13°C)*	
Fruit crops	Apple, pear, plum, cherry, labrusca grape, strawberry, blueberry, raspberry, cranberry
Vegetable crops	Asparagus, rhubarb, spinach, lettuce, the cabbages, the root crops, pea, potato
Flower and ornamental crops	Aster, calceolaria, carnation, cineraria, cyclamen, geranium, petunia, snapdragon, verbena, vinca
Crops which require a moderately high temperature range, 55–65°F (13–18°C)*	
Fruit crops	Peach, apricot, vinifera grape, rotundifolia grape, fig, persimmon, blackberry, pecan, tung
Vegetable cops	Tomato, capsicum pepper, eggplant, the cucurbits, bean, protepea
Flower and ornamental crops	Asparagus plumosus and sprengeri, African violet, azalea, begonia, Cattleya orchids, coleus, gardenia, hydrangea, kalanchoe, lily, poinsettia, rose
Crops which require a high temperature range, 65–75°F (18–24°C)*	
Fruit crops	Banana, Brazil nut, cacao, cashew, citrus, coffee, date, fig, olive, papaya, pomegranate
Vegetable crops	Sweet potatoes, yams, cassava, chayote, okra, the cucurbits, sweet corn
Flower and ornamental crops	Philodendron, dieffenbachia, croton, pothos, ficus, peperomia, dracaena, aspidistra, gloxinia

*Night temperature. The day temperature may be 5–20°F (30–12°C) higher depending on the amount of light available.

comparatively high rates will take place within the upper half and moderately high rates will take place within the lower half. The high rate of cell division makes for a rapid development of stems, leaves, and absorbing roots and a rapid utilization of carbohydrates; and the moderately high rates make for a moderately rapid development of stems, leaves, and absorbing roots and a moderately rapid utilization of carbohydrates. Thus, with other factors in favorable supply and with high rates of photosynthesis and normal rates of respiration, the maintenance of temperature within the upper half of the optimum range promotes vigorous vegetative growth, whereas the maintenance of temperature on the lower half induces moderately vigorous vegetative growth. Therefore, at temperatures within the upper half of the optimum range, most of the carbohydrates which the plant is making will be used and very few will be stored. On the other hand, at temperatures within the lower half, lesser quantities of carbohydrates will be used and greater quantities will be stored. The influence of night temperature within the optimum range is briefly summarized as follows:

Optimum Night Temperature Range

Lower half	Upper half
Moderately rapid rates of cell division	Rapid rates of cell division
Moderately vigorous vegetative growth	Vigorous vegetative growth
Moderately rapid usage of carbohydrates with moderate quantities available for storage	Rapid usage of carbohydrates with low quantities available for storage

The marked effect of temperature on the growth and development of crop plants is manifested by the care and attention given to the maintenance of night temperatures in glass and plastic houses. In general, growers of greenhouse crops maintain night temperatures within the upper half of the optimum range when their crops are in the vegetative phase. During this phase, the rate of gross photosynthesis and the rate of new cell formation is high and the rate of respiration is correspondingly high. As a result, large quantities of free energy are liberated. On the other hand, growers maintain temperatures within the lower half of the optimum range when their crops are in the reproductive phase. During this phase, the rate of gross photosynthesis is moderately high, the rate of new cell formation is relatively low, and the rate of respiration of the plant as a whole is accordingly low. As a result, relatively large quantities of free energy are stored. More importantly, the quality of the crop is enhanced accordingly, for example, the color, fragrance, attractiveness, and keeping quality of cut flowers, and the color, fragrance, and taste of fruits. In other words, the intensity of metabolism varies with the phase of growth. During the vegetative phase, it is relatively high, since large quantities of protoplasm and many new cells are being made, and the rate of respiration is accordingly high. On the other hand, during the reproductive phase, the intensity of metabolism is relatively low since only small quantities of protoplasm and small numbers of new cells need to be made and the rate of respiration is accordingly low.

Optimum Day and Night Temperature Range in Greenhouse and Plastic-House Culture As previously explained, the rate of net photosynthesis for any given day determines the amount of carbohydrates for cell division and cell elongation during the following night. In greenhouse culture, the water supply and the essential-element supply are usually not limiting factors in growth, and the rate of photosynthesis is dependent on the light supply. Thus, during sunny weather the optimum day temperature range can be relatively high, since the rate of gross photosynthesis is high due to the abundant light and, despite the relatively high rates of respiration, abundant carbohydrates will be available for growth and development. However, during cloudy weather the rate of gross photosynthesis is low. If the same temperature range is used as during sunny weather, the rate of respiration will be relatively high; as a result, relatively few carbohydrates would be available for growth and development. Thus, to conserve the carbohydrate supply, the temperature during the day is lowered somewhat in order to lower the rate of respiration without any marked reduction in the rate of gross photosynthesis. This explains why the optimum day temperature range during cloudy weather for any given plant is lower than the optimum day temperature range during sunny weather. In fact, growers of greenhouse crops have found that the optimum day temperature range during cloudy weather is not much higher than the optimum night temperature range, in face, not much higher than 5°F (2.7°C) at most.

Optimum Day and Night Temperature Ranges in Outdoor Culture Although the grower of outdoor crops cannot control temperature in precisely the same way as the grower of greenhouse crops, he can utilize favorable temperatures by taking advantage of southeasterly or northeasterly slopes, by using various soil types and elevations, and more particularly by planting at the right time. Study the curves of Figure 5-2. Note that plants of the Porto Rico sweet potato set in the field at the earliest practical date, usually one month after the last killing frost in the spring, produced more marketable roots than plants set at later dates. In fact, the data show that for each day's delay in planting, the grower lost 2.4 bushels of marketable roots per acre. Examples for other crops could be cited.

Temperature, in combination with other factors, markedly influences the localization of horticultural crop industries. In general, cool-season crops grow best in regions characterized by relatively cool weather, and warm-season crops thrive best in regions characterized by relatively warm weather. For example, the temperature range of the summer season in Holland, Denmark, and Great Britain is particularly suitable for the growing of asparagus, spinach, lettuce, cabbage, Brussels sprouts, cauliflower, garden pea, and other cool-season crops. In these

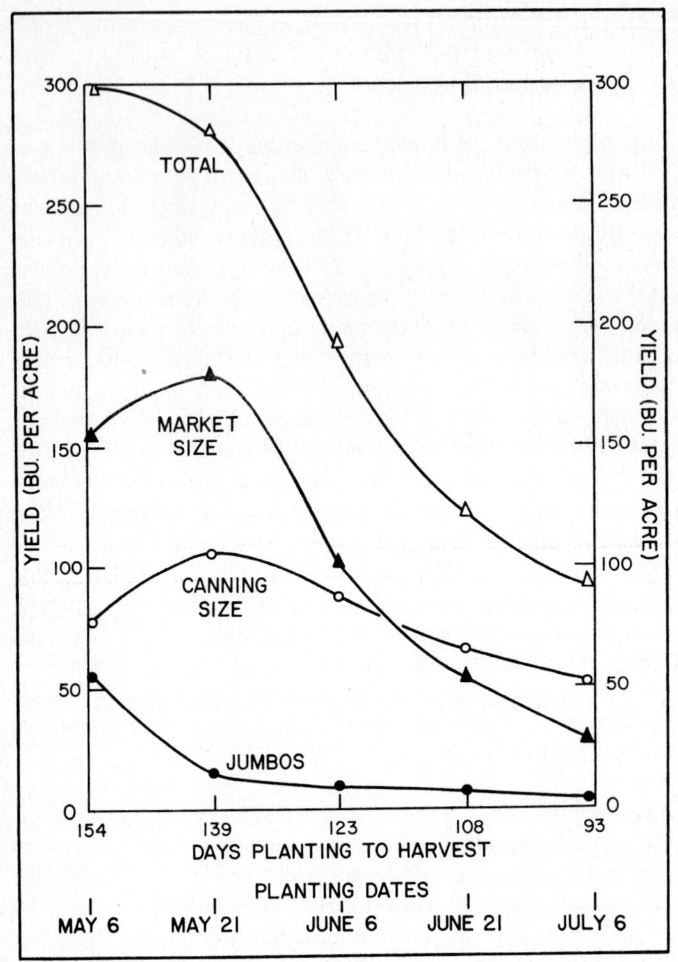

Figure 5-2 Effect of planting at the right time. *(Redrawn from Figure 1, Md. Agr. Ext. Serv. Fact Sheet 24, 1955.)*

countries warm-season crops, such as oranges, grapefruit, lemons, limes, peaches, apricots, and cantaloupes, are raised in glass houses because the heat of summer is insufficient to bring these crops to full maturity. In the continental United States the cool-season vegetables are grown during the summer in the northern regions of the country and during the fall, winter, and early spring in the southern regions. In general, the apple and labrusca grape are confined to the northern regions, whereas citrus fruits, dates, European and muscadine grapes, and other warm-season crops are grown in the southern regions.

Temperature also influences the adaptability of varieties within a given kind or species. For example, the Winesap and York Imperial varieties of apple require a comparatively high optimum temperature range, whereas Baldwin and McIntosh require a low optimum temperature range. The Earliana and Bonny Best tomato varieties thrive best in the northern regions, and Marglobe thrives best in the southern regions. The Tom Watson variety of watermelon thrives well in the southern region but fails to produce profitable crops in the northern region.

UNFAVORABLE EFFECTS OF TEMPERATURE

The unfavorable effects of temperature are classified as follows: (1) growing-season temperatures above the optimum night temperature range, (2) growing-season temperatures below the optimum night temperature range, and (3) unfavorable effects of winter temperature.

Growing-season Temperatures above the Optimum Night Temperature Range

Plants subjected to night temperatures *above* their optimum range, particularly during the later stages of growth, generally produce low yields. How are these low yields produced? As discussed in Chapter 1, the yield for any given plant equals the amount of carbohydrates made per unit of time minus the amount used. Since the initial carbohydrates are made by photosynthesis and are used in respiration, the yield equals the rate of gross photosynthesis minus the rate of respiration. In general, when the plants of any given crop are grown at comparatively high day temperatures combinedwith high night temperatures, that is, with night temperatures above the optimum range, the rate of gross photosynthesis remains at a high level but the rate of respiration increases markedly. As a result, the amount of carbohydrates available for growth and yield of any given crop becomes increasingly less. Thus, the higher the night temperature above the optimum range, the lower is likely to be the yield.

The relation of temperature to the rate of photosynthesis and the rate of respiration to the yield is shown in Figure 5-3. Note that at high temperatures the rate of gross photosynthesis remains at a high level, *but the rate of respiration increases quite markedly*. Also note that at point D the rate of respiration equals the rate of photosynthesis and that no carbohydrates are available for growth and development. Thus, in order for plants to grow and develop rapidly, the rate of photosynthesis must always be greater than the rate of respiration; and the greater the difference between the rates of these processes, the greater is likely to be the yield. This effect of temperature above the optimum range for growth explains why potatoes and other cool-season crops are grown in the southern United States during the spring and fall only; why carnations produce relatively large, fragrant flowers in cool weather and relatively small, nonfragrant flowers in warm weather; why garden peas fail to produce high yields during hot weather; why they produce abundant yields during cool weather; why the flowers of *Primula sinensis* var. *rubra* are red at comparatively low temperatures and white at high temperatures; why the apple industry, for the most part, is located in the northern regions of the country and in the elevated regions in the upper South; and why the cool-season vegetable and flower crops are grown in the winter-garden areas in the South and during the summer in the North. Figure 5-4 shows the effect of temperature above the optimum range for the development of potato tubers. Note that as the plant's soil temperature increases above the optimum range, the yield of tubers decreases.

Growing-season Temperatures below the Optimum Night Temperature Range

Plants subjected to temperatures *below* their optimum night temperature range for growth also produce low yields. How are these low yields produced? Here again both photosynthesis and respiration are involved. The relation of temperature below the optimum range to the rate of each of these processes and the amount of carbohydrates available for growth is shown in Figure 5-3. Note that at temperatures below the optimum range, the rate of photosynthesis and the rate of respiration decrease, *but the rate of photosynthesis decreases to a greater extent than the rate of respiration.* Also note that at point A, the rate of each of these

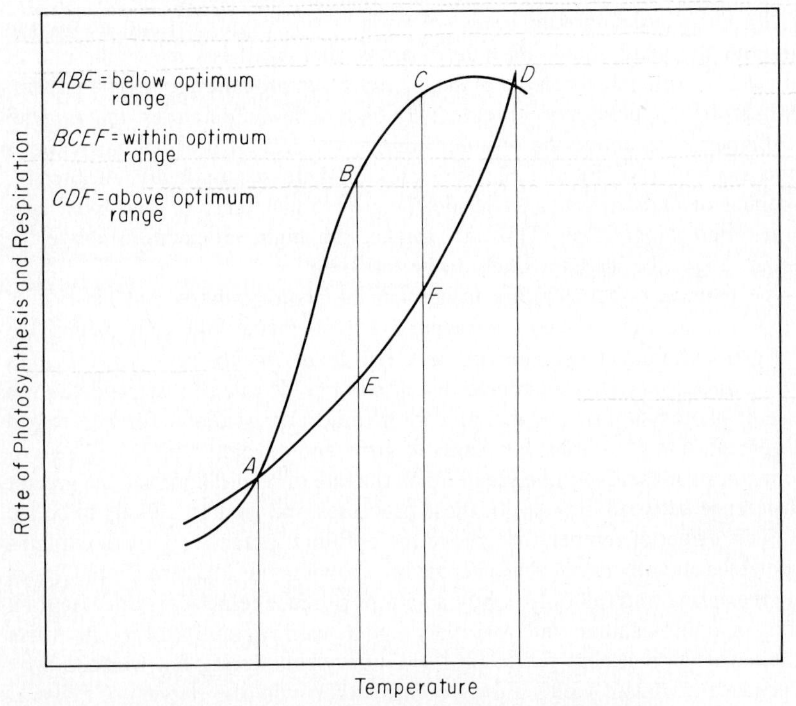

Figure 5-3 Relation of photosynthesis and respiration to the available carbohydrate supply for growth and development of crop plants.

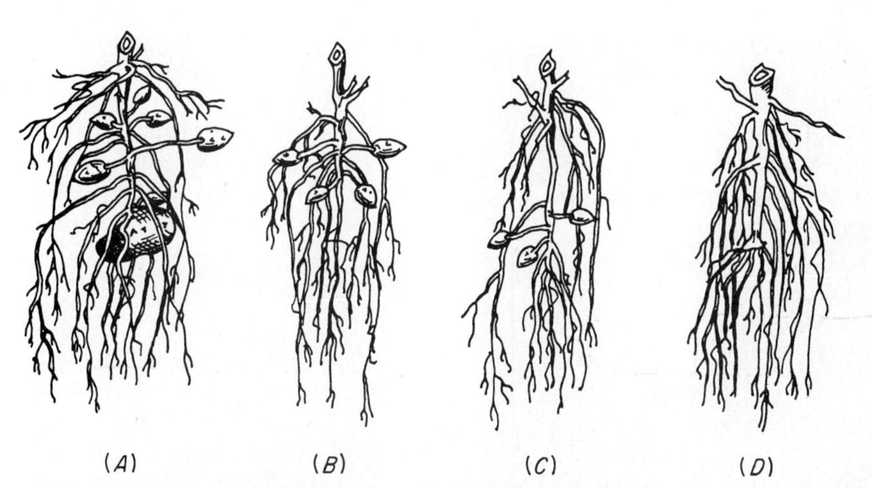

Figure 5-4 Effect of temperature on the development of potato tubers. A: 68°F, 20.9 gm. B: 73°F, 5.0 gm. C: 79°F, 1.6 gm. D: 84°F, 0 gm. *(Redrawn from Figure 9,* Minn. Agr. Exp. Sta. Tech. Bul. *34, 1925.)*

processes is equal and no carbohydrates are available for growth and development. Further, at temperatures below the optimum night temperature range for any given plant, *the rate of protein formation in the making of new cells is low.* As a result, the rate of growth is slow and the yields, if any, are accordingly low. This effect of temperature below the optimum range for growth explains why warm-season vegetable crops thrive well only in favorable locations during the summer in the coastal regions of Alaska; why the northern limit of the commercial sweet potato industry extends from New Jersey to southern Ohio, southern Indiana, and Iowa; why the tung industry in the United States is confined to a narrow belt extending from southern Louisiana to northern Florida; why the date industry in the United States is limited to the Coachella Valley of southern California; why an unusually cool growing season in apple and peach orchards is likely to make the plants susceptible to winter injury; and why peaches and apricots in northern Europe are grown on the south side of high walls.

Temperatures just below the optimum range also induce a condition called *bolting* in certain valuable vegetable crops, e.g., cabbage, collards, celery, onions, and beets. These kinds are biennials. Under favorable temperatures the plants develop the storage organ during the first year and flowering stalks, flowers, fruit, and seed during the second. If seedling plants of these crops are subjected to temperatures somewhat below the optimum range, that is, between 40 and 50°F (4 to 10°C) for four to eight weeks depending on the kind and variety, they produce flowering stalks and seeds instead of producing the storage organ, e.g., the head in cabbage, the fleshy leaves in collards, the fleshy petioles in celery, the bulb in onion, and the fleshy root in beets. Why does bolting take place at temperatures just below the optimum range? In general, the rate of photosyn-

Figure 5-5 Nonbolted and bolted cabbage plants. *(Courtesy, E. L. Moore, Mississippi State University.)*

thesis is fairly high and at the same time the rate of respiration and the rate of cell division are low. As a result, relatively low quantities of carbohydrates are used and large quantities of carbohydrates are stored. This more or less rapid accumulation of carbohydrates is associated with the formation of flower-forming hormones and the initiation of flower buds which, with the onset of warm weather, develop into flowering stalks which bear the fruit and seed. Thus, these biennials become annuals when the seedling plants are subjected to relatively low temperatures for a sufficient period of time. Note the two cabbage plants in Figure 5-5.

QUESTIONS

1. What is meant by the optimum temperature range for plant growth and development?
2. Compare cool-season and warm-season crops from the standpoint of their optimum temperature ranges for photosynthesis and for cell division.
3. With other factors favorable, night temperatures within the upper half of the optimum range favor vigorous vegetative growth and the utilization of carbohydrates. Explain.
4. How does the greenhouse grower use night temperatures to regulate the vegetative and reproductive phases of plant growth?

5 With other factors favorable, temperatures within the lower half of the optimum range are more favorable for the development of large, highly colored flowers than temperatures within the upper half. Explain.
6 In general, the red varieties of tomatoes develop lycopin (the red pigment of tomatoes) more readily in the North in July, August, and September than in the South during the same months. Explain.
7 At what time of the growing season in the South would you expect tomato fruits to develop their highest color? Explain.
8 The Puget Sound district of the United States and the climate of England are particularly favorable for the growing of garden peas. Explain.
9 Cool-season vegetable crops are grown in the summer in the northern United States and in the fall, winter, and early spring in the southern United States. Explain.
10 How do night temperatures above the optimum range for growth produce low yields?
11 Investigations have shown that crops produce low yields when the rate of respiration approaches the rate of gross photosynthesis. Explain.
12 For the growing of greenhouse crops, the optimum cloudy day temperature range should be not much higher than the optimum night temperature range. Explain.
13 Carnations grown in greenhouses in the southern United States and harvested in February and March are likely to have larger flowers and thicker stems than those harvested from the same plants in April and May. Explain.
14 Garden peas planted three or four weeks before the average date of the last killing frost in the southern United States generally produce greater yields than those planted two or three weeks later. Explain.
15 In your opinion can satisfactory crops of apples be produced in the southern United States? If so, where? Explain.
16 How do temperatures below the optimum range for growth produce low yields? Explain.
17 In the British Isles cucumbers, cantaloupes, and tomatoes are grown more satisfactorily in glass and plastic houses than outdoors. Explain.
18 Sweet potatoes are grown in regions which have a growing season of at least 150 to 170 days during which the temperature is relatively high. Explain.
19 How do temperatures just below the optimum range induce bolting in biennial vegetables?

Winter Temperatures

Harmful effects of winter temperatures are classified as follows: (1) injuries due to immaturity of the tissues, (2) injuries due to unseasonably high temperatures followed by low temperatures, (3) injuries associated with winter drought, and (4) injuries due to insufficient cold.

Injuries Due to Immaturity of the Tissues The ability of woody plants to withstand the low temperatures of winter is dependent on the ability of the protoplasm to bind water. This, in turn, is related to the ability of the protoplasm to make compounds known as *hydrophilic colloids*. These hydrophilic colloids have a large surface in proportion to their size and adsorb large quantities of water on these surfaces. The molecules of water are drawn to each other, and they act more like a solid than a liquid. This water is called *bound water* and cannot be frozen at the usual subfreezing temperatures. In general, the hydrophilic colloids are made from accumulated carbohydrates, and mature tissues contain more carbohydrates than immature tissues. Thus, the rate of photosynthesis, the relative rate of carbohydrate usage and storage, particularly during the latter part of the growing season, and the ability of the plant to make hydrophilic colloids are all concerned. This explains (1) why a water deficit within woody plants during the latter part of the growing season is likely to predispose the tissues to winter injury; (2) why weakly vigorous trees are likely to be injured to a greater extent than moderately vigorous trees; (3) why excessively vigorous trees are susceptible to winter injury; (4) why unseasonably high temperatures combined with moist soil during the latter part of the growing season are likely to make woody plants susceptible to winter injury; and (5) why heavy applications of available nitrogen to moderately vigorous trees during the latter part of the season should be avoided. The high supply of available nitrogen combines with the sugars in the formation of proteins for the making of the protoplasm. This, of course, depletes the carbohydrate supply for the formation of the hydrophilic colloids. In cases (1) and (2) the rate of photosynthesis is low and very few carbohydrates are made; and in cases (3), (4), and (5) the rate of photosynthesis is high, but most of the carbohydrates have been used. Very few are left for making the hydrophilic colloids.

In general, under favorable conditions horticultural plants vary in their ability to withstand low winter temperatures. Some are resistant, but others are not. As previously stated, resistant, or hardy, plants possess the ability to bind their water to a greater extent than nonresistant, or nonhardy, plants. Examples of relatively hardy and nonhardy horticultural crops are shown in Table 5-3.

Injury due to the immaturity of the tissues may be severe or mild, depending not only on the degree of carbohydrate accumulation in trees and shrubs, but also on the time and severity of the low temperature. In general, a sudden drop in temperature of 30, 40, or 50°F (−1, 4, or 10°C) is more severe than a drop of 10, 15, or 20°F (−12, −9, or −6°C). For example, the late summer and fall of 1940 in the middle western United States were characterized by unseasonably high temperatures, abundant light, and relatively high rainfall. As a result, the trees and shrubs in this region were moderately to vigorously vegetative and, as would be expected, comparatively few carbohydrates were stored in the tissues. On November 11, the temperature in that area dropped from a high of 50 to 60°F (10 to 16°C) to a low of 10°F (−12°C) within a very short period. The sudden drop in temperature severely injured the wood of many deciduous-tree fruits and ornamental shrubs. As a result, thousands of fruit trees were killed and the productive capacity of many of the trees which survived was considerably reduced. Thus, environmental conditions during the growing season and the time, duration, and

Table 5-3 Hardy and Nonhardy Horticultural Plants

Fruits	Vegetables	Ornamentals
	Hardy	
Apple, plum, pear, peach, cherry, blueberry, cranberry, labrusca grape	Cool-season crops (potato excepted)	Flowering peach, quince, hawthorn, rose, spirea, barberry, lilac
	Nonhardy	
Citrus, date, olive, fig, muscadine grape, raspberry, blackberry	Warm-season crops	Magnolia, nandina, cape jasmine, crepemyrtle, camellia

intensity of cold determine, to some degree at least, the localization of fruit-crop industries. Note the ornamental tree and shrub hardiness zones presented in Figure 5-6 and the winter injury to old and young apple branches shown in Figure 5-7.

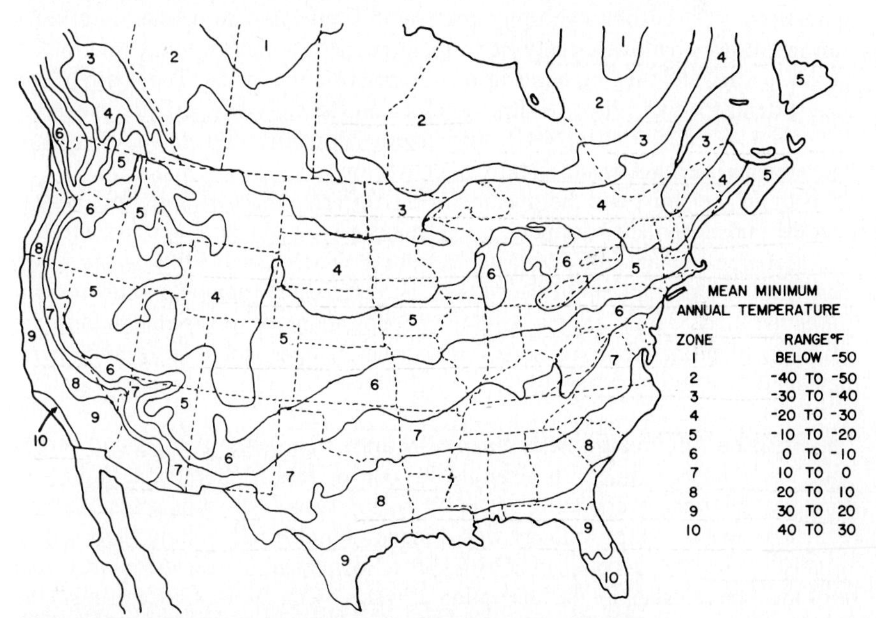

Figure 5-6 Ornamental tree and shrub hardiness zones for the 48 contiguous United States and southern Canada. *(Adapted from U.S. Dept. Agr. Misc. Pub. 814.)*

Figure 5-7 Examples of winter injury. Left: bark on trunk of Golden Delicious apple tree. Center: freezing cracks on main branch. Right: freezing cracks on young branches. *(Courtesy, A. B. Groves, Virginia Polytechnic Institute and State University.)*

Injuries Due to High Temperatures Followed by Low Temperatures Unseasonably high temperatures during late winter and early spring followed by low, usually freezing, temperatures are frequently injurious in peach, pecan, and tung orchards of the southern United States. The warm period may extend for two, three, or four weeks, and the daily temperature may fluctuate from 40 to 75°F (4 to 24°C). With internal conditions favorable for growth, the temperature range is sufficiently high to promote the absorption of water, cell division of the buds, and the opening of the flowers. What is more important is that the buds lose their hardiness, or resistance to low temperatures. As a result, the near-freezing or subfreezing temperatures which usually follow the warm period kill the protoplasm of the cells of the pistil(s) and stamens. Thus, for any particular kind of fruit, early blooming varieties are more susceptible to this type of injury than late blooming sorts, and northern slopes would seem to be more desirable than southern slopes.

Injuries Associated with Winter Drought The tops of trees and shrubs lose water in the winter. In general, the rate of loss is more or less directly proportional to the evaporating power of the air. Thus, a dry atmosphere accompanied by winds of high velocity would induce a higher rate of transpiration than a moist atmosphere accompanied by winds of low velocity. The water lost by the tops must be replaced by the absorption of water by the roots. Consequently, the relative rates of water absorption and transpiration are involved. If the soil contains very little available water and if the rate of transpiration is very high, the protoplasm gradually becomes desiccated and the tissues finally die. This type of winter injury is more common in regions characterized by light summer rains, a dry atmosphere, and intense winter cold than in regions characterized by moderate rains, a humid atmosphere, and mild cold. How can this type of injury be reduced or avoided? Manifestly, practices which increase the water content of the soil and the water absorption of the tree and which reduce the rate of transpiration would be beneficial. Accordingly, the application of irrigation water and the judicious use of windbreaks are likely to be helpful.

Injuries Due to Insufficient Cold To appreciate the effects of insufficient cold on deciduous trees and shrubs during the winter, the student should know the meaning of the rest, or physiologic dormant period. As an example, let us study the behavior of three comparable lots of two-year-old moderately vigorous peach trees. Let us assume that these trees are grown in tubs (small, wooden containers), that they have shed all their leaves by the last week of October, and that on November 1 they are divided into three comparable lots—A, B, C—and placed under the following conditions: lot A is kept out of doors from November 1 to February 1 and then placed in a warm greenhouse; lot B is kept in a warm greenhouse from November 1 to May 1; and lot C is kept outdoors throughout this period. Let us also assume that the temperature outdoors varied between 20 and 65°F (−6 and 18°C) from November 1 to December 31 and between 15 and 60°F (−12 and 16°C) from January 1 to March 21 and that the temperature in the greenhouse varied from 55 to 85°F (13 to 29°C). Thus, the different temperature treatments are as follows: for lot A, a low temperature level for a period of three months followed by a high temperature level; for lot B, a high temperature level for the entire period; and for lot C, a low temperature level for the entire period. How would the buds of lot A and lot B appear on February 15? The buds of the plants of lot A would be open, and those of lot B would be inactive and show little, if any, signs of growth. What is the difference in temperature treatments between lots A and B? Note that lot A was subjected to a period of cold, but lot B was not. In other words, the trees of lot B failed to grow, even though environmental conditions were favorable for growth. Evidently, conditions within the trees were unfavorable for the growth of buds.

Now let us compare lots A and C. The trees of lot C would show no signs of bud development, but since they had received the same amount of cold as those of lot A, the trees would be out of their rest, or physiologic dormant, period and in an environmental dormant period. In other words, *when the internal conditions are unfavorable for the growth of the leaf and flower buds, particularly for the process of cell enlargement, the trees and shrubs are in the rest, or physiologic dormant, period, and when internal conditions are favorable for the growth of the leaf and flower buds and the buds show no sign of normal growth, the trees and shrubs are in the environmental dormant period.*

In general, trees and shrubs in temperate climates enter their rest, or physiologic dormant, period during the latter part of the growing season—in

August or September in the northern hemisphere and in February or March in the southern hemisphere. Recent results indicate that the length of the light period initiates the rest, that the buds enter the state of rest gradually, that buds of vigorously vegetative trees enter the rest period somewhat later than those from moderately vigorous trees, and that the hormones associated with cell division and enlargement decline, while those associated with very little or no cell division increase. In other words, when the buds enter their rest period, the change in rate of metabolism is quite marked—from one which is quite intense to one which is quite low in intensity.

Reasons for the Rest Period Why do trees and shrubs go into a rest, or physiologic dormant, period? It seems to be a matter of survival of the fittest. In the warmer part of temperate regions, winter temperatures are mild. Quite often, periods exist during which the temperatures are sufficiently high to promote growth of leaf and flower buds, and quite often these bud-swelling temperatures are followed by leaf and flower bud-killing temperatures. Thus, with these alternating high and low temperatures, the buds of trees and shrubs with no rest period would be killed, while those with a rest period would remain dormant during the warm periods.

Amount of Cold Required As shown in Table 5-4, the amount of cold required varies with the kind of plant and with the variety within a given kind. Note that the apple requires more cold than the peach, and the peach requires more than the vinifera grape. Also note that the apple varieties McIntosh, Rome Beauty, and Wagner require more cold than Rhode Island Greening, Wealthy, and Winter Banana; the pear varieties Bartlett, Bosc, and Tyson require more cold than Kieffer, Garber, and LaConte; and the peach varieties Mayflower, Salway, and Dixie Red require more cold than Golden Jubilee, Belhaven, and Hiley.

Table 5-4 The Chilling Requirement of Certain Horticulture Plants

Crops with Short Chilling Requirements
Apricot; fig; Japanese plum; peach var. Babcock; persimmon; quince; vinifera grape; muscadine grape; currant; flowering peach; forsythia; wisteria
Crops with Moderately Long Chilling Requirements
Peach (most varieties, e.g., Golden Jubilee, Belhaven, Hiley); pear var. Kieffer, Garber, LaConte; plum (most varieties); sweet cherry; walnut (northern varieties); gooseberry
Crops with Long Chilling Requirements
Apple var. Wealthy, Rhode Island Greening, Winter Banana; pear var. Bartlett, Bosc, Tyson; sour cherry; peach var. Mayflower, Salway, Dixie Red
Crops with Very Long Chilling Requirements
Apple var. McIntosh, Rome Beauty, Wagner, Northern Spy

Source: W.H. Chandler, et al., Chilling requirements for opening of buds of deciduous orchard trees and some other plants in California, *Calif. Agr. Exp. Sta. Bul.*, 611, *1937*.

Symptoms of Prolonged Physiologic Dormancy The flower buds located near the tips of branches may develop into blossoms in late January or early February while the basal buds are still dormant. Two months later the basal buds may blossom as a result of a very prolonged blossoming season. Frequently, some fruits may develop to the size of a walnut, while adjacent flower buds are just opening. In such situations, the blossoms are small and deformed, and the pollen production is poor. The development of leaves is also irregular. Initially the first buds are two-year-old or older buds located in the centers of the trees, forming clusters of foliage. Later, occasional lateral buds and finally tip buds may start. Foliage is scant well into the growing season and sufficiently lacking to maintain fruit growth. Growth from delayed opening buds appears weak at first, but later it is normal in all respects including bud set.

Of the many deciduous fruit crops, prolonged dormancy is most severe with the peach in commercial districts adjacent to the subtropical regions of the world—in northern Florida, southern Georgia, southern Mississippi, and southern Louisiana; in Italy; in parts of South Africa and South America; and in Queensland, Australia. Figure 5-8 shows peach trees in and out of their physiologic dormant period. Note the degree of foliation on the tree which received sufficient cold to break the rest and that on the tree which did not receive sufficient cold. To reduce the problems pertaining to prolonged physiologic dormancy, research workers of agricultural experiment stations have attacked the problem in two ways: (1) by testing chemicals which bring partially dormant buds out of their rest period, and (2) by breeding varieties which combine short rest periods and consistent production of marketable fruit. Of these two methods, past experimental evidence indicates that the breeding of adapted varieties is the more practical.

In addition to deciduous trees and shrubs, the buds of the storage organs of certain herbaceous crops go into a physiologic dormant period. In general, these

Figure 5-8 Effect of sufficient and insufficient cold on the behavior of two varieties of peach trees. The Southland tree on the left requires about 750 hours of cold, whereas the Redhaven tree on the right requires about 950 hours. Both trees were growing in the same orchard and were photographed on the same day, May 18, 1950. *(Courtesy, J. P. Overcash, Mississippi State University.)*

organs are the fleshy rhizomes of rhubarb; the head of cabbage; the bulbs of onion, lily, and tulip; and the corms of gladiolus. All of these organs require a period of exposure to cold, usually between 35 and 45°F (1.7 and 7.2°C), to break the rest. The means by which this period of exposure is attained varies with the kind of plant.

Specific Temperature Effects

Temperature and Starch-Sugar Transformations of Harvest Products The quality of many horticultural products is determined by the degree of sweetness. In crops where the principal reserve carbohydrate is starch, the following reactions are involved:

(enzymes)

1. sugar ⟶ starch
2. starch ⟶ sugar
3. sugar ⟶ CO_2 + H_2O + heat and other forms of energy

Low temperature decreases the rate of all three reactions, but the rate of decrease is not uniformly the same. Investigations have shown that reactions (1) and (3) are decreased more than (2). In this way sugars in the plant product accumulate and remain at a high level. High temperatures increase the rate of all three reactions, but (3) increases more than (1) or (2). Therefore, the sugars remain at a low level. This effect of temperature on starch-sugar transformation of certain harvested products explains why potato tubers become sweet when stored at temperatures of 32 to 40°F (0 to 4°C); why sweet corn raised in Maine remains sweet for a relatively long time; why garden peas are the sweetest if cooked immediately after they are harvested; why apples contain more sugar when stored at 40°F (4°C) rather than at 55°F (13°C); why sugar flows in the hard maple in the spring; and why sugarcane becomes sweet after the first frost.

Temperature and Diseases As with horticultural crops, many plant pathogens—fungi and bacteria which cause disease—have optimum temperatures for growth. In general, these organisms may be placed in either of two groups: (1) those which thrive best in cool weather and (2) those which thrive best in warm weather. Examples of cool-weather pathogens are the fungi which cause apple scab and peach leaf curl. Examples of warm-weather pathogens are fungi which cause bitter rot and black rot of apple, smut on corn, and fusarium wilt on tomato. These differences in temperature requirements explain why a certain disease is prevalent in one region of the country and nonexistent or of little consequence in another. For example, peach leaf curl is more prevalent in northern peach-growing districts. In fact, this disease does not exist in central Georgia, central South Carolina, and central Mississippi. On the other hand, black rot and bitter rot of apples are more prevalent in southern apple-growing districts because of the higher temperature requirements of the organisms responsible for these diseases.

Insect pests are also influenced by temperatures. In general, there are more generations in the South than in the North simply because of the longer period of favorable temperature for reproduction. For example, the codling moth caterpillar, a serious insect pest of the apple, has only one generation in Maine, three or four in New Jersey, and seven or eight in western South Carolina; the cucumber beetle has two or three generations in Michigan and four or five in south Mississippi. In addition, the cold winter temperatures of the North destroy many insect pests. The mild winter temperatures of the South favor their hibernation.

QUESTIONS

1 Fruit trees and other woody plants low in carbohydrates at the end of the growing season are more susceptible to winter injury than trees high in carbohydrates. Explain.

2 Fruit trees and other woody plants with heavy crops are more likely to be injured by low temperatures during the winter than trees with moderate crops. Explain.

3 Partial defoliation, particularly during the latter stages of growth, makes woody plants susceptible to winter injury. Explain.

4 In general, the Great Plains region of the United States is unfavorable for the development of satisfactory crops of the deciduous fruits. Explain.

5 In the northern United States peach trees often have the past season's growth injured during the winter, and in the southern United States the flower buds are often the only part of the tree injured. Explain.

6 Many shrubs adapted to the northern United States do not thrive well in the southern United States. Explain.

7 Moderate cold following a period of rapid vegetative growth in the fall is likely to cause more winter injury than severe cold following a period of moderate vegetative growth. Explain.

8 What is meant by the *rest*, or *physiologic dormant, period?*

9 Under the usual conditions prevailing in a greenhouse, a peach tree, an apple tree, or a grape vine, which has grown there for one season, would not come out of its rest period in March or April (the usual time) but in late spring or early summer. Explain.

10 Explain why commercial apple orchards do not exist along the Gulf Coast or in Florida.

11 A warm November and December and a cold January and February in the southern United States are usually followed by satisfactory crops of peaches. Opposite conditions often result in smaller crops. Explain.

12 In general, the chilling requirements of peach varieties for orchards in warm temperate districts should be considered. Explain.

13 Explain why citrus orchards do not exist in northern regions.

14 Garden peas become hard, yellow, and unpalatable if they are exposed to relatively high temperatures after being picked. Explain.

15 Peas and sweet corn stay comparatively sweet when kept cool after they are harvested. Explain.
16 Peas canned immediately after they are picked contain more sugar than those canned two hours after being picked. Explain.
17 Harvested sweet corn and garden peas cease to remain sweet if subjected to high temperatures. Explain.
18 How would you expect high temperatures to affect quality and length of the harvest period of sweet corn?
19 Experiments have shown that sweet corn stored at 86°F (30°C) loses its sweetness six times faster than sweet corn stored at 32°F (0°C). Explain.
20 One ton of husked green corn during the first 24 hours of storage at 86°F (30°C)loses about 3 pounds of sugar. Explain. Can the loss of sugar be entirely prevented? Can it be retarded? If so, how?
21 When sweet corn is kept in storage, the loss of sugar is about two to four times as fast at 50°F (10°C) as at 30°F (−1°C). Explain.
22 How should garden peas, lima beans, and protepeas be handled immediately after the pods are picked in order to retain their sweetness and quality?
23 A home gardener wants information on the best way to harvest garden peas and sweet corn. Outline two methods for each.
24 Parsnips subjected to chilling temperatures become sweet. Explain.
25 Spinach canned in the fall is sweeter than spinach canned in the spring. Explain.
26 Certain diseases and insects are more prevalent in some regions than in others. Explain.

SELECTED REFERENCES FOR FURTHER STUDY

Thompson, H. C. 1939. Temperature in relation to vegetative and reproductive development in plants. *Proc. Amer. Soc. Hort. Sci.* 37:672–679. A concise and comprehensive review of the early literature on the effect of temperature and photoperiods on the vegetative and reproductive phases, with special reference to the biennial vegetables.

Weinberger, J. H. 1950. Prolonged dormancy of peaches. *Proc. Amer. Soc. Hort. Sci.* 56:129–133. An interesting and first-hand account of the symptoms of prolonged physiologic dormancy in peach trees, and their practical implications presented by an authority on the dormancy of woody plants.

Went, F. W. 1957. Climate and agriculture, *Sci. Amer.* 196:83–94. A concise statement of man's dependence on agriculture, the development and use of the phytotron at the Earhart Plant Research Laboratory, and the behavior of plants under controlled conditions.

Chapter 6

The Light Supply

And God said, Let there be light; and there was light.

Genesis 1:3

NATURE AND PROPERTIES OF LIGHT

Light is a form of kinetic energy. Its properties may be examined from two different points of view: the particulate and the electromagnetic. According to the particulate concept, light comes from the sun in the form of discreet units, or tiny particles, called *quanta,* or *photons.* These tiny particles travel at an average speed of 186,191 miles per second, they have weight, and, as such, they exert a pressure of 5×10^{-11} atmosphere.[1] Scientists have estimated that the weight of light emitted from the sun is about 250 million tons per minute and that about 3.3 pounds of light fall each year on each square mile of the earth's surface.

According to the electromagnetic concept, the tiny photons, or quanta, exhibit the properties of waves, properties of length and frequency. According to

[1]An atmosphere at sea level exerts a pressure of 14.67 pounds per square inch.

the scientists who study the sun, this life-giving body is a slow-burning hydrogen furnace. In its center and for every second, 564 million tons of hydrogen change to 560 million tons of helium with the release of 4 million tons of kinetic energy. This energy consists mostly of the relatively short, high frequency x-rays. As these rays travel to the surface of the sun, some remain as x-rays; other change to short, high frequency cosmic rays; others to moderately long, moderately frequent ultraviolet and "visible" rays; and others to the long, low frequency, infrared and very long radio waves. Not all of the rays which are directed toward the earth reach its surface. Certain rays are filtered out or absorbed by various components of the atmosphere. Note the three filters presented in Figure 6-1. What wave lengths are filtered out by carbon dioxide, by ozone, and by water, respectively? What wave lengths reach the surface of the earth at sea level?

Measurement of Light

In general, crop-plant scientists measure the amount of light in terms of foot-candles or thermal units. *A foot-candle is the amount of light impinging on a surface one foot away from a standard wax candle.* A commonly used instrument

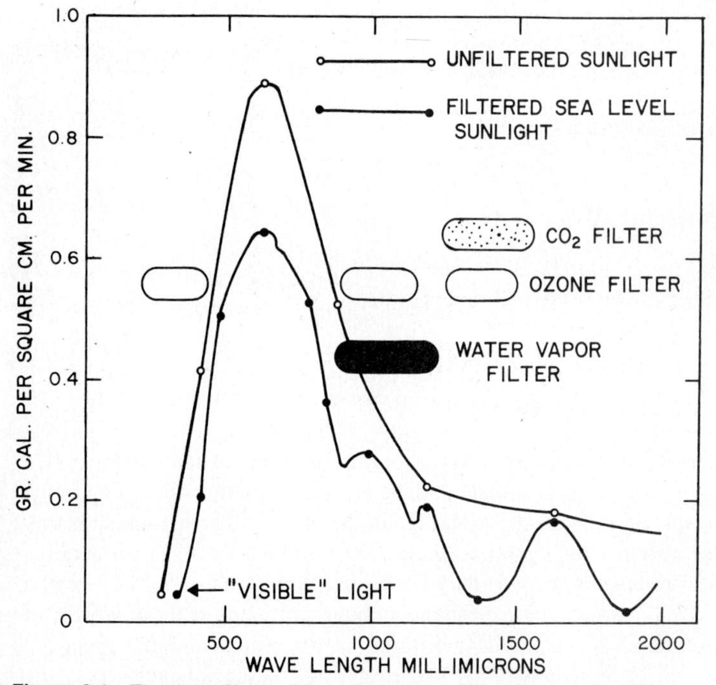

Figure 6-1 The three filters of the atmosphere. Note that not all of the radiation directed toward the earth arrives at the surface of the earth. *(Courtesy, A. C. Leopold, Purdue University and McGraw-Hill Book Company, Inc.)*

is the Weston Illuminometer. This instrument consists of a light receptor cell, a sensitive galvanometer, and a volt meter from which the number of foot-candles can be read directly. It is quite compact, portable, and inexpensive, and can be carried from place to place for use in the laboratory and field.

Thermal units are the British thermal unit, abbreviated BTU, and the gram calorie, abbreviated gm-cal. *A BTU is the amount of heat required to raise one pound of water 1°F; a gm-cal. is the amount of heat required to raise one gram of water from 3.5 to 4.5°C (38 to 40°F).* The standard instrument consists of a pyrheliometer and a continuous recording potentiometer. In general, the pyrheliometer is placed in the open, usually on top of the building which contains the potentiometer and its accessories.

Total Amount of Light

Four factors are involved: differences between seasons in the same place; differences between places during the same season; differences between arid and humid locations within the same latitude; and differences between high and low elevations within the same latitude. Differences between seasons in the same place and differences between places during the same season for four locations in the humid section of the United States are shown in Figure 6-2. Note that for each location the light supply is at a minimum in January, increases to a maximum about June 21, and decreases to its January minimum by December 21. In

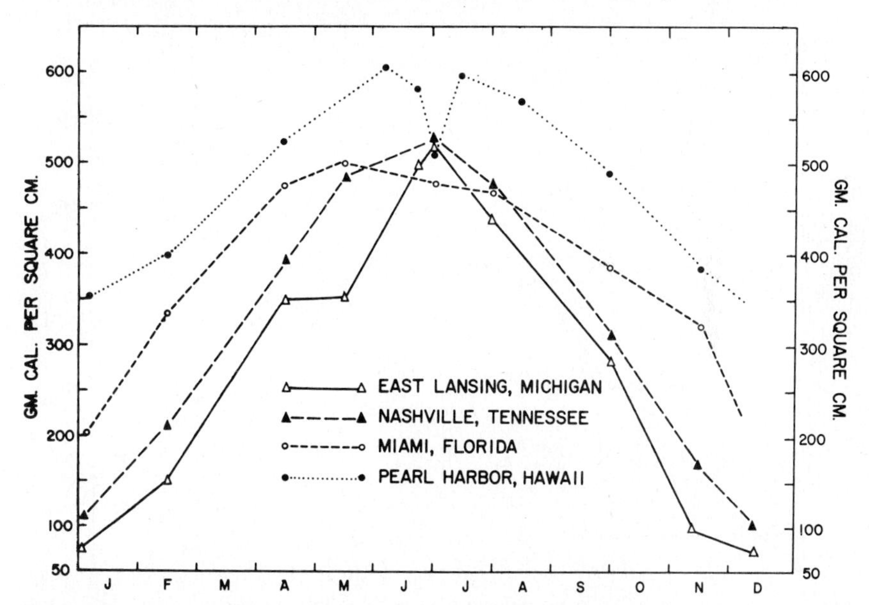

Figure 6-2 The total amount of light at four locations in the United States. *(Adapted from Figures 14, 16, 17 and 23,* Mich. Agr. Exp. Sta. Tech. *Bul. 222, 1950).*

other words, for each of the four places the light supply is relatively low during November, December, January and February, moderately high during March, April, September and October, and high during May, June, July and August. Would you say this statement is generally true for all places in the northern Hemisphere? What are the months of low, moderately high, and high supplies of light in the southern Hemisphere?

Note also the differences between places during the same season, that is, the differences due to latitude. In general, and from one year to the next, southerly locations in the northern Hemisphere have a greater light supply than northerly locations. This is particularly the case in the winter, spring, and fall months but not necessarily true in the summer. For example, the light supply in East Lansing, Michigan, the most northerly location of the four, is somewhat greater than the light supply in Miami, Florida from June 25 to July 15, and equal to the light supply in Nashville, Tennessee at the time of the summer solstice. Why is it that the light supply in the northerly locations in June, July, and August is comparable to the light supply in the southerly locations? Would you say that the same situation holds for places in the southern Hemisphere?

Differences in the light supply due to differences in the moisture content of the atmosphere for two comparable locations are shown in Figure 6-3. Note that the light supply is greater at the location with the relatively dry atmosphere. This is to be expected since water vapor has the capacity to absorb light. Thus, since

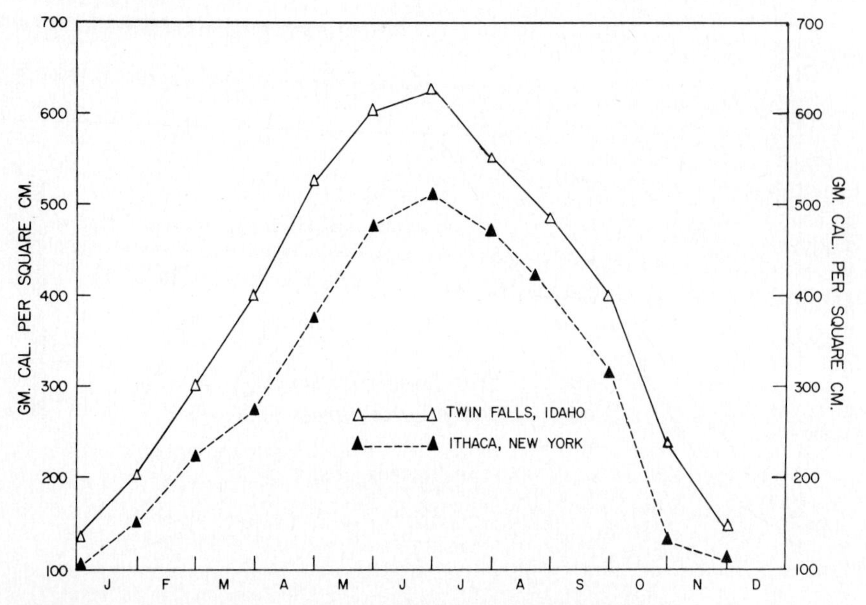

Figure 6-3 The total amount of light at an arid and a humid location in continental United States. *(Adapted from Figures 41 and 42,* Mich. Agr. Exp. Sta. Bul. *222, 1950.)*

more life-giving sunshine reaches the farmer's fields in the arid sections, as compared to those in the humid sections, with other factors favorable, there is greater opportunity for high rates of net photosynthesis and the production of acceptable, high-market yields.

Differences in light supply between places with high and low elevations at the same latitude within the humid area of the United States are exemplified by the differences in elevation between Columbia, Missouri at 730 feet (223 meters) and New York at 55 feet (18 meters). With the exception of the latter part of October and the first part of November, Columbia has a greater supply of available light than New York City has. However, the differences between the two places are not strictly comparable because of the wide distance between the two places and the fact that the climate of Columbia is continental, whereas that of New York City is maritime. Nevertheless, the data indicate that places at high elevations have a greater supply of light than places at low elevations. Thus, with other factors favorable, places of higher elevation will have a greater rate of net photosynthesis and a correspondingly greater yield.

Primary Function

As shown on page 6, light is an integral part of the photosynthetic reaction in that it provides the energy for the combination of carbon dioxide and water in the formation of the first manufactured compounds. Since this energy, either directly or indirectly, comes from the sun, the greater the amount of light available, with other conditions favorable, the greater the rate of gross photosynthesis and the amount of carbohydrates available for plant growth and development. Thus, with other factors favorable, regions with an abundant light supply should produce a greater quantity of the products of photosynthesis from one season to the next than regions with a low light supply.

In addition to the necessity of light energy for photosynthesis, light has other effects on plant growth. To obtain a working knowledge of these effects, a study of light intensity, light quality, and relative length of the light and dark periods for any given 24-hour day is necessary.

LIGHT INTENSITY

Light intensity refers to the number of quanta, or photons, impinging on a given area, or to the total amount of light which plants receive. In general, for any given location, the intensity varies with the day, with the seasons, and with the distance from the equator. It gradually increases from sunrise to the middle of the day and gradually decreases from the middle of the day to sunset; it is high in summer, moderately high in spring and fall, and low in winter; and it is highest at the equator and gradually decreases from the equator to the two poles. Other variations in light intensity are due to dust particles and water vapor in the atmosphere, to differences in slope of the land, and to differences in elevation.

How does light intensity influence plant growth and development? Its effect is discussed from three standpoints: (1) compensation and light saturation points of sun and shade plants, (2) intensity within the optimum range, (3) intensity below the optimum range, and (4) intensity above the optimum range.

Compensation and Light Saturation Points of Sun and Shade Plants

Just before and at sunrise, light intensity is very low and the rate of gross photosynthesis is correspondingly low. Usually, under these conditions the rate of gross photosynthesis is lower than the rate of respiration. As the sun appears above the horizon and climbs toward its zenith, the light intensity increases and the rate of gross photosynthesis correspondingly increases. The light intensity at which the rate of gross photosynthesis equals the rate of respiration is called the *compensation point*. Thus, at this intensity the rate of apparent, or net, photosynthesis is zero. In general, the compensation point varies with the kind of plant. In fact, investigations have shown that plants which require high light intensities have a higher compensation point or range than plants which require low light intensities. Note the compensation points for the sun and shade plants presented in Figure 6-4. What is the average compensation point for the eight sun plants and for the five shade plants? Under conditions of low light intensity, why do shade plants survive and why do sun plants finally die?

As the sun approaches its zenith, light intensity increases and the rate of net, or apparent, photosynthesis correspondingly increases. However, this increase in the rate of net photosynthesis does not take place over the entire range of light intensity. Usually at some point within this range, the rate of net photosynthesis

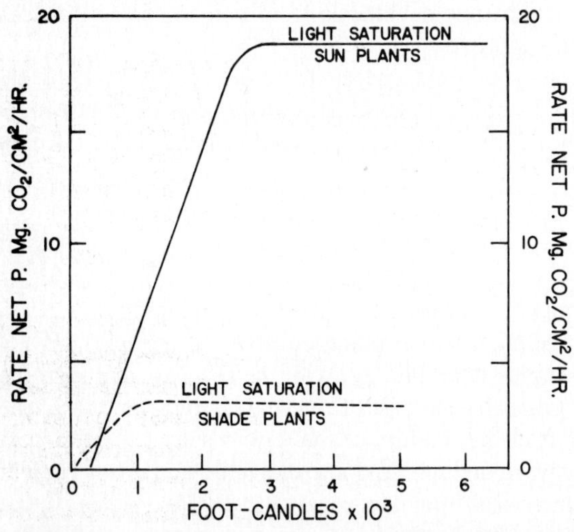

Figure 6-4 The compensation and light saturation points of sun and shade plants. *(Adapted from Figure 16,* Amer. Jour. Bot. *43:560, 1956.)*

levels off. In general, the light intensity at which there is no further increase in the rate of net photosynthesis is called the *light saturation point*. As with the compensation point, the light saturation point varies with the kind of plant. Note the average light saturation points for the eight sun plants and the five shade plants presented in Figure 6-4. With the sun plants, the rate of net photosynthesis increased from about 100 to 150-foot-candles to about 2800 foot-candles, at which point the rate of photosynthesis leveled off. With the shade plants, the rate of net photosynthesis increased from about 50 foot-candles to about 960 foot-candles, at which point it leveled off. Thus, sun plants have a decidedly higher light saturation point than shade plants. These differences in light-intensity utilization explain, partially at least, why certain plants thrive in full sun or sunny locations and would not survive in shade, and why other plants thrive in shade and would not survive in full sun.

Optimum Light Intensity Range

The effect of the optimum range may be illustrated by the use of the line AB.

A ———————————————————————————— B

The rate of photosynthesis is high.
The rate of respiration is normal.
Abundant carbohydrates are available for growth.

Note that within this range, with other factors favorable, the rate of photosynthesis is high, the rate of respiration is normal, and, as a result, the amount of carbohydrates available for growth and development is high. Thus, if the plants of any given crop are handled properly, particularly with respect to the vegetative and reproductive phases, the marketable yields are likely to be high.

The optimum, or favorable, light intensity range is not the same for all crops. For example, fern, many foliage plants, and African violet require relatively low light intensity; carnation, chrysanthemum, and rose require relatively high light intensity; and dogwood, nandina, and arborvitae thrive well over a wide range of light intensity. In other words, the favorable light intensity range varies with the kind of plant. Some kinds have their highest rate of photosynthesis and make more foods per unit time at relatively low intensity, whereas other kinds have their highest rate of photosynthesis and make more foods per unit time at relatively high intensity. Although the optimum light intensity range for many crops is not definitely known, experience, particularly with ornamental plants, indicates that plants may be classified as follows: (1) plants which require low light intensity, the so-called shade plants; (2) plants which require moderately high light intensity, the partial shade and sun plants; (3) plants which require high light intensity, the so-called sun plants; and (4) plants which thrive well over a wide range of light intensity, the shade or sun plants. Examples are presented in Table 6-1.

Table 6-1 Examples of Crop Plants with Various Light Intensity Requirements*

Shade only and no direct sun 500–1000 ft-c	Shade and direct sun for short time daily 1000–3000 ft-c	Direct sun mostly 3000–8000 ft-c	Slight shade and direct sun tolerant 2000–8000 ft-c
		Fruit Crops	
	Cacao, coffee, tea	Banana, citrus, coconut, date, fig, olive, papaya, pecan	Apple, pear, peach, plum, South American rubber tree
		Herbaceous crops	
Ginseng Tobacco (certain varieties	Vanilla, pepper	Capsicum pepper, cereal grains, cotton, corn, cucurbits, eggplant, flax, rice, pineapple, sweet potato, tobacco (certain varieties)	Cabbage, potato, peanut
		Ornamentals	
African violet, Aspidistra, begonia, diffenbachia, ferns (most species), ficus, mahonia, philodendron, pothos, schefflera	Calladium, dogwood, English ivy, gloxinia, ilex (certain species), peperomia (species), periwinkle, orchids (species), sansevieria (species)	Carnation, chrysanthemum, crapemyrtle, deutzia, gladiolus, kalanchoe, lantana, lily, poinsettia, rose, snapdragon	Abelia, berberis, buxus, forsythia, gardenia, ilex (certain species), magnolia, nandina

*Data from various sources.

Intensity below the Optimum Range

The effect of light intensity below the optimum range is illustrated by the data in Table 6-2. This study was made at the Michigan Experiment Station with the Grand Rapids Forcing tomato, a variety adapted to greenhouse culture. The plants were grown in a greenhouse from February 15 to June 15, and at the beginning of the test, they were divided into three groups. The first group was grown in full sunlight, the second under one layer of cheesecloth, and the third under two layers. The one layer of cheesecloth admitted 50.4 percent of the light, and the two layers admitted 25.0 percent. Note that the plants in full sunlight had a greater chlorophyll content and produced a greater yield than those under one layer of cheesecloth, and the plants under one layer of cheesecloth had, in turn, a greater chlorophyll content and produced a greater yield than those under two layers. Thus, the plants grown in full sunlight had the highest rate of photosynthesis, and those grown under two layers of cheesecloth had the lowest.

Similar results were obtained in a study of the daily rate of photosynthesis of an eight-year-old McIntosh apple tree at Cornell University. For the entire tree the rate of photosynthesis increased with increases in light intensity up to full sunlight. For example, on five cloudy days the average rate of carbon dioxide assimilation was 5.2 milligrams per hour per 100 square centimeters of leaf surface, and on five sunny days the average rate of photosynthesis was 20.6 milligrams. During the middle part of clear days there was more light than the leaves needed. In other words, some factor other than light intensity became limiting in photosynthesis. However, on the cloudy days there was insufficient light to carry on a high rate of photosynthesis. Figure 6-5 shows the relative thickness and number of rows of palisade cells of the leaves of unshaded and shaded apple trees. Note the greater thickness and larger number of palisade cells of the leaves grown in the sun.

Why are growth, development, and yield relatively low at deficient light intensities? As previously stated, light energy is needed for the combination of carbon dioxide and water in the photosynthetic reaction. With deficient light intensity the amount of energy available for the union of carbon dioxide and

Table 6-2 Effect of Low Light Intensity on the Growth of Tomatoes

Treatment	Relative amount of light admitted	Average daily intensity, ft-c	Yield, lb of fruits	Relative chlorophyll content	Relative efficency
Plants in full sunlight	100	1,140	65	High	High
Plants under one layer of cheesecloth	50	583	51	Moderately high	Moderately high
Plants under two layers of cheesecloth	25	261	32	Low	Low

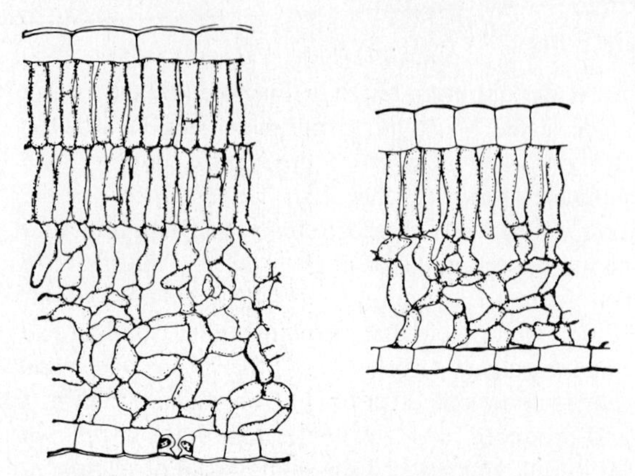

Figure 6-5 Diagrams of sun and shade leaves of apple. Left: sun leaf. Right: shade leaf. *(Courtesy, F. W. Emerson,* Basic Botany, *New York: McGraw-Hill Book Company, Inc.)*

water is low and, as a result, the rate of production of the initial carbohydrates for the making of other compounds is comparatively low. Thus, growth and development are accordingly slow and, in general, yields are likely to be low. This effect of deficient light is shown diagrammatically by line XA.

X **A**

Decreasing rate of photosynthesis with normal rates of respiration
Decreasing supplies of carbohydrates for growth and yields

Note that as light intensity decreases from A to X, the rate of photosynthesis decreases. At point X the rate of photosynthesis equals the rate of respiration and the amount of carbohydrates available for growth is zero. This effect of light intensity below the optimum range explains, partially at least, why crop plants are planted at definite distances; why growth and development of crops in greenhouses during the winter are relatively low; why greenhouse roofs are washed in early fall; why the interior of greenhouses is painted white; why plants growing in dense shade have relatively shallow root systems; and why ornamental hedges should have a broad base and a relatively narrow top.

Intensity above the Optimum Range

The effect of light intensity above the optimum range is shown by the data in Table 6-3. This study was made at the Mississippi Experiment Station with the Stokesdale tomato, a variety adapted to certain parts of the southern United States. The three levels of light intensity were obtained by growing the plants in full sun, under nylon, and under unbleached muslin. Note that the yield of the plants in full sun is relatively low.

Table 6-3 Effect of Excess Light Intensity on Yield of Tomatoes

Treatment	Average daily intensity, ft-c	Relative amount of light admitted, %	Yield, lb/10 plants	
			Sept. 29-Oct. 19	Sept. 29-Nov. 3
Plants in full sun	7,725	100	2.2	16.5
Plants under nylon	3,440	45	5.4	24.1
Plants under muslin	2,132	28	4.9	19.4

Source: Adapted from Table 3, *Proc. Am. Soc. Hort. Sci.* 60:293, 1952.

The question arises: Why are yields low in excess light intensity? Three explanations are set forth. The first explanation concerns the chlorophyll content. With certain plants excess light intensity reduces the chlorophyll content and the leaves become yellowish-green. As a result, the rate of absorption of light is low and the rate of photosynthesis is correspondingly low. This effect is sometimes called *solarization.* The second explanation concerns the water supply. The excess light intensity markedly increases the temperature of the leaves. This, in turn, induces a high rate of transpiration, and the rate of absorption of water does not keep up with it. As a result, the guard cells lose turgor, the stomates partially or entirely close, and diffusion of carbon dioxide into the leaves is slowed down. Thus, the rate of photosynthesis is slowed down while respiration continues, and the supply of carbohydrates available for growth and development is low. The third explanation concerns enzyme activity. The excess light increases the temperature of the leaves. This, in turn, inactivates the enzyme system which changes sugars to starch. As a result, the sugars accumulate and, according to the law of mass action, the rate of photosynthesis slows down. This effect of light intensity above the optimum range explains, partially at least, why shading materials are placed on greenhouse roofs in late spring and summer; why cloth or plastics are suspended above plants in greenhouses during the summer; why certain ornamental plants are grown in lathhouses during the summer, particularly in the southern United States; and why companion cropping is practiced in many parts of the world. In companion cropping, two or more kinds are grown in the same area at the same time. One kind is tall and withstands high light intensities. The other kind is short and is injured by high light intensities. The tall, upright crop shades the low-growing crop. This reduces the light intensity for the low-growing crop, which, in turn, promotes high rates of photosynthesis and high yields.

LIGHT QUALITY

As previously stated, photons have weight and exert a pressure. They also exhibit the properties of waves, and *quality of light* refers to the length of the waves. These wave lengths are set forth as follows:

Invisible	Visible spectrum									Invisible
Ultraviolet	V	B	B-G	G	Y	O	R	F-R		Infrared
				Millimicrons						
15–390	390	460	490	530	585	610	660	730		760+

Note the parts: the visible, the invisible, and the composition, or color spectrum, of visible light. The question arises: How does light quality influence plant growth and development?

The Visible Spectrum

The composition of visible light affects the rate of growth, as measured in terms of dry weight, and the vegetative and reproductive phases. With reference to the rate of growth, investigations at the California Institute of Technology have shown that tomato plants grown in red or blue light produce a greater dry weight over the same period of time than comparable plants grown in green light, and that plants grown in red or blue light produce a greater dry weight than plants grown in "white" light. This indicates that green light actually inhibits plant growth and may explain why light bulbs which transmit large quantities of red and blue light are more satisfactory in greenhouse culture than those which transmit relatively small quantities.

The Significance of Phytochrome

Phytochrome is a pigment, therefore, it absorbs light. According to our present knowledge phytochrome is widely distributed in crop plants and exists in very minute quantities in any given plant, a matter of parts per billion. Further, phytochrome exists in two reversible forms: the red light absorbing, designated as P_R, and the far red light absorbing, designated as P_{FR}. Of these forms, P_R is considered the stable, inactive, or storage form, and P_{FR} is considered the unstable, or biochemically active, form. In other words, the protein part of the molecule of P_{FR} acts as an enzyme or part of an enzyme system in changing the type of metabolism from one which is associated with the formation of vegetative buds to one which is associated with the formation of flower buds. According to recent research, the behavior of the phytochrome system is as follows. During the light period of any given 24-hour day, the P_R form changes to the P_{FR} form, and during the dark period, the P_{FR} form gradually changes to the P_R form. With short-day–long-night plants, the biochemically active form required a relatively long dark period to change the metabolism from the vegetative phase to the reproductive phase, whereas with long-day–short-night plants, a relatively short dark period is required, provided that optimum night and day temperatures are maintained in both cases. This reversible biochemical system also explains other types of crop plant behavior. For example, in the germination of the seed of the grass family, the epicotyl elongates while the seedling leaves remain undeveloped, but as soon as the light strikes the seedling at the surface of the soil, the epicotyl stops elongating and the leaves begin to grow and expand. In the germination of the seed of the legume family, the hypocotyl elongates while the seedling leaves remain small and protected by the cotyledons. Here again, as soon as light strikes the seedling, the hypocotyl stops elongating, the cotyledons open, and the leaves begin to expand. In both cases, while the seeds are in the dark (in the soil), the phytochrome system is in the P_R inactive form; but when light strikes the plant, the P_R form changes to the active P_{FR} form which, in turn, triggers the changes that are necessary for the expansion of the leaves, the shortening of the internodes, and the growth of the stems. The reversibility of the phytochrome system is illustrated as follows:

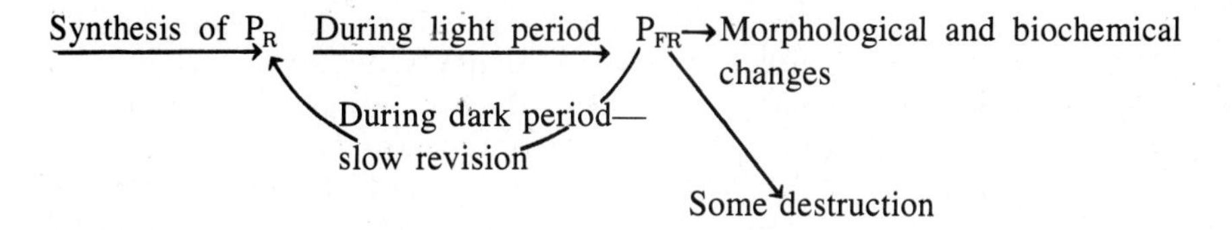

The Invisible Spectrum

As shown on page 120, the invisible spectrum has two parts: ultraviolet and infrared. What is the influence of the ultraviolet? Despite the popular opinion that ultraviolet light is beneficial to the growth of plants, investigations have shown that ultraviolet light near the visible end of the spectrum has no advantage in increasing dry weight or in hastening the time of flowering. For example, at the United States Plant Industry Station, Beltsville, Maryland, tomato and capsicum pepper plants were started in hotbeds covered with ultraviolet-transmitting glass and hotbeds covered with greenhouse glass. There were no practical differences in behavior between the two lots of plants. The plants started under greenhouse glass grew as rapidly and produced the same yields as those started under glass admitting ultraviolet light.

What is the influence of the infrared? Although definite information on the influence of infrared light is meager, recent research shows that infrared light has very little, if any, beneficial effect on plant growth. Since excellent crops are produced in greenhouses and since greenhouse glass absorbs most of the ultraviolet and infrared light, the invisible part of the spectrum seems to be unnecessary for plant growth and development. Further, the high-elevated, 4,000 to 5,000 feet (1,219 to 1,524 meters), potato-producing districts of eastern Idaho are known to produce substantially lower yields of marketable tubers than comparable low-elevated, 1,000 to 1,300 feet (305 to 396 meters), producing districts. According to the investigative authorities, the lower yields at the higher elevation are due to a greater amount of ultraviolet light impinging on the leaves of the potato plants. Thus, the results indicate that the relatively large amount of ultraviolet light at high elevations in arid regions may actually be detrimental to the growth and marketable yield of crop plants.

RELATIVE LENGTH OF THE LIGHT AND DARK PERIODS

The length of the light and dark periods refers to the period of light and the period of dark for any given 24-hour ''day'' for any given location. To obtain a working knowledge of the variation in the length of the light and dark periods, the length of these periods on the following dates should be kept in mind. These dates are March 21, September 21, December 21, and June 21. On March 21 and September 21, the spring and fall equinox respectively, the sun is directly over the equator and rises exactly in the east and sets exactly in the west. As a result, the length of the light and the length of the dark periods are the same at all places in both the northern and southern hemispheres. In other words, 12 hours of light and 12 hours of dark occur at all places on the surface of the earth on March 21 and September 21. On December 21, the sun is farthest south of the equator; the shortest light and the longest dark periods take place for all locations in the northern hemisphere, and the longest light and the shortest dark periods take place for all locations in the southern hemisphere. Conversely, on June 21, the sun is farthest north of the equator; the longest light and the shortest dark periods take place at all locations in the northern hemisphere, and the shortest light and longest dark periods take place at all places in the southern hemisphere. The length of the light and dark periods at representative locations in the northern and southern hemispheres are set forth in Table 6-4.

Note that the longest light period and the shortest dark period vary with distance from the equator. Thus, between northern and southern locations in the northern hemisphere, the length of the light period is longer in the northerly locations from March 21 to September 21, and the length of the light period is longer in the southerly locations from September 21 to March 21. Study the daylight curves for the six locations in Figure 6-6.

How does the relative length of the light and dark periods influence growth and development? In general, these periods influence crop plants in three ways: (1) in relative amount of carbohydrates made by all crops, (2) in time of flower-bud formation of many crops, and (3) in development of storage organs of such crops as onions and potatoes.

Amount of Carbohydrates

Naturally, the relative length of the light and dark periods influences photosynthesis and respiration. In general, the longer the light period, provided other factors are favorable, the greater will be the amount of foods made in photosynthesis; and the shorter the dark period, the lower will be the amount of foods used in respiration. Thus, a plant exposed to a favorable light period of 17 hours and to a favorable night period of 7 hours will have a greater quantity of carbohydrates

Table 6-4 Relative Length of Light and Dark Periods in Hours in the Northern and Southern Hemispheres

Dates	Degrees Latitude North											
	10[a]		20[b]		30[c]		40[d]		50[e]		Arctic Circle	
	Light	Dark	Light	Dark	Light	Dark	Light	Dark	Light	Dark	Light	Dark
Mar. 20–22 and Sept. 20–22	12.0	12.0	12.0	12.0	12.0	12.0	12.0	12.0	12.0	12.0	12.0	12.0
Dec. 20–22	11.7	12.3	11.0	13.0	10.2	13.8	9.3	14.7	8.1	15.9	0	24.0
June 20–22	12.7	11.3	13.3	10.7	14.1	9.9	15.0	9.0	16.4	7.6	24.0	0

Dates	Degrees Latitude South											
	10[f]		20[g]		30[h]		40[i]		50[j]		Antarctic Circle	
	Light	Dark	Light	Dark	Light	Dark	Light	Dark	Light	Dark	Light	Dark
Mar. 20–22 and Sept. 20–22	12.0	12.0	12.0	12.0	12.0	12.0	12.0	12.0	12.0	12.0	12.0	12.0
Dec. 20–22	12.7	11.3	13.3	10.7	14.1	9.9	15.0	9.0	16.4	7.6	24.0	0
June 20–22	11.7	12.3	11.0	13.0	10.2	13.8	9.3	14.7	8.1	15.9	0	24.0

[a]Saigon; Kwajalein Atoll; San Jose; Addis Ababa.
[b]Haiphong; Wake Island; Mauna Loa, Hawaii; Mexico City; Bombay.
[c]Shanghai; New Orleans; Jacksonville; Alexandria.
[d]Darien; Kansas City; Philadelphia, Rome.
[e]Vancouver; Winnipeg; Calais; Stalingrad.
[f]Port Moresby; Guadalcanal Island; Maccio, Brazil; Luanda.
[g]Bowen, Queensland; Tonga; Belo Horizonta; Beira.
[h]Graften, New South Wales; Porto Alegro, Brazil; Durban.
[i]Bass Strait; Cook Strait; Bahia Blanca.
[j]Santa Cruz, Kergulen Islands.

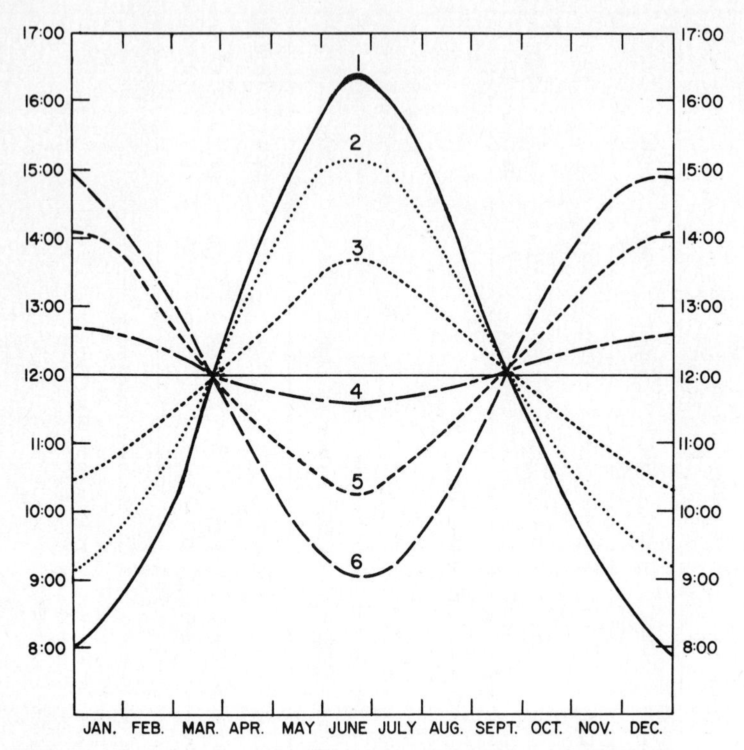

Figure 6-6 Day-length curves for three latitudes in the northern hemisphere and for three latitudes in the southern hemisphere. [1]50° N: Vancouver, Winnipeg, English Channel, Karkov, Karaganda. [2]42° N: Hackodate, Chicago, Rome, Istanbul, Tashkent. [3]24–26° N: Taipei, Midway, Miami, Karachi. [4]10° S: Port Moseby, Maraquersa Islands, Maceio, Luanda. [5]30° S: Graften, Porto Alegro, Durban. [6]40° S: Bass Strait, Cook Strait, Bahia Blanca.

for growth and development than a plant exposed to a favorable light period of 14 hours and a favorable night period of 10 hours. This explains, partially at least, why high marketable yields of many crops are secured in the more northerly latitudes during the summer months. During the long light period, relatively large quantities of carbohydrates are made in photosynthesis; and during the short dark period, small quantities are used in respiration. As a result, large quantities of carbohydrates are available for growth and yield.

Time of Flower-bud Formation

The relative length of the light and dark periods also determines the time at which flower buds are formed in many crops. Many investigators have studied the effects of various lengths of light and dark periods. In general, they have found that some plants, e.g., spring radish and delphinium, require long light periods combined with short dark periods for the formation of their flower buds; other plants, e.g., poinsettia and kalanchoe, require short light periods combined with long dark periods for the formation of their flower buds; and still other plants, e.g., tomato and cotton, form flower buds during both long and short light periods. Plants which require long light and short dark periods for the formation of their flower buds are called *long-day–short-night plants;* those which require short light and long dark periods are called *short-day–long-night plants*; and those which form flower buds in both long light and dark periods and short light and long dark periods are called *day-night neutral plants*. In general, within any given 24-hour period, most long-day–short-night plants require 8 to 10 hours of continuous dark, and most short-day–long-night plants require 10 to 14 hours of continuous dark for the formation of their flower buds.

When short-day–long-night plants are grown during long days and short nights, they manufacture abundant carbohydrates and make abundant proteins. These, in turn, are used for the development of the stems, leaves, and absorbing roots. Thus, short-day–long-night plants grown during long days and short nights are vegetative, nonflowering, and nonfruitful. On the other hand, when long-day–short-night plants are grown during short days and long nights, they manufacture very little carbohydrates and, in turn, make few proteins. Thus, long-day–short-night plants grown during short days and long nights, because of lack of light, are weakly vegetative and nonflowering.

The Critical Photoperiod

The critical photoperiod varies with the photoperiodic response of plants. With short-day plants, it is the maximum length of the light period at which the plant differentiates its flower buds and develops its flowers, fruit, and seed. In other words, when the length of the light period exceeds the critical photoperiod, the vegetative phase of the crop plant is promoted and the reproductive phase is suppressed. In general, the critical period for short-day plants seems to vary from 11 to 14 hours. For example, the large headed chrysanthemum differentiates its flower buds at a light period of 13.5 hours. If, after the plant has differentiated its flower buds, it is subjected to a photoperiod longer than 14 hours, its vegetative phase is promoted and its flowers do not develop normally.

With long-day plants, the critical photoperiod is the minimum length of the light period at which the plant differentiates its flower buds. In other words, and in sharp contrast to the response of short-day plants, when the length of the light period is less than the critical photoperiod, the vegetative phase of the plants is promoted and the reproductive phase is suppressed. In general, the critical photoperiod for long-day plants seems to be from 12 to 14 hours. For example, spinach is a long-day plant. It differentiates its flower buds at a photoperiod of 13 hours. If, after the plant has differentiated its buds, it is subjected to a photoperiod shorter than 13 hours, its vegetative phase is promoted and its reproductive phase is suppressed. The reaction of short-day–long-night plants and long-day–short-night plants is presented in Figure 6-7. Just how the length of the light and dark periods controls the flowering of long-day–short-night plants and short-day–long-night plants is not definitely known. One explanation is that the length of the light and dark periods controls the time when protein synthesis takes place. For example, certain species of salvia are short-day–long-night plants. These plants accumulate nitrates and carbohydrates in their tissues during short days and long nights. Apparently, the short days combined with long nights limit the synthesis of proteins, which, in turn, limits the making of protoplasm and the development of new cells. Because of the limited protein synthesis, the carbohydrates accumulate. This, in turn, results in the formation, or the laying down, of flower-bud primordia. Another and more widely accepted explanation is that the length of the light and dark periods determines the formation and translocation of flower-bud forming hormones, and that these hormones are made in the leaves. For example, if a young, physiologically active leaf of the cockleburr, a short-day–long-night plant, is exposed to 10 hours of dark, the plant does not form flower buds. Further, if a single leaf of a flowering plant is grafted on a nonflowering plant, the nonflowering plant forms flower buds and blooms even though it is growing under long-day and short-night periods. A brief list of each of the three groups of plants is presented in Table 6-5.

What practical use can be made of the differential response to the length of the light and dark periods? In general, fruit, vegetable, and flower growers select varieties which are adapted to the relative length of the light and dark periods in any given location, and florists, particularly the growers of commercial varieties of chrysanthemums, actually control the length of the light and dark periods.

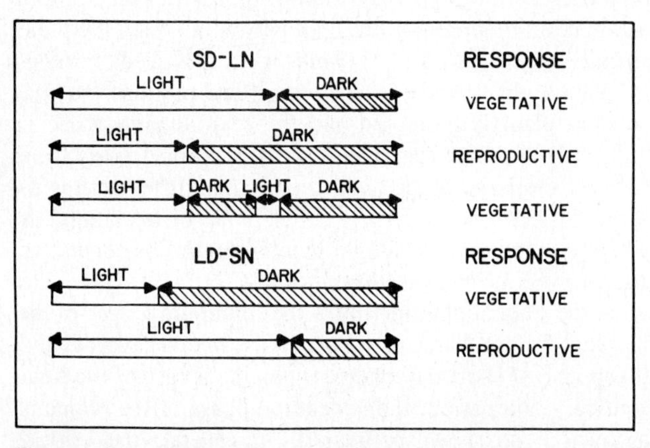

Figure 6-7 Response of short-day–long-night plants and long-day–short-night plants to the vegetative and reproductive phases. Note that a short exposure of light during the dark period promotes the vegetative phase in short-day–long-night plants and the reproductive phase in long-day–short-night plants.

Table 6-5 Some Short-day–long-night, Long-day–short-night, and Day-night Neutral Plants*

Group	Short-day–long-night	Long-day–short-night	Day-night neutral
Fruit Crops	Coffee, strawberry (certain varieties)	Strawberry (certain varieties)	Strawberry (everbearing)
Herbaceous Crops	Potato (certain varieties), snap bean (dwarf varieties), sweet potato, winter rice, soybean (certain varieties), sugar cane	Chinese cabbage, peanut, lettuce, radish, spinach, peppermint, barley (certain varieties), soybean (certain varieties), cereals	Tomato, eggplant, pepper, cucurbit, cotton, polebean, var. snap bean
Ornamentals	Chrysanthemum (large headed varieties), bouvardia, cosmos, kalanchoe, poinsettia, Japanese morning glory, gardenia	Tuberous rooted begonia, China aster, feverfew delphinium, stock, hibiscus, pyrethrum	African violet, carnation, petunia, zinnia, cyclamen

*Data from various sources.

Under natural conditions the greenhouse chrysanthemum flowers in October. By giving the plants additional light during the winter and shortening the light period during the summer, growers are able to place greenhouse varieties of this crop on the market at all times of the year. Figure 6-8 shows how the light period is lengthened during the winter, and Figure 6-9 shows how the light period is shortened during the summer.

Development of Storage Organs

The length of the light and dark periods also influences the time of formation of certain storage organs, e.g., bulbs of the onion, and tubers of the potato and Jerusalem artichoke. Investigations have shown that comparatively short light periods favor the development of tubers of certain varieties of the potato, that comparatively short light periods (10 to 11 hours) are required for the formation of bulbs of the Bermuda onion, and that long light periods (14 to 16 hours) are necessary for the formation of bulbs of American types. This explains, partially at least, why certain northern varieties of onions are unadapted to the length of the light and dark periods of the South and why the southern varieties are unadapted to the length of the light and dark periods of the North.

Figure 6-8 An installation showing the use of electric lights to lengthen the period of light. *(Courtesy, Alex Laurie, Sarasota, Florida.)*

Figure 6-9 An installation showing the use of black cloth to shorten the period of light. *(Courtesy, Alex Laurie, Sarasota, Florida.)*

QUESTIONS

1 As related to the crop plant, state three outstanding properties of light.
2 State the primary function of light in crop-plant production.
3 Calculate the mean weight of light which falls on each acre or hectare of your community (1) from one year to the next and (2) during the frost-free growing season.
4 What are the months of low, moderately high, and high supplies of light in the Northern Hemisphere? In the Southern Hemisphere? Give reasons.
5 Under conditions of low light intensity, why do shade plants survive and why do sun plants finally die?
6 Under conditions of high light intensity, why do sun plants survive and why do shade plants finally die?
7 Within limits, increases in light intensity increase the rate of gross photosynthesis. Explain.
8 Light is generally the limiting factor in growing plants in greenhouses during the winter months. Explain.
9 Plants grown in dense shade have less extensive root systems than plants grown in the sun. Explain in terms of the light supply.
10 Holly trees growing in shade have a lesser rate of growth than those growing in sun. Explain.
11 The covering of ornamental shrubs with opaque material in late fall is a bad practice. Explain.
12 The sugar content of many fruits (apples, peaches, grapes, cantaloupes) is associated in a positive way with the amount of sunshine during the period of fruit growth. Explain.
13 Under what conditions is light intensity likely to be harmful? Explain.
14 When raising certain crops in greenhouses, lime or special shading material is placed on the roof during spring and summer and removed during fall and winter. Explain.
15 In general, the propagating houses of florists are more heavily shaded than other houses. Explain.
16 How do shade houses promote the production of high-quality flowers during periods of excessive light intensity?
17 A bright, sunny fall is more favorable for the winter conditioning of woody plants than a cloudy fall. Give two reasons.
18 When do the equinoxes occur? How long are the light and dark periods on these dates?
19 For any given location, on what date does the longest light period occur? The shortest?
20 What is a short-day–long-night plant? A long-day–short night plant? A day-night neutral plant?
21 A home gardener in continental United States writes: "I have been growing a certain variety of pole bean for a number of years. I find it never produces pods until October. I have tried to make it produce pods earlier by growing the plants in different soils and in different locations on the farm but with little success." Can you give any explanation?

22 In home gardens, spinach is grown during late fall, winter, and early spring. Explain why in terms of the light supply.

23 How does the commercial florist lengthen the marketing season of large-headed varieties of chrysanthemums? Explain.

24 A variety of onions requires a minimum of 16 hours of light during a 24-hour day for the development of satisfactory bulbs. With other factors favorable, would this variety be adapted to your particular location? Explain.

SELECTED REFERENCES FOR FURTHER STUDY

Butler, W. L., and R. J. Downs. 1960. Light and plant development. *Sci. Amer.* 203:56–63. A technical discussion of the effect of red and far red light on the behavior of plants.

Garner, W. W., and H. A. Allard. 1923. Further studies in photoperiodism, the response of the plant to relative length of day and night. *Jour. Agr. Res.* 23:871–920. A classic paper by the two scientists who discovered photoperiodism in crop and ornamental plants.

Parker, H. M., and H. A. Borthwick. 1942. Day length and crop yields. *U.S. Dept. Agr. Misc. Pub.* 507. A popular, nontechnical, and well-illustrated bulletin on the relation of photoperiodism to crop-plant production.

Chapter 7

The Essential Elements

Crop plants are God's gift to all mankind.

ESSENTIAL ELEMENTS AND ESSENTIAL RAW MATERIALS

The green plant is a biochemical factory. Certain raw materials are used, either directly or indirectly, in the making of the all-important foods, fibers, enzymes, hormones, and vitamins. These raw materials meet at least two requirements: (1) they contain one or more essential elements for growth and development, and (2) they exist in a form which plants can absorb and use. For example, nitrogen is part of the molecule of all proteins and part of the molecule of both chlorophyll a and chorophyll b. Nitrogen is therefore an essential element. Although nitrogen exists in many types of compounds, crop plants absorb and use nitrogen from the soil mostly in two relatively simple forms: the nitrate ion and the ammonium ion. Because these ions are absorbed and used by plants in making the many nitrogenous organic compounds, they are called essential raw materials. Thus, *essential raw materials are chemical compounds or parts of chemical compounds which contain one or more essential elements for plant growth and are absorbed and utilized by plants.*

The essential elements necessary for plant growth and development are carbon, oxygen, hydrogen, nitrogen, phosphorus, potassium, sulfur, calcium, magnesium, manganese, iron, boron, zinc, copper, and molybdenum. Since any one of these elements may become a limiting factor in growth and development, the student should learn what the role of these elements is in plant life, how to recognize symptoms of essential-element deficiency, and when and how best to replenish the supply.

CARBON AND OXYGEN

Properties and Role

As they affect the crop plant, the properties of carbon are unique. Firstly, the carbon atom has the remarkable ability to combine with itself and with other elements. Thus, compounds of carbon may exist where the carbon atoms form straight chains, rings, or combinations of the two. Secondly, the carbon atom has the ability to form both symmetric and asymmetric compounds; that is, one compound is the mirror image of the other. One compound turns the plane of polarized light to the right, and the other turns the plane of polarized light to the left, hence, the terms *dextrorotary* and *levorotary*. Thirdly and most importantly, crop plants are able to use compounds which are either dextrorotary or levorotary, but not both. For example, all the amino acids made by crop plants are levorotary. In other words, the enzyme systems of crop plants have the capacity to change the L amino acids to proteins and vice versa, but not the ability to change the dextrorotary, or D forms.

Oxygen is a constituent of all carbohydrates, lipids, proteins, and related substances. Of these compounds the carbohydrates are relatively high in oxygen, the proteins are moderately high, and the lipids are relatively low.

The Carrier

The carrier of carbon and oxygen is carbon dioxide. In other words, of all the many compounds which contain carbon and oxygen, plants obtain these two essential elements only as carbon dioxide. Thus, carbon dioxide is an essential raw material for plant growth and development.

As is well known, the supply of carbon dioxide for crop plants is the atmosphere—the huge canopy of air which surrounds the surface of the earth. In this canopy carbon dioxide exists as a gas in relatively low concentration, about 3 to 4 parts in 10,000 parts by volume (0.03 to 0.04 percent) in open fields and in somewhat larger amounts in greenhouses and hotbeds. Although this concentration by volume is low, the actual amount of carbon dioxide in the atmosphere is tremendous. In fact, scientists have estimated that the total quantity of this gas in the troposphere—the lower part of the earth's canopy—is on the order of 600 billion tons and that about 70 billion tons are used each year by plants in photosynthesis.

Despite this enormous quantity used in photosynthesis, the total amount of carbon dioxide in the atmosphere remains relatively constant. How is the carbon dioxide supply replenished? Carbon dioxide is released into the atmosphere for the most part by two similar processes: (1) the respiration of all living organisms and (2) the combustion of all organic compounds. The living organisms include both animals and plants on the land, in the soil, and in the waters of the earth. Of particular importance are the organisms which decompose organic wastes. In general, these organisms give off by far the greater quantities of carbon dioxide, and, in this way, they keep the amount of organic wastes at a minimum.

Carbon dioxide gets into the manufacturing tissues by the process of diffusion. As long as photosynthesis is taking place, the gas diffuses from the atmosphere through the stomates into the intercellular spaces of the chlorophyll-containing cells. On the surface of these cells carbon dioxide enters into solution with water, diffuses to the surface of the chloroplasts, and finally unites with water in the formation of the initial food substances and related compounds.

When is the concentration of carbon dioxide likely to be the limiting factor in photosynthesis and in plant growth and development? In general, when other factors influencing photosynthesis are in optimum supply, carbon dioxide is the limiting factor. This condition is likely to occur when the water supply within the plant is such that the guard cells are fully turgid and the stomates are open, when the temperature and light intensity are within their respective optimum ranges, when the periods of light and dark are favorable for the type of growth the plant should be making, and when all other essential raw materials are in optimum supply. With these optimum conditions, investigations have shown that an enrichment of air with carbon dioxide increases plant growth. The tests with this compound were conducted in greenhouses or in light-transmitting cabinets where the carbon dioxide supply could be more or less confined, and when other environmental factors were not limiting growth. In one experiment, cucumbers supplied with air consisting of 31.3 parts of carbon dioxide in 10,000 parts by volume increased 60.0 percent in dry weight over those supplied with 3.0 parts of this gas in 10,000 parts, the usual concentration. Study the data in Table 7-1. In the three tests with lettuce and the test with tomatoes, the addition of carbon dioxide during the day to a maximum concentration of 2,000 parts per million (ppm) increased yields, even under the relative low light intensities in East Lansing, Michigan. Increasing the concentration of carbon dioxide in experiments with the same or other crops has increased either the yields, the quality, or both, of roses, carnations, and snapdragons at the Colorado Station, of potatoes, beets, and carrots at the Vermont Station, and of tomatoes at the National Vegetable Research Station in England and the Royal Agricultural and Veterinary College at Copenhagen, Denmark. In general, increasing the concentration from 0.03 percent to 0.3 percent by volume increased the rate of photosynthesis

Table 7-1 Effect of Increasing the CO_2 Content of the Air in a Greenhouse at Michigan State University, 1961 and 1963*

Crop	No. of varieties	Period of growth	CO_2 not added		CO_2 added	
			ppm	lbs	ppm	lbs
Lettuce	7	Jan. 18–Feb. 26 '61	250–350	1.9†	300–600	2.5†
Lettuce	3	Feb. 12–Mar. 26 '63	125–500	1.4†	800–2000	2.4†
Lettuce	4	Nov. 20–Dec. 19 '63	125–500	1.3†	800–2000	1.7†
Tomatoes	9	Feb. 15–June 1 '63	125–500	9.8‡	800–2000	14.0‡

**Source:* Adapted from Tables 1, 2, 3 and 4. *Econ. Bot.* 18: 42–48, 1964.
†lbs/ 10 heads.
‡lbs/ plant.

and the rate of growth, until some other factor became limiting in growth. At present, the addition of carbon dioxide to air of greenhouse crops is widely practiced throughout the world. However, research is needed to determine how best to use the carbon dioxide, particularly with respect to variations in temperature and light intensity. The use of carbon dioxide on crop plants in the field is in the experimental stage.

Air Pollution and Crop-Plant Photosynthesis

The natural components of the atmosphere are nitrogen (79 percent), oxygen (20 percent), water vapor, carbon dioxide, and argon. All of these with the exception of argon have a definite role in crop plant production. Hence, they are beneficial. In addition to the natural components and as a result of man's activities, the atmosphere contains a wide variety of other substances. In general, these substances are called *pollutants,* and they may be divided into two groups: (1) the tiny particles of solids called *particulates;* examples are fly ash and soot in thick black smoke, and (2) gases, principally sulfur dioxide, carbon monoxide, the oxides of nitrogen, and incompletely oxidized hydrocarbons. All of these pollutants are inimical to crop plant production. The fly ash and soot get between the crop plant and the life-giving sun. This reduces light intensity and is likely to reduce the rates of photosynthesis, growth, and yield. The gases, particularly those which are the components of smog, destroy chlorophyll and render the plant incapable of making fresh supplies. Of these gases sulfur dioxide is particularly injurious. This gas combines with water vapor in the air and forms sulfuric acid which, after precipitating on the plant, rapidly oxidizes the tissue of the leaves. Thus, the rate of photosynthesis is reduced or entirely stopped, and growth and yield are reduced or entirely stopped. Note in Figure 7-1 the loss of chlorophyll on the leaves of the apple. According to plant scientists at the Air

Figure 7-1 Effect of injurious concentrations of sulfur dioxide on the chlorophyll content and photosynthesis of apple leaves. *(Courtesy, J. M. Skelly, Virginia Polytechnic Institute and State University.)*

Pollution Research Center, Riverside, California, smog injures all kinds of plants—plants both inside and outside the home, plants in commercial gardens and orchards, and trees in the forest. For example, in 1970, commercial growers in California sustained a loss of 25 million dollars and over one million Ponderosa pines in the San Bernadino Forest due to the exhaust from automobiles. Air pollution has damaged crops in other states also—in Florida, Michigan, New York, New Jersey, Oregon, and Washington. As is well known, the gases which pollute the air are injurious to human health. They are also injurious to the health of crop plants and are likely to make the production of food and fiber more difficult. For this reason, anything that stands in the way of the crop plant and the life-giving sun should be removed, and the air in the vicinity of all crop plants should be kept continually clean.

HYDROGEN

Role

Hydrogen, like carbon, is a constituent of almost all organic compounds made by plants. Hydrogen is therefore an essential element. In addition, the relative proportion of hydrogen to oxygen in the molecule of a given food determines the stored energy content of that food. For example, fats are relatively high in hydrogen in proportion to oxygen. They, therefore, contain relatively large quantities of chemical energy. For this reason, fats are excellent energy reserve materials.

The Carrier

The carrier of hydrogen is water. Water is therefore an essential raw material. In photosynthesis, water is decomposed into hydrogen and oxygen; the hydrogen is used in the formation of the manufactured compounds, and the oxygen is given off as the by-product.

However, the student should remember that of the total amount of water absorbed, the amount which combines with carbon dioxide in the photosynthetic reaction is very small, usually less than 1 percent. Thus, most of the water absorbed is needed to maintain turgor of the living cells, particularly the guard cells, for the diffusion of carbon dioxide through the stomates. This important role of water is discussed more fully in Chapters 4 and 13.

NITROGEN

Role

Nitrogen enters into the formation of many compounds made by plants. It is part of the molecule of all proteins and enzymes, of chlorophyll a and chlorophyll b, of certain acids of the nucleus and certain hormones. For these reasons nitrogen is an essential element.

The Carriers

As previously stated, plants absorb most of their nitrogen in the form of nitrate-nitrogen or ammonium-nitrogen. For this reason these ions are called the carriers of nitrogen, or, preferably, the available forms of nitrogen. In acid soils plants absorb more nitrate ions, and in slightly alkaline soils they absorb more ammonium ions. For example, scientists at the New Jersey Experiment Station supplied apples and tomatoes with equal amounts of nitrate-nitrogen and ammonium-nitrogen growing on very acid (pH 4.0), moderately acid (pH 6.0), and neutral (pH 7.0) soils. On the acid soil (pH 4.0), the apple trees and tomato plants assimilated nitrate ions to a greater extent than ammonium ions; on the neutral soil (pH 7.0), they assimilated ammonium ions to a greater extent than nitrate ions; and on the moderately acid soid (pH 6.0), they assimilated both forms in practically equal amounts. Apparently, the relative amount of nitrate-nitrogen or ammonium-nitrogen which is assimilated depends to some extent on the soil reaction. Each form behaves differently in the soil. In general, *the nitrate ion leaches readily,* and the ammonium ion is fixed according to the exchange capacity of the soil. Which form is more readily lost by heavy rains or excessive irrigation?

Symptoms of Deficiency and Excess

Since relatively large or small quantities of available nitrogen may exist in the soil solution and since the relative amount of available nitrogen has marked effects on the vegetative and reproductive phases, the student should become familiar with the symptoms of nitrogen deficiency and the symptoms of nitrogen excess. Symptoms of nitrogen deficiency are as follows: *with monocots the middle portion of the leaf blade becomes yellowish-green, but the margins remain green; and with dicots the leaf blade becomes uniformly yellowish-green.* In both groups, the old leaves show symptoms first, and the chlorophyll content of the plant as a whole is relatively low. Thus, with the low content of chlorophyll, relatively low quantities of light are absorbed. As a result very few carbohydrates are made per unit time, and growth and yields are likely to be low. (See Fig. 7-2)

Symptoms of nitrogen excess with most plants are as follows. The vegetative phase proceeds rapidly; in other words, there is a rapid development of stems and large, dark green leaves. These large, dark green leaves contain large quantities of chlorophyll. As a result,relatively large quantities of light are absorbed per unit time, and large quantities of carbohydrates are made. Since under these conditions large quantities of available nitrogen are present, most of the sugars are used in making cells of the stems, leaves, and absorbing roots, and very few carbohydrates are left for thickening the cell walls, developing the fibers, storing starch, and forming flowers, fruit, seed, and storage organs. Thus, the stems are

Figure 7-2 Nitrogen-, phosphorus-, and potassium-deficient leaves of gardenia. Upper left: healthy. Upper right: nitrogen-deficient. Lower left: potassium-deficient. Lower right: phosphorus-deficient. *(Courtesy, Alex Laurie, Sarasota, Florida.)*

soft and succulent, the cell walls are thin, the development of fibers is limited, flowering and fruiting are often delayed or entirely suppressed, and the yield of storage organs is likely to be low.

The Available Nitrogen Supply and the Vegetative-Reproductive Balance

As previously stated, sugars and available nitrogen combine in the formation of the initial proteins. These, in turn, combine with each other in the formation of the complex proteins, and these combine with certain carbohydrates and fatty substances in the formation of the living substance, the protoplasm. It follows, therefore, that with other factors favorable for growth, the available nitrogen supply determines the rate at which new protoplasm and new cells are made and the rate at which sugars are used. Thus, when a plant has a high rate of photosynthesis and a normal rate of respiration, and absorbs and *uses large quantities of available nitrogen,* and is subjected to temperatures within the upper half of the optimum range, particularly during the night, the sugars will be used almost entirely for the production of stems, leaves, and absorbing roots. When, on the other hand, the plant has a high rate of photosynthesis, a normal rate of respiration, and absorbs and *uses moderately large quantities of available nitrogen,* the development of stems and leaves will be less rapid; and, as a result, there will be less rapid utilization of sugars for stem, leaf, and root growth. Some sugars will be left for the development of flower-forming hormones and flowers, fruit, seed, or storage structures. Thus, the amount of available nitrogen may regulate the type of growth to a marked degree, provided other factors are not limiting. In general, with abundant carbohydrates and abundant available nitrogen, vegetative processes are dominant over reproductive processes; and with abundant carbohydrates and moderately abundant available nitrogen, vegetative processes are less dominant and reproductive processes are more evident.

Nitrogen in the Atmosphere

Nitrogen in the atmosphere exists as a stable, inert gas with two atoms combined to from one molecule of the gas, $N \equiv N$. Further, since this gas is the chief component of the huge canopy of air which surrounds planet earth, the actual amount almost staggers the imagination. Based on the fact that the atmosphere exerts a pressure of 14.67 pounds per square inch and that 79 percent of this pressure is due to the gaseous nitrogen, 36,348 tons of nitrogen gas exist over every acre of the earth's surface.

Unfortunately, crop plants cannot use this raw gaseous form directly. Before they can use it, the gaseous nitrogen is changed to protein nitrogen or some other form of organic nitrogen. Fortunately for the farmer and the welfare of the human race, certain lower forms of plants have acquired the ability to make this change. In general, these organisms are the blue-green algae, certain species of bacteria, and a group of fungi called *mycorrhiza.* They use the gaseous nitrogen in the formation of their protoplasm, which contains large quantities of proteins. The process is called *nitrogen fixation.*

The blue-green algae are tiny, single-celled, almost circular microscopic plants. They tend to form colonies and are found in warm, moist places—on the surface of warm soils and on the surface of ponds and lakes. In general, these tiny plants reduce the gaseous nitrogen by using the energy from the sugars of their own photosynthesis. According to Van Overbeek, they change large quantities of atmospheric nitrogen above rice fields in the tropics into a form which the rice plants can absorb and use.[1]

The nitrogen-fixing bacteria belong to two distinct groups: the nonsymbiotic and the symbiotic.[2] The nonsymbiotic live independently of crop and other plants. In other words, they have ability to use the energy of light in the reduction of the gaseous nitrogen in order to form amino acids and proteins. In sharp contrast, the symbiotic bacteria live in conjunction with a certain group of higher plants which, because of their characteristic fruit, are called *legumes.* These bacteria have been placed in the genus *Rhizobium.* According to the latest research, the roots of the crop plants actually fix the nitrogen, but, for reasons not fully understood, the roots cannot do so without the aid of the rhizobia. In general, these rhizobia are specific in action. For example, the strain that grows on alfalfa roots does not grow on the roots of peas or beans, and vice versa. Thus, legumes are inoculated with proper cultures of bacteria. Agricultural experiment stations have developed quick and economical methods of mixing seeds of crops of the legume family with these bacteria—a practice known as *inoculation.* In general, growers have found that inoculation pays.

Mycorrhizae are fungi. As pointed out in Chapter 15, the body of a typical fungus consists of mycelia and spore cases. Since the beginning of the twentieth century, tree fruit horticulturists have observed a rather intimate association between the mycelia and the root systems of many valuable horticulture plants—apples, pears, peaches, plums, pecans, walnuts, filbert, citrus, cacao, cranberry, gooseberry, currant, and strawberry, and ornamentals such as azalea and rhododendron. This intimate association has stimulated the curiosity of horticulturists. Is the mycelia entirely parasitic on the host plant or does it act as a symbiont, a benefactor, in return for the food it receives from the host? Do the mycorrhizae actually fix nitrogen and/or do they make other essential elements available? In other words, is the association beneficial or harmful, or neither? The research to date suggests that, in general, the association is beneficial, but the degree of benefit varies with the essential element. With respect to nitrogen, the research indicates that the mycorrhizae do not fix nitrogen directly but that they assist in some way in increasing the nitrogen content of the soil, since the total amount of nitrogen in forest soils is greater than the amount of nitrogen contributed by rain and snow and by the activities of *Azotobacter* and *Clostridium.* With respect to phosphorus, calcium, and magnesium, the results

[1]J. Van Overbeek, *The lore of living plants,* New York: Scholastic Book Services, Chap. 3, 1964.

[2]The word *symbiosis* means living with other kinds of organisms to mutual advantage.

show that mycorrhizae make the essential ions of these elements more available. This is doubtless due to the respiration of the fungi, with the carbon dioxide combining with water in the formation of carbonic acid. With respect to the essential ions of manganese, iron, and copper, the organic matter resulting from the dead bodies of the mycorrhizae form chelates (pronounced keylates) which are more available to crop plants than the nonchelated forms.

Nitrogen in Organic Matter

The nitrogen in organic matter occurs largely in the form of protein nitrogen. As such, crop plants cannot absorb it; it must be changed into the ammonium or nitrate form. Two groups of minute plants effect this change. The first group comprises certain fungi and bacteria which are called the *ammonifiers;* the process is called *ammonification.* The second group consists of bacteria only which are called the *nitrifiers;* the process is called *nitrification.* Vast numbers of these organisms exist in fertile soils. Bacteriologists have estimated that as many as 46 billion bacteria exist in a gram of decomposing material and that under average conditions a fertile soil contains from ½ to 1 ton per acre (1.1 to 2.2 metric tons per hectare), or about 100 to 250 billion per pound (454 grams) of soil. Obviously these organisms have a significant role in the growing of crop plants.

The changes which gaseous nitrogen and protein-nitrogen undergo in the formation of the available forms of nitrogen is illustrated as follows:

$$\text{Gaseous N} + \text{sugars} \xrightarrow{\text{(nitrogen fixers)}} \text{protein N}$$

$$\text{Protein N} \xrightarrow{\text{(ammonifiers)}} \text{amino acids} \longrightarrow NH_4 + 0_2$$

$$\xrightarrow{\text{(nitrifiers)}} NO_2 + 0_2 \longrightarrow NO_3$$

If the fungi and bacteria which make ammonium-nitrogen and nitrate-nitrogen were nonexistent, the farmer's bill for nitrogen fertilizers would be enormously increased. In other words, these fungi and bacteria make large quantities of available nitrogen, a quantity which certain scientists have estimated as varying from 50 to 300 pounds per acre (55 to 336 kilograms per hectare) per year. Their activity is markedly affected by the soil environment. Principal factors are (1) temperature, (2) water supply, (3) oxygen supply, and (4) acidity.

Temperature Temperature affects the activity of the ammonifiers, nitrifiers, and nitrogen fixers in much the same way it affects higher plants. When the soil is cold or extremely hot, their activity is retarded. In general, temperatures for their greatest activity are moderately high (60 to 85°F or 15.6 to 29.4°C). Consequently, the curve of nitrate-nitrogen formation, other factors being equal, varies with the temperature of the season, the locality, and the soil. Thus, for any given location and soil, the rate of nitrate-nitrogen formation is more rapid in warm soils than in extremely hot or cold soils.

Water Supply Moist soils are more favorable than dry soils. The effect of water in the soil is twofold: (1) it aids in the decomposition of organic matter, and (2) it brings essential raw materials into solution. Since bacteria and fungi make protoplasm and new cells, they require essential elements for growth in much the same way as do higher plants. In other words, the carriers of the essential elements must be in solution before their absorption into the cells can take place.

Oxygen In general, abundant supplies of oxygen are necessary for the respiration of the ammonifiers, nitrifiers, and nitrogen fixers. For this reason, soils should be well drained. In fact, nonaerated, poorly drained soils contain certain bacteria called *Pseudomonas denitrificans*. These organisms live in the absence of free oxygen. Fortunately, these organisms do not thrive in well-drained soils. The decomposition of nitrate-nitrogen to gaseous nitrogen is illustrated as follows:

$$2HNO_3 \rightarrow 2N + H_2O + 2\frac{1}{2}O_2$$

Note that this reaction depletes the soil of nitrogen.

Acidity of the Soil Investigations have shown that the nitrifying and nitrogen-fixing organisms do not thrive well in very acid soils. For these organisms, the optimum range seems to be from pH 6.0 to 7.0. Thus, satisfactory nitrate formation does not take place in very acid or alkaline soils.

To summarize the effect of the soil environment on the activity of the ammonifiers, nitrifiers, and nitrogen fixers, these organisms thrive best in warm, moist, well-aerated, well-drained, and slightly acid soils. Any deviation from these desirable conditions may correspondingly reduce the available nitrogen supply.

Nitrogen in Commercial Fertilizers

Despite the relatively large amounts of available nitrogen made by soil organisms, only in exceptional cases are the quantities sufficient for the satisfactory growth of crop plants. Thus, applications of nitrogen-carrying fertilizers are necessary. From the standpoint of their effect on the soil and plants, nitrogen fertilizers are classified as follows: (1) physiologically acid and (2) physiologically alkaline. Regarding physiologically acid carriers, plants absorb the cation to a greater extent than the anion. Regarding physiologically alkaline carriers, plants absorb the anion to a greater extent than the cation. The continued use of the former makes soils more acid, and the continued use of the latter makes soils more alkaline. Consequently, the choice of either type of fertilizer depends to some extent on the soil reaction. Several acid- and basic-forming nitrogen carriers are shown in Table 7-2.

PHOSPHORUS

Role

In general, phosphorus is part of the molecular structure of several vitally important compounds: deoxyribonucleic acid (DNA), the two forms of ribonucleic acid (mRNA and tRNA), the enzyme systems necessary for the energy transformations in photosynthesis and respiration, and adenosine diphosphate (ADP) and adenosine triphosphate (ATP). As pointed out previously, the energy released from the high energy bonds of ATP is used by all living things to maintain the living condition and to perform work. Since the high energy bonds are made available by crop plants, the vital importance of phosphorus to the nutrition of crop-plant, animal, and human life is obvious.

Symptoms of Deficiency

Symptoms of phosphorus deficiency are not as frequent as those of nitrogen deficiency. A lack of phosphorus is indicated by a slowing down of growth and by late maturity. *In monocots, the leaves of plants in the vegetative stage usually show reddish or purplish areas instead of the desirable dark green. In dicots, the main veins of old leaves frequently become reddish or purple, while the young leaves are dark green or grayish-green.* In fruit trees the blossoms drop off, the fruit is small, unattractive in color, matures slowly, and few flower buds are formed for the next year's crop. Note the phosphorus-deficient leaf blades in Figure 7-2.

Available Form and Carriers

Plants absorb phosphorus only as phosphate ions. These ions are dihydrogen phosphate (H_2PO_4), monohydrogen phosphate (HPO_4), and phosphate (PO_4). Of these ions the H_2PO_4 form is absorbed the most because of its greater solubility. However, the availability of the dihydrogen ion depends on the degree of acidity of the soil solution. In general, the range of maximum availability seems to be

Table 7-2 Acid and Basic Nitrogen Carriers

Acid-forming carriers	N, %	Basic-forming carriers	N, %
Ammonium nitrate	33	Anhydrous ammonia	82
Ammonium phosphate	11	Calcium nitrate	16
Ammonium sulfate	21	Cyanamide	22
Urea	46	Sodium nitrate	16

between pH 5.5 and 6.8. In very acid soils, especially where abundant iron or aluminum silicates exist, the monocalcium phosphate changes to insoluble iron and aluminum phosphate; and in alkaline soils, especially where abundant calcium exists, the monocalcium phosphate changes to the relatively insoluble tricalcium phosphate. *Unlike nitrate ions, the phosphate ion is readily "fixed" by soils.* Consequently, very little movement, or leaching, of phosphorus takes place and excessive amounts in the soil solution do not exist. Therefore, phosphates may be applied when convenient, preferably close to the plant's roots. For most crops it is usually applied at the time of soil preparation and as side dressings during the early states of growth.

The principal carrier is superphosphate. This material is made by treating raw rock phosphate with sulfuric acid or phosphoric acid. Note the equations.

$$Ca_3(PO_4)_2 + 2H_2SO_4 \rightarrow Ca(H_2PO_4)_2 + 2CaSO_4$$

$$Ca_3(PO_4)_2 + 4H_3PO_4 \rightarrow 3Ca(H_2PO_4)_2$$

In the first case, a mixture consisting of monocalcium phosphate and gypsum is formed. The product is called superphosphate and varies from 16 to 20 percent P_2O_5, depending on the relative amount of phosphate and gypsum in the mixture. In the second case, only monocalcium phosphate is formed. The product is called *double* or *treble superphosphate* and varies from 40 to 50 percent P_2O_5.

QUESTIONS

1. The green plant is essentially a chemical factory. Explain.
2. What is right and what is wrong with the following statement? Plants make foods and plants absorb foods from the soil.
3. State the role of carbon dioxide in plant growth.
4. How is the supply of carbon dioxide maintained?
5. In general, the average concentration of carbon dioxide for any given 24-hour period is 0.03 percent. At what part of the 24-hour day would you expect the concentration to be greater than 0.03 percent, and at what part of the day would you expect the concentration to be less? Explain.
6. Calculate the amount of carbon dioxide above 1 acre or 1 hectare of the earth's surface.
7. Under what conditions is carbon dioxide likely to be the limiting factor in plant growth?
8. In general, soils high in organic matter give off more carbon dioxide than soils low in organic matter. Explain.
9. Water is an essential raw material for plant growth. Explain.
10. State the role of nitrogen in plant growth.

11 Calculate the amount of N_2 above each acre of the earth's surface, assuming an atmospheric pressure of 14.67 pounds per square inch.
12 Available nitrogen is needed for the development of meristematic tissues. Explain.
13 Distinguish between nitrogen-deficiency symptoms in the leaf blade of monocots and in the leaf blade of dicots.
14 In general, plants growing in soil containing low quantities of available nitrogen store carbohydrates. Explain.
15 For the same period of time a plant growing in a small pot produces a smaller top than a plant growing in a large pot. Explain from the standpoint of the nitrogen supply.
16 In general, cotton in the Mississippi Delta produces a larger top and a greater yield than cotton in the hills. Explain.
17 Outline the changes taking place in organic matter from protein-nitrogen to nitrate-nitrogen and from gaseous nitrogen to nitrate-nitrogen.
18 What is Ammonification? Nitrification? Nitrogen fixation?
19 Name the soil factors influencing the activity of the ammonifiers, nitrifiers, and nitrogen fixers.
20 The natural nitrate supply is low in winter and comparatively high in summer. Explain.
21 Experiments at the Louisiana Station have shown that a complete fertilizer is necessary for the fertilization of spring crops of snap beans and that a fertilizer without nitrogen is necessary for fall crops. Explain.
22 Side dressings of readily available nitrogen to crops in the seedling stage in the southern United States are frequently necessary during cold, wet weather. Give two reasons.
23 Decomposition of organic matter in winter in the greenhouse is more rapid than outdoors. Explain.
24 Show how the continual use of a physiologically acid fertilizer makes the soil more acid and how the continual use of a physiologically alkaline fertilizer makes the soil less acid and more alkaline.
25 In what form do plants absorb phosphorus?
26 State four roles of phosphorus in plant growth.
27 Distinguish between phosphorus deficiency in monocots and dicots.
28 Phosphorus becomes unavailable in soils with greater than pH 8.5. Explain.
29 Phosphorus becomes unavailable in soils with less than pH 5.0. Explain.

POTASSIUM

Role

Potassium differs from carbon, hydrogen, oxygen, and other essential elements in that it is not a constituent of the manufactured compounds or a part of any living tissue. Nevertheless, plants do not grow in the absence of potassium. In general, this element appears to be necessary for the synthesis of amino acids, since plants growing in cultures high in ammonium ions and low in potassium ions accumulate large quantities of ammonium in the tissues.

Symptoms of Deficiency

The symptoms of potassium deficiency vary somewhat with different crops. With fruit crops, e.g., apple, pear, plum, gooseberry, and black and red currant, *potassium deficiency is first evidenced by a faint yellowing of the leaf margins.* Later, the yellowing proceeds along the veins and the margins turn dark brown. This condition is known as *scorch.* It is very serious in Great Britain, Quebec, and Nova Scotia and occurs to some extent in the northern and western United States. With cucumbers, the margins of the old leaves become yellow, but the midrib and veins remain green. With potatoes and sweet potatoes, the leaves become rough and puckered, and the margins curl downward, turn yellow, and finally turn brown. With many dicotyledonous flower crops, the margins of the leaves turn yellow and finally dark brown, but the remainder of the leaves remain green. With sweet corn and other monocots, yellowing begins at the tips and goes down the edges, leaving the center green. With both the dicots and monocots, *the old leaves usually show the symptoms first.* Note the appearance of the margins of the leaves in Figure 7-2.

Available Form and Carriers

Plants absorb potassium as potassium ions, and the principal source seems to be the ions attached to the surface of colloidal particles. These ions are exchanged for hydrogen ions, hence, the term exchangeable potassium. Important carriers are potassium chloride (KCL) and potassium sulfate (K_2SO_4), and potassium nitrate (KNO_3). In general, these compounds are quite soluble in water, and they dissociate into potassium ions and chloride, sulfate, and nitrate ions, as the case may be. Most soils fix potassium ions to a lesser degree than they fix phosphate ions. Consequently, if comparatively large quantities are applied, leaching may take place and excessive amounts in the soil solution may exist.

SULFUR

Role

Sulfur is part of the molecular structure of three essential amino acids—cysteine, cystine, and methionine—and of B_1, or thiamin, a coenzyme of respiration. Thus, sulfur is necessary for the making of new cells and for most of the catabolic activities of crop plants. In addition, sulfur is part of certain volatile compounds, such as mustard oil and the allyl sulfites characteristic of the onion group.

Symptoms of Deficiency

Investigations have shown that, with many flower crops, *the veins of the leaves are lighter green than the tissue between the veins.* Note this condition is exactly opposite to that for magnesium, manganese, and iron deficiency. In general, the leaves on the upper part of the plant show sulfur-deficiency symptoms first. This indicates that sulfur is not readily transferred from one part of the plant to another.

Available Form and Sources

Plants absorb sulfur only in the form of sulfate ions (SO_4), a highly oxidized form of sulfur. Important sources are sulfur in organic matter, sulfur in the atmosphere, and sulfur in commercial fertilizers.

The sulfur in organic matter is in the form of protein-sulfur. Since plants cannot use it as it is, it must be changed into the sulfate form. As in the formation of nitrates from organic matter, this change is effected by certain soil organisms. In general, the protein-sulfur is changed into hydrogen sulfide. This, in turn, is oxidized into sulfuric acid, which combines with minerals in the soil solution and forms a salt. This change from protein-sulfur to sulfate-sulfur is illustrated as follows:

$$\text{Organic matter containing sulfur} \xrightarrow{\text{(soil organisms)}} H_2S$$

$$+ 2O_2 \xrightarrow{\text{(sulfur bacteria)}} H_2SO_4$$

$$H_2SO_4 + MOH^3 \longrightarrow MSO_4 + H_2O$$

The sulfur in the atmosphere is washed into the soil by rain. In the soil the sulfur is oxidized. This oxidized sulfur combines with water in the formation of sulfuric acid which, in the presence of a mineral, forms a salt. In regions near certain industrial establishments the atmosphere contains relatively large quantities of sulfur. The change of elemental sulfur into sulfate-sulfur is illustrated as follows:

$$S + O_2 \rightarrow SO_2 + O \rightarrow SO_3 + H_2O \rightarrow H_2SO_4 + M \rightarrow MSO_4$$
(rain)

Commercial fertilizers supplying sulfur are crude sulfur, or flowers of sulfur; ammonium sulfate; potassium sulfate; land plaster, or gypsum; and superphosphate containing calcium sulfate. As shown by the above equation, elemental sulfur, or sulfur dioxide, is oxidized to sulfate—the only form of sulfur absorbed by crop plants.

Sulfur deficiencies are rather widespread throughout the United States and the world. In districts where deficiencies exist, experiments have shown that applications of sulfur-supplying materials have restored crop plants to health, resulting in normal growth, acceptable quality, and profitable yields—in orange

[3]M stands for *minerals in soil solution* in this case, as well as in the compound MSO.

groves and in legume and grassland areas in California; in legume-growing districts in Oregon, Washington, and British Columbia; in white clover fields in south Mississippi and Alabama; in cotton fields in Louisiana; in tea plantations in Java; in tea and tobacco plantations in Nyasaland; and in pastures in Australia.

CALCIUM

Role

Calcium is necessary for the formation of calcium-pectate which, together with magnesium-pectate, binds the cellulose chains together in cell wall formation. Hence, in healthy, rapidly growing plants, large quantities of calcium are necessary for the meristem of the roots, the stems, and the cambia. Calcium also appears to be necessary for the absorption of nitrate-nitrogen since investigations have shown that calcium-deficient plants accumulate sugars and starches in the tissues and are unable to absorb nitrate-nitrogen. Immediately after calcium-starved plants are supplied with this element, nitrates, if available, are found within the tissues in a relatively short time.

Symptoms of Deficiency

Unlike nitrogen, phosphorus, and potassium, but like sulfur, calcium is relatively immobile in plants. Thus, *young leaves show symptoms of deficiency before old leaves.* Investigations at the New Jersey Experiment Station have shown that calcium-deficient plants of the Marglobe tomato become stunted, stiff, and woody. The upper leaves become yellowish-green, and the lower leaves remain normal green. The margins are usually less dark green than the central portion. Roots become short, bulbous, and brownish at the tips. Fruits are few. Calcium-deficiency symptoms of the apple are somewhat similar. The young leaves are yellowish-green, and mature leaves are dark green. On young trees, the leaves may be abnormally small. On large trees, the leaves may be normal in size, except those near the tips of the twigs. In severe cases, the margins of the leaves turn brown, and the growth of the roots is seriously retarded. The roots may form, but they live for a short time only. Similar deficiency symptoms have been noted on coffee, cacao, and tea. In many herbaceous flowering plants, the symptoms are similar to those of vegetable plants: *the upper leaves become yellowish green, the terminal buds fail to develop, and the roots die in the area of the tips in a relatively short time.*

A lack of adequate supplies causes certain physiologic diseases, prevalent in many parts of the world—bitter pit of fruits of the apple in South Africa and Australia, black heart of celery in many producing districts of the United States, blossom end rot of tomato and capsicum pepper in practically all producing districts in the United States. In all cases, applications of calcium chloride have cured the disorder.

Available Form and Carriers

Plants absorb calcium as calcium ions. The principal carriers are the three forms of lime: calcium oxide, calcium hydroxide, and calcium carbonate. In soils, calcium oxide and calcium hydroxide gradually change to calcium carbonate, according to the following equations:

1. $CaO + HOH = Ca(OH)_2$
2. $Ca(OH)_2 + CO_2 = CaCO_3 + H_2O$

Growers usually use the hydrated and carbonate forms.

In addition to supplying calcium for the specific roles of this element, the various forms of lime are used to make soils less acid. The degree of acidity of the soil solution has marked effects on the growth and yield of many economic crops. If the soil solution is either very acid or very alkaline, growth and development of many crops are markedly reduced. In addition, crops vary in acidity requirements. Some thrive well in markedly acid soils; others thrive best in moderately acid or slightly acid soils; and others thrive best in slightly acid or neutral soils. Since soils vary in degree of acidity and since plants vary in acidity requirements, the student should have a working knowledge of the factors which cause acidity, how the acidity of the soil influences growth and yields, and how the acidity may be modified to produce high marketable yields.

Nature and Measurement of Acidity and Alkalinity

The soil solution, like any aqueous solution, always contains a certain concentration of hydrogen ions and hydroxyl ions. The acidity of the solution is due to hydrogen ions, and its alkalinity is due to hydroxyl ions. Water ionizes, forming hydrogen and hydroxyl ions according to the following equation:

$$H_2O \rightleftharpoons \overset{+}{H} + \overset{-}{OH}$$

Thus, the soil solution contains water molecules, hydrogen ions, and hydroxyl ions. Scientists have developed a scale called the pH scale, which measures the degree of acidity of solutions. This scale has the following characteristics: (1) it is divided into 14 main divisions from 0 to 14, and the value of 7.0, the midpoint, is the reference point; (2) at this value, the solution is neutral, at values lower than 7, the solution is acid, and at values greater than 7, the solution is alkaline; (3) the values on the scale are negative logarithms to the base of 10, so that for every decrease in unit pH there is a tenfold increase in H-ion concentration and a corresponding tenfold decrease in the OH-ion concentration. For example, a solution of pH 6.0 is ten times more acid than a solution of pH 7.0, and a solution of pH 4.0 is one hundred times more acid than a solution of pH 6.0.

Soil Reaction and Essential-element Availability

A definite relation exists between the degree of acidity of soils, essential-element availability, and plant growth. In very acid soils, usually less than pH 5.0, nitrate-nitrogen formation is low since both the nitrifying and nitrogen-fixing bacteria do not thrive well in very acid soils; monocalcium phosphate changes to iron and aluminum phosphate, which are relatively insoluble; and compounds which supply available calcium and magnesium are usually present in insufficient quantities for rapid plant growth. On the other hand, in alkaline soils, usually greater than pH 8.5, nitrate-nitrogen formation is also low since the nitrifying bacteria do not thrive well in alkaline soils; monocalcium phosphate changes to tricalcium phosphate, which is relatively insoluble; and compounds which supply potassium, manganese, iron, and boron are likely to become unavailable. Thus, high acidity and high alkalinity should be avoided. In fact, most crops thrive best in moderately acid or slightly acid soils. The optimum pH range of certain horticultural crops based on growers' experience and the latest experimental evidence is presented in Table 7-3.

Experiment stations have developed comparatively simple tests to determine the degree of acidity of the soil solution. In general, these tests involve the use of an indicator—a compound allowed to percolate slowly through a small mass of soil. The change in color of the indicator denotes a certain degree of acidity. These tests are fairly accurate, reliable, and inexpensive and permit making many determinations in a comparatively short time.

Reducing Acidity Many soils in many districts in humid sections are very acid, usually less than pH 5.5. To reduce the acidity of these soils for the growing of crops which thrive best in moderately acid or slightly acid soils, lime

Table 7-3 Soil-acidity Requirements of Certain Crop and Ornamental Plants*

Optimum pH range	Crop or plant
Very acid (pH 4.5–5.5)	Blueberry, cranberry, huckleberry, azalea, camellia, gardenia, hydrangea, and foliage plants, e.g., philodendron, peperomia, and pothos
Moderately to slightly acid (pH 5.5–6.8)	Apple, pear, peach, plum, pecan, cherry, orange, lemon, grapefruit, blackberry, raspberry, strawberry, cabbage, carrot, radish, cucumber, tomato, pepper, eggplant, chrysanthemum, carnation, snapdragon.
Slightly acid to slightly alkaline (pH 6.5–7.5)	Asparagus, spinach, lima bean, beet, celery, lettuce, onion, garden pea, sweet pea, alfalfa, clover

*Data for this table have been obtained from various sources.

is added. The amount required depends largely on (1) soil type and (2) acidity requirements of the crop. In general, sands and sandy loams are sensitive to changes in acidity because of low buffer capacity. Hence, comparatively light applications are needed. Crops which thrive best on slightly acid soils require more lime than those which thrive best on moderately acid soils. In districts where the soils are naturally deficient in magnesia-bearing rock, as for example the Atlantic Coastal Plain, dolomitic limestone, a form of limestone containing large quantities of magnesium carbonate mixed with calcium carbonate, is used to a greater extent than ordinary limestone.

Increasing Acidity Certain soils in humid sections are moderately or slightly acid. To increase the acidity of these soils for crops which thrive best in very acid soils, ammonium sulfate or flowers of sulfur is added. The amount required depends on the exchange capacity of the soil and the change in pH that should be made. In general, applications vary from 50 to 500 pounds per acre (56 to 560 kilograms per hectare).

MAGNESIUM

Role

Magnesium has at least two important roles. This element is the center of the molecule of both chlorophyll a and chlorophyll b, and it is part of the molecule of magnesium-pectate which together with calcium-pectate binds the cellulose chains in the formation of the cell walls. Thus, magnesium is necessary for photosynthesis and for cell division.

Symptoms of Deficiency

The outstanding symptom for practically all crops is the loss of chlorophyll of the tissues between the veins of the leaves. In other words, *the intervascular areas change from dark green to light green and from light green to yellow, but the veins remain green.* This change from green to yellow starts at the tips and gradually proceeds to the center of the blades. As with a deficiency of nitrogen, phosphorus, and potassium, *the old leaves show the symptoms first.* Thus, when magnesium is deficient, it is transferred from the old to the young tissues. Because of the low chlorophyll content, there is a low rate of light absorption and a corresponding low rate of photosynthesis; and as a rule the tissues are low in sugars and starch, the cells are thin-walled, the strengthening tissue is poorly developed, and the roots are small and few in number. Figure 7-3 shows healthy and deficient leaves of chrysanthemum. Note the chlorotic condition of the interveinal areas.

Magnesium deficiencies have appeared in many parts of the world—in the apple orchards of New Zealand and the states of Maine and New York; in the citrus groves of Brazil, Palestine, South Africa, and the states of California and Florida; in the grape vineyards of Alabama and North Carolina; and in the

Figure 7-3 Healthy and magnesium-deficient leaves of chrysanthemum. Leaf on left is from a plant grown with a normal supply of magnesium; leaves on right are from a magnesium-deficient culture. *(Courtesy, Alex Laurie, Sarasota, Florida.)*

tobacco and vegetable crop plantings of the coastal plain area of the United States.

Available Form and Carriers

As with potassium and calcium, crop plants absorb magnesium as magnesium ions. Principal carriers are magnesium nitrate, magnesium sulfate, and dolomitic limestone. In general, magnesium nitrate and magnesium sulfate are used in plantings where deficiency symptoms are evident since they are quite soluble and are absorbed by the leaves. Thus, they may be applied either as sprays or in solid form. On the other hand, dolomitic limestone is used to maintain adequate supplies in the soil since it is less soluble and longer lasting than either of the other two compounds.

MANGANESE AND IRON

Manganese and iron are discussed together because their role in plant growth, symptoms of deficiency, and behavior in the soil are quite similar.

Role

Recent research has shown that manganese is part of an enzyme system for condensing soluble amino compounds into proteins and that iron has two roles: as part of an enzyme system for the making of chlorophyll and as part of the molecular structure of the heme pigment. This pigment, in turn, is needed for the terminal phase of respiration, the liberation of hydrogen from organic compounds, and the joining of this hydrogen to oxygen in the formation of water.

Thus, manganese is necessary for photosynthesis, and iron is necessary for both photosynthesis and respiration.

Symptoms of Deficiency

In general, *the interveinal areas of young leaves become light green and yellow, but the veins remain green.* Since both these essential elements are relatively immobile in the plant, *the young leaves show symptoms first.* As the deficiency becomes severe, necrotic, or dead, areas appear in the interveinal areas. According to deficiency-symptom experiments conducted at the Ohio Experiment Station, manganese deficiency induces relatively small dead areas along the middle of the leaf only, whereas iron deficiency induces relatively large dead areas in all of the interveinal tissues.

Available Forms and Behavior in Soil

As with other essential cations, plants absorb manganese and iron as manganese and iron ions. The degree of acidity of the soil solution has marked effects on the availability of these ions. In general, excessive supplies are usually present in very acid soils, moderate and adequate quantities are present in slightly acid soils, and deficient supplies exist in very slightly acid, neutral, and alkaline soils. In fact, certain tests have shown that manganese is almost insoluble in soils which have pH values from 6.2 to 8.0. In other words, the optimum pH range for manganese and iron availability seems to be from 5.5 to 6.2. Thus, the degree of acidity of the soil should be determined to confirm any suspected deficiency or excess of these two elements.

Carrier of Manganese

The common carrier is manganese sulfate. When artificial applications are necessary, this material may be applied either as a foliage spray or directly to the soil. At present applications are required in certain citrus orchards in California and Florida, in many apricot, peach, and cherry orchards in Utah, in the truck-crop areas near Charleston, South Carolina, and Norfolk, Virginia, and in certain ornamental nurseries and plantings in Florida. In some cases, particularly where the soils are slightly acid or neutral, spraying the foliage may be more economical than applying the material to the soil. For example, experiments in Florida have shown that spraying manganese sulfate at the rate of 10 pounds per acre (11 kilograms per hectare) on manganese-deficient snap beans was as effective in correcting the deficiency as a soil application of 20 to 25 pounds per acre (22 to 28 kilograms per hectare). In general, the rate of application varies from 5 to 100 pounds per acre (6 to 112 kilograms per hectare) depending on soil type, degree of acidity, and manganese requirements of the crop.

Carriers of Iron

Common carriers of iron are ferrous sulfate, an inorganic compound, and chelates containing iron, a group of organic compounds, e.g., ethylenediaminetetraacetic acid, called EDTA, and diethylenetriaminepentaacetic acid, called DTPA. These compounds hold iron in an available form and facilitate the entrance of iron into the plant and the distribution of iron within the plant. In addition they are not broken down in soils. The chelates may be applied either in solution with water or in the dry form. They have corrected chlorosis in many areas: in citrus and vegetable areas of central Florida, in ornamental plant-growing districts in Pennsylvania and Massachusetts, and in certain high-lime soil areas in California. Compare the healthy and deficient leaves in Figures 7-4 and 7-5.

BORON

Role

Although the exact physiologic role of boron is not known, recent research indicates that this element is necessary for cell division, the germination of pollen, the movement of sugars through protoplasmic membranes, the development of phloem, and the transport of certain hormones. In the absence of adequate supplies, the middle lamella of new cells develops poorly and the phloem tubes break down. Thus, calcium and boron have related roles in plant growth. In fact, investigations have shown that if the calcium content of the growing tissues is high, the boron content should be high also.

Figure 7-4 Manganese-deficient and healthy leaves of tung. *(Courtesy, R. D. Dickey, University of Florida.)*

Figure 7-5 Healthy and iron-deficient leaves of rose. Left: minus iron. Right: with iron. *(Courtesy, Alex Laurie, Sarasota, Florida.)*

Symptoms of Deficiency

Since rapid cell division does not take place in the absence of boron, the meristems are profoundly affected. As would be expected, *vegetative and flower buds fail to develop, leaves are small, lopsided, and misshaped, and the flesh of storage organs breaks down.* Since boron, like calcium, is immobile, *the younger parts of the plants show symptoms of deficiency first.* Note the boron-deficient cabbage head in Figure 7-6.

Available Form and Carriers

Plants absorb boron as the negatively charged borate ion, and they require this element in relatively low concentration, usually from 0.1 to 1.0 ppm in the soil solution. If the concentration becomes greater than 1.0 ppm, toxic effects are likely to occur.

Factors influencing the amount of available borate are (1) soil type and (2) soil reaction. Soils most deficient are old soils, sandy soils, mucks and peats, those derived from igneous rocks, and alkaline and neutral soils. Applications are made in many commercial crop industries. Examples are apple and pear orchards in Australia, New Zealand, Europe, Canada, and the New England states; apricot orchards in the light sandy soils of New Zealand; grapefruit and orange groves in California; cauliflower districts in the Catskill Mountains of New York; turnip and rutabaga fields in Great Britain and Ireland, Iceland, the maritime provinces of Canada and northeastern United States; celery districts in southern Michigan, central Florida, Oregon, and certain areas in California; cabbage fields in central Mississippi; and chrysanthemum districts in Florida and southern Alabama. Figure 7-7 shows boron deficient areas in the United States. Thus, boron deficiencies are widespread throughout the United States and other parts of the world.

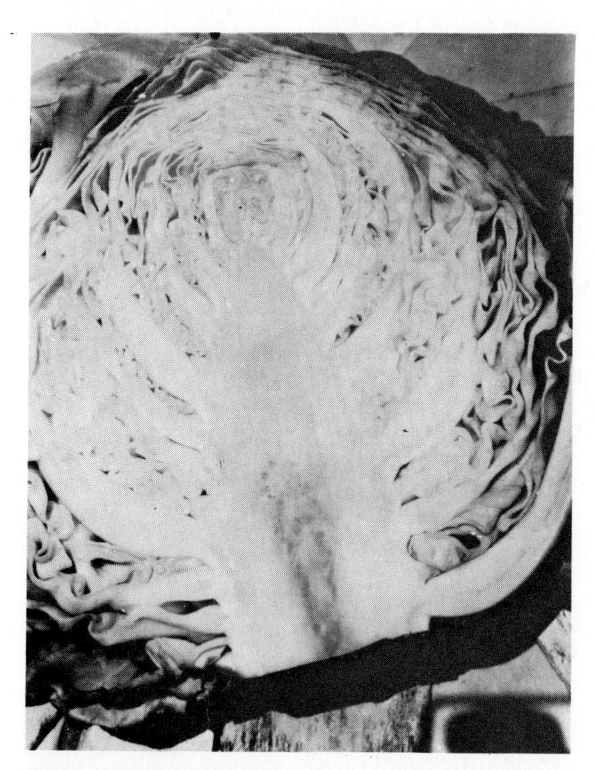

Figure 7-6 A boron-deficient head of cabbage. Note the dark area in the basal portion of the stem. *(Courtesy, J. A. Campbell, Mississippi Agricultural Experiment Station.)*

The principal carrier is fertilizer borate ($Na_2B_4O_7 \cdot 4H_20$). In general, when applications are necessary, fruit trees are given from 2 to 16 ounces of borax per tree, and herbaceous crops from 5 to 30 pounds of borax per acre. This form is highly soluble and subject to rapid leaching. A less soluble form, colemanite ($Ca_2B_6O_{11} \cdot 5H_2O$), has been tested with satisfactory results at the South Carolina Experiment Station. Thus, with the use of the calcium form, losses due to leaching may be avoided.

ZINC

Role

Zinc is necessary for the formation of tryptophane, the precursor of indoleacetic acid, and is a component of at least two enzyme systems: (1) the glyco-glycine dipeptidases necessary for the formation of proteins and (2) the dehydrogenases necessary for the glycolysis of sugar in the terminal phase of respiration. Thus, zinc is necessary for cell elongation, for the formation of proteins, and for the oxidation phase of respiration.

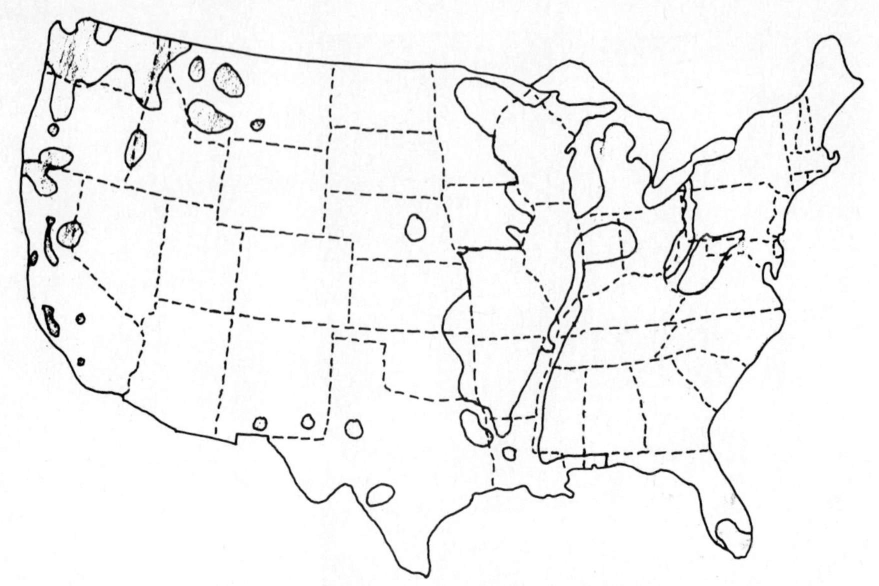

Figure 7-7 Boron-deficient areas in continental United States. *(Courtesy, D. A. Russell,* U.S. Dept. Agr. Yearbook, *1957.)*

Symptoms of Deficiency

Zinc-deficiency symptoms vary somewhat with the crop. With apples, peaches, and pecans, *deficient branches have small, narrow, mottled leaves,* and they usually die back each year; with citrus, *the leaves are small and extremely mottled;* and with herbaceous plants, *the leaves have yellow interveinal areas and green veins.* Note that in all cases inadequate quantities of chlorophyll develop. As a result, the rate of photosynthesis of zinc-deficient crops is low and the yields and quality are low. Note the small, narrow leaves of the zinc-deficient peach tree in Figure 7-8.

Zinc deficiencies have been found in both tree and herbaceous crop industries. The tree crop industries include deciduous fruit districts in Oregon, California, Washington, and British Columbia; orange groves in southern California; peach orchards in southeastern Arizona and southern New Mexico; apple, pear, and peach orchards in Utah and Colorado; citrus, peach, pecan, and tung orchards in the southeastern United States; and cacao plantations of the Gold Coast. The herbaceous crop industries include bean, sweet corn, and tomato plantings in the Pacific Coast states; onion-growing districts on the high-lime mucks in Michigan; and pea bean-growing districts of slightly alkaline soils also in Michigan.

Available Form and Carrier

Plants absorb zinc in the ionic form, and the usual carrier is zinc sulfate ($ZnSO_4 \cdot 7H_2O$). This compound is quite soluble in water and is applied as a

Figure 7-8 A zinc-deficient peach tree. *(Courtesy, R. D. Dickey, University of Florida.)*

foliage spray or directly to the soil. Of these methods, spraying is more widely used, since zinc is readily fixed by soil colloids and becomes unavailable in neutral and alkaline soils. In fact, these soils usually contain large quantities of zinc which are unavailable to crops. Thus, the fixing power of the soil for zinc and the degree of acidity are important. In general, the available concentration varies from 1 to 10 ppm with the midpoint of this range as the optimum concentration. The use of zinc chelates, organic carriers of zinc, is the experimental stage.

COPPER

Role

Copper is an essential constituent of certain oxidizing enzymes, e.g., tyrosinase, the enzyme which in the presence of oxygen darkens the flesh of the potato tuber; and ascorbic acid oxidase, the enzyme which oxidizes ascorbic acid. Since with deficient supplies *the interveinal areas of leaves become yellow,* copper is necessary for the formation of chlorophyll.

The need of certain soils in the continental United States for copper was discovered in the Florida Everglades in 1927. Crops sprayed with Bordeaux mixture, a suspension consisting of copper sulfate and calcium hydroxide, or dusted with copper sulfate produced satisfactory yields, whereas crops not sprayed or dusted with the copper-containing compound made scant growth. Subsequent tests showed that applications of 50 pounds of copper sulfate per acre changed the soil, a saw grass peat, from a low to a remarkably high state of fertility. Since that time, tests conducted in various regions have shown that copper has a beneficial effect on the growth of many plants. Applications of available copper have cured the "frenching," or spotting, of leaves in citrus and almond trees in California, corrected copper chlorosis of corn and citrus in Florida and of deciduous fruits in South Africa, increased the thickness of the scales and improved the color of onions grown on the muck soils in New York and Michigan, and increased the yields of potatoes, sweet potatoes, onions, and snap beans in Delaware.

Available Form and Carrier

Here again, plants absorb copper in the ionic form. The usual carrier is copper sulfate. It may be applied either as a foliage spray or directly to the soil. As with certain other compounds, the amount applied depends largely on the amount of available copper in soils. In general, acid soils, or those usually less than pH 5.5, the low-lime mucks, and sands require more copper than the moderately acid or slightly acid soils. Applications vary from 10 to 250 pounds per acre (11 to 280 kilograms per hectare), and usually one application supplies sufficient copper for a period of several years.

MOLYBDENUM

Role

Molybdenum is part of the molecular structure of an enzyme, a riboproteinase which is essential for reducing the inert nitrogen gas of the atmosphere to an ammonium compound both in crop plants and in the nitrogen-fixing bacteria, *Azotobacter* and *Rhizobia*. Note the equation:

$$N \equiv N + 6(H) \longrightarrow 2NH_3 + H_2O \longrightarrow 2NH_4OH + \begin{pmatrix} \nearrow ATP \searrow \\ p \\ \nwarrow ADP \swarrow \end{pmatrix} \longrightarrow \begin{matrix} \text{Amino acids} \\ \\ \text{Proteins} \end{matrix}$$

Energy is needed to break the triple bonds between each of the two nitrogen atoms. This energy comes from organic compounds and is released by respiration. Since all of the nitrogen in the many kinds of compounds on the earth originally came from nitrogen of the air and since protein deficiency is quite prevalent among humans throughout the world, the relative importance of this reaction is obvious. In fact, many crop-plant scientists consider that nitrogen fixation is the most important chemical reaction in the world next to photosynthesis and respiration.

Symptoms of Deficiency

Deficiency symptoms have been described for many economic crop plants. These plants include members of the citrus family, the cabbage family, the legume family, spinach, lettuce, onions, tomatoes, cucumbers, and cantaloupes. With oranges and grapefruit, *the leaves have clearly defined yellow areas.* With cauliflower, *the leaves are long and narrow with cupped or wavy margins;* and with tomatoes, cucumbers, and snap beans, *the leaves turn brown or yellow at the margins, which turn upward.* In all cases growth is slow, the plants become stunted, and yields and quality are low.

Molybdenum deficiency has been found in many parts of the world, e.g., in the Coastal Plain area of eastern United States, in the Serpentine soils of central California, in the Palouse district of Washington, and in Great Britain, Ireland, the Netherlands, France, India, Japan, Australia, and New Zealand. Molybdenum deficiency usually occurs in very acid soils. In fact, scientists have found a close positive correlation between the degree of acidity of the soil solution and the presence of available molybdenum. In general, very acid soils have less available molybdenum than slightly acid or neutral soils.

Available Form and Carriers

Crop plants absorb molybdenum as the negatively charged molybdate ion, and in crop-plant production the carriers are ammonium molybdate and sodium molybdate. These materials may be applied in solution with water as a seed treatment or as a liquid spray when the plants are young or dry, either alone or with other fertilizing materials, or even with pest control materials. In general, applications vary from ¼ ounce (7 grams) to 1 pound (454 grams) per acre.

CHLORINE

Recent investigations have shown that chlorine is essential for the growth of 11 species of plants. According to research workers at the University of California, about 100 ppm of the dry weight of the species is the minimum requirement.

The natural source of supply is rain. In general, the amount varies with the distance from oceans. In other words, the rainfall along the coasts contains a greater amount of chlorine than that some distance away.

The behavior of the chlorine ion in soils is similar to that of the nitrate and borate ions. In other words, it is not fixed by soils and it is susceptible to leaching.

OTHER ELEMENTS

From the time agriculture research began and until 1920 only 10 of the original 92 elements were considered essential for the growth of the crop plants. These elements are carbon, hydrogen, oxygen, nitrogen, phosphorus, potassium, sulfur, calcium, magnesium, and iron. Since that time, investigations have shown that six other elements are essential—boron, manganese, copper, zinc, molybdenum, and chlorine. With continuous improvements in techniques for determining element essentiality, other elements will doubtless be discovered as essential. Elements which suggest themselves as essential are aluminum, cobalt, iodine, sodium, silicon, and vanadium.

ESSENTIAL ION UPTAKE

As previously stated, crop plants require specific raw materials for the making of manufactured substances. With the exception of carbon dioxide and water, these raw materials exist as ions in the soil solution. Examples are the ammonium, nitrate, phosphate and potassium ions. The questions arise: How do crop plants obtain these essential ions, and how do these ions move to the site of manufacture? Recently, plant scientists have discovered that the concentration of any given ion is frequently greater in the living cells of the root and even within the xylem of the root than the concentration of that ion in the soil. This means that crop plants have the ability to move essential ions against a concentration gradient of that ion. As explained in Chapter 1, this is similar to pushing an object uphill. This requires work and the expenditure of a certain amount of free energy. In other words, when crop plants are absorbing essential ions against a concentration gradient, they are working, and the energy comes from the high energy phosphate bonds of ATP through respiration. In contrast to simple diffusion, the process is called *active absorption,* or *dynamic transport, of ions*. In general, this explains why the absorption of water and the uptake of essential ions are independent processes, why the absorption of essential ions may take place in the absence of the absorption of water, why the uptake of ions and rates of photosynthesis and respiration are intimately associated, and why any reduction in photosynthesis of the tops and respiration of the roots is likely to correspondingly reduce the rate of absorption of essential ions by the roots.

QUESTIONS

1. State the role of potassium in plant growth.
2. What part of the leaf is first affected by potassium deficiency?
3. How would you distinguish potassium deficiency from nitrogen deficiency in the leaf blade of monocots?
4. Name two commercial sources of potassium.
5. State the roles of sulfur in plant growth.
6. In what form do plants absorb sulfur?
7. Outline the changes taking place from protein-sulfur to sulfate-sulfur and from sulfur to sulfate-sulfur.
8. Name two commercial fertilizers containing sulfur.
9. State the roles of calcium in plant growth.

10 Calcium-deficient plants store carbohydrates. Explain.
11 Young leaves show calcium-deficiency symptoms before the old leaves do. Explain.
12 State the three principal carriers of calcium.
13 What are acidity and alkalinity due to?
14 Soil with a pH 3 is acid; with a pH 8, alkaline; with a pH 7, neutral. Explain.
15 Yields of most crops are low when the plants are grown in very acid soils, usually less than pH 5.0. Give three reasons.
16 State the roles of magnesium in plant growth.
17 Describe the outstanding symptoms of magnesium deficiency.
18 In what section of the United States is magnesium naturally deficient in soils? Give two reasons.
19 How does dolomitic limestone differ from ordinary limestone?
20 Dolomitic limestone is used to a greater extent in the southeastern United States than ordinary limestone. Explain.
21 How would you distinguish magnesium deficiency from manganese and iron deficiency?
22 Chlorotic plants are lower in sugars and carbohydrates and grow more slowly than healthy plants. Explain.
23 Chelates containing iron are more effective in controlling iron chlorosis than are inorganic compounds containing iron. Explain.
24 State the outstanding symptoms of boron deficiency.
25 How is boron deficiency corrected? State the substances which are used and how they are applied.
26 How is zinc deficiency corrected?
27 Although large quantities of zinc may be present in the soil, it is not always available. Explain.
28. In your opinion, how do applications of copper sulfate benefit onions growing on muck soils in New York, Michigan, and Florida?
29 State the role of molybdenum in plant growth.

SELECTED REFERENCES FOR FURTHER STUDY

Hooker, W. J., et al. 1972. Air pollution effects on potatoes and beans in Southern Michigan. *Mich. Agr. Exp. Sta. Res. Rept.* 167. An excellent report on the effect of ozone, oxides of nitrogen, sulfur dioxide, and carbon monoxide on the growth and yield of crop plants.

Springer, J. K. 1972. Is air pollution choking your profits? *Amer. Veg. Grower.* 20 (3):74–76. An account of the effects of harmful concentrations of air pollutants on the growth of vegetable crops in various parts of the United States.

Van Overbeek, J. 1964. *The lore of living plants.* New York: Scholastic Books Service. Chap. 3. A distinguished crop-plant scientist discusses the discovery of and the world need for commerical fertilizers and the importance of the nitrogen supply in crop-plant production.

Wallace, J. 1961. *The diagnosis of mineral deficiencies in plants*. New York: Chemical Color photographs of the mineral deficiency symptoms of a large number of crop plants.

Part Two

Principal Horticultural Practices

Chapter 8

Basic Crop-Plant Genetics in Relation to the Development of Superior Varieties

The greatest service that can be rendered to any country is to add a useful plant to its culture.

Thomas Jefferson

THE WORK OF MENDEL

Modern crop-plant breeding is based on modern genetics. Modern genetics began with the work of Gregor Mendel, a research worker and teacher at the Augustinian monastery at Brunn, Austria, in what is now Czechoslovakia. Mendel worked continuously and assiduously for eight years with the garden pea, *Pisum sativum* L. He presented a paper on his results at a meeting of the Brunn Society of Natural History in February, 1862, and published in 1866 a monograph entitled, *Experiments in Plant Hybridization.*

Mendel's work differed from the work of previous investigators in that he crossed races of the garden pea which ''bred true''; he studied the inheritance of the seven contrasting traits listed in Table 8-1, usually by examining one pair of

Table 8-1 Results of Mendel's Experiments With Garden Peas, *Pisum sativum* L., 1854-1868

			F_1 generation	F_2 generation		
Trait	Parents		All plants	Dominant Trait	Recessive Trait	Ratio
Seed coat structure	smooth	wrinkled	smooth	smooth 5474	wrinkled 1850	2.96:1
Cotyledon color	yellow	green	yellow	yellow 6022	green 2001	3.01:1
Seed coat color	gray	white	gray	gray 705	white 224	3.15:1
Pod shape	smooth	constricted	smooth	smooth 882	constricted 299	2.95:1
Pod color	green	yellow	green	green 428	yellow 152	2.82:1
Inflorescence	lateral	terminal	lateral	lateral 651	terminal 207	3.14:1
Plant height	tall 6–7 feet	dwarf 3/4–1 foot	tall	tall 787	dwarf 277	2.84:1

Source: Gregor Mendel, *Experiments in plant hybridization,* Cambridge, Mass.: Harvard University Press, 1866.

contrasting traits at a time; and he carried his crosses to the first, the second, and sometimes the third filial generation (written F_1, F_2, F_3 respectively). More importantly, Mendel made careful observations of the F_1 and succeeding generations, and he counted the plants which possessed any of the contrasting traits in the F_2 and succeeding generations. The contrasting traits, the appearance of the F_1, and the appearance and number of plants with any of the contrasting traits in the F_2 are presented in Table 8-1. Note that all of the plants of the F_1 had only one contrasting trait and that both traits appeared in the F_2. Mendel called the trait which appeared in the F_1 the *dominant trait* and the one which did not appear the *recessive trait*. In this way, Mendel discovered the principle of dominance; that is, some traits can entirely mask the appearance of another trait. Note also that the dominant and recessive traits were passed on from one generation to the next in a definite mathematical ratio. Mendel concluded that some material substance which could be associated with the inheritance of characters, or traits, was being passed from one generation to the next. He called this substance a *determinator,* or *factor*.

In 1886 Hugo De Vries, a Dutch botanist, found a group of evening primrose plants growing in a corner of an old potato field. This group of primroses differed from the evening primrose growing in other parts of the field. DeVries collected seed from each of these groups and found that the characters of the plants of each different group remained relatively constant. He concluded that "attributes of organisms consist of distinct, separate and independent units." About the same time Karl Correns, a German scientist, studied the inheritance of characters of garden peas and obtained practically the same results as those of Mendel. Somewhat later Eric von Tschermak, an Austrian scientist, repeated Correns' experiments with garden peas and found constant ratios in the F_2 generation. All three scientists presented their results before the German Botanical Society in 1900: De Vries on March 24, Correns on April 24, and Von Tschermak on June 24. Their results not only confirmed Mendel's work but also the work of William Spillman, who was experimenting with wheat in the United States, and William Bateman, who was experimenting with animals in England. Like Mendel, these scientists concluded that some material substance is passed on from generation to generation that is associated with the inheritance of characters. Like Mendel, they called this substance a *factor,* or *determinator.* Later, about 1910, this factor became known as the *gene* and more recently as the definite chemical compound called *deoxyribonucleic acid,* abbreviated DNA.

Gene Designation, Homozygous and Heterozygous Genes

To explain his results, Mendel assumed that the hereditary factors, or genes, in the somatic tissue (tissue of the body) of the garden pea always existed in pairs, with one factor coming from the ovule, or mother parent, and the other coming from the pollen, or father parent. He used letters of the alphabet either in the upper or lower case to indicate each member of a given pair. This type of designation has continued to be used since Mendel's time. If the members of a given pair of genes are the same, as for example, AA or aa, they are called *homozygous genes,* and if they are unlike or are not the same, as for example, Aa, they are called *heterozygous genes.*

Genes and the Chromosomes

In 1831, Robert Brown, a plant anatomist, found that nearly every living plant cell had a nucleus, and with the development of the low-powered microscope, plant scientists drew diagrams of the dividing cells. The rodlike bodies—the chromosomes—caused great speculation. Many scientists believed they carried the hereditary factors, or genes. In 1887 August Weisman, a German zoologist, postulated that if the hereditary factors are carried on the chromosomes, there should be two kinds of cell division. He called the first kind *equal division* of the nucleus, which today it is called *mitosis.* He called the second kind *reduction division* of the nucleus, which today it is called *meiosis.* The essential features of mitosis and meiosis are shown in Figure 8-1. Note that in mitosis a given chromosome splits longitudinally and duplicates itself, with each duplicated chromosome separating into daughter cells. The splitting actually occurs at the

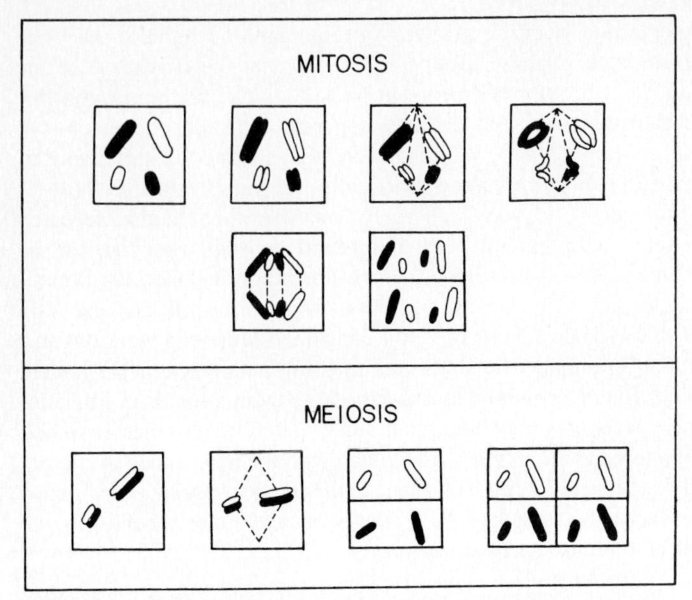

Figure 8-1 Mitosis and meiosis. Note that in mitosis the chromosomes split longitudinally and reduplicate themselves; in meiosis they do not split longitudinally.

junction of each pyrimidine-purine base, and although a variety of nitrogen bases exist in the cell sap, adenine (A) always bonds with thymine (T), and guanine (G) always bonds with cytosine (C). In this way, the DNA strands make exact copies of themselves, and unless changes take place in the chemical structure of DNA, the DNA, or genetic constitution, remains the same throughout the life of the plant. Note also that in meiosis homologous chromosomes come together in pairs, that they may or may not exchange complementary material, and that each homolog separates into daughter gametes. Here again, unless mutations take place in the chemical structure of DNA, the genetic constitution remains constant throughout the life of the species.

Mendel's Laws

Because Mendel carefully selected his material (he worked with 22 homozygous strains of the garden pea) and because he made careful observations and counts of the F_1 and succeeding generations, he was able to isolate two laws: (1) the law of segregation of the genes and (2) the law of independent assortment of the genes.

The law of segregation involves the behavior of both the chromosomes and the genes. At meiosis, homologous chromosomes migrate to opposite poles of the cell, and the members of a given pair of genes also migrate to opposite poles. Note the seven crosses in Table 8-1. All the plants of the F_1 displayed the dominant trait only, and the plants of the F_2 segregated in the proportion of three dominants to one recessive. Thus, the gene for any given recessive character was present in the cells of the F_1 and separated from the dominant gene in some of the plants of the F_2. Thus, the law states that the members of a given pair of genes separate in the formation of the gametes, or sex cells.

The law of independent assortment also involves the behavior of both the chromosomes and the genes. At meiosis, the chromosomes of nonhomologous pairs separate independently from each other, and the genes on nonhomologues assort independently during sex cell formation. For example, Mendel crossed two homozygous races; one race had smooth and yellow seed, and the other had wrinkled and green seed. From his work Mendel found that smooth is dominant over wrinkled and yellow is dominant over green. Thus, all of the F_1 plants had smooth and yellow seed. Mendel allowed the F_1 plants to self-pollinate; and he found that some of the plants of the F_2 had smooth and yellow seed, others had smooth and green seed, some had wrinkled and yellow seed, and the remainder had wrinkled and green seed. In this way, Mendel found that two new combinations occurred: smooth and green, and wrinkled and yellow. When Mendel considered the inheritance of each pair of contrasting traits alone, 423, or 76.1 percent, had smooth seed and 133, or 23.9 percent, had wrinkled seed, 416, or 74.8 percent, had yellow seed and 140, or 25.2 percent, had green seed. Thus, with independent assortment of each of these two pairs of genes, ¾ × ¾ of the plants had both dominants in the F_2, ¾ × ¼ had one dominant and one recessive, ¼ × ¾ had the other dominant and the other recessive, and ¼ × ¼ had both recessives. Mendel's actual counts are shown in Table 8-1. Note that despite the relatively small number of plants involved, the counts are well within the limits of experimental error. Numerous investigations have substantiated Mendel's results, and they show that genes on different chromosomes assort independently in sex cell formation.

Meaning of the Terms *Genotype* and *Phenotype*

Consider Mendel's cross between homozygous tall and homozygous dwarf plants of the garden pea. At meiosis, one-half of the eggs of the female parent have the T gene and one-half contain the t. In like manner, one-half of the sperms of the male parent have the T gene and one-half contain the t. The eggs and sperms combine according to the following diagram.

♂ \ ♀	T	t
T	TT	Tt
t	Tt	tt

From the standpoint of genetic constitution, or genotypes, there are three distinct classes, (TT), (Tt), and (tt); and from the standpoint of outward appearance, there are two distinct groups, or phenotypes, (TT and Tt) and (tt), since the homozygous tall and the heterozygous tall cannot be distinguished from each other. Thus, the *genotype* refers to the genetic constitution, and the *phenotype* refers to the outer appearance of any given population.

Types of Hybrids, Genotypic and Phenotypic Ratios

As previously stated, Mendel worked with homozygous races of the garden pea. He made numerous crosses between these races, and the F_1 progeny of any given cross he called a *hybrid*. The question arises: What is a hybrid? If two parents are crossed and one parent has three dominant genes and the other has the corresponding three recessive genes, the F_1 will have three pairs of heterozygous genes. For example, AA BB CC dd EE x aa bb cc dd EE = Aa Bb Cc dd EE. Thus, *a hybrid is an individual or group of individuals of any given population which contains one or more heterozygous genes with the remaining genes being, presumably, homozygous.*

The Monohybrid The monohybrid is obtained by crossing two individuals which differ in one pair of genes only or in one pair of contrasting traits. For example,

AA BB cc x aa BB cc
F_1 = Aa BB cc

Therefore, *a monohybrid is an individual or a group of individuals which contain one pair of heterozygous genes with the remaining genes being, presumably, homozygous.* With complete dominance, as in the preceding example with tall and dwarf garden peas, there are two distinct classes, or kinds, of gametes in the F_1 and three genotypes and two phenotypes with a phenotypic ratio of 3:1 in the F_2. However, not all genes are completely dominant. Some are partially dominant, as for example, the inheritance of the color in the flowers of snapdragon or morning glory. If a race bearing red flowers is crossed with a race bearing white flowers, all of the plants of the F_1 produce pink flowers. The plants of the F_2 segregate in the proportion of one red, two pink, and one white. Thus, with partially dominant genes, the number of genotypes equals the number of phenotypes.

The Dihybrid The dihybrid is obtained by crossing two individuals which differ in two pairs of genes only or in two pairs of contrasting traits. For example,

AA BB cc DD x aa bb cc DD
F_1 = Aa Bb cc DD

Therefore, *a dihybrid is an individual or group of individuals which has two heterozygous genes with all remaining genes being, presumably, homozygous.* With complete dominance, there are four distinct kinds, or classes, of female and male gametes in the F_1, nine distinct genotypes and four distinct phenotypes with the phenotypic ratio of 9:3:3:1 in the F_2. However, there are many cases in which the two pairs of genes interact differently. In other words, the typical, 9:3:3:1 ratio becomes modified. Some modifications are 9:7, 9:3:4, and 13:3, the explanation of which is beyond the scope of this text.

The Trihybrid *The trihybrid is obtained by crossing two individuals which differ in three pairs of heterozygous genes with all remaining genes being, presumably, homozygous.* With complete dominance, the trihybrid produces eight kinds of female and male gametes in the F_1, 27 distinct genotypes and eight distinct phenotypes in the F_2, with a phenotypic ratio of 27:9:9:9:3:3:3:1.

The three types of hybrids with the number of gametic classes, genotypes, and phenotypes are presented in Table 8-2. Note that as the number of heterozygous genes increases the number of gametic classes, genotypes, and phenotypes increase in geometric proportion. This explains, partially at least, why geneticists and plant breeders usually work with only one or two pairs of contrasting traits at a time. In this way, the laws of segregation and independent assortment have full opportunity to express themselves.

Linkage and Crossing-over

Mendel found that the law of independent assortment did not operate fully in all of his crosses. For example, he crossed two races of the garden pea, one with red flowers and gray seed coats and one with white flowers and white seed coats. All the plants of the F_1 had red flowers and gray seed coats. In the F_2, he expected some of the plants would have either one of the two new combinations, red flowers and white seed coats or white flowers and gray seed coats. In actuality, the plants had either red flowers and gray seed coats or white flowers and white

Table 8-2 Types of Hybrids, Number of Gametic Classes, Genotypes, and Phenotypes*

Type of hybrid	No. of heterozygous genes	No. of gametic classes	No. of genotypes in F_2	No. of. phenotypes in F_2	Least number required in F_2
Monohybrid	1	2	3	2	4
Dihybrid	2	4	9	4	16
Trihybrid	3	8	27	8	64
General Formula		†2^n	3^n	2^n	4^n

*With dominance complete
†n = number of heterozygous genes

seed coats. Mendel noted this observation in his paper on plant hybridization, but he did not advance any explanation.

The explanation was developed by Bateson and Punnett in their studies on the inheritance of traits of the sweet pea *Lathyrus odoratus* L. in 1906. They crossed two races, one with purple flowers and relatively long pollen grains and one with white flowers and round pollen grains. All of the plants of the F_1 had purple flowers and long pollen. Thus, purple was dominant to white and long pollen dominant to round. Their results with the F_2 are shown in Table 8-3. Note that the genes for purple flowers and long pollen and the genes for white flowers and round pollen tended to stay together. Bateson and Punnett called this phenomenon *coupling of the genes*. About 1910, T. H. Morgan and his associates, a team of research workers at Columbia University, New York, studied intensively the inheritance of characters of the fruit fly *Drosophylla melanogasta*. In general, they found that the insect had four groups of linked characters and hence, four groups of linked genes. From this work, they concluded that linked genes are on the same chromosome. Geneticists working with other organisms found that the number of linkage groups corresponded to the number of homologous chromosomes. Thus, with diploids, at least, the number of linkage groups equals the number of paired chromosomes. Note the number of linkage groups of the diploid crop plants listed in Table 8-4.

Linked genes do not always stay together. A phenomenon known as *crossing-over* takes place. As previously stated, at meiosis homologues come together in pairs and, as shown in Figure 8-2, breaks may occur and the homologues may exchange comparable parts. In this way, new combinations of genes and traits may take place with genes of the same linkage group or on the same chromosome. In general, the amount of crossing-over between any two genes depends on the distance between them. For example, if three genes, A, B, and C, are on the same chromosome and A and B are next to each other and A and C are some distance from each other, breaks are more likely to occur between A and C than between A and B. In fact, the percentage of crossing-over between any two genes is directly proportional to the relative distance between them and provides the basis for the making of chromosome maps. In general, the phenomenon of linkage and crossing-over is the hope and despair of the crop-

Table 8-3 Results of a Cross Between Two Races of Sweet Peas *(Lathyrus oderatus* L.)

F_2 generation	Purple flowers, long pollen	Purple flowers, round pollen	White flowers, long pollen	White flowers, round pollen	Total
Expected ratio	9	3	3	1	
Expected numbers	3910	1304	1304	434	6952
Actual numbers	4831	390	393	1338	6952

Table 8-4 Diploid and Haploid Numbers and Linkage Groups of Various Crop Plants

Crop	Diploid number (2*n*)	Haploid number (*n*)	Number linkage groups
Peach	14	7	7
Raspberry (certain species)	14	7	7
Plum (certain species)	18	9	9
Bean	22	11	11
Cabbage	18	9	9
Onion	16	8	8
Potato	24	12	12
Corn	20	10	10
Sweat pea	14	7	7
Chrysanthemum (certain species)	18	9	9
Petunia (certain species)	14	7	7
Primrose (certain species)	18	9	9

plant breeder. If two superior genes are linked, the breeder hopes that no breaks will occur. However, if a superior gene is linked with an inferior gene, the breeder hopes that breaks will occur and an exchange will be made with a superior gene.

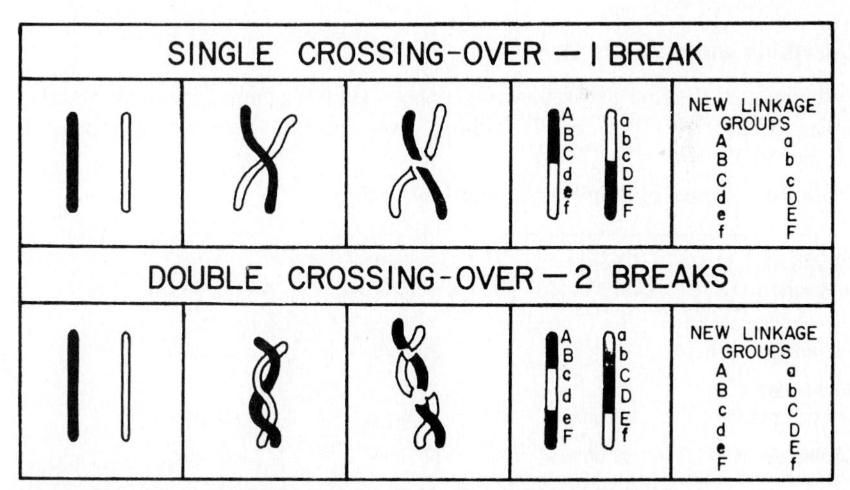

Figure 8-2 Single and double crossing-over. Note the formation of new linkage groups.

QUALITATIVE AND QUANTITATIVE CHARACTERS

In general, qualitative characters are distinct and can be readily distinguished from each other. Examples are the various colors of fruit, vegetables, and flowers. In sharp contrast, quantitative characters are indistinct, continuous, and cannot readily be distinguished from each other. In other words, qualitative characters differ in kind, whereas quantitative characters differ in degree. Examples of quantitative characters are the all-important marketable yield, the shape of individual fruits, and the various components of quality. In general, these differences in degree are due to the combined effect of the genetic constitution of the crop plant and the environment under which it is grown. Consequently, in the inheritance of quantitative characters, the effect of the genes and the various factors of the environment must be separated. This is done by using appropriate experimental techniques and appropriate statistical methods in the analysis of plant performance. Further, in quantitative inheritance a large number of genes are involved, and each gene contributes a definite quantity to the inheritance. In general, dominant genes contribute more than recessive genes, and dominant genes interact with each other either arithmetically or geometrically. In general, there are three types of dominant gene interaction: (1) arithmetic, (2) arithmetic and geometric, and (3) geometric. An example of each type of interaction is shown in Table 8-5. Each dominant gene, in either the homozygous or the heterozygous condition, contributes eight units to the inheritance and each recessive gene contributes two units.

However, in the actual inheritance of quantitative characters, the situation is not as simple as the three cases presented. A large number of dominant genes are involved. These genes may interact with each other and with the environment to determine the inheritance of a single quantitative character.

F_1 Hybrids and Male Sterility

F_1 hybrids are the first generation crosses between two parents; the ovule and the pollen. These crosses are usually within a given species—between two strains of

Table 8-5 Types of Dominant Gene Interaction

Type of interaction	Parent AA BB cc dd	Parent aa bb CC DD	F_1 Aa Bb Cc Dd
Arithmetic	20 units	20 units	30 units
Arithmetic and geometric	68 units	20 units	80 units
Geometric	68 units	68 units	256 units

Source: Poole, C. F., *Vegetable breeding,* Davis, Calif.: University of California, 1935.

the same variety, between a strain and a related variety, and between two varieties of the same species. Naturally, with this type of cross, the ovule parent must not be allowed to produce pollen. The discovery of male sterility in several crop plants has greatly facilitated the development of F_1 hybrids. There are two types of male sterility: (1) chromosomic and (2) chromosomic-cytoplasmic. With the chromosomic, the genes responsible for male sterility are located on the chromosomes only. In general, male fertility is dominant to male sterility. Thus, MM or Mm produces fertile stamens and mm produces sterile stamens. This type of sterility has been found in raspberry, tomato, and cantaloupe. With the chromosomic-cytoplasmic, the male sterile genes, mm, are recessive to the dominant fertile genes, MM and Mm, and they interact with two types of cytoplasm: normal N, which makes for the development of fertile pollen, and sterile S, which makes for the development of shriveled, nonfunctional pollen. Thus, crops with S cytoplasm and the mm genes will produce shriveled anthers and nonfunctional pollen, and crops with N cytoplasm and the MM genes or Mm genes will produce normal anthers and functional pollen. Crops with this type of male sterility are corn, sweet corn, and onion. The student should remember that thc cytoplasmic gene systems N and S are inherited through the ovule parent only, since the pollen parent does not contribute any cytoplasm in the development of F_1. The formation of male sterile and male fertile F_1 hybrids of the onion is presented in Table 8-6.

Ploidy

Ploidy refers to the number of chromosome sets, or genomes, in crop plants. As shown in Table 8-7, each crop plant has a basic, or x, number, a gametic, or *n*, number, and a somatic, or 2*n*, number. The basic, or x, number constitutes a set, or a genome. Each chromosome within a set is different from any other chromosome in that set in size, shape, and, more particularly, in the genes it carries.

Table 8-6 The Production of Male Fertile and Male Sterile F_1 Hybrids in Onions

Ovule parent	Pollen parent	F_1	
mm S	mm	mm S pollen sterile	
mm S	MM	Mm S pollen fertile	
mm S	Mm	Mm ½S pollen fertile	mm ½S pollen stertile

Table 8-7 Basic and Somatic Chromosome Numbers of Certain Crops

Crop	Basic number (x)	Somatic numbers (2*n*)	
Malus (apple)	17	34, 51	
Prunus (peach, plum, cherry)	8	16, 24, 32	
Citrus	9	18, 27, 36	28
Vitrus (grape)	19	38, 40, 76	
Fragaria (strawberry)	7	14, 42, 56	
Solanum (potato)	12	24, 36, 48, 60, 72, 96	
Ipomea (sweet potato)	15	90	
Lycopersicum (tomato)	12	24, 48	
Brassica (cabbage)	9, 10	18, 36, 20, 56	
Zea (corn)	10	20	
Rosa (rose)	7	14, 28, 35, 42, 56	
Chrysanthemum	9	18, 36, 54, 72, 90	
Lilium (lily)	12	24, 36, 48	45
Dahlia	8	32, 64	36
Lathyrus (sweet pea)	7	14	

The *n*, or gametic, number is usually one-half of the 2*n* number.

Also note that with most of our crop plants the somatic, or 2*n* number is a complete multiple of the x number; with a few other crops, the 2*n* number is not a complete multiple. Since the list in Table 8-7 is representative of crop plants as a whole, the 2*n* number of most of our crop plants is a complete multiple of the x number. In general, crop plants with complete sets are called *euploids*, or *homoploids*, and those with incomplete sets are called *aneuploids*, or *heteroploids*.

Euploids

Note the examples of euploids in Table 8-7. Some kinds have two and three sets, for example, apple and pear; other kinds have two, three, and four sets, for example, citrus and grape; and other kinds have five or more sets, for example, potato, chrysanthemum, and rose. The names of these kinds, along with the corresponding number of sets are listed in Table 8-8. In general, kinds with two sets, the diploids, are quite fertile since at meiosis normal pairing of homologs

Table 8-8 Number of Chromosome Sets in Various Euploids

No. of sets	Name	No. of sets	Name	No. of sets	Name
1	Haploid	4	Tetraploid	7	Septaploid
2	Diploid	5	Pentaploid	8	Octoploid
3	Triploid	6	Hexaploid	9	Nonoploid

takes place, whereas kinds with three sets are quite sterile since normal pairing of homologs does not take place. Of special significance is the double diploid, also referred to as *amphidiploid* or *allopolyploid*. In general, two closely related kinds are crossed with an almost sterile hybrid resulting. This is followed by the doubling of the chromosomes within each line of descent and the development of a fertile hybrid. Note the scheme presented in Figure 8-3.

Mutations, Bud Sports and Chimeras

With crop plants, a *mutation* refers to a marked change in a specific character, or trait, in the phenotype. In general, this trait is the end result of a change in the chemical makeup of a specific gene or portion of DNA. This gene, in turn, directs the formation of a specific enzyme, which results in the development of the new trait. In other words, there are changes in the biochemistry of the crop plant and in the making of definite chemical compounds. For example, consider the formation of scarlet red, rose red, and delphinium blue in Figure 8-4. Note that the formation of any one of these pigments is due to the degree of oxidation of the benzene ring. Oxidation of the number 4, the numbers 4 and 5, and the numbers 3, 4, and 5 results in the formation of scarlet red, rosc rcd, and delphinium blue, respectively. Also note that a specific enzyme is needed for the formation of each of these pigments. Thus, the development of a given trait requires the formation and activity of a specific enzyme system which, in turn, is formed under the direction of a specific portion of DNA.

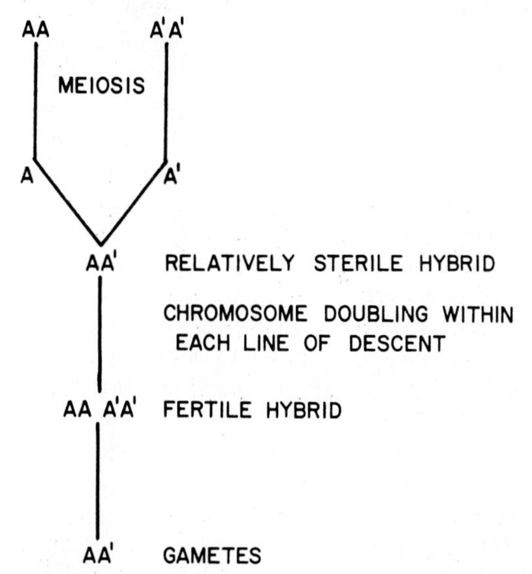

Figure 8-3 The formation of a double diploid, amphidiploid, or fertile hybrid.

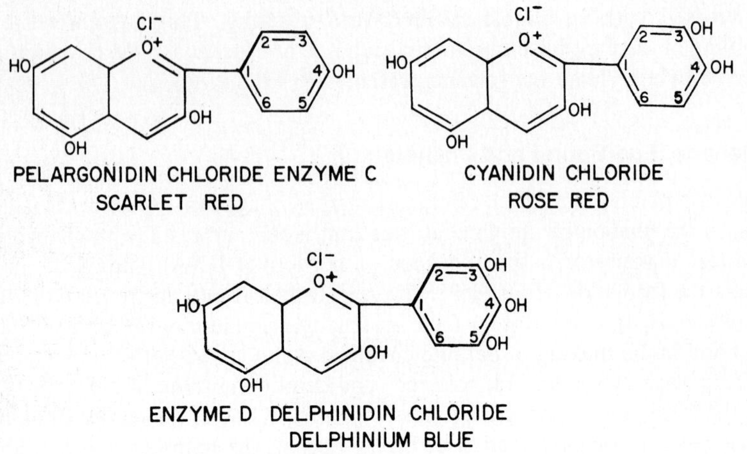

PELARGONIDIN CHLORIDE ENZYME C CYANIDIN CHLORIDE
SCARLET RED ROSE RED

ENZYME D DELPHINIDIN CHLORIDE
DELPHINIUM BLUE

Figure 8-4 Three anthocyanin pigments. Note that the differences in color are due to differences in degree of oxidation of the benzene ring.

Types of Changes From the standpoint of the genotype and the phenotype, there are three types of changes in genes: from the dominant to the recessive, from the recessive to the dominant, and from the viable to the lethal. In the first case, if the crop plant is self-pollinated, the mutation will first appear in the F_2 generation, and if it is cross-pollinated selfing of the F_1, or the application of a plants' pollen to its pistils will be necessary for the mutation to appear. In the second case, the mutation will appear immediately after it is formed. In the third case, since most lethals are recessives, they will be carried in perpetuity in the heterozygous condition and will become lethal when the genes become homozygous. An example of lethal genes are those which result in a lack of chlorophyll formation. They have been found in corn, sweet corn, asparagus, cowpeas, snapdragons, and many other crops. Examples of horticulturally important gene mutations in seed-propagated crops are presented in Table 8-9.

Bud Sports and Chimeras. Two types of genetic change can take place: all of the tissues of a given bud may change in genetic constitution, or one or more tissues or part of a tissue may change. The former change is referred to as a *bud sport,* in this text this term refers to genetic changes in the buds of asexually propagated crops only and the latter is referred to as a *chimera*. Thus, *a bud sport is a living entity which has the same genetic makeup in all of its tissues* whereas *a*

Table 8-9 Examples of Horticulturally Important Gene Mutations

Crop	Type of gene change	Horticultural significance
Field corn	Dominant to recessive	Development of a new plant, sweet corn
Snap beans and lima beans	Dominant to recessive	Development of new varieties
Carrot	Dominant to recessive	Development of varieties for human use
Garden Pea, Tomato	Recessive to dominant	Development of fusarium wilt-resistant varieties
Peach, Capsicum pepper, Protepea	Recessive to dominant	Development of nematode-resistant varieties
Sweet pea	Dominant to recessive	Appearance of several new flower types and new varieties
Cyclamen	Recessive to dominant	Appearance of purple flowers

chimera is a living entity which has tissues with different genetic makeups. In general, there are three main kinds of chimeras: the periclinal, the mericlinal, and the sectorial. These are illustrated in Figure 8-5. Note the specific tissues which differ in genetic constitution from that of the remaining tissues. In the periclinal only, the outer cover has changed, in the mericlinal only part of the outer cover has changed, and in the sectorial, a sector of all the tissues has changed.

The Frequency and Practical Significance of Bud Sports and Chimeras. During the first half of the twentieth century, many horticulturists studied the frequency and nature of bud sports and chimeras in asexually propagated crops—in apple, pear, peach, plum, cherry, orange, lemon, grape, potato,

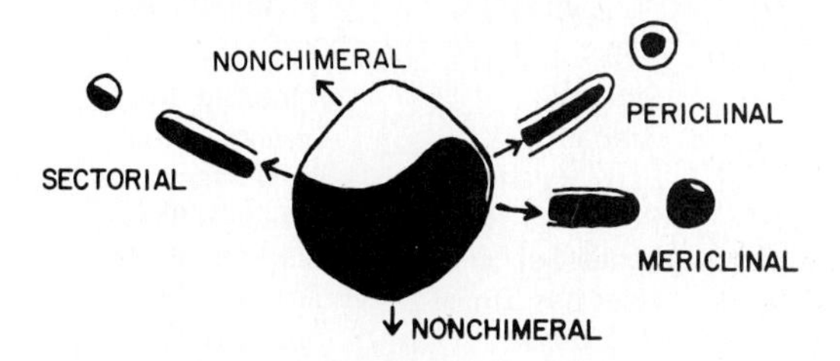

Figure 8-5 Three types of chimeras and two types of nonchimeras. *(Adapted from W. N. Jones* Bot. Rev. *3:545-562, 1937.)*

sweet potato, rose, chrysanthemum, dahlia, coleus, and marigold. In general, their studies show that the rate of appearance of bud sports and chimeras is relatively high. In fact, the progeny of asexually propagated crops is no more uniform than that of seed-propagated crops. This dispelled a previously prevailing idea that asexually propagated crops bred true. In addition, their data show that most chimeras are in the inferior direction and that relatively few appear in the superior direction. Thus, rigid and continuous selection of propagation stock is necessary for at least two reasons: the elimination of inferior types, and the probable discovery and dissemination of superior types. Examples of superior bud mutations and chimeras are presented in Table 8-10.

BREEDING METHODS

In general, there are two distinct methods: (1) selection and (2) combination breeding. Selection is limited to the genes within a given strain, variety, or species. These superior or inferior genes may be dominant or recessive, qualitative or quantitative, resistant or susceptible to specific environmental factors or to specific disease and insect pests. Thus, in the development of superior varieties by selection, the procedure consists largely of isolating and maintaining superior phenotypes and discarding inferior phenotypes. There are two types: mass selection and single family or single plant.

In contrast to selection, combination breeding consists of placing superior genes from one or more closely related kinds in one individual. In other words, superior genes displace the inferior genes of that individual. In general, each major crop has its own unique methods of selection and combination breeding. This is largely due to each crop's distinct characteristics and its environmental and cultural requirements.

Table 8-10 Examples of Superior Bud Sports and Chimeras

Crop plant	Parent variety	Mutant variety	Practical significance
Apple	Delicious	Starking	Attractive fruit
	Northern Spy	Graham	Distinct variety
Peach	Halehaven	Early Halehaven	Early strain
Orange	Washington Navel	Robertson Navel	Attractive flesh
Grapefruit	Thompson Seedless	Texas Seedless	Attractive flesh
	Emperor	Seedless Emperor	Seedless flesh
Potato	De Sota	Red Desota	Attractive red skin
	Burbank	Russett Burbank	Attractive russet skin
Sweet potato	Little Stem Jersey	Orlis	High carotene
	Nancy Hall	Red Nancy	High carotene
	Centennial	Rose Centennial	Light pink skin
Hybrid tea rose	Briarcliff	Better Times	Dark red petals

PLANT PATENTS

In 1930, the Congress of the United States passed a plant patent law. This law provided for the granting of patents to new and distinct varieties of asexually propagated crops only, provided that the part used for propagation was not used for human consumption. Thus, patents could be extended to all new varieties of asexually propagated crops, except varieties of potato, Jerusalem artichoke, and sweet potato.

In 1970, the Congress extended the law to include all new varieties of asexually propagated crops and all new varieties of seed-propagated crops, except those of carrot, celery, cucumber, okra, capsicum pepper, and tomato. Thus, patents are now available for all new varieties of fruit crops, all new varieties of vegetable crops, except those listed above, and all new varieties of flower crops.

The purpose of the act is similar to that of the Inventors Patent Act of 1787 in that it provides a means by which the breeder or originator of the new variety may receive a profitable return from his investment—an investment involving knowledge, time, facilities, and money. Thus, in this way the development of new varieties may be stimulated.

According to the Act of 1970, each new variety should meet at least two requirements: (1) it must be distinct, e.g., clearly different in at least one character from any other variety; (2) it must be reasonably uniform, e.g., the variation in any one character should have the same characteristics as the parents. According to the Act, any new variety may be patented for a period of 17 years. During this period, the originator may control the amount of seed for sale by growing the seed himself or by authorizing others to grow the seed for him. In both cases, the seed is grown to meet the certification requirements of the state in which the seed is grown. In this way, the originator has some control of the parent stock of each new variety. Plant patents and certificates are secured through and administered by the Plant Variety Protection Office, U.S. Department of Agriculture, Plant Industry Station, Beltsville, Maryland. Plant patent laws are also in operation in other countries of the world, such as Great Britain, the Netherlands, Australia, and New Zealand.

QUESTIONS

1 Cite three ways in which Gregor Mendel's work differed from that of previous investigators.
2 Show how mitosis maintains a constant number of chromosomes in the life of a given individual and how meiosis maintains a constant number in the life of a given species.
3 Demonstrate how two pairs of heterozygous genes, Aa Bb, assort independently in sex cell formation.

4 Determine the number of phenotypes in the following crosses, assuming complete dominance: AA BB cc Dd Ee ff x AA Bb cc Dd ee ff, and Aa BB Cc x Aa BB Cc.
5 What is a monohybrid? A dihybrid? A trihybrid?
6 Breeders of horticulture crops generally work with only one or two pairs of genes or characters at a time. Can you think of any reason for this?
7 In general, linked genes require larger populations for adequate segregation than nonlinked genes. Explain.
8 Differentiate between qualitative and quantitative characters.
9 In any breeding program involving quantitative characters, the effect of the various factors of the environment should be measured. Explain.
10 Show how male sterility is particularly advantageous in the development of F_1 hybrids.
11 What is the meaning of the x number, the *n* number, and the 2*n* number of crop plants?
12 In general, amphidiploids, or allotetrapoloids, are usually more fertile than autotetraploids. Explain.
13 From the standpoint of vegetative propagation, sectorial and mericlinal chimeras are unstable, whereas periclinal chimeras are relatively stable. Explain.
14 How does selection differ from combination breeding?
15 Single plant selection is more effective than mass selection. Explain.
16 In combination breeding, the new combinations need not be made homozygous in asexually propagated crops, but they should be made homozygous in sexually propagated crops. Explain.
17 Show how the plant patent law extended by the Congress of the United States stimulates the development of new varieties.

SELECTED REFERENCES FOR FURTHER STUDY

1. Mendel, G. J. 1866. *Experiments in plant hybridization*. Cambridge, Mass.: Harvard University Press. An English translation of Mendel's original paper.
2. Separates of the *U.S. Dept. Agr. Yearbook of 1937*. Each separate contains an excellent summary of research and related information on horticulture subjects up to 1937.

Title of separate	No. of separate	No. of literature citations
Breeding apples and pears	1586	26
Improvement in stone fruits	1588	34
Nut breeding	1590	52
Improvement in subtropical fruits	1589	212
Breeding small fruits	1583	122
Breeding vegetable crops	1581	335
Breeding and genetics in potato improvement	1582	74
Improvement of flowers by breeding	1591	564

Chapter 9

Propagation

METHODS OF PROPAGATION

Propagation of crop plants involves the formation and development of new individuals. These new individuals are used in the establishment of new plantings. In general, two methods are employed: (1) the using of seed and (2) the using of vegetative parts of plants, called *vegetative,* or *asexual,* propagation. Table 9-1 presents examples of crops which are propagated by seed, by using vegetative parts, and by both of these methods.

PROPAGATION BY SEED

Behold a sower sent forth to sow.—*Matthew 13:3*

What Is a Seed?

Essentially, a seed consists of an embryo with nourishing and protecting tissue. The embryo is a minute plant. Principal parts are plumule, radicle, hypocotyl, epicotyl, and cotyledons. The plumule is the first growing point of the stem; the

Table 9-1 Examples of Crops Propagated by Seed, by Vegetative Parts, and by Both Seed and Vegetative Parts

Type of propagation	Fruit crops	Vegetable crops	Flowers and ornamentals
By seed	Cocoanut, tung	Asparagus, cabbage, celery, onion, spinach, tomato, vine crops	Aster, calendula, centauria, delphinium, forget-me-not, marigold, pansy, salvia, sweet pea
By vegetative parts	Banana, pineapple, raspberry, strawberry	Potato, sweet potato, rhubarb, globe artichoke	Carnation, chrysanthemum, geranium, poinsettia, irls, peony, phlox
By both seed, for rootstocks, and vegetative parts, for cions	Apple, peach, pecan, citrus, tung		Dogwood, flowering peach, flowering cherry, crabapple, juniper

radicle is the first growing point of the root; and the hypocotyl and epicotyl together constitute the first, or original, stem of the plant. The nourishing tissues are endosperm, and cotyledons. In well-developed mature seed, these tissues are packed with stored food—starch, hemicellulose, reserve proteins, or fats—depending on the kind of plant. For example, sweet corn stores starch and dextrin; asparagus, onion, and date store hemicellulose; pea and bean store reserve proteins and carbohydrates; and pecan, walnut, lettuce, okra, cucurbits,[1] and ornamental sunflowers store comparatively large quantities of fat. These stored, or reserve, materials change to soluble forms for the respiration of the embryo in storage and for the respiration and growth of the embryo in germination. The protecting tissue is the seed coat.[2] In general, the coat retards the rate of transpiration; in some kinds of plants, it retards the rate of respiration while the seeds are in storage and protects the delicate embryo from mechanical injury to some extent. Thus, *a seed may be defined as a minute plant with nourishing and protecting tissues.*

Principles and Practices in Growing Seed for Sale

In general, new varieties and strains are bred or developed in the locality or region where they will be grown for home-garden or commercial production. Usually after a plant breeder has developed a new strain or variety, he has only a small quantity of seed, a handfull or several pounds of that strain or variety. This small

[1]The cucurbits include cucumber, cantaloupe, pumpkin, squash, and watermelon.

[2]In some crops, e.g., carrots, strawberries, and sweet corn, part of the fruit becomes attached to the seed coats, and the entire structure is commonly called *seed*.

quantity should be multiplied as rapidly as possible to meet the demands of all growers. To supply this demand, the seedsman selects an environment which is most favorable for the production of high yields of high-quality seeds, consistent with low costs of production.

In general, this selection requires a working knowledge of the genetic constitution of the new variety, its climatic requirements, its mode of pollination, and its susceptibility to specific diseases. The genetic constitution refers to the ever-changing dynamism of the genes. In common with the genes of other living things, the genes of crop plants are never static, they are always dynamic, and the changes are more likely to take place in the inferior direction than in the superior. Thus, the seedsman is always alert to the necessity of roguing out off-types, or inferior plants. The *climatic requirements* refer to the rate of apparent, or net, photosynthesis of the crop. As with home- and commercial-garden crops, the rate of net photosynthesis of the seed crop should be high. Thus, with other factors favorable, regions blessed with abundant sunshine and with an available supply of high-quality water should be well adapted to the production of seed for commercial stocks. Coffee and other crop plants which require low light intensities are exceptions. The mode of pollination is related to the degree of isolation necessary. If, for example, the new variety is largely self-pollinated and flowers at the same time as a sister variety, the distance between the two varieties can be relatively short. If, however, the new variety is largely cross-pollinated and flowers at the same time as a sister variety, its distance should be relatively wide. This distance depends on the direction of insect flight, if the varieties are insect-pollinated, or the direction of the prevailing wind, if the varieties are wind-pollinated. In general, plant breeders by virtue of their past experience on intervarietal pollination have a working knowledge of the minimum distance which is required.

Finally, and for obvious reasons, the seedsman selects sites which are free from diseases to which the new variety is susceptible. This is particularly true for diseases which are carried by the seed. For example, anthracnose is a fungus disease of the snap bean. The causal organism has the ability to infect the embryo of the seed. Thus, if infected seed is multiplied and placed in retail outlets throughout any given country and if the many home and commercial gardeners plant the diseased seed, the disease would be rapidly disseminated throughout the country. This, in turn, would seriously reduce both the quantity and quality and the reputation of the edible product—the relatively thick-walled pods. Thus, it is necessary for the seedsman to multiply the seed of any new variety in locations free from its seed-borne diseases.

Harvesting, Processing, and Drying

When the seed is severed from the parent plant, its source of food is cut off, and since respiration continues, the food supply within the seed begins to decline. In other words, with severance from the parent plant the process of aging, or senescence, sets in, and the vigor of the seedling, as measured by its rate of growth, begins to decline. Further, harvesting, processing, and drying require mechanical manipulation of the seed. This manipulation of the seed, if violent, is likely to damage the little plant within the seed. Thus, in all harvesting, processing, and drying operations the rate of respiration should be low, the seed should be handled with care, and the time interval between harvesting and drying should be reduced to a minimum.

Methods of harvesting, processing, and drying vary with the kind of crop and more particularly with the type of fruit. In general, for fleshy fruits the principal steps in sequence are (1) running ripe fruit on corrugated rollers, e.g., tomatoes, or cutting the fruit with knives, e.g., cantaloupes, (2) allowing the juice and seed with its gelatinous coat to ferment in large containers or vats, (3) washing the seed in running water, and (4) drying the seed either naturally in dry climates or artificially in humid climates, e.g., tomato and cucumber seed.

For dry dehiscent fruits of peas and beans, the principal steps are (1) cutting the base of the plants when they are nearly dry but before the pods begin to dehisce, (2) running the plants through a separator to separate the seeds from the pods and, (3) running the seed through a milling machine to remove unwanted plant debris. For these crops special separators have been developed, and the rotary screens are run at moderate speeds usually not more than 350 revolutions per minute (rpm), since investigations have shown that speeds greater than 350 rpm irrevocably damage the embryos. Examples of irrevocable damage are seedlings without plumules and seedlings without one or both of the cotyledons. For the siliques of cabbage and the capsules of onion, the sequential steps are similar to those for peas and beans with the exception that large pieces of canvas are usually used for drying the seed-bearing stalks. This is necessary to catch any seed which may have dehisced.

For dry indehiscent fruits, e.g., the "seed" balls of beet, the schizocarp of carrot and celery, the achene of lettuce, and the utricle of spinach, the sequential steps are (1) cutting the base of the plants either by hand or machine or cutting the seed heads only when the so-called seed is mature, (2) placing the plants in shocks, small piles, or windrows for four days to three weeks depending on relative humidity of the area, (3) threshing to separate the fruit from the remainder of the plant, and (4) milling to get rid of the remaining debris. In addition, carrot "seed" is run through rubbing machines to remove the spines from the seed.

Annuals and Biennials in Seed Production

As the term suggests, annuals require one growing season to complete their life cycle. In other words, it is simply a case of crop going from seed to seed within one growing season. In general, these plants may be divided into two groups: those which develop their absorbing system, stems, and leaves in the first part of the growing season and develop flowers, fruit, and seed during the latter; and

Table 9-2 Examples of Seed Producing Districts in the World

Country	Crops	Seed-producing areas
Australia	Both cool- and warm-season vegetable and flower crops	New South Wales, Victoria, Tasmania, North and South Island, New Zealand
Canada	Beets, crucifers, spinach, carrots, onions, lettuce, peas, snapbeans, flower crops	Lower coastal area of British Columbia, inland valleys of British Columbia
Europe	Beets, crucifers, spinach, carrots, onions, flower crops, lettuce, endive, flower crops	Denmark, Netherlands, southern Sweden, southern England, southern France, Italy, Spain, Canary Islands
Japan	Crucifers, tomatos, eggplants, onions, peas	Agriculture areas of Honshu
United States	Beets, crucifers, spinach	Puget Sound, Washington
	Beets, crucifers, snapbeans, spinach, onions, cucumbers	Willamette Valley, Oregon
	Peas, snapbeans, corn	Palouse of Washington and Idaho
	Most vegetable crops	San Joaquin Valley, California
	Cucumbers and cantaloupes	Arkansas River Valley
	Tomatoes	Canning districts of Middle-West
	Collards, pimiento pepper, okra, watermelon	Georgia, Florida, Louisiana
	Flower crops	Coastal areas of central and southern California

Data from various sources

those which develop stems and leaves simultaneous with the development of flowers and fruit. Examples of the first group are the fleshy leaves of spinach and mustard, the head of lettuce and Chinese cabbage; and examples of the second group are the fleshy fruits of tomatoes, capsicum peppers, and the various members of the cucumber family.

In sharp contrast to the annuals, the biennials require two consecutive growing seasons to complete their life cycle. During the first season, the plants develop roots, stems, leaves, and storage organs; and during the second season, they develop their flowering stems, fruit, and seed from the storage organs. Examples are the head and supporting stalk of cabbage, the bulb of onion, and the fleshy roots, called *stecklings*, of garden beets, carrots, turnips, and rutabaga. Thus, growing seed of the biennial vegetables is somewhat more complicated than growing seed of the annuals, in that the storage organs have to be maintained in a viable condition from harvest time in the fall to planting time in the following spring and the genetic constitution of the original seed has to be maintained for a period of two seasons instead of one.

Home Saving of Seed

Under what conditions is the home saving of seed likely to produce satisfactory results? To answer this question three important facts should be kept in mind. (1) Varieites and strains of a given kind of crop cross-pollinate, and the amount of cross-pollination may vary from less than 1 percent to 50 percent or more. Even the so-called self-pollinated crops, e.g., garden peas and tomatoes, cross from 0.1 to 5.0 percent. (2) In general, this uncontrolled or random cross-pollination does not promote uniformity in plant performance. (3) For effective isolation from unwanted pollen, varieties of the same kind in flower at the same time should be separated by a distance of at least 1 mile (1.6 kilometers). This explains why the saving of seed from home gardens in a city or village or from varieties of the same kind in flower at the same time in the same home garden is likely to lead to disappointing results; and why the saving of seed from isolated gardens, especially when only one variety of a given kind is grown, is likely to result in the development of superior sorts. An example of a new variety developed under isolated conditions is the Alabama No. 1 pole bean. The original lot of seed was obtained from a farm family which had been saving seed from superior plants for numbers of generations. Tests at the Alabama station showed that this variety was resistant to the root-knot nematode—a serious pest in many home and commercial gardens in the South. Another example is one of the parents of Clemson Spineless okra. This parent has spineless, light green pods, and the original lot of seed was obtained from a farm family which had been saving seed from spineless podded plants for numbers of generations. This plant was crossed with a spiney, dark green podded sort, and selections were made for dark green, nonspiney pods for six generations. The new variety was released as Clemson Spineless okra.

Storing Seed The principal process concerned is respiration. The stored food combines with water (absorbed from the air) in the formation of soluble food which, in turn, combines with oxygen in the formation of carbon dioxide and water and usable forms of kinetic energy and the liberation of heat. Note the equation:

$$\text{Stored food} + \text{absorbed } H_2O \rightarrow \text{soluble food} + O_2 \xrightarrow{\text{(hydrolyzing and oxidizing enzymes)}} CO_2 + H_2O + \text{usable energy} + \text{heat}$$

Temperature of the Air versus Dryness of the Seed While seeds are stored, their rate of respiration should proceed very slowly. Principal environmental factors in the storage of seed of most crops are the temperature of the air surrounding the seed and the water content of the seed. Note in the preceding equation the unique position of the water supply to the rate of respiration. If the relative humidity of the air is high, large quantities of water will be absorbed and, as a result, large quantities of insoluble foods will be hydrolyzed to soluble forms. This will result in an increase in the respiration rate and a decrease in storage life. If, however, the seed is allowed to absorb only small quantities of water, only small quantities of insoluble foods will be hydrolyzed into soluble forms. As a result, the rate of respiration will be low even in the presence of high temperatures, and the storage life will be prolonged.

Numerous investigations with artificially dried seed stored in moisture-proof or moisture-resistant containers show that this is the case. For example, workers at the Seed Testing Laboratory of the Asgrow Seed Company reduced the water content of freshly harvested seed of onion, tomato, capsicum pepper, and watermelon to about 5 percent, stored the seed in moisture-proof containers, and kept them in a cabinet held at a constant temperature of 90°F (32°C). At the end of three years, the germination of the seed in percent was as follows: onion 80, tomato 85, pepper 75, and watermelon 90. Work at other laboratories—the Boyce Thompson Institute, Yonkers, New York, the U.S. Department of Agriculture Research Station, Beltsville, Maryland, the Seed Testing Laboratory at Mississippi State, Mississippi, and the Agricultural Experiment Station, Davis, California—has been equally encouraging. Study the data in Table 9-3. The onion seed stored in the sealed can was protected from the high moisture content of the air and had a high germination percentage throughout the period of the experiment, whereas the seed stored in the cloth bag was exposed to a near saturated atmosphere and lost its ability to germinate by the end of a storage period of 32 weeks. As a result of this work, moisture-resistant, or moisture-proof, containers are now used in the packaging, storing, and merchandising of vegetable and flower crop seed. Examples of 100 percent moisture-resistant, or moisture-proof, containers are sealed tin cans; sealed aluminum cans; hermetically sealed glass jars; and sealed pouches made of aluminum foil, laminated to mylar or polyethylene. Examples of 80 to 90 percent moisture-resistant containers are aluminized polyethylene pouches; properly sealed multiwalled paper bags with the inner wall

Table 9-3 The Beneficial Effect of Protecting Onion Seed from High Moisture in the Air in the Presence of a High Temperature.

Container	Weeks in storage							
	4		8	16	16		32	
	M[a]	G[b]	M	G	M	G	M	G
Sealed can[c]	7.0	80[e]	7.1	83	7.2	80	7.1	81
Cloth bag[d]	23.0	72[e]	30.0	65	40.0	43	59.0	0

Source: Table 4, *Calif. Agr. Exp. Sta. Bul.* 792, 1963.

[a]Percent moisture in seed.

[b]Percent germination.

[c]Seed protected from near saturated atmosphere.

[d]Seed exposed to near saturated atmosphere.

[e]The original germination was 82 percent.

laminated with aluminum; and high density polyethylene pouches, 3 mil or more in thickness. Examples of nonresistant packages are paper bags; cloth bags; and burlap sacks.

Longevity of Seed

Some plants develop seed which retains its vitality longer than the seed of others. In fact, vegetable and flower crop seed may be placed in three groups: (1) kinds dependable for a short period (one or two growing seasons), (2) kinds dependable for a moderately long period (two or three growing seasons), and (3) kinds dependable for a long period (three to five growing seasons). Examples are shown in Table 9-4.

Testing Seed

The prime object of testing seed is to determine the rate of planting. As is well known, thinning due to excess rates of planting or poor stands due to nonviable seed always increases the cost of production. In general, the rate of planting is determined more accurately by the vitality of the seed than by its percentage of germination. Vitality refers to the rate of growth of the seedling plant. Two lots of seed may have the same percentage of germination, but the sprouts of one lot may make a greater amount of growth during the same period of time. The lot with the more vigorous sprouts is said to have the greater vitality.

At least two methods are available for testing seed: (1) the planting of seed in soil or sand and (2) the germination of seed between folds of blotting paper,

Table 9-4 Longevity of Vegetable and Flower Seed

Dependable for:		
1 or 2 seasons	**2 or 3 seasons**	**3 or 4 seasons**
Onion, sweet corn, celery, parsnip, primrose, lantana viola, verbena, begonia, petunia, dephinium	Asparagus, pea, bean, cabbage, lettuce, spinach, okra, petunia, salvia, calendula, gypsophila, gailardia	Cucurbits, dahlia, dianthus, lupine, carnation, amaranthus, gourds, zinnia

cotton flannel, or burlap. Each method has advantages and disadvantages. Modern seed houses conduct their own germination tests, and they record the results for any given lot of seed on the package. In addition, the U.S. Department of Agriculture and many states have laboratories for the testing of seed of horticultural and agronomic crops.

Treating Seed

The prime object of treating seed is to protect seed and seedling plants from certain parasites. A *parasite* is an organism which cannot make its own foods and feeds on living tissues. Parasites attacking seed are divided into two groups: (1) those which attack seeds and seedlings of most crops and (2) those which are specific and attack seed and seedlings of certain crops only.

Parasites which attack seed of most crops are present in many garden soils. They usually attack the plants in the seedling stage, particularly when they are growing in greenhouses, hotbeds, and cold frames. Because of the high humidity frequently maintained in these structures, these parasites, usually a group of fungi, produce a condition which is characterized by growers as seed rot or "damping off".

In general, the germinating seedlings may be attacked and killed before they emerge from the soil, or they may be attacked after they emerge. Many chemicals, called *protectants,* e.g., Captan, Diclone (Phygon), and Chloronil, have been tested to control this disease. These chemicals are applied in the dust or slurry form; directions for their application are usually printed on the container and should be carefully followed.

Parasites which attack seed or seedlings or specific crops are fungi and bacteria. As in the case of damping off, scientists have developed methods of controlling these pests. Agricultural experiment stations throughout the United States and other parts of the world have developed tables and charts on the use of chemicals for treating seed and seedstock of major local crops. As with the use of protectants, directions for their application should be carefully followed.

Germination of Seed

Germination is essentially a quickening of the growth of the embryo or seedling plant. Before germination begins, the young plant is relatively small and dormant. As germination proceeds, the growing points of the radicle and plumule divide rather rapidly. Usually, the radicle emerges from the seed coat first, proceeds downward, and develops into the root system; the plumule proceeds upward and develops into the shoot system. The fundamental process concerned is respiration. The stored insoluble foods are changed to soluble foods, and auxinic hormones are made in the endosperm and cotyledons. These soluble foods and hormones are translocated to the rapidly dividing meristems where they are utilized for the making of new cells and for the liberation of kinetic energy. *Germination, therefore, is entirely a food utilization process.*

Processes going on in seed during germination are (1) absorption of water, (2) secretion of enzymes and hormones, (3) hydrolysis of stored foods into soluble forms, and (4) translocation of soluble foods and hormones to the growing points. These processes are either wholly or in part influenced by the following factors: (1) food reserves, (2) hormone supply, (3) water supply, (4) oxygen supply, and (5) temperature level.

Food Reserves The main function of the reserve food is to nourish the young seedling until it can make its own foods, enzymes, and hormones. Thus, if the reserve food supply is low, the young seedling is likely to become weak and stunted. In extreme cases it may not have sufficient food to provide the free energy to push its radicle very far into the soil or its plumule above the surface of the soil. In general, relatively small, shriveled, immature seed is usually low in food reserves. This seed is separated from plump, nonshriveled seed during seed processing.

The Hormone Supply The main function of the hormones is to give the cell walls the ability to stretch. The stretching of the cells takes place in the region of elongation. The seat of production of the cell-stretching hormones is the endosperm and the cotyledons. Thus, if the endosperm or cotyledons are not developed fully or have been injured during harvesting, processing, or storing, the supply of hormones necessary for cell elongation is likely to be low and growth of the seedling is accordingly retarded.

The Water Supply Functions of water in germination are (1) to soften the seed coat, (2) to combine with stored foods in the formation of soluble foods, (3) to serve as the transportation medium of soluble foods and hormones to the meristems, and (4) to serve with the hormones in enlarging the new cells. Seeds germinate slowly in comparatively dry soil. In fact, tests have shown that soils should be maintained at or near field capacity for the rapid germination of seed of many crops. Quite often the use of special devices is necessary to facilitate the germination of small seed which, of necessity, must be planted shallowly. For example, celery and lettuce seed are firmed into the surface of finely pulverized, carefully leveled soil, and burlap is spread on the surface to prevent washing of the seed. Recently, glass-wool wick was tested for its ability to supply water for the germination of certain flower crop seed in flats. The plant wick is in sections 5 to 6 inches (12.5 to 15 centimeters) long, and one end is placed through a hole in the center of the bottom of a flat and flared out like the spokes of a wheel. The flat is filled with a sterilized mixture, one-third consisting of sand, and two-thirds of soil. Seed is planted in rows and watered thoroughly. The flat is then placed on a shallow pan containing water and covered with a pane of glass for one or two days. With this method only one watering is necessary. Excellent results have been secured with delphinium, snapdragon, stock, cineraria, begonia, petunia, and calceolaria.

The Oxygen Supply Functions of oxygen in germination are (1) to oxidize fats and other reserve compounds in the formation of sugars and other soluble compounds and (2) to oxidize the sugars in the process of respiration. This need of most kinds of seed for oxygen explains why soil should be moist but not wet; why the germination media should be loose and friable; why, in most cases, greater percentages of germination are obtained in sand or sandy loams than in clays; and why certain seeds are planted shallowly.

Temperature Level In general, temperature markedly influences the rate of many germination processes: absorption of water, translocation of soluble forms and hormones, respiration, and cell division and elongation. Since germination should proceed rapidly in the establishment of new plantings, high temperature levels, usually within the upper half of the optimum range for the growth of any given plant, should be utilized or maintained.

Light Investigations have shown that light stimulates the germination of seeds of some horticultural crops and reduces the germination of others. For example, the germination of freshly harvested seed of lettuce, celery, and primrose is markedly stimulated by exposure to weak light. On the other hand, the germination of onion, garlic, and chive seed is retarded by light. However, the germination of most kinds of vegetable and flower crop seed is not influenced by light; they germinate equally well in the light or in the dark.

Rest Period of Seed Although the seed of most vegetable and flower crops germinates immediately after it is harvested, the seed of many fruit and ornamental crops requires a rest or afterripening period before it germinates. In other words, certain physiologic and biochemical changes must take place within the seed before the young plant will grow. These changes may be related to the secretion of enzymes, the production of hormones, the absorption of water, the diffusion of oxygen into the seed, the diffusion of carbon dioxide away from the seed, or some other process. In general, the rest period is broken by storing the seed in moist media at temperatures from 40 to 45°F (4.4 to 7.2°C) for a period of one to three months, depending on the kind of seed. The student will note that this temperature range is quite similar to that required for breaking the rest of buds of the deciduous-tree fruits. Plants which have seeds with a rest period are apple, pear, peach, plum, pecan, some of the alpines, and many woody ornamental shrubs.

Pelleting consists of coating individual seeds with very finely divided adhesive materials, e.g., montmorillonite and powdered vermiculite. Each pellet is round, smooth, and uniform, regardless of the size, shape, and roughness of the individual seed. For example, carrot and lettuce seeds are long and narrow and rough on their surfaces, whereas onion seeds are blocklike and angular. Carrot and lettuce seeds are actually fruits and are commonly called *seed.* In general, tests have shown that benefits from using pelleted seed vary with the nature of the crop. With head lettuce grown in the irrigated districts of California, the use of pelleted seed as compared to nonpelleted seed has markedly reduced the cost of thinning out seedling plants; with carrots and onions, it has resulted in a more uniform spacing of plants in the row, and this, in turn, has increased the number of marketable roots or bulbs per unit area. With all of these crops, the use of a well-prepared seed bed is basically necessary.

Planting Seed

Seed is planted by hand or by machine. Seed planted in containers (pots, pans, flats) and in greenhouse beds or benches is usually planted by hand. Seed planted in gardens, fields, and orchards is usually planted by machine. Machines which plant seed are called *drills.* Seed drills may be operated by hand or by tractor. In general, they plant seed at the required depth, at a uniform rate, and firm the soil around the seed. Some drills are equipped to place commercial fertilizer at varying rates on the side of the seed. Note the six-row seed drill in Figure 9-1.

The depth of planting depends on (1) type of germination and (2) moisture and oxygen content of the soil. In general, seedlings with cotyledons which emerge from the soil usually require more shallow planting than seedlings with cotyledons which stay in the soil. For example, the cotyledons of snap bean and lima bean emerge from the soil, and although the seed is large, it should be planted shallowly. Note the differences in behavior of the hypocotyl and epicotyl of certain plants in Figure 9-2.

As previously stated, the water and oxygen supply exists in the pore spaces of the soil. Thus, if the pore spaces on the upper level of soil are nearly saturated, the oxygen supply is likely to be the limiting factor and relatively shallow planting is required. On the other hand, if the pore spaces of the upper level contain low quantities of available water, the water supply is likely to be the limiting factor and relatively deep planting is required. This explains why seeds

Figure 9-1 A tractor-drawn–six row seed drill in operation. Note the well-prepared seedbed and row marker on each side of the machine. *(Courtesy, International Harvester Co., Chicago, Ill.)*

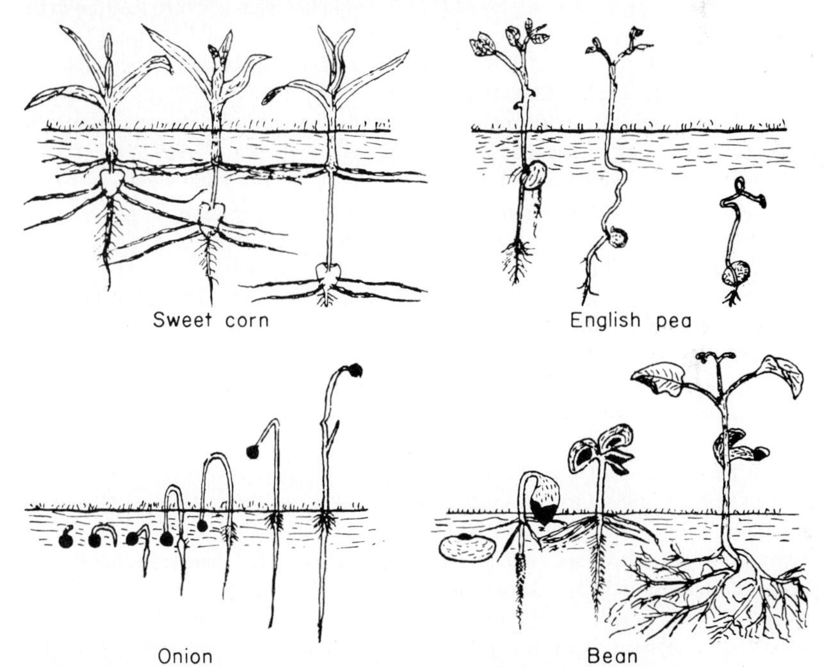

Figure 9-2 Relation of type of germination to depth of planting. Top: cotyledons remain in the soil. Bottom: cotyledons "lifted" above the soil.

are usually planted at a greater depth in the summer than in the fall, winter, and early spring.

Time of planting seed in the garden and field depends largely on the temperature requirements of the crop and the temperature of the locality. Since warm-season crops cannot withstand comparatively low temperatures, their seeds are not planted until the soil is sufficiently warm for at least moderately rapid germination. Consequently, the time of planting the seed of these crops is determined largely by the average date of the last killing frost in the spring and the first killing frost in the fall. On the other hand, since many cool-season crops withstand temperatures as low as 20°F (−6.7°C), seed of these crops may be planted before the last killing frost in the spring and after the first killing frost in the fall. This is true especially in regions characterized by mild winter temperatures.

QUESTIONS

1. Name the two methods by which new plantings are established.
2. What is a seed?
3. In general, new varieties and strains are developed in the regions where they are to be grown. Explain.
4. In general, most vegetable and flower crop seed are produced in regions favorable for high yields of high-quality seed. Explain.
5. Under what conditions would you recommend the saving of seed from crops grown in the flower or vegetable garden?
6. How would you obtain reliable information on adapted strains and varieties of horticultural crops for your location?
7. State the fundamental process concerned in the storage of seed.
8. Seeds while stored gradually decrease in dry weight. Explain.
9. Seed stored in a warm room and protected from high moisture of the air is likely to have a longer storage life than seed not so protected. Explain.
10. In the absence of refrigeration how would you store for a period of one year small lots of garden seed in a warm, humid climate? Give reasons.
11. Outline a practical and effective method of storing a small quantity of okra or watermelon seed.
12. What is the primary purpose of testing seed?
13. Growers should always try to obtain a full stand at the first planting. Give two reasons.
14. Why are many kinds of seed treated with certain chemicals called protectants?
15. Germinating seeds quickly decrease in dry weight. Explain.
16. The germination of seed is primarily a food utilization process. Explain.
17. Name four important processes going on in seeds during germination. State the principal environmental factors conditioning these processes.
18. Open, porous, well-drained soils facilitate germination to a greater extent than closed, nonporous, poorly drained soils. Give two reasons.

19 Saturated soils may retard or entirely prevent the germination of most vegetable and flower crop seed. Explain.
20 Seed of cool-season flower and vegetable crops may be planted earlier in the spring than seed of warm-season crops. Explain.
21 In general, firming of the soil immediately after planting the seed facilitates germination. Explain.
22 For any given kind of vegetable and flower crop seed, depth of planting varies with type of germination and with the moisture and oxygen supply of the soil. Explain.
23 Flats or seed pans containing germinating seed should always be set level. Explain.
24 How does the use of glass-wool wick provide water for germination?
25 How does pelleting provide for precision in machine planting?

VEGETATIVE PROPAGATION

> A good tree bringeth forth good fruit, but an evil tree bringeth forth corrupt fruit.—*Matthew 7:17*

As stated before, vegetative propagation consists of using vegetative structures—stems, leaves, or roots. These structures contain or develop buds which grow and develop into new individuals. This method of propagation is essential for the raising of many economic crops, e.g., the deciduous fruit crops, evergreen fruit crops, nut fruits, many flowering and ornamental crops, and certain vegetable crops.

Reasons for Vegetative Propagation

The prime reason for vegetative propagation is that *many crops, if propagated by seed, would not resemble the parents which produced the seed.* For example, if seed from a Winesap, Delicious, or Baldwin apple is planted, the trees which would develop from this seed would bear apples quite unlike those of the parent. They would vary greatly in size, shape, color, quality, season of maturity, keeping ability, chemical composition, and other characteristics. On the other hand, if a vegetative bud from a Winesap, Delicious, or Baldwin tree is grafted on the stem of a young apple tree, the tree that would grow from this bud would eventually bear apples exactly like those of the tree from which the bud was taken. The same situation exists with many other fruit crops, with many flowering and ornamental plants, and with certain vegetable crops. Obviously, *a primary advantage of vegetative propagation is that valuable varieties or individuals are perpetuated, which, in turn, makes possible the production of a standardized high-quality product.*

Other reasons for vegetative propagation are as follows: (1) certain valuable plants produce little or no seed, e.g., flowering cherry, flowering almond, flowering peach, and gardenia; (2) other plants produce seed which germinate with difficulty, e.g., holly, some of the viburnums, and rose; (3) some plants are more resistant to diseases, others are more resistant to nematodes, and others are more vigorous when they are grown on roots of related kinds, for example, certain species of the American grape are resistant to the root louse, certain species of peach are resistant to nematodes, and certain species of the labrusca grape impart vigor to the tops; (4) some plants are propagated more economically by vegetative means, e.g., strawberry, blueberry, and potato. A discussion of four principal methods of vegetative propagation follows.

Cuttage

Cuttage consists of producing new individuals *after* the piece of stem, whole leaf, piece of leaf, or piece of root has been severed from the parent plant. Detached stems with or without leaves are called *stem cuttings,* and detached roots are called *root cuttings.* Of these types, stem cuttings are the more widely used. They are classified as follows: (1) cuttings which require leaves and (2) cuttings which do not require leaves at the time they are severed from the parent plant. Since these two types differ in maturity and carbohydrate content of the tissues, they are discussed separately.

Cuttings Which Require Leaves Cuttings which require leaves are taken from herbaceous plants or from woody plants when the wood is immature. With these cuttings, rapid healing of the wounded surface and rapid production of roots are indispensable to the welfare of the cutting. If the cut surface heals slowly or not at all, most of the all-important water within the cutting escapes, and rot-producing organisms are likely to invade the tissues. How does the cutting heal the wounded surface? In general, immediately after the cut is made the intercellular spaces and the cells just beneath the cut become filled with sap. The sugars in the sap change to unsaturated fatty acids, and these, in turn, combine with oxygen of the air in the formation of the skinlike, varnish-like layer of material, or suberin. Suberin possesses the remarkable property of keeping the water within the cutting and resisting the attacks of rot-producing organisms. However, this layer is effective for a short time only since it is very shallow and nonelastic and cannot adjust itself to changes in water pressure within the cutting due to the intake and outgo of water. For these reasons a more permanent layer is formed. How does the cutting develop this layer? In general, in very young dicotyledonous herbaceous stems the permanent layer develops from the pericycle or cortex. These tissues have the ability to change into meristem and thus produce new cells. On the other hand, in relatively old dicotyledonous herbaceous stems and in immature woody stems, the permanent layer develops from the cambium. In both cases the walls of the new cells are impregnated with suberin, tannin, and other materials and they are corky in nature. Since this layer is being renewed constantly, it is durable; since it is several cells thick, it is deep-seated; and since it is elastic, it withstands the stress and strain due to changes in water absorption and transpiration. How does the cutting develop the root system? In general, the pericycle in young stems and the cambium in the somewhat older stems develop growing points, and these growing points develop into individual roots.

Is there anything that can be done to facilitate the development of the temporary and permanent protective layers and the production of roots with rapidity? A discussion of important factors follows.

Temperature Since with cuttings which require leaves the problem consists in producing roots from shoots, growth of the tops is retarded and growth of the roots is accelerated. The problem, therefore, is to keep the tops cool and the basal end of the cuttings relatively warm. In general, this is done by maintaining a relatively low air temperature and by applying artificial heat to the medium in which the cuttings are placed. The low air temperature, combined with high humidity of the air, maintains a low rate of transpiration. This low rate of transpiration keeps the guard cells turgid and the stomates open. As a result, carbon dioxide diffuses into the leaves and carbohydrate and hormone manufacture takes place. The relatively high temperature at the base of the cuttings promotes rapid oxidation of the fatty acids in the formation of suberin and speeds up the rate of cell division in the formation of the corky layer and the development of the root system.

The application of heat to the basal portion of cuttings is known as *bottom heat*. Bottom heat is supplied in various ways: by lead-covered electric resistance wire, by steam in pipes, and by hot water in pipes. Many experiments have shown that high rooting-media temperatures, combined with relatively low air temperatures, facilitate rapid root production. For example, tests at the Ohio Experiment Station have shown that chrysanthemum cuttings kept in sand at 60°F (16°C) produced a satisfactory root system in 10 days, whereas a comparable lot kept in sand at 50°F (10°C) required 20 days.

Relative Humidity and Light Intensity These factors affect both transpiration and photosynthesis. The student will recall that relative humidity and light intensity have opposite effects on the rate of transpiration. In general, high relative humidity promotes low rates of transpiration and high light intensity promotes high rates. Since low rates of transpiration are needed and since light is needed for the making of the carbohydrates and the hormones, the higher the relative humidity, the greater will be the amount of light the leaves can absorb without wilting. For this reason a high relative humidity is maintained.

Oxygen and Moisture Supply The formation of suberin requires abundant oxygen, and the rapidly dividing meristem requires both abundant oxygen and water. Hence, in the rooting of cuttings, media are used which will enable the growing points to obtain abundant oxygen and, at the same time, sufficient moisture for rapid root production. In general, washed, sharp, silica sand, mixtures of sand and peat moss, vermiculite, and mixtures of sand and perlite are satisfactory propagating media for herbaceous and softwood cuttings. These materials are porous, easily drained, and hold sufficient moisture for rapid root development.

The Leaf Area The healing of the cut surface and the production of roots require a supply of carbohydrates and auxinic hormones. These substances are made in nonwilted leaves. To prevent continuous wilting, the number of leaves, particularly on stem cuttings, is frequently reduced. However, if reduction of the leaf area of any given lot of cuttings is necessary, it is reduced only enough to prevent continuous wilting, and a high relative humidity is maintained to keep the leaves turgid.

Activity of the Root-producing Tissue The tissues of both herbaceous and woody plants vary in ability to form growing points. For example, certain herbaceous plants have an active, well-developed pericycle in their stems, and certain woody plants have active cambia. These active tissues form growing points very readily. This may explain, in part at least, why cuttings of some kinds root relatively easily and why cuttings of other kinds root slowly or with difficulty.

Types of Cuttings In general, cuttings which require leaves may be divided into three groups: (1) the herbaceous stem cutting, (2) the leaf and leaf-bud cutting, and (3) the softwood cutting.

The herbaceous stem cutting usually consists of the terminal portion of stems of herbaceous plants. In general, terminal portions of stems of moderately vigorous plants are preferred. With most kinds considerable reduction of the leaves is necessary, and the cuttings are prepared just before they are to be placed in the rooting medium. Many valuable herbaceous plants are propagated by stem cuttings. Examples are set forth in Table 9-5 and illustrated in Figures 9-3 and 9-4.

Leaf and leaf-bud cuttings consist of whole leaves with or without the petiole or of whole leaves with a piece of stem supporting the petiole. In general, they are usually taken from plants which develop thick, fleshy leaves and are prepared and rooted in much the same way as the herbaceous stem cutting. In fact, herbaceous stem cuttings and leaf and leaf-bud cuttings are frequently

Table 9-5 Kinds of Cutting Used in Propagating Crops

Kind of cutting	Horticultural crops
Herbaceous stem	Coleus, carnation, chrysanthemum, geranium, lantana, tomato
Leaf and leaf bud	Begonia, bryophyllum, cacti, gloxinia, Saintpaulia
Softwood	Dogwood, Japanese barberry, lilac, spirea, weigela, azalea, boxwood, holly, privet, arborvitae, Japanese yew, juniper
Hardwood	Arizona cypress, camellia, gardenia, currant, fig, grape, gooseberry, philadelphus
Root	Blackberry, bouvardia, wisteria

Figure 9-3 Paired cuttings of geranium and coleus. The left-hand cutting was treated with a growth regulator; the right-hand cutting was not treated.

Figure 9-4 Paired cuttings of sultana, camellia, and peach. The left-hand cutting was treated with a growth regulator; the right-hand cutting was not treated.

placed side by side in the propagation house. Examples of plants propagated by leaf or leaf-bud cuttings are shown in Table 9-5.

Softwood cuttings are taken from both deciduous and evergreen woody plants. Invariably, the terminal portion of stems is used, and the cuttings are severed from the parent plant when the wood is immature. Since the wood is immature, low quantities of carbohydrates have been stored in the tissues and, as would be expected, leaves are necessary not only for making additional carbohydrates, but also for making auxinic hormones. Thus, as with the herbaceous cutting, the leaves are maintained in the turgid condition.

The time of taking softwood cuttings varies somewhat with the kind of plant. In general, cuttings from deciduous plants are taken before or immediately after the new shoots have ceased to elongate; cuttings from broad-leaved evergreens are usually taken in the late summer; and cuttings from coniferous evergreens are taken in the fall or early winter. Examples of plants propagated by softwood cuttings are presented in Table 9-5.

Mist Propagation of Herbaceous and Softwood Cuttings Mist propagation consists of maintaining a film of water on the leaves of the cutting and a high relative humidity of the ambient air. In this way, the rate of transpiration is reduced to a minimum, and as a result the guard cells remain turgid, the stomates remain open, and the manufacture of carbohydrates and related substances proceeds unabated even in the presence of high light intensity. Further, with the high light intensity, the evaporation of water from the leaves keeps the tops relatively cool, and this, in turn, lowers the rate of respiration. Thus, with the low rate of transpiration combined with the low rate of respiration, high rates of apparent photosynthesis take place and abundant carbohydrates and other manufactured substances become available for the initiation and growth of the root system. Sample data on the effectiveness of mist propagation are presented in Table 9-6. Note that of the six kinds of cuttings four responded quite well to mist propagation while the remaining two kinds responded equally well to both the mist and the nonmist propagation. Similar results have been secured in other tests. In general, most but not all kinds of cuttings respond positively to the misting technique.

There are two kinds or types of misting: continuous and intermittent. The continuous involves applying the mist continuously during the light period, whereas the intermittent system involves applying the mist at definite intervals during the light period. Each system has advantages and disadvantages. The main advantage of the continuous system is that a hand valve installed in the water line is all that is necessary to control the system. Its main disadvantage is that excessive leaching of soluble compounds from the leaves is likely to take place and, unless the media is well drained, waterlogging of the media is likely to occur. The main advantage of the intermittent system is that excessive leaching of soluble compounds from the leaves and waterlogging of the media is not likely to take place. Its main disadvantage is that a timing mechanism is required to control the misting cycle. Thus, the advantages of the one become the disadvantages of the other. However, a research survey of systems throughout the world shows that the intermittent system is the more widely used. Of a total of 1,105 installations, 729, or 66 percent, used the intermittent system, and 376, or 34 percent, used the continuous system.

In general, five systems have been developed to control the misting cycle of the intermittent system. Of these five systems, three are used in commercial installations. These are the misting-cycle time clock, the electronic leaf, and the Misto-matic. The electric time clock consists of a day-night clock and a misting-cycle clock connected with a solenoid valve. The day-night clock turns on the system in the morning, usually just after sunrise, and off in the evening, usually just after sunset; the misting-cycle clock turns the system on and off for definite intervals during the day. For example, a commonly used misting cycle is one minute with the clock set to mist the cuttings for three, four, or five or more seconds per minute. The main advantage of this system is that it has a relatively low installation and maintenance cost. Its main disadvantage is that manual operation is required to change the misting cycle.

Table 9-6 The Response of Terminal Softwood Cuttings under Mist and No Mist

Kind of plant	No. of cuttings	Days to root	Mist				Ordinary greenhouse no mist			
			A*	B	C	Index†	A*	B	C	Index†
Deutzia lemoinei	100	15	43	26	28	321	36	20	27	267
Euonymus europaens	100	61	32	22	12	238	10	15	25	120
Pachysandra terminalis	120	29	50	44	7	389	48	40	30	390
Philadelphus coronarius	100	30	34	21	33	226	3	3	32	56
Spiraea vanhouttei	100	61	24	9	22	169	28	7	7	168
Viburnum lantana	100	33	30	18	23	227	1	3	17	34

Source: Table 1, *Mich. Exp. Sta. Quart. Bul.* 32 (2): 245–49, 1949.
*A = well rooted, B = moderately well rooted, C = poorly rooted
†Index = no. under A × 5 + no. under B × 3 + no. under C × 1

The so-called electronic leaf consists of a piece of plastic with two electrodes and a control unit connected to a solenoid valve. It is based on the assumption that the evaporation of water from the plastic is the same as the evaporation and transpiration of water from the leaves. When a continuous film exists between the two electrodes, misting is considered to be unnecessary and the solenoid valve is then in the closed position. With the evaporation of water between the electrodes, the film evaporates, contact between the electrodes is broken, and the solenoid valve opens and misting begins. Thus, the main advantage of this system is that the period of misting varies with the rate of evaporation and transpiration of water from the leaves, which varies according to temperature and light intensity. In addition no manual control is necessary. Its main disadvantage is that mineral salts may gradually accumulate on the electrodes, eventually forming a continuous film between the electrodes which would close the solenoid valve.

The Misto-matic system consists of a small stainless steel screen or grid and a control unit connected to a solenoid valve. Here again, this system operates on the assumption that the rate of evaporation of water from the screen or grid is practically the same as the rate of transpiration and evaporation of water from the leaves. When the mist is falling on the cuttings, it is also falling on the grid, which gradually declines to a point where the mercury switch closes the solenoid valve and misting stops. With loss of water from the leaves and corresponding loss of water from the grid, the grid gradually rises to a point where the mercury switch opens the solenoid valve and misting begins. Thus, the operation of the Misto-matic system is similar to that of the electronic leaf system, and its main advantages and disadvantages are practically the same.

Misting propagation structures include structures which were used before the misting technique was developed and structures especially designed for the misting technique. Examples of the former are greenhouses, plastic houses, and lathhouses, and examples of the latter are low-lying, quonset-type plastic tents, close or open-topped rectangular frames, mist boxes and the so-called bubbles. In all of these types, the rooting media are placed on level and well-drained soil to prevent waterlogging and to provide adequate aeration of the media, and the covers are moveable to regulate the temperature within the enclosure and, more particularly, to regulate the light intensity impinging on the leaves. For example, when a lot of cuttings is first placed in the media, the light intensity should be relatively low. As the cuttings become established, that is, as they develop their root systems, the light intensity can be safely increased. In addition, the frames and boxes have sides consisting of light-transmitting plastic to protect the cuttings from strong winds and to provide uniform application of the mist. Further, in case bottom heat is necessary, a thermostatically controlled heating cable is placed directly under the rooting media.

Cuttings Which Do Not Require Leaves Cuttings which do not require leaves are taken from woody plants when the current season's wood is mature. At this time the tissues are well supplied with carbohydrates. In general, there are two types: (1) the deciduous hardwood cutting and (2) the root cutting.

The Hardwood Cutting Principal factors concerned are (1) activity of the buds and (2) activity of the vascular and wound cambium in the healing of the cut surface and the formation of roots. Buds produce root-forming hormones. These hormones are translocated from the buds to the base of the cutting, where they are needed for cell division and cell elongation. Both the vascular cambium and the wound cambium possess the ability to form new roots. In general, roots which develop at the base of the cut arise from the wound cambium and those which develop from the nodes arise from the vascular cambium. Since both types of cambia are more active at the nodes than at the internodes, the basal cut is made just below the bud. Operations in the handling of this type of cutting are (1) securing the cuttings during the nongrowing season (usually, medium-sized wood from 6 to 10 inches, or, 15 to 25 centimeters long is used). (2) healing over the cut surface, during the winter, (3) planting in the field in the spring, and (4) allowing the cuttings to grow for a season or two before they are transplanted. Figure 9-5 shows hardwood cuttings of four kinds: grape, mock orange, forsythia, and currant.

The Root Cutting Root cuttings are pieces of roots. New shoots develop from adventitious buds, and the new roots develop from the old root or from the base of the new individual. In general, root cuttings are made 2 to 6 inches (5 to 15 centimeters) long from roots about the size of an ordinary lead pencil. There

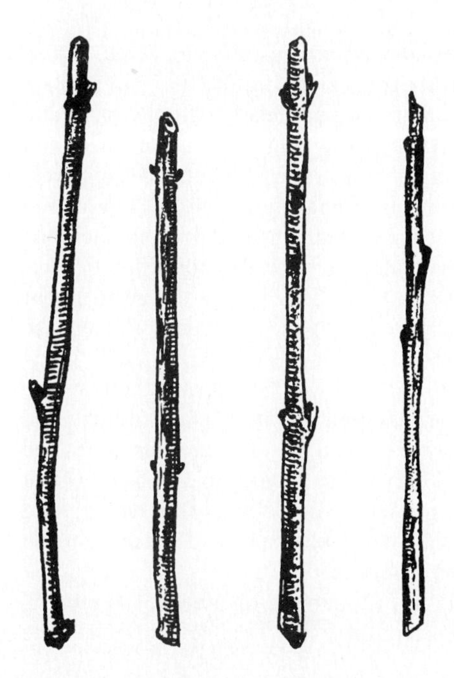

Figure 9-5 Examples of hardwood cuttings. Left to right: grape, mock orange, forsythia, and currant. (*Redrawn from* Maryland Agr. Exp. Sta. *Bul. 335, 1932.)*

are two methods by which the cuttings are handled. The first consists of taking the cuttings in early winter, storing them in sand, and planting them the following spring. The second consists of starting the cuttings in hotbeds or greenhouses in winter and transplanting the young plants to the field the following spring. The cuttings may be planted either horizontally or vertically. If they are planted vertically the end next to the crown of the plant is placed uppermost. Plants which develop "suckers" readily are propagated easily by root cuttings. Table 9-5 lists examples of plants propagated by hardwood and root cuttings.

Chemical Treatments and Rooting of Stem Cuttings Many investigations have shown that the application of certain chemicals promotes the development of roots of stem cuttings. Of the numerous chemicals which have been tested indoleacetic acid (IAA), indolebutyric acid (IBA), and naphthaleneacetic acid (NAA) have produced the most striking results. These chemicals not only speed up the healing of the wound and the production of roots, but they also induce the development of a large number of roots and are now used throughout the world in the propagation of many kinds of crop plants. These chemicals act in much the same way as the auxinic hormones. They are effective in very dilute concentration and are usually applied as a dust. In general, the cuttings are placed in groups or small bundles, and the basal ends of the cuttings are dipped in water and then in the chemical dust. The cuttings are then ready for placing in the propagation bed.

Layerage

Layerage involves the production of new individuals, usually on stems, *before* they are severed from the parent plant. Its advantages are that the parent plant supplies the new individual with water and food, particularly carbohydrates and proteins, and with hormones, particularly the auxins, until it makes its own food and hormones and that the expensive equipment necessary for cuttage is unnecessary. Two disadvantages are that this form of propagation is usually limited to plants which form growing points readily and that this method does not facilitate the production of a large number of individuals in a relatively short time. In other words, the number of individuals which can be produced from any given parent plant by layerage is relatively few compared with the number which can be produced by cuttage.

Principal factors influencing the production of roots and shoots on layered stems are (1) temperature, (2) moisture and oxygen supply, (3) lack of light, and (4) age of wood. In general, as previously explained, temperature, moisture, and oxygen directly influence the rate of cell division and enlargement, lack of light stimulates the production of roots, and growing points form and develop more readily in young wood. When the new plant has developed an adequate stem and root system, it may be severed from the parent plant.

The main kinds of layerage are (1) tip, (2) simple, (3) trench, (4) mound, and (5) pot, or air. In the discussion of each of these methods, note how adequate moisture and oxygen are provided, how light is excluded, and that young wood is used. (See Fig. 9-6).

Tip layering involves covering the tips of stems with moist soil. During the latter part of the growing season the stems bend and the tips come in contact with and grow downward in the soil for a short distance and then bend upward. In due time the meristem at the tips develops roots and shoots, and the following spring the young plants may be severed from the parent stem and planted in a new location. In general, this method is limited to plants which have flexible stems, e.g., black and purple raspberries, trailing blackberries, loganberries, and dewberries.

Simple layering involves covering a stem just back of the tip with moist soil or other appropriate media. The layered portion of the stem usually is slit or notched, and the top of the stem with its leaves is allowed to remain above the surface of the soil. Slitting or notching of the stem promotes the development of growing points of the root system, and the leaves manufacture food and hormones for the development of the root system. Forsythia, yellow jasmine, and climbing roses may be propagated by this method.

Trench layering consists in placing the basal and middle portions of young stems in a shallow trench and covering these sections 2 to 4 inches (5 to 10

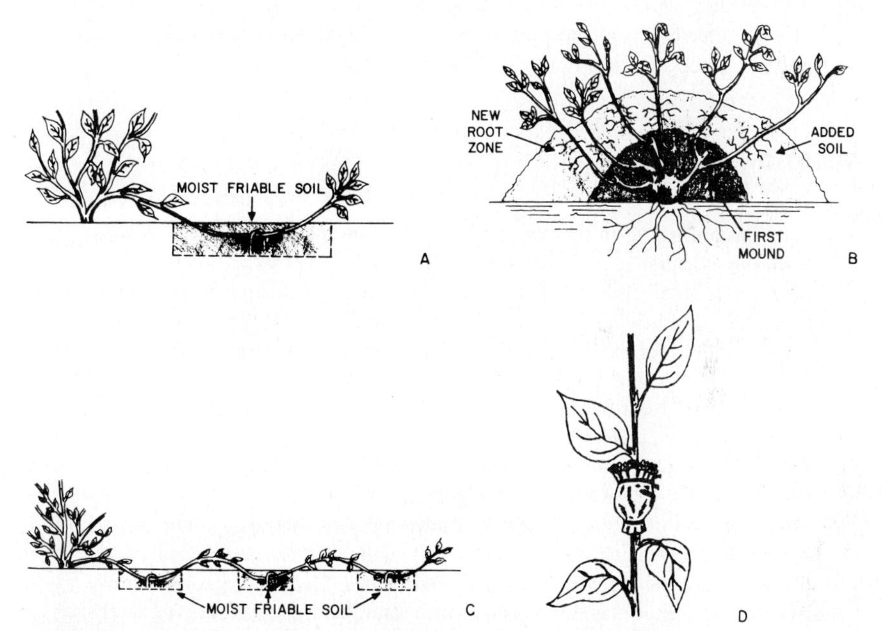

Figure 9-6 Top left: tip layering. Top right: mound layering. Bottom left: trench layering. Bottom right: air layering. *(Adapted from* Cornell Agr. Ext. Bul. *1006, 1958.)*

centimeters) deep with moist soil. The terminal portion is left exposed to manufacture foods and hormones for the developing plants. A modification of this method is discontinuous, or serpentine, layering. With this method only the nodes of the basal and middle portions are covered with soil. Muscadine grapes are propagated by these methods. The canes are placed in trenches during the late fall or winter, and shoots develop from the nodes during the following spring or summer. In late fall or early spring the new plants may be removed from the parent stem. With some plants wounding is necessary to induce root formation. For example, rhododendron stems are notched, ringed, or slit. Other plants propagated by this method are rose, spirea, and other deciduous shrubs.

Mound layering consists in cutting back the stems of the plant during the nongrowing season and covering the young stems with a mound of soil. These stems produce roots in the soil and are removed the following fall or spring and set out as separate plants. Currants, gooseberries, quinces, certain ornamental shrubs, and certain root stocks of the apple are propagated by this method.

Air layering consists in surrounding stems of the previous season's growth with moist peat moss held in place by a split pot, a wooden box, or sheets of plastic film, as for example polyethylene. In general, these films have high permeability to carbon dioxide and oxygen and low permeability to water vapor and they withstand weathering for long periods. Usually, the stem is girdled to facilitate the production of roots just above the girdle. When the roots are well developed, the stem is severed from the parent plant. Air layering is practiced usually with such plants as codiaeum, ficus, litchi, and Persian lime.

QUESTIONS

1. State the prime reason for vegetative propagation. State three other reasons.
2. State the essential steps in healing of the cut of the herbaceous cutting.
3. Show how relatively high humidity, in the presence of high rooting temperature, promotes rapid healing of the cut surface of herbaceous and softwood cuttings.
4. What is mist propagation? How does mist propagation promote the production of roots of cuttings with leaves?
5. Herbaceous and softwood cuttings, while being prepared, should not be placed in the hot sun or in wind. Explain.
6. Investigations have shown that herbaceous cuttings of many flower crops root faster at 70°F (21°C) than at 60°F (16°C). Explain.
7. Given two lots of cuttings. Lot A is placed in dry air (40 percent humidity). Lot B is placed in moist air (95 percent humidity). Other factors being equal, which lot is likely to have the greater percentage of survival? Give two reasons.
8. Florists and nurserymen use washed silica sand, peat, and mixtures of sand and peat instead of soil for propagating herbaceous and softwood cuttings. Explain.
9. Leaves are necessary for the production of roots in herbaceous and softwood cuttings. Give two reasons.

10 The leaf area of herbaceous and softwood cuttings should not be reduced if wilting can otherwise be avoided. Explain.
11 Given two lots of softwood cuttings. Lot A has two leaves; lot B has three leaves. With environmental factors favorable, which lot is likely to produce more roots in a given time? Give your reasons.
12 Sweet potato vine cuttings have a very active pericycle. Under favorable conditions they root easily. Explain.
13 In your opinion, how do growth regulators promote the rooting of cuttings?
14 In the hardwood cutting, why is the cut made just below a bud? Explain.
15 Crops which produce suckers readily can be propagated by root cuttings. Explain.
16 How does layerage differ from cuttage?
17 Name two advantages and one disadvantage of layerage.
18 How does girdling a layered stem facilitate root formation?

Graftage

Graftage involves the union of two separate, usually woody, structures—the union of a root and a stem or, more frequently, the union of two separate stems. In general, the upper part of the union is called the *cion* (scion) and the lower part is called the *stock*.

Reasons for Graftage Graftage makes possible (1) the changing of the tops of trees, usually from an undesirable variety to a desirable variety; (2) the growing of several kinds of flowers or fruit on one tree or plant; and (3) the utilization of stocks which influence the growth of the cion. These effects may be divided into four categories: (1) rootstocks which permit the cion variety to thrive on light or heavy, poorly drained or salinaceous soil, e.g., seedlings of the rough lemon, which are well adapted to sandy soil; (2) rootstocks which are resistant to relatively low temperatures, e.g., *Citrus trifoliata;* (3) rootstocks which are resistant to certain pests, e.g., certain rootstocks of the apple which are resistant to the root louse and certain rootstocks of the peach which are resistant to nematodes; (4) rootstocks which impart a differential effect on the ultimate size of the tree of the cion variety of the apple, pear, peach, and plum and which are of prime importance. As result of research at two British research stations, the East Malling Research Station and the John Innes Horticultural Institute, four types of rootstocks are now available, particularly for the apple industry. These types and their effect on the behavior of the tree are presented in Figure 9-7. Note the differences in height and ultimate spread of the trees, the age at which the trees begin bearing fruit, and the relative productive period of these types. In your opinion, which type of tree is most conveniently pruned and sprayed and from which type is the fruit to be most conveniently harvested? Which size of tree will produce the greatest yield per unit area?

Various hypotheses have been set forth to explain the differential effect of the East Malling and Malling-Merton stocks. One hypothesis is that the conduct-

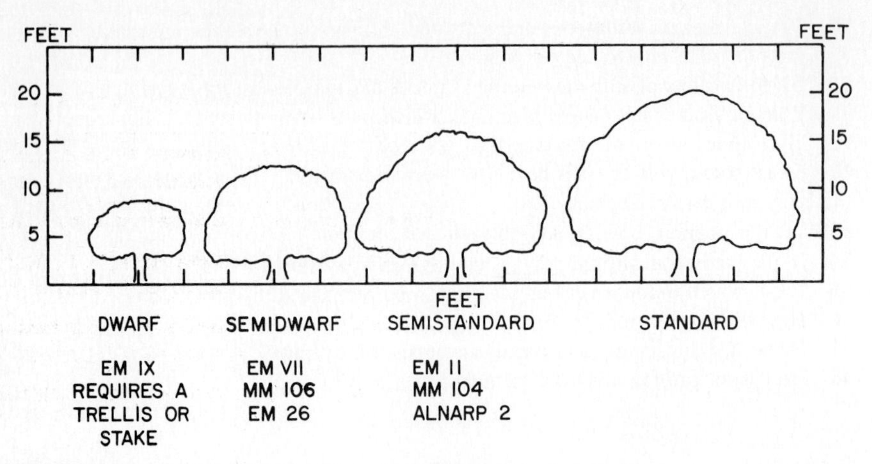

Figure 9-7 Influence of the stock on the ultimate size of apple trees. *(Adapted from Mich. Agr. Ext. Serv. Bul. 432, 1964.)*

ing tissues of the union restricts the movement of the manufactured substances from the tops to the roots. This hypothesis is supported by the relatively rapid development of the stems just above the cut, the sparse development of the root system of the stock, and the short period required to bring the tree into bearing. Another hypothesis is that the tissues of the union interfere with the normal hormone relations of the top to the root and the root to the top. Thus, research is needed to test these and other hypotheses so that the physiologic and biochemical basis of the dwarfing effect of the rootstock on the cion can be elucidated.

Limitations of Graftage In general, the principal tissues concerned in graftage are the cambia, particularly the vascular cambium. As previously stated, cambia have the ability to make new cells, and the formation of cells is necessary for the unification, or the growing together, of the cion and stock. Thus, graftage is limited to plants which develop the secondary plant body: the conifers and the dicots. In other words, monocots cannot be grafted very readily. Another limitation is that even within the conifers and the dicots, only structures which are closely related botanically will grow together. In general, the wood of two horticultural varieties within the same botanical species forms successful unions, and in some instances the wood of each of two species within the same genus grows together satisfactorily, e.g., apple and crabapple, pecan and hickory, garden rose and wild rose.

Formation of the Graft Union How do the cion and the stock grow together? In general, the two structures are prepared in such a way that the vascular cambium of each structure is placed close to or in contact with each other and held together until the two structures grow together. Principal steps in the formation of the union are as follows: (1) the exposed cells of the cambium of each of the two structures produce a mass of parenchyma cells; (2) these parenchyma cells intermingle and interlock with each other; (3) certain cells of this parenchyma tissue become meristematic and form a cambium which connects with that of the cion and that of the stock; and (4) this cambium divides and forms secondary phloem and secondary xylem which, in turn, connect with the secondary phloem and secondary xylem of the cion and the stock. These tissues which form the union are called *callus*. Thus, the callus forms a bridge of living tissue between the cion and the stock. In this way, water, hormones, and essential raw materials pass from the stock to the cion, and the manufactured foods and hormones pass from the cion to the stock.

Types of Grafting Two types are recognized: budding and grafting. Budding consists of uniting a vegetative bud to a seedling tree or to a mature tree. Two important kinds are T-budding and patch budding.

T-budding consists of making an incision in the bark of the stock in the form of a T and inserting the bud under the bark. Raffia, rubber bands, or adhesive tape are used to hold the bud tightly to the stock. T-budding is commonly used in the propagation of apples, peaches, nectarines, apricots, almonds, and plums. In the southeastern United States, the budding of peaches is done in June or early July, and the trees, called *June buds,* are ready for sale at the end of the growing season. In the northern United States, the budding is usually done in August or early September. At this time the buds are in the rest period and are inactive. The following spring they develop rapidly, and the trees are ready for sale in the fall. In both cases, the stock consists of young trees raised from seed of the wild peach, the cion consists of a single bud of a current season's growth of the desired variety, and the budding is done in the nursery row. In the preparation of the buds, the leaves are removed by cutting the petiole just below the leaf blade. In this way, the piece of the petiole which remains serves as a handle to facilitate the insertion of the bud beneath the bark of the stock. Note the operations set forth in Figure 9-8.

Patch budding consists of removing a square or rectangular piece of bark from the stock and replacing it by a similar patch of bark which includes the desirable bud. The wrapping material usually used is waxed cloth or budding tape. Patch budding is commonly used in the propagation of thick-barked trees, such as pecan and walnut. In the budding of the pecan, nuts from seedling trees or standard varieties are planted in rows, the cion buds are inserted in August of the second summer, and the trees are headed back 6 inches above the bud. When the shoot no longer requires the support of the stub, it is removed. Note the steps presented in Figure 9-9.

Grafting consists of uniting a piece of twig with two or more vegetative buds to a seedling tree or to a mature tree. The many kinds of grafting are classified

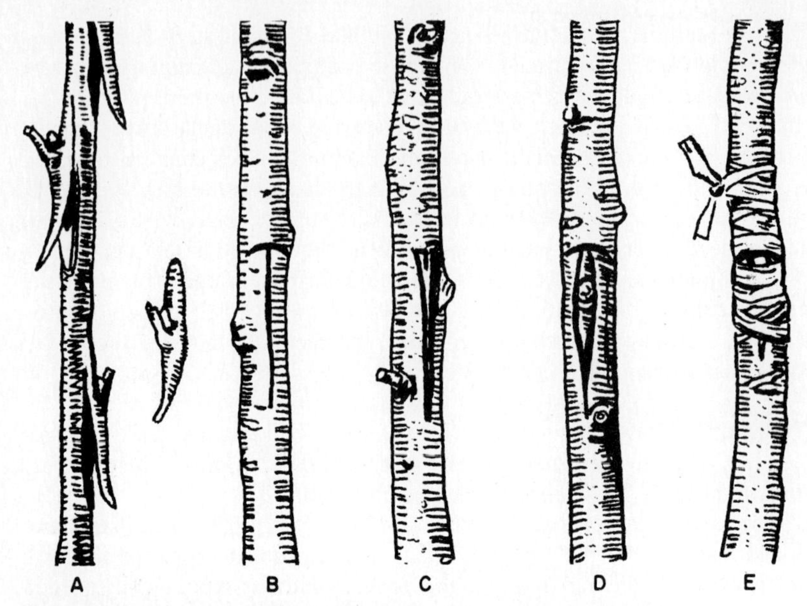

Figure 9-8 T-budding. A: bud stick. B: T-cut through bark. C: bark raised to admit bud. D: bud in place. E: bud wrapped with raffia. *(redrawn from* U.S. Dept. Agr. Farmers' Bul. *1567, 1932.)*

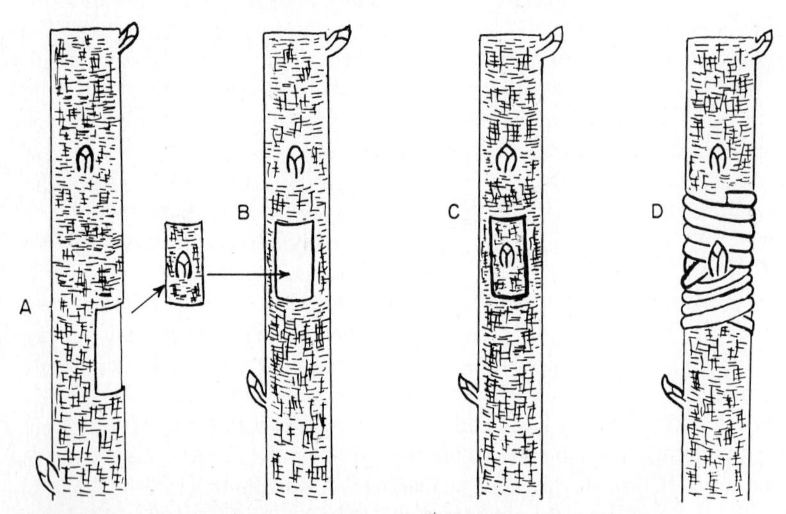

Figure 9-9 Patch budding. A: cion removed from bud stick. B: patch of bark removed from stock. C: cion from A inserted in B. D: cion wrapped with waxed cloth. *(Redrawn from* Mississippi Agr. Exp. Sta. Bul. *375, 1943.)*

according to the relative diameter of stock and cion. On this basis two kinds are recognized: (1) those of which the diameters of the stock and the cion are similar; and (2) those of which the diameter of the stock is greater than that of the cion.

Diameters of Stock and Cion Similar Naturally, the stock consists of relatively young wood and is practically the same age as that of the cion. Principal kinds are (1) tongue, root, or whip grafting; and (2) approach grafting. The tongue, whip, or root graft is used extensively in the propagation of apples and pears. In general, one-year-old seedling trees are dug in the fall and the grafting is done in the winter. Sloping cuts about 1½ inches (3.8 centimeters) long are made at the base of the cion and at the top of the stock, and a reverse cut is made on each piece about ⅓ inch (.8 centimeters) from the tip and ½ inch (1.3 centimeters) in depth. The two pieces are then fitted together and wrapped. Note the steps set forth in Figure 9-10.

Approach grafting consists in joining the stems of plants growing on their own roots. This method is used in the propagation of coniferous evergreens. The plants used as stocks are potted and placed close to the plants which are to be used as cions. The bark on one side of the stock is sliced away for a distance of 1 or 2 inches (2.5 or 5 centimeters). A similar cut is made on the side of the cion.

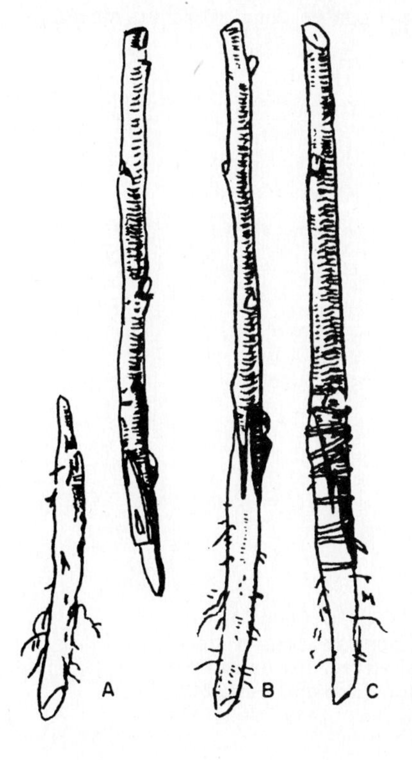

Figure 9-10 Root, tongue, or whip grafting. A: cion and stock prepared. B: cion and stock fitted together. C: cion and stock wrapped with raffia. (*Redrawn from* U.S. Dept. Agr. Farmers' Bul. *1956, 1932.*)

The two wounded surfaces are then pressed together and bound with waxed cloth, waxed strips, or tape to prevent the cut surface from drying out and to hold the cambia close to each other. The stock and cion are allowed to remain undisturbed throughout the growing season, during which time the two plants grow together.

Diameter of Stóck Greater than that of Cion Naturally, the stock is older and has a diameter larger than that of the cion. Common kinds of grafts are (1) cleft, (2) bark, (3) notch, and (4) wedge. Cleft grafting consists in (1) splitting the stock branch down the center, (2) holding the wedge open while the cions are inserted, (3) placing the cion in the cleft in such a way as to ensure contact of the cambia, and (4) waxing over the cut surface. The main advantage of cleft grafting is that it can be done during the dormant season. Its main disadvantage is that wood-decaying organisms may get into the graft. Note the operations illustrated in Figure 9-11. Bark grafting consists in (1) splitting the bark of the stock, (2) nailing the cion in place, and (3) waxing over the cut surface. Bark grafting can be done only in the spring when the bark begins to slip, and the cion wood must be gathered when dormant and stored until needed. Notch grafting consists in (1) notching the bark of the stock, (2) cutting the cion to fit the notch, (3) nailing the cion in place, and (4) waxing over the cut surface. Notch grafting combines the advantages of cleft and bark grafting, since it can be done over a considerable

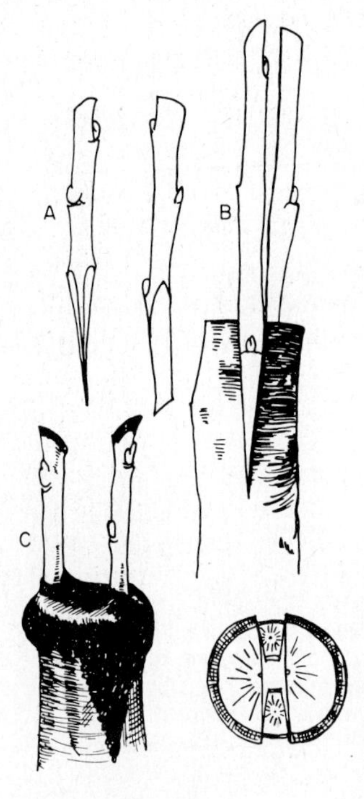

Figure 9-11 Cleft grafting. A: cion prepared. B: cions inserted in stock. C: cut surface waxed. (*Redrawn from* U.S. Dept. Agr. Farmers' Bul. *1567, 1932.)*

period of time and the stock is not split. However, notch grafting, when properly done, requires considerable time. Wedge grafting consists in (1) removing a V-shaped wedge of tissue down the middle of the stock, (2) making a tapering wedge at the base of the cion, and (3) inserting the cion in the wedge of the stock so that the cambia of the stock and cion are in contact with or close to each other. Camellias are frequently propagated by this method.

Bridge Grafting Bridge grafting is not a method of asexual propagation. It is done to save valuable trees. Each year the base of the trunk of many valuable fruit trees is girdled either partly or completely by rodents or by mechanical means. The question arises: How does girdling the trunk of a tree kill the tree? The tissues of the stems of coniferous and dicotyledonous trees may be divided into two parts: the bark and the wood. The bark contains the layer of cork, cork cambium, secondary phloem, and vascular cambium; and the wood contains secondary xylem only. When such a tree is girdled, the secondary phloem is severed. As a result, the foods, hormones, and vitamins made in the tops are no longer translocated to the roots. Thus, when the carbohydrates and other compounds of the root system, which were present before girdling, are exhausted, the root system dies and the top dies. This explains why a girdled tree may live for one or two or more seasons before death finally takes place. In general, bridge grafting consists in inserting one-year-old cions above and below injured areas of trunks or limbs. The cions usually are collected during the dormant season and are inserted in the tree when the bark begins to slip. Note the operations illustrated in Figure 9-12. If young suckers are present at the base of the tree, they may be used to bridge over the injured area. This method is sometimes called *inarching*.

Closely related to girdling is the effect taking place within a tree which develops a heavy crop of fruit and seed and then dies. For some reason, the growth of the root system has been retarded or entirely prevented. As a result, large quantities of carbohydrates are stored in the stems and an excessively large number of flower buds are layed down. During the following year, these buds develop into flowers and the ovaries of the flowers grow and become ripe and contain mature seed. With the development of the flowers and the growth of the fruit, large quantities of carbohydrates are used, and with the ripening of the fruit and maturation of the seed, large quantities of carbohydrates are stored. As a result, insufficient carbohydrates are available for making the water-retaining colloids in the fall, for respiration and maintenance of the living tissues during the winter, and for growth and development of the stems and leaves and absorbing system the following spring.

Grafting Waxes and Tools The primary function of waxes is to protect the wounded surfaces from wood-decaying fungi, to retard transpiration from the tissues, and to permit the normal exchange of oxygen and carbon dioxide in

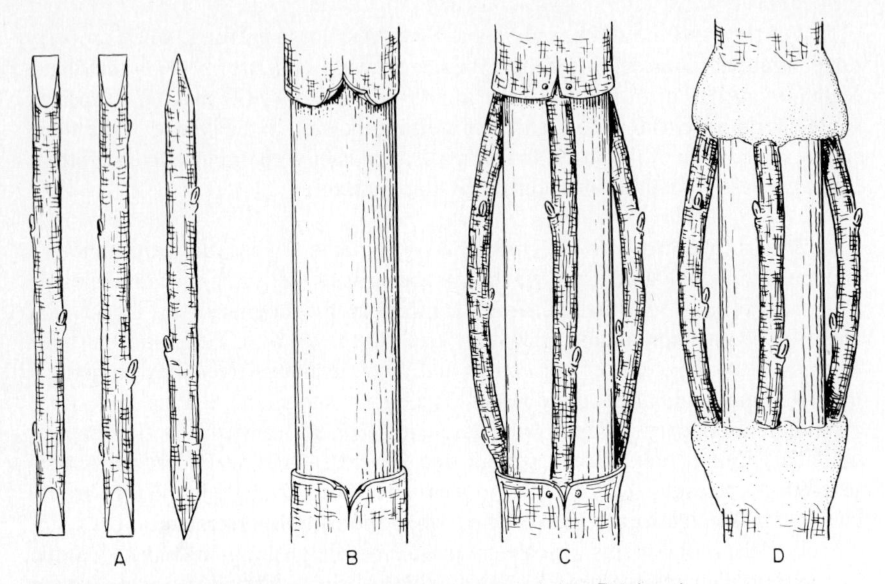

Figure 9-12 Bridge grafting. A: three views of prepared cion. B: injured area prepared to receive cions. C: cions inserted. D: cions and bark waxed. *(Adapted from* Calif. Agr. Ext. Circ. *96, 1936, and* Mississippi Agr. Exp. Sta. Bul. *375, 1943.)*

respiration. For these reasons, suitable waxes must not crack or melt at ordinary temperatures and must be free from substances toxic to the tissues.

Various kinds of waxes are hot, or hard, wax; hand, or soft, wax; and emulsified asphalt. The ingredients, preparation, and application and the advantages and disadvantages of each type may be obtained from any standard text on plant propagation or from the horticultural department of your local land-grant college or university.

Using Storage Organs

The use of storage organs is limited to plants which develop specialized storage structures of organs. In general these storage organs are bulbs, corms, tubers, rhizomes, and fleshy roots.

Bulbs are produced by monocots and, specifically, by certain members of the lily family. In general, an individual bulb consists of the basal portion of leaves, called *scales,* which are attached to a small disk-shaped platelike stem. There are two types: (1) tunicate and (2) scaly. Tunicate bulbs have concentrically arranged, thin, membranous outer scales and concentrically arranged, thick, fleshy inner scales, e.g., the bulbs of onion and tulip. On the other hand, nontunicate, or scaly, bulbs are not surrounded by thin membranous scales and the fleshy scales are not concentric but separated from each other. An example is the scalelike bulbs of lily. Easter lilies, tulips, and narcissus are propagated by using young bulbs. In general, small bulbs form around, above, or at the base of the mother bulb. When mature plants are removed from the soil, the young bulbs are separated from the mother plant; and when planted separately, they develop into new individuals. (See Fig. 9-13)

Corms are produced by certain monocots also. In general, an individual corm consists of the enlarged base of a stem which is surrounded by dry, scalelike leaves. Like bulbs, roots develop from the lower portion of a corm, but only one stem develops from the upper part. In general, small corms, called *cormels,* develop just above the mother corm. These, in turn, are separated from the parent plant and, when planted in a new location, develop into new individuals. Gladiolus and crocus produce corms and are propagated by the planting of cormels.

Tubers are produced by certain dicots. In general, an individual tuber is a short, thick, fleshy underground stem with scalelike leaves subtending nodes, commonly called *eyes,* e.g., the tubers of the potato, the Jerusalem artichoke, and caladium. Shoots arise from the nodes and develop into independent plants. In the vegetative propagation of the potato and Jerusalem artichoke, large tubers are cut into several pieces. Each piece has one or more nodes and stored food, chiefly starch, for the nourishment of the seedling plant.

Rhizomes are produced by certain monocots and dicots. In general, an individual rhizome is a stem with well-developed nodes and internodes and which grows in a horizontal direction along the surface of the soil or in the soil. Rhizomes may be divided into two groups: (1) short, thick, and fleshy and (2) relatively long, slender, and nonfleshy. Thick, fleshy rhizomes are packed with stored food, particularly at the end of the growing season, and are cut or divided into pieces for the production of new individuals. In general, cutting into pieces,

Figure 9-13 Young and old bulbs of lily and tulip. (*Redrawn from* Calif. Agr. Ext. Cir. *132, 1947 and* U.S. Dept. Agr. Cir. *372, 1937.*)

called *division,* is done usually in the spring or late fall and seldom in the summer. Many herbaceous perennials are propagated by using pieces of rhizomes—lily of the valley, aster, hollyhock, violet, rudbeckia, gaillardia, coreopsis, chrysanthemum, Shasta daisy, campanula, phlox, delphinium, and rhubarb. Slender, nonfleshy rhizomes produce new individuals at the nodes, particularly when the new individual is connected with the parent plant. Certain grasses, e.g., bent grass and Bermuda grass, are propagated through the use of slender rhizomes.

Fleshy roots are produced by certain dicots. In general, an individual fleshy root is short and thick, has the anatomy typical of roots, and, unlike stems, does not have nodes and internodes, e.g., the fleshy roots of sweet potato and dahlia. With the sweet potato, adventitious buds arise from the root itself and the buds develop into seedling plants. With the dahlia, a piece of stem containing at least one bud is allowed to remain attached to the fleshy root.

Using Apomictic Seed

Apomictic seed is formed without the benefit of the male parent. In other words, no actual fertilization takes place. Thus, only the reproductive tissue of the female is involved. In general, the new individual may develop (1) from a $2n$ nonreduced megaspore mother cell, (2) from one or more cells of the $2n$ nucleus, or (3) from a nonreduced egg. Thus, apomixis tends to promote homozygosity and the development of pure lines. Although this form of asexual reproduction is rather widespread throughout the plant kingdom, it is of practical importance in only a few crop plants—Kentucky blue grass, citrus, and mango.

QUESTIONS

1. The cambia are the chief tissues concerned in graftage. Explain.
2. Grafting is limited to coniferous and dicotyledenous plants. Explain.
3. Describe briefly the essential steps in the formation of the callus.
4. In general, the callus serves as a living bridge between the cion and the stock. Explain.
5. Differentiate between budding and grafting.
6. In general, buds in the middle portion of one-year-old wood make the best unions in budding and grafting. Explain.
7. In the budding of fruit crops the buds are wrapped tightly until they form a union with the stock. Explain.
8. Why are tongue-grafted trees usually not planted immediately after the graft is made?
9. The cut surface of a cleft graft in tops of trees is usually covered with grafting wax. Give two reasons.
10. How does girdling the base of the trunk of a tree kill the tree?
11. How does bridge grafting save girdled trees?

12 A tree develops an excessively large amount of fruit and then dies. Explain from the standpoint of the carbohydrates.
13 Give the similarities and differences between a bulb and a corm.
14 What is the main difference between a tuber and a fleshy root?
15 In cutting potatoes for seed each piece should contain at least one eye. Explain.
16 In propagating dahlias each piece of stem with the root should contain at least one bud. Explain.

SELECTED REFERENCES FOR FURTHER STUDY

Carlson, R. F. 1964. Dwarf fruit trees. *Mich. Agr. Ext. Serv. Bul.* 432. A general discussion on the effect of the East Malling and Malling-Merton rootstocks on the behavior of cion varieties of the apple and pear.

Harrington, J. F. 1961. The value of moisture-resistant containers in vegetable seed packaging. *Calif. Agr. Exp. Sta. Bul.* 792. An excellent report on how an investigator solves a problem, benefiting the vegetable seed industry and the grower.

Row-Dutton, P. 1959. *Mist propagation of cuttings.* Farnham Royal, Bucks, Eng: Commonwealth Agricultural Bureau. An excellent review of the literature on the use of mist propagation throughout the world, including 160 references.

Stefferud, A. 1961. Seeds. *U.S. Dept. Agr. Yearbook.* A discussion based on recent experimental data on the characteristics, production, processing, treating, storing, and marketing of vegetable and flower crop seed. (Since U.S. yearbooks are widely distributed, high school and college libraries are likely to have a copy.)

Chapter 10

Climates, Sites, and Soils

May the blessings of abundant, high-quality food spread to all the people of the earth.

CLIMATES

Climate refers to meteorological conditions at any given place or within any given region. More specifically, these meteorological conditions refer to changes in the atmosphere, the huge blanket of air which surrounds planet earth. From the standpoint of crop-plant production, there are four climates, or climatic zones: (1) the tropical, (2) the subtropical, (3) the warm-temperate, and (4) the cool-temperate. These zones can be delineated according to their latitude, or distance north and south of the equator. In general, the tropical zone lies between 0 and 20° north and south, the subtropical between 20 and 30° north and south, the warm-temperature between 30 and 40° north and south, and the cool-temperature between 40 and 60° north and south. These differences in latitude make for differences in degree of heat, that is, in temperature level, in length of the frost-free growing season, and in intensity and total amount of light between any two zones. Note the differences set forth in Table 10-1. The degree of heat, or

Table 10-1 Climates and Their Relation to Latitude, Rainfall, Temperature Level, and Amount of Light

Type of climate	Location and latitude	Mean annual rainfall* (in.)	Mean temperature* (degrees F) Jan.	July	Frost-free growing season* (days/year)	Mean daily amt. light	Total amt. light Frost-free season†
						Gm-cal	
Tropical	Honolulu Hawaii 20°N	25.3	71	76	365	505	184,248
Subtropical	Gainesville Fla. 29°N	49.1	58	81	285	430	114,339
Warm-temperate	Nashville Tenn. 36°N	44.8	40	79	215	411	88,284
Cool-temperate	Lansing Mich. 42°N	30.5	23	71	158	386	61,004
Warm-humid	Gainesville Fla. 29°N	49.1	58	81	285	430	114,229
Warm-dry	Riverside Calif. 32°N	11.5	52	76	265	466	123,434

*From *U.S. Dept. Agr. Yearbook*, 1941.
†From *Mich. Agr. Exp. Sta. Tech. Bul.*, 222, 1950.

temperature level, the length of the frost-free growing season, and the daily supply of light vary more or less directly with the distance from the equator. As a result, the tropical zone has the highest temperature level, the longest frost-free growing season, and the greatest amount of light, followed by the subtropical, the warm-temperate, and the cool-temperate zones, in descending order.

Within each climatic zone there are certain features of the physiographic environment which influence the water supply, the temperature level, and the light supply. In general, these physiographic features pertain to the amount and distribution of the annual rainfall, the elevation of the land above and below sea level, the direction of mountain slopes, the presence of relatively warm and cold rivers or currents of water in the oceans, and the presence of large bodies of fresh water.

Amount and Distribution of the Annual Rainfall

As is well known, the annual rainfall varies widely within each climatic zone. Note in Figure 10-1 that some areas receive relatively small quantities each year, whereas others receive large supplies. In general, crop-producing areas which receive insufficient supplies of rain and require the application of irrigation water for the production of high yields are known as the *arid* and *semiarid* areas; crop-producing areas which receive sufficient supplies annually and require the application of irrigation water as a form of insurance for the production of high yields are known as *humid* areas. From the standpoint of net photosynthesis and the production of high yields, the essential difference between arid, semiarid, and humid areas is the difference in the dryness of the atmosphere. In arid and semiarid areas, the air is relatively dry, the sky is relatively clear, the light intensity is relatively high, and abundant sunshine prevails. In humid areas the air is relatively moist, the sky contains more clouds, the rains are more frequent, and lesser quantities of sunshine prevail. Study the data in Table 10-1 for Riverside, California, a semiarid location, and for Gainesville, Florida, a humid location. Note that the amount of light each day during the frost-free growing season is greater in the semiarid location than in the humid location. Further, dry air is less favorable for the development of certain foliage diseases of crop plants than is moist air. Thus, the dry air combined with the greater amount of sunshine in arid and semiarid areas promotes high rates of net photosynthesis and high marketable yields. This explains, partially at least, why many vegetable and flower crop industries are situated in the arid and semiarid districts of the western United States and why the same situation exists between dry and humid areas in other parts of the world.

The Elevation of the Land

The *elevation of the land* refers to the altitude of the surface of the land above or below sea level. In general, for each 300 feet in elevation there is an average decrease of 1°F (0.6°C). Thus, with an elevation of 4,000 feet (1,220 meters) a

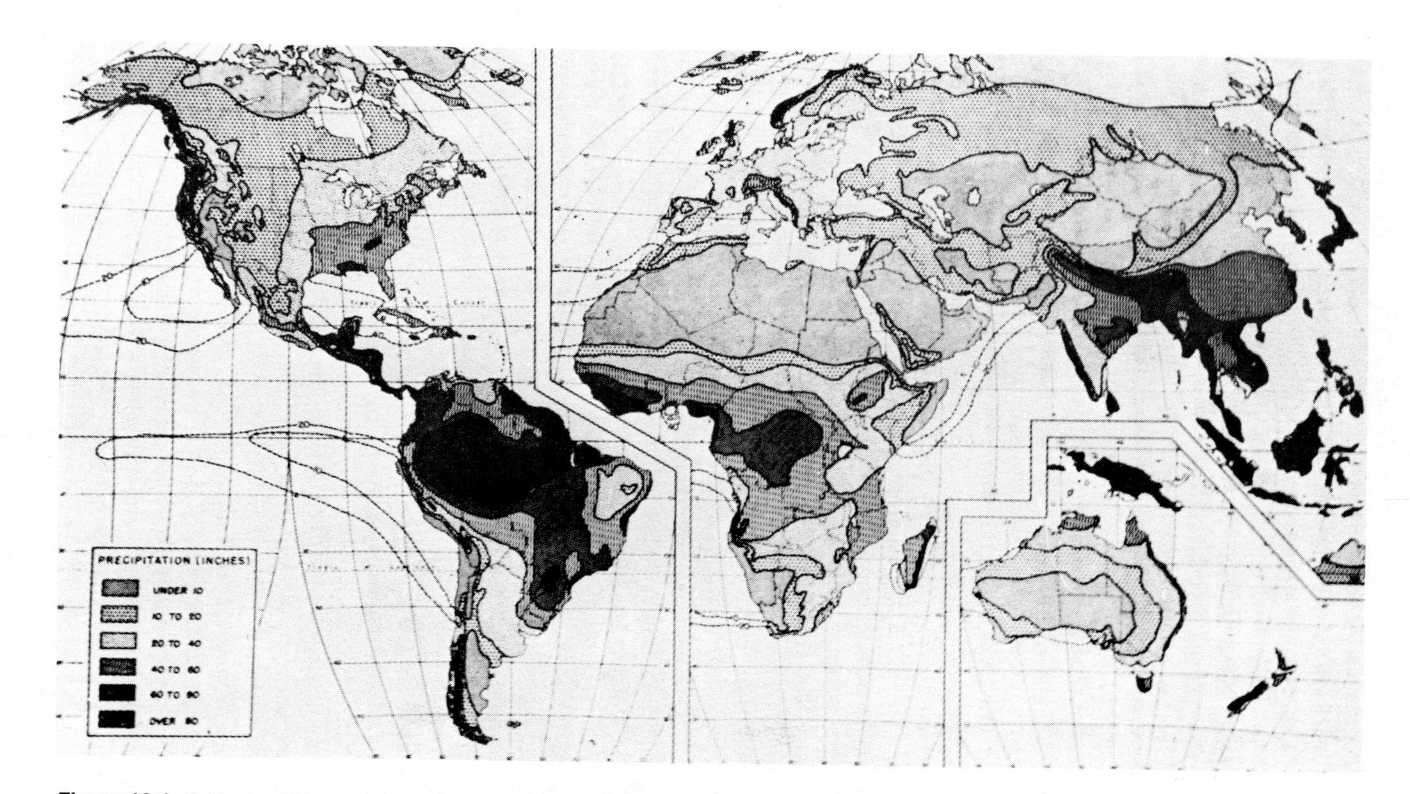

Figure 10-1 Arid, semiarid, and humid areas, based on the annual average precipitation. *(Courtesy, U.S. Department of Agriculture, Washington, D.C.)*

decrease of about 13°F (7.8°C) is obtained. These differences in elevation within any particular region are chiefly responsible for differences in the length of the frost-free growing period and the maximum temperatures that occur. For example, at Clemson, South Carolina, the elevation is 774 feet (236 meters); at Highlands, North Carolina, 50 miles west of Clemson, the elevation is 4,500 feet (1,372 meters). These differences in elevation make for marked differences in temperature between the two places. At Clemson the length of the frost-free growing period is 215 days, and the summer temperature is sufficiently high to permit the successful production of long-season, warm-temperature crops, such as sweet potato, but it is disastrous for the growing of cool-season crops, such as cabbage, potato, or garden pea. On the other hand, at Highlands the mean length of the frost-free growing period is 178 days, and the summer temperature is favorable for the production of the cool-season crops, such as cabbage and potato, but it is too low for the growing of warm-season crops, such as sweet potato. As another example, consider any one of the five major islands of the state of Hawaii. Tropical crops, e.g., cocoanut palm, banana, and papaya, thrive at the low elevations; subtropical crops, e.g., pineapple, sugar cane, and sweet potato, thrive at the intermediate elevations; and warm-temperate and even cool-temperate crops thrive at the high elevations. A similar situation exists in Mexico, Costa Rico, and Kenya, as well as in other parts of the world.

The Direction of Mountain Slopes

The *direction of mountain slope* refers to its exposure to the direction of the prevailing wind. In general, mountain slopes facing the direction of the prevailing wind, the windward slope, have a higher rainfall and a lesser amount of sunshine than slopes not facing the prevailing wind, the leeward side. For example, in Hawaii the rainfall of the windward slopes is quite high, varying from 100 to 200 inches (254 to 508 centimeters), whereas the rainfall on the leeward side is comparatively low, varying from 20 to 50 inches (51 to 127 centimeters) per year. Further, for any given mountain slope the rainfall is higher from November 1 to April 30 than it is from May 1 to October 31, and for any 24-hour period the rainfall is higher during the dark period than it is during the light period. Thus, windward slopes have more rain and lesser amounts of sunshine, and, unless the soils are well drained, excess water is likely to be a problem. Leeward slopes have relatively dry air and relatively greater amounts of sunshine, and, unless irrigation is practiced during periods of water deficits within the plants, deficient water is likely to be a problem. A similar situation exists in the western part of the contiguous 48 states. The west to east winds as they flow across the Pacific Ocean pick up large quantities of water. When they arrive at the series of mountain ranges just west of the Continental Divide, the winds are forced upward. This, in turn, lowers their temperature, and, as a result, they drop most of their moisture and flow across the plains in a relatively dry condition.

Influence of Ocean Streams

Ocean streams are rivers or currents of water which have a different temperature from that of the adjacent water. A striking example is the Gulf Stream, a current of warm water, which flows alongside currents of relatively cold water. According to authorities, its color is blue while adjacent currents are green. Opposite Miami, it extends from the shore of Florida to the Burmuda Banks and flows at the rate of four knots per hour. This unique stream originates in the Gulf of Mexico, flows northward along the east coast of Florida and in a northeasterly direction across the north Atlantic to the shores of France, Ireland, England, Scotland, Iceland, and finally Norway. (See Fig. 10-2).

In general, the Gulf Stream has a marked and beneficial effect on the climate of these countries. During the winter, the prevailing west to east winds pick up heat as they cross the Gulf Stream, and their temperature is likely to be raised above that which kills plant tissue. For example, if the temperature of the wind is 20°F (−6.7°C) when it reaches the Gulf Stream, it is likely to be 10 to 15°F higher (1.1 to 1.7°C) just after it crosses the stream. During the summer, the prevailing winds lose some of their heat as they cross the stream and become relatively cool. Accordingly, they help maintain cool conditions in adjacent lands. In general, the modifying influence of the Gulf Stream explains why the winters of Florida are relatively warm, permitting the successful production of oranges and lemons and many herbaceous crops throughout the Florida penin-

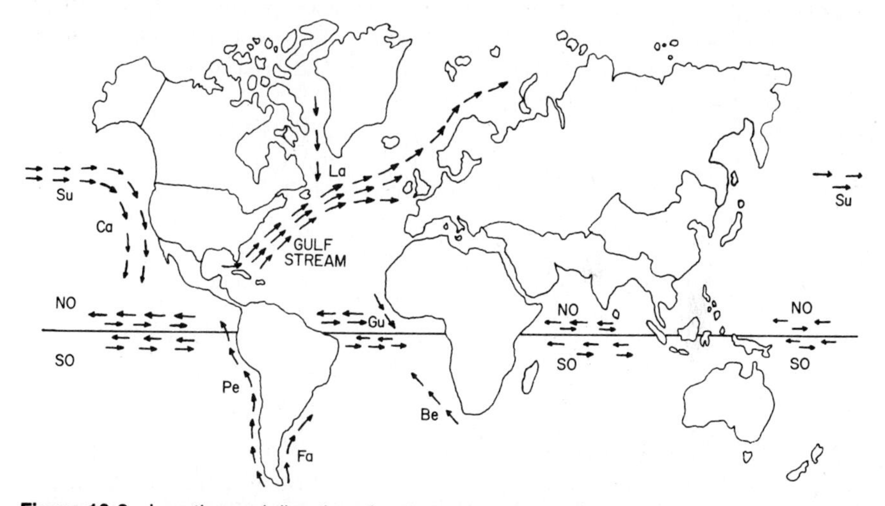

Figure 10-2 Location and direction of main ocean currents throughout the world. Be = Benquela Current. Fa = Falkland Islands Current. GS = Gulf Stream. Gu = Guinea Current. La = Labrador Current. No, So = North and South Equatorial Currents. Su, Ca = Subarctic and California Currents. *(Adapted from the world map of the National Geographic Society, Washington, D.C.)*

sula; and why home garden and commercial production of cool-temperature crops are possible in the British Isles and in latitudes as far north as the Arctic Circle.

Another important ocean stream is the sub-arctic current. In general, this current originates just south of the Aleutian Islands, flowing from west to east to the shores of Washington and Oregon and south along the coast of California. In sharp contrast to the Gulf Stream, the subarctic current is a river of relatively cold water. As the wind flows over this current, its temperature is lowered, and as it continues to flow over the land between the shore and the coastal range, it modifies the temperature accordingly. In this way, the modifying effect of the current makes possible the successful production of many high-income crops, for example, peas and snapbeans in Washington and Oregon, head lettuce in the Salinas Valley of California, seed of many flower crop plants near Lompac, California, and dried lima beans in southern California.

Influence of Large Bodies of Fresh Water

A striking example is the influence of the Great Lakes. These five lakes, or inland seas, comprise an area of about 96,000 square miles and vary in depth from a maximum of about 130 feet (40 meters) for Lake Erie to about 900 feet (274 meters) for Lake Superior. Thus, they contain an enormous quantity of water. As pointed out in Chapter 5, water has the capacity to absorb and give off large quantities of heat. In general, these large bodies of water absorb heat during the summer and give off heat during the winter. The prevailing wind over the lakes flows from northwest to northeast. Thus, as the wind flows over the lakes during the summer, its temperature decreases keeping adjacent offshore areas relatively cool. Conversely, as the wind flows over the lakes during the winter, its temperature increases keeping adjacent offshore areas relatively warm. This modifying influence of the Great Lakes, combined with the direction of the prevailing wind, explains why many deciduous tree fruit industries—apples, pears, peaches, plums, cherries, and labrusca grapes—and small fruit industries—strawberry, raspberry, and blueberry—are found in districts as far as 42° north. It also explains why these crops are more advantageously produced in districts on the east coast rather than the west coast of Lake Michigan and on the south coast rather than on the north coast of Lake Erie and Lake Ontario.

Climate and Changes in the Weather

As is well known, air flows above the surface of the earth in much the same way as water flows in rivers and small streams on the surface. For the most part there are two types of air-mass flow: (1) the rivers, or masses, of cold-dry air, and (2) the rivers, or masses, of warm-moist air. In general, masses of cold-dry air are relatively dense and as such exert a relatively high barometric pressure. For this reason they are called *high pressure areas*. In the northern hemisphere, they originate in the Arctic Circle and flow clockwise from west to east at rates varying from less than 1 to 20 miles (1.6 to 32.2 kilometers) per hour or more. In the southern hemisphere, they originate in the Antarctic Circle and flow counterclockwise, also at rates varying from less than 1 to 20 or more miles (1.6 to 32.2 or more kilometers) per hour. In both instances, their outer edge, or front is usually indicated on weather maps as shown in Figure 10-3.

In sharp contrast, masses of warm-moist air are relatively light in weight and as such exert a relatively low pressure. For this reason, they are called *low pressure areas*. They originate in the tropical and subtropical zones and flow counterclockwise from south to north in the northern hemispheres and clockwise from north to south in the southern hemispheres. In either case, they flow at rates varying from less than 1 to 20 or more miles (1.6 to 32.2 or more kilometers) per hour. Their outer edge, or front, is usually indicated on the weather map as shown in Figure 10-3.

In general, the dominance of either high or low pressure masses and their interaction determines the type of weather in any given locality. When a high pressure mass dominates in any given locality, the temperature of the area is relatively low, the air is clear, and the days are sunny. Conversely, when a low pressure mass dominates in any given locality, the temperature of the area is relatively high, the air contains large quantities of water vapor, and the days are partly cloudy, cloudy, or rainy.

Frequently, the high and low pressure areas clash. Manifestations are increasing cloudiness followed by total clouds, rain, sleet, or snow. If the mass of high pressure is greater than that of the low, the mass of low pressure is likely to disappear and that of high pressure will be reduced in size since a portion of the cold-dry air will have been used to condense some of the water vapor of the warm-moist air. If the mass of low pressure is greater than that of the high, the mass of high pressure is likely to disappear and that of the low pressure will be reduced in size since a portion of the warm-moist air will have been used to get rid of the cold-dry air. Finally, if the masses of cold-dry air and those of warm-moist air lack dominance, the rate of flow of the one is likely to stop the rate of

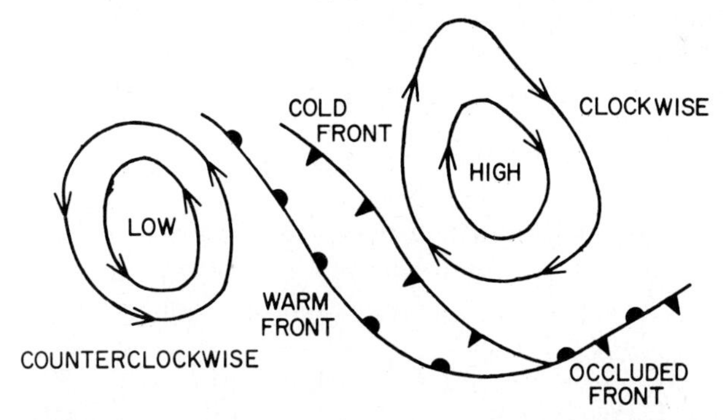

Figure 10-3 Direction of flow of high and low pressure areas in the northern hemisphere, and designation of cold fronts, warm fronts and occluded fronts.

flow of the other; in other words, the two fronts will become joined and stationery. This type of front is called an *occluded front,* as indicated on the weather map as shown in Figure 10-3. Thus, a working knowledge of the nature and behavior of high and low pressure masses provides a basis for understanding changes in the weather in a specific area and a means by which a grower can plan his operations.

SITES

Sites refer to places where any given horticulture product is produced. Examples are an apple orchard, a grape vineyard, a field of tomato or potato plants, or a greenhouse range given over to the production of roses or chrysanthemums. Sites include two factors: (1) topography and (2) soil. In horticulture, *topography* refers to the contour of the land, its elevation or depth, and similar features of the terrain. As previously stated, the soil is the home of the plant's roots and the reservoir for the plant's essential elements and water.

Kinds of Sites

In general, vegetable crops and herbaceous flowering plants are grown on level and slightly sloped sites. Most vegetable and flower crops require cultivation and pest control, and harvesting operations can be performed more efficiently on level land than on sloped land. On the other hand, deciduous- and evergreen-tree fruits are grown on sloped land. With these crops air drainage requires more consideration than it does for herbaceous crops. Note the contour lines of the young peach orchard in Figure 10-4 and the straight rows and the level land of the field in Figure 10-5.

Kind and Degree of Soil Erosion In many parts of the world soil erosion, the removal of soil particles from crop-plant fields, is a serious problem. In fact, many scientists believe that soil erosion is responsible for more of the losses in soil fertility than any other single factor. Clearly, soil erosion should be reduced to a minimum. Three types of erosion are considered: (1) erosion by wind, (2) erosion by water, and (3) erosion by gravity.

Erosion by wind is due to the carrying power of the wind. In fact, this carrying power is directly proportional to the square of its velocity. Thus, decreases in velocity will correspondingly reduce the erosive action. In general, this is done by growing cover crops and using windbreaks. Types of windbreaks are trees, shrubs, fences of boards, slats, or cloth, and strips of grain or forage crops.

Erosion by water is due to the carrying power of streams of water. Here again the carrying power is directly proportional to the square of its velocity. Thus, as with the action of the wind, a decrease in the rate of water flow correspondingly decreases the rate of erosion. In general, this decrease is

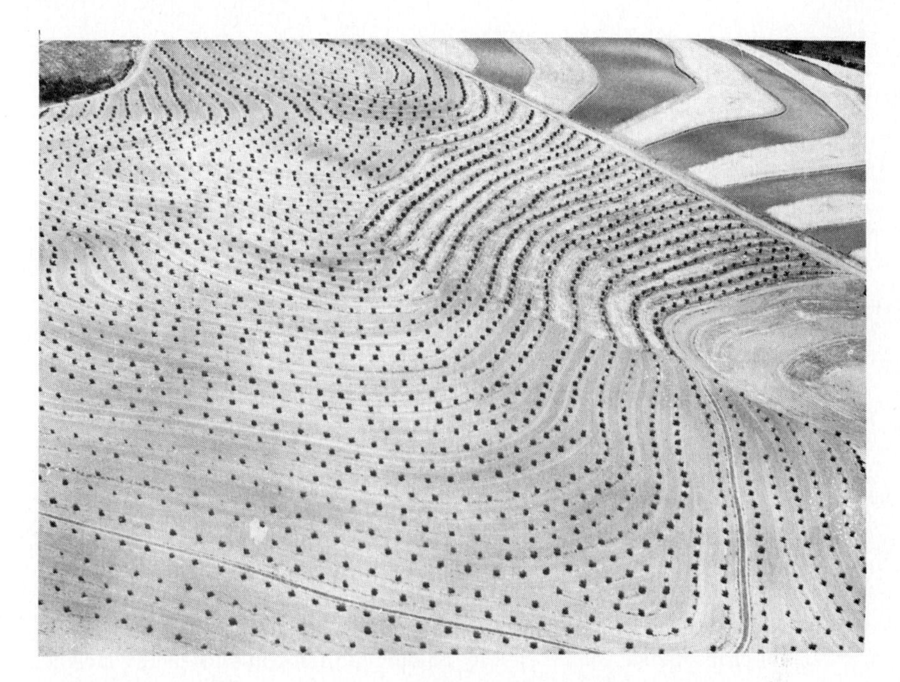

Figure 10-4 A peach orchard in the Spartanburg-Inman district of South Carolina planted on the contour. *(Courtesy, John T. Bregger, Soil Conservation Service, U.S. Department of Agriculture.)*

Figure 10-5 Part of a 640-acre field of celery in 1949, near Belle Glade, Fla. The narrow rows are 24 inches apart; the wide rows are 48 inches apart. The wide rows are necessary to accommodate the wheels of the spray rig. In this field, the height of the water table is controlled to maintain adequate supplies of water. *(Courtesy, J. C. Hoffman, Vegetable Breeding Laboratory, U.S. Department of Agriculture.)*

achieved by terracing the land, contour planting, stripcropping, and growing cover crops.

Terracing the land involves following the contours of the land and dividing the area into separate drainage systems. The function of each terrace is to reduce the rate of water flow and to hold the water within each drainage area. This allows the water which has fallen within any given area to gradually soak into the soil. Broad, relatively flat terraces are most commonly used.

Contour planting involves planting crops with the contour. In this way, the cultivation of a particular row is done on level land rather than up and down or at right angles to the slope. Contour tillage retards erosion by holding water in the small terrace made by the cultivator.

Strip cropping involves planting cultivated and noncultivated crops alternately in narrow or broad strips with the contour of the land. The noncultivated crops absorb the water falling on the noncultivated area and the runoff from the adjacent cultivated area. Thus, soil erosion is reduced. The width of each cultivated or noncultivated strip is determined by the steepness of the slope, the rate of percolation of water through the soil, the water-holding capacity of the soil, and the amount and character of the rainfall. Obviously, the steeper the slope, the lesser the rate of percolation and water-holding capacity, and the greater the rainfall, the narrower the cultivated and noncultivated strips.

Cover crops reduce soil erosion by holding the soil particles and by increasing the rate of percolation and the water-holding capacity of the soil.

Erosion by gravity is due to the force which draws all soil particles toward the center of the earth, and this force, in turn, is directly proportional to the steepness of the slope. Thus, this type of erosion is likely to be severe on steep slopes such as the slopes of highway cuts during heavy rains. Erosion by gravity is reduced by establishing flat terraces at intervals up the slope and using retaining materials or walls along the steep portion, by establishing plants which root quickly and easily in relatively infertile soil, and by providing naturally controlled diversion channels.

Air Drainage Air drainage involves the flow of air from areas of high elevation to areas of low elevation. Cold air flows from the high lands—the hills and mountains—into the low lands—the valleys. How does this flow of air take place? During the night radiation takes place, and the earth's surface cools. This, in turn, cools the layer of air next to the soil, and since cold air is heavier than warm air, the cold air flows downward and the warm air flows upward. This phenomenon is called *temperature inversion,* and when it occurs, the high lands are warmer than the low lands. (See Fig. 10-6)

For the successful production of the many tree fruits, adequate air drainage is particularly necessary. The cold air must be carried away from the immediate vicinity of the orchard, particularly during the blossoming season. Frequently during the period of flowering, the temperature of the cold air is from 1 to 5°F (0.6 to 2.8°C) below the killing temperature of the flowers and buds. At the same

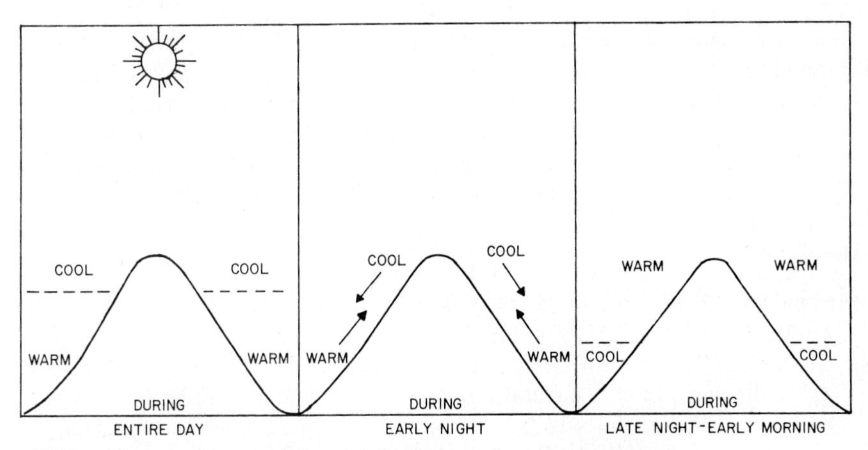

Figure 10-6 Flow of cold and warm currents of air in the absence of wind showing inversion of temperature and need for air drainage during the blossoming season.

time the temperature of the warm air is from 1 to 5°F (0.6 to 2.8°C) higher. In many orchards, observations have shown differences of from 15 to 20°F (8.4 to 11.1°C) between the top and bottom of moderate slopes, and that variations in temperature much less than these often mean the difference between live and dead trees. In general, the higher the land in relation to adjacent areas and the greater the area of low land which receives the cold air, the greater the freedom of air flow and the freedom from injury. Thus, before orchards are planted, the frost hazard of proposed sites is carefully evaluated.

Exposure *Exposure* refers to the direction of the slope. In general, a southern or eastern slope (1) promotes early blooming and early maturity of the crop, (2) offers some protection against north or west winds, and (3) permits early planting of herbaceous crops. On the other hand, a northern slope (1) delays blooming of orchard trees and (2) offers some protection against sunscald in regions of high light intensity.

SOILS

Without photosynthesis there would be no soil.

In general, soils comprise the natural mantle of decomposing minerals and organic matter on the surface of the earth. Soils came about and developed through the activities of green plants. Green plants originated and evolved in the water. In time, certain kinds established themselves on the barren rocks of the seashore. As these plants and their successors grew, they secreted compounds which decomposed and dissolved the mineral matter, and, when they died, their remains-

—organic matter—mixed with the minerals. In this way, soil came into existence. Thus, two important components of soils are (1) decomposed mineral matter and (2) decomposed organic matter. Soils which are derived from and contain large quantities of mineral matter are called mineral soils. Soils which are derived from and contain large quantities of organic matter are called muck or peat soils. Of these, mineral soils have the wider adaptation and distribution.

Mineral Soils[1]

Mineral soils consist of mixtures of sand, silt, and clay. Sand particles are relatively large (from 0.15 to 1.0 millimeter in diameter); silt particles are small (from 0.002 to 0.05 millimeter); and clay particles are very small (from 0.0001 to 0.002 millimeter). In fact, mineral soils can be classified according to the proportion of sand, silt, and clay which they contain. Principal classes are (1) sands, (2) sandy loams, (3) loams, (4) silt loams, and (5) clay loams. In horticultural crop production, each class has advantages and disadvantages.

Sands In general, sands are coarse-textured, well drained, relatively infertile, and vary in acidity. They contain 80 to 95 percent sand and 5 to 20 percent silt and clay, with 0.1 to 1.0 percent organic matter. The large proportion of sand facilitates rapid drainage, aeration, and rapid decomposition of organic matter. On the other hand, the low proportion of silt and clay and the low content of organic matter provide for relatively small quantities of available water and small quantities of potential nitrate-nitrogen. In addition, many sands are deficient in certain essential elements, particularly calcium and magnesium. Thus, the crop-producing capacity of sands is increased largely by reducing acidity, adding deficient elements, and increasing the organic matter. Despite the relatively low fertility of sands, with good management they produce satisfactorily such crops as peach, raspberry, asparagus, sweet potato, and watermelon.

Sandy Loams In general, sandy loams are moderately coarse-textured, well drained, moderately fertile, and slightly acid or slightly alkaline. They contain 50 to 80 percent sand and 20 to 50 percent silt and clay, with 0.1 to 3.0 percent organic matter. This proportion of sand is sufficient to permit rapid drainage, abundant aeration, and a moderately rapid oxidation of organic matter. At the same time, the proportion of silt and clay and the higher content of organic matter provide for a moderately high water-holding capacity and fairly high quantities of potential nitrate-nitrogen. Despite the somewhat high fertility of sandy loams as compared to sands, the crop-producing capacity is increased in much the same way as that of the sands, by reducing acidity, applying sufficient fertilizers, and maintaining adequate supplies of organic matter. Sandy loams have a much wider adaptation than sands. Most ornamental plants, small fruits,

[1]This discussion is limited to soils in the humid regions. Since the amount of organic matter in mineral soils varies greatly, some soils naturally fall outside the ranges of organic matter which are presented.

peaches, plums, nuts, market-garden crops near large centers of population, and truck crops grown in the winter-garden areas of the South thrive well on these soils.

Loams In general, loams are moderately fine-textured, moderately well drained, moderately fertile, and slightly acid or slightly alkaline. They contain 30 to 50 percent sand and 50 to 70 percent silt and clay, with 1.0 to 4.0 percent organic matter. As compared with sandy loams their lesser proportion of sand and greater proportion of silt and clay, combined with their greater organic matter content, permit moderately rapid drainage, fairly abundant aeration, a moderately rapid oxidation of organic matter, and a moderately high water-holding capacity and nitrate-nitrogen supply. In general, loams have practically the same crop adaptation as sandy loams.

Silt loams In general, silt loams are fine-textured, fairly well drained, fertile, and slightly acid or alkaline. They contain 20 to 30 percent sand and 70 to 80 percent silt and clay, with 1.0 to 4.0 percent organic matter. This greater proportion of silt and clay permits moderately slow drainage, fair aeration, and a moderately slow oxidation of organic matter. The greater amount of organic matter provides for a high amount of available water and an abundant supply of potential nitrate-nitrogen. In these soils the maintenance of good tilth is more difficult than in loams and sandy loams. In marked contrast to the sands and sandy loams, the crop-producing power of silt loams is increased by improving the drainage and aeration.

Clay Loams In general, clay loams are very fine-textured, poorly drained, fertile, slightly acid, or slightly alkaline. They contain 20 to 50 percent sand, 20 to 60 percent silt, and 20 to 30 percent clay, with 1.0 to 6.0 percent organic matter. This large proportion of silt and clay markedly decreases aeration and drainage, and the large amount of organic matter provides for a large amount of available water and potential nitrate-nitrogen. As with silt loams, the maintenance of good tilth is difficult, and the crop-producing capacity is increased by drainage and aeration.

Like the other types of soils, silt and clay loams have their place in crop production. In general, they are adapted to crops when large yields are more important than early yields. For example, most tomatoes and sweet corn for the cannery are grown on silt and clay loams. On the other hand, silt and clay loams are not well adapted to crops grown for their fleshy roots or to tree fruits, unless the subsoil is well drained and permits deep and extensive penetration of the root system. Figure 10-7 shows the proportion of sand, silt, and clay in each of six soil classes.

Soil Layers, or Horizons A vertical section of any given mineral soil shows distinct layers, or horizons. In general, these layers are the topsoil, or A

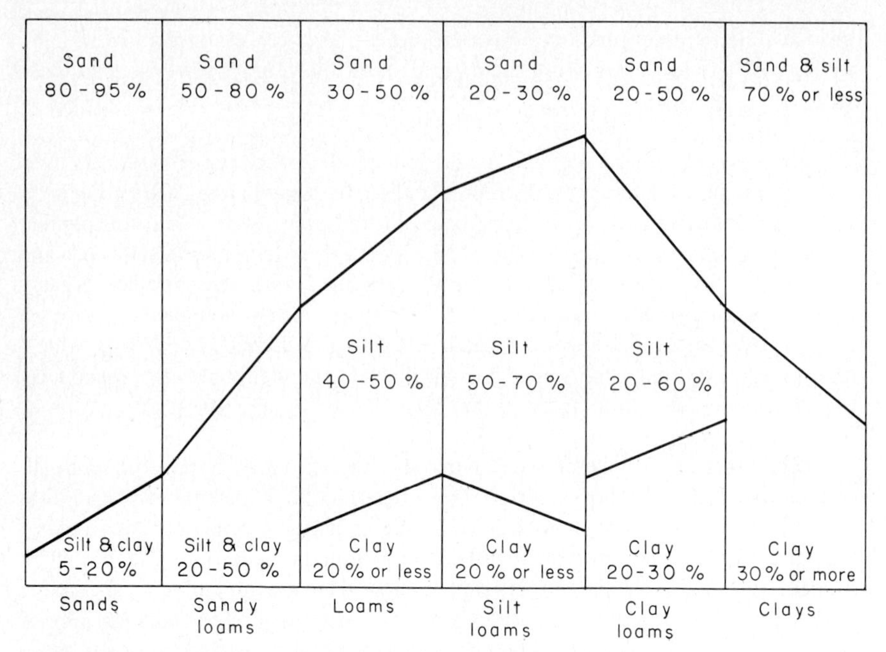

Figure 10-7 The proportion of sand, silt, and clay in various soil classes. *(Courtesy, The late Jackson B. Hester, Elkton, Md.)*

horizon, the subsoil, or B horizon, and the substratum, or C horizon. Of these, the topsoil is more fertile, more subject to weathering and tillage operations, and contains more organic matter than the other horizons. Soil organisms are, therefore, more abundant in this layer. The lower layers usually have a finer texture and a more compact structure. In some sections the subsoil becomes very compact, forming an impermeable layer called *hardpan* which retards drainage and resists root penetration.

The texture and acidity of the subsoil have an important bearing on the depth and penetration of the root system and the ability of plants to withstand drought. As previously stated, plants with deep root systems can obtain water and essential ions in the lower levels of the soil. In this way, the plants possess the capacity to withstand drought during hot weather, particularly when the transpiration rate is high. Tight, impervious subsoils resist root penetration, and highly acid subsoils are unfavorable for root growth. In many areas subsoils are limed in order to make them more favorable for the growth and deep penetration of the roots.

Muck or Peat Soils

Muck or peat soils are composed of partially or highly decomposed plant remains which have accumulated through the ages in low, wet places and contain from 20 to 70 percent organic matter. If the soil has a fine texture and the plant remains are in an advanced state of decomposition, it is called *muck*. On the other hand, if the vegetable remains are still coarse and fibrous so that they mat together, the soil is called *peat*. In general, peat is used for mulching ornamental trees and shrubs, for growing crops in greenhouses, and for shipping herbaceous and ornamental plants. Mucks, however, are more valuable for horticultural crop production. They are classified according to their acidity or lime content. There are three kinds: (1) strongly acid, or low-lime, (2) moderately to slightly acid, or high-lime, and (3) alkaline.

Low-lime Mucks In general, low-lime mucks are strongly acid (pH 3.0 to 5.5). Because of this high acidity, comparatively few horticultural crops are adapted to them, e.g., cranberry and blueberry. Cranberries are grown extensively on the muck soils of New Jersey and Wisconsin.

High-lime Mucks High-lime mucks are moderately to slightly acid (pH 5.5 to 6.8). In sharp contrast to the low-lime mucks, the acidity of these mucks is within the optimum reaction range for the growth of many horticultural crops, e.g., celery, onion, cabbage, lettuce, and carrot. In fact, most of the commercial crops of onions and celery grown in New York and Michigan are produced on these so-called high-lime mucks.

For any particular crop the management of mucks differs greatly from that of mineral soils. In general, mucks require packing when prepared for planting, maintenance of the water table close to the surface, application of large quantities of potash, and protection from wind erosion. Packing of the soil is necessary to close air spaces in the muck and to facilitate capillary action. The high water table, usually kept within 2 to 4 feet (.6 to 1.2 meters) from the surface, is necessary to supply the plants with adequate water, to help prevent injury from summer frost in northern climates, and to reduce wind erosion. Applications of large quantities of potash are necessary since muck soils are notably deficient in potassium. Mixtures frequently used are 0-8-24, 0-10-20, 2-8-16, 3-9-18, and 3-12-12.

Alkaline, or Sapric, Mucks As the name suggests, alkaline mucks are alkaline in reaction (pH 7.1 to 8.2). Because of this alkalinity, these mucks are generally unadaptable for the growing of horticultural crops. In some instances, slightly alkaline mucks have been made productive by applications of flowers of sulfur.

QUESTIONS

1. From a horticultural standpoint, state briefly the advantages and the limitations of each of the four climatic zones.
2. List the arid, semiarid, and humid areas shown in Figure 10-1.

3 The dry summer climates of the Pacific coast of the United States and the Mediterranean coast of France are particularly adapted to the raising of a large number of horticulture products. Explain.

4 The vinifera grape thrives in southern France and northern Italy. It does not thrive in England and northern France, except in glasshouses. Explain.

5 Many cool-temperate trees and small fruits thrive in England and Scotland. Explain.

6 Deciduous-fruit production is more extensive and varied on the east side of Lake Michigan and on the south side of Lake Ontario than it is on the opposite side of each. Explain.

7 How do large bodies of water and the direction of the prevailing wind influence the climate of the land located nearby?

8 A sloped topography is most desirable for tree fruits. Explain.

9 How do sloped sites protect the blossoms of fruit trees—apple, peach, and tung—from killing temperatures?

10 Open sites for fruit trees are generally more satisfactory than closed sites. Explain.

11 For fruit trees, a northern slope may be more desirable in one section of a country and a southern slope more desirable in another. Explain.

12 Level sites are generally used for vegetable and flower crops. Explain.

13 What is the main difference between sand, silt, and clay?

14 Which type of soil, a sandy loam or a clay loam, would result in earlier reproductive growth in the tomato or capsicum pepper? Explain.

15 The texture and drainage of the subsoil is particularly important to the grower of tree fruits. Explain.

16 What properties do muck soils and the heavy loams have in common?

17 What properties do muck soils and the light loams have in common?

18 Onions, lettuce, celery, cabbage, and carrots thrive exceptionally well on slightly acid muck soils. They do not thrive on very acid or very alkaline mucks. Explain the different responses of these crops?

SELECTED REFERENCES FOR FURTHER STUDY

Hambridge, G. 1941. Climate and man. *U.S. Dept. Agr. Yearbook.* Discussions on climates of the world, climate in relation to fruit and vegetable production, and the scientific basis of modern meteorology.

Stefferud, A. 1957. Soil. *U.S. Dept. Agr. Yearbook.* A nontechnical and easily understandable discussion by 143 authorities on soils and their relation to all types of crop-plant production.

Stout, G. J. 1958. *Successful truck farming.* New York, N.Y.: Macmillan, chap. VI. An instructive account by an outstanding teacher on the relation of frosts and the selection of sites to vegetable crop production.

Chapter 11

Soil Organic Matter and Commercial Fertilizers

We have received the world as an inheritance. None of us has a right to damage it–and everyone has the duty to leave it in an improved condition.

Adapted from Joseph Joubert

SOIL ORGANIC MATTER

Organic matter in the soil is derived from plant and animal remains and is a mixture of these materials at various stages of decomposition. The compounds present are those which were part of living tissues: the carbohydrates, lipids, proteins, and related substances. In the process of decomposition, some of these compounds are oxidized to their end products and others to an intermediate product called *humus*. Therefore, organic matter in soils consists of a mixture of compounds which become completely oxidized or are changed to humus.

COMPLETELY OXIDIZED COMPOUNDS

In general, compounds which become completely oxidized are the relatively simple forms—the sugars, starches, hemicelluloses and related compounds, and simple proteins. These compounds are decomposed by the heterotropic organisms in soils—most bacteria, all fungi, and all actinomycetes. Like all living things, these organisms must have a source of energy for their vital needs. They obtain this energy from the organic compounds originally made by green plants. This decomposition is biochemical and is illustrated as follows:

Sugars + O_2 → CO_2 + H_2O + heat and other forms of kinetic energy
Simple proteins + O_2 → CO_2 + H_2O + NH_3 + heat and other forms of kinetic energy
Complex proteins containing sulfur + H_2O + O_2 → CO_2 + H_2O + NH_3 + H_2S + minerals + heat and other forms of kinetic energy

Note that oxygen is needed for the decomposition of the sugars and simple proteins. Both oxygen and water are needed for the decomposition of the complex proteins, and carbon dioxide, ammonia, hydrogen sulfide, and certain minerals are the end products. As previously explained, carbon dioxide is one of the essential raw materials for crop plants. Ammonia is changed to nitrate-nitrogen; hydrogen sulfide is changed to sulfate-sulfur; and the minerals, particularly calcium, magnesium, and potassium combine with certain anions in the formation of salts or remain in the soil solution as dissociated ions. Compounds which are completely oxidized have at least two important uses in crop production: (1) they provide the heterotropic organisms in soils with a source of energy; and (2) they supply crop plants, either wholly or in part, with essential elements.

FACTORS AFFECTING THE RATE OF DECOMPOSITION

Soil Temperature

The effect of temperature on the rate of respiration of heterotropic soil organisms and the consequent rate of decomposition of organic matter follows rather closely the van't hoff's law. This law states that for every increase of 10°C from 0 to 35°C the rate of a process increases two or more times. Thus, the rate of decomposition of organic matter will be at least twice as rapid at 30°C (86°F) as at 20°C (68°F), and twice as rapid at 20°C (68°F) as at 10°C (50°F). How can this temperature effect be applied to the rate of decomposition of organic matter under various climatic conditions? In general, higher soil temperatures exist in warm climates than in cool climates, thus the rate of decomposition is greater in warm climates than in cold climates. Numerous investigations have shown this to be the case. Consequently, with other factors comparable, applications of organic matter in warm climates should be more frequent than in cool climates, and in the same climate, more frequent in greenhouses than outdoors.

Soil Aeration and Moisture

The decomposition of organic matter in soils requires free oxygen and water. The oxygen is needed for the respiration of the heterotropic organisms, for the nitrifiers in changing ammonia to nitrate-nitrogen, and for the sulfur bacteria in changing hydrogen sulfide to sulfate-sulfur. Water is needed for cell division of these organisms and for the absorption of essential raw materials, such as nitrate-nitrogen and phosphate-phosphorus. As previously explained, oxygen is a component of the soil air, and this air, together with water, occupies the pore spaces of the soil. Thus, if the pore spaces are full of water, the oxygen supply will be low and the rate of decomposition will be accordingly low. On the other hand, if the pore spaces contain very little available water, the rate of decomposition will be low also. In general, saturated soils have inadequate quantities of oxygen and dry soils have inadequate quantities of water, whereas moist soils have adequate quantities of both oxygen and water.

Soil Reaction

Soil reaction refers to the degree of acidity and is expressed by pH. Soils may be very acid, moderately acid, slightly acid, slightly alkaline, or very alkaline. As stated in Chapter 7, the degree of acidity of the soil solution markedly influences the growth and activity of many kinds of soil organisms. For example, the heterotropic soil organisms and the chemosynthetic soil organisms, such as the nitrifying bacteria, are more active and abundant in moderately acid and slightly acid (pH 6.0 to 6.5) soils than they are in strongly acid (pH 4.0 to 4.5) or alkaline (pH 8.0) soils. Thus with other factors favorable, the rate of decomposition is most rapid in moderately acid or slightly acid soils.

The Chemical Composition of Organic Matter

Green-manuring crops, crop residues, and organic mulches vary greatly in chemical composition. In general, young green-manuring crops contain relatively large quantities of sugars, starches, and simple proteins—compounds which decompose rapidly; and green-manuring crops in the mature stage contain relatively small quantities of sugars and simple proteins and relatively large quantities of cellulose—a compound which decomposes less rapidly. Crop stubble, stover, or sawdust, however, consists largely of lignocellulose and lignin—compounds which decompose slowly. Thus, succulent crops or residues may be expected to decompose more rapidly under favorable conditions than mature green-manuring crops, crop stubble, stover, or sawdust.

NATURE AND VALUE OF HUMUS

Humus is an intermediate compound of decomposition; that is, it is made by the action of soil organisms on the organic residues contained in or on the soil. The individual particles of humus are in the colloidal state, they are relatively stable, and they are closely related chemically to the lignoproteins. Humus is important as a source of cation exchange, as it affects phosphate availability and improves water-holding capacity and soil aeration.

Cation Exchange

The humus particles are adsorbed on the colloidal clay particles and form a humus-clay colloidal complex. These particles have a large external and internal surface which contains numerous negative charges. As would be expected, these charges attract positively charged ions, and most of these ions are essential for plant growth. Thus, the humus-clay colloids serve as a storehouse for certain essential ions.

The essential ions adsorbed on the surface of the colloids include calcium, magnesium, potassium, ammonium, and others. In fertile soils, these essential ions are present in optimal amounts for plant growth. This situation is shown in simplified form in Figure 11-1. As plants absorb these essential ions, they exchange them for hydrogen ions. For example, for the exchange of one calcium ion or one magnesium ion, two hydrogen ions are needed; and for the exchange of one potassium ion or one ammonium ion, one hydrogen ion is required. Thus, as the plants absorb these essential cations, the surfaces of the colloidal particles contain more and more hydrogen ions. This explains why the removal of cations by crops tends to make soils more acid and why these essential ions are supplied in the form of commercial fertilizers and lime.

Plants require free energy in order to exchange hydrogen ions for essential cations on the colloidal surfaces. This energy comes from the sugars which are decomposed in respiration. Thus, both gross photosynthesis and respiration are needed for the absorption of essential ions from colloidal surfaces. This explains why plants with high rates of gross photosynthesis and normal rates of respiration have a greater capacity for the absorption of essential ions than plants with low rates of gross photosynthesis and normal rates of respiration; why plants with dark green or healthy leaves have a greater capacity for the absorption of essential ions than plants with light green or diseased leaves; and why the root system of most crops require well-drained soils, due to the fact that carbon dioxide given

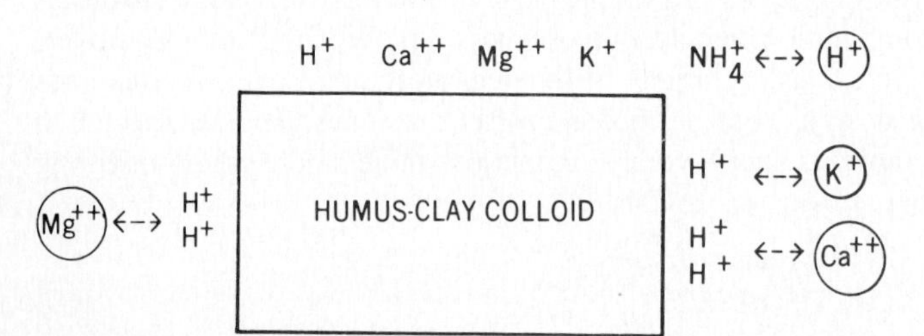

Figure 11-1 The adsorption of essential cations from colloidal surfaces. Note that hydrogen ions are exchanged for these cations.

off by respiration should diffuse rapidly from the roots and oxygen required by respiration should diffuse rapidly to the roots.

Phosphate Availability

In general, phosphate ions remain in an available form between pH 5.5 and 7.0. In very acid soils, usually less than pH 5.0, compounds containing aluminum, iron, and sometimes manganese become soluble and may result in highly toxic concentrations of these ions. Humus decomposes, or degrades, forming humic acids which, in turn, combine with aluminum, manganese, and iron to form humates of these ions. These substances are relatively stable. In this way, the aluminum, manganese, and iron ions are taken out of solution, and the phosphate ions, which would otherwise form the relatively insoluble aluminum manganese and iron phosphates, remain available to plants. In other words, under the foregoing conditions, humus reduces the amount of phosphorus that is combined in an unavailable state in the very acid soils; hence, greater quantities are available for the crop. Therefore, under these conditions humus is beneficial for crops which require very acid soil.[1]

Aluminum Toxicity

When a relatively large concentration of aluminum ions exists in the soil solution, horticultural plants grow slowly and yield poorly. As stated previously, the concentration of these ions is dependent largely on the degree of acidity of soils. In many soils aluminum ions begin to appear in toxic concentration when the pH is at or slightly below 5.0. Studies at the Virginia Truck Station show that humus reduces the concentration of aluminum ions in very acid soils. For example, Norfolk fine sand at pH 5.0 with 1 percent organic matter had 875 milligrams of AL_2O_3 per 100 grams of soil, whereas the same soil with 5 percent organic matter had only 0.2 milligrams. Plants growing in the soil with 1 percent organic matter failed to grow normally, whereas those growing in the soil with 5 percent organic matter made satisfactory growth. Evidently, humus is very helpful in reducing or entirely preventing aluminum toxicity. Manifestly, the humus unites with the aluminum ions in the formation of a relatively stable colloid. In this way, toxic concentrations of aluminum ions are taken out of solution and rendered harmless to crop plants.

Water-Holding Capacity

Humus particles possess a large surface in proportion to their weight. This surface is hydrophilic. Thus, humus possesses the ability to adsorb considerable quantities of water. In fact, humus has approximately four times the water-holding capacity of some clay colloids. This explains why soils high in humus

[1]In the potato belt of eastern Virginia, the soils are kept at pH 5.0 to 5.5 because of the presence of scab.

have a relatively high water-holding capacity; why the addition of humus, particularly to sandy soils, is likely to increase the water-holding capacity; and why crop plants growing in soils high in humus withstand short periods of drought without the application of irrigation water.

Aeration

Aeration refers to the amount of air in the soil. As stated in Chapter 4, the air and water of a soil occupy the space between the soil particles. If a soil is fully saturated, no air is present; hence, the quantity of air in the soil is inversely proportional to the water content. The principal gases of soil air are nitrogen, oxygen, and carbon dioxide. This air is comparatively high in carbon dioxide (0.5 to 10.0 percent) and low in oxygen (5.0 to 19.0 percent). The oxygen supply must be constantly renewed since root systems and most soil organisms require abundant oxygen for their respiration. Organic matter increases the pore space of soils and thereby improves the rate of exchange of gases in the soil. Thus, mineral soils high in organic matter are likely to have improved aeration as compared to mineral soils low in organic matter.

THE CARBON-NITROGEN RATIO OF ORGANIC MATTER

When organic matter is incorporated into soil, the ratio of carbon to nitrogen is usually quite wide (approximately 50:1). As decomposition proceeds, relatively large quantities of carbon dioxide are liberated, and relatively small amounts of ammonium-nitrogen and nitrate-nitrogen are formed. Decomposition continues until the ratio of carbon to nitrogen is about 10 parts carbon to 1 part nitrogen. When this ratio has been attained, further decomposition results in parallel rates of carbon and nitrogen production. In other words, organic matter in its more advanced stage of decomposition has a ratio of 10 parts carbon to 1 part nitrogen.[2]

How do green-manuring crops which vary in the proportion of carbon to nitrogen differ in the amount of carbon dioxide given off during their decomposition? Suppose two crops of the same tonnage but with different ratios are turned under. Lot A has a ratio of 80:1, and lot B has a ratio of 20:1. Which lot will give off the greater amount of carbon dioxide? Since the soil organisms will reduce each lot to the same ratio of approximately 10:1, much more carbon dioxide will be released from lot A than from lot B. Hence, lot B will supply more humus than lot A. In general, organic matter possessing narrow ratios will be more effective in building up humus content of soils than will material with wide ratios. A classification of organic matter based on the C:N ratio is shown in Table 11-1. Note that various types of organic matter possess different C:N ratios and that the same crop has a narrow ratio at the early stages of growth and a wide ratio at the later stages.

[2]Some authorities state 12:1.

Table 11-1 Organic Matter of Various C:N Ratios

Organic matter low in C, high in N (a very narrow ratio)	Organic matter high in C, high in N (a narrow ratio)	Organic matter high in C, low in N (a wide ratio)	Organic matter very high in C, low in N (a very wide ratio)
Liquid manure (10:1) Legumes in early stages of growth (15:1–20:1)	Legumes in late stages of growth (20:1) Nonlegumes in early stages of growth (20:1)	Rotted straw and rotted leaves (60:1) Nonlegumes in late stages of growth (60:1)	Straw (80:1) Stubble (80:1) Strawy manure (80:1) Leaves (80:1) Sawdust (400:1)

Soil Organisms and Plant Growth

Soil organisms compete with higher plants for available nitrogen and other essential elements. In fact, plant scientists have shown that the needs of most soil organisms are met first. This ability of soil organisms to successfully compete with crop plants for available nitrogen may or may not be beneficial. This fact should be kept in mind in management decisions regarding organic matter additions. If the plants have a slow growth rate due to an inadequate available nitrogen supply, the utilization of nitrate-nitrogen by soil organisms will further reduce the already short supply of nitrogen. However, if the plants are making unduly vigorous vegetative growth because the available nitrogen supply is in excess, the utilization of the nitrogen by soil organisms will be beneficial. Such behavior by soil organisms explains why symptoms of nitrogen deficiency may occur on crops which have been planted immediately after a mature green-manuring crop has been turned under; why mulching crops with raw straw or undecomposed sawdust is likely to retard vegetative growth and, in extreme cases, produce symptoms of nitrogen deficiency; why applications of readily available nitrogen will facilitate decomposition of mature green-manuring crops; and why a mulch of straw slows down vigorous vegetative growth of trees in an orchard or plants in a vineyard in the late summer or fall. Note the effect of applying available nitrogen to a green-manuring crop of mature rye on nitrate formation in Figure 11-2.

QUESTIONS

1 Name the two components of organic matter in mineral soils.
2 Write the equation illustrating the decomposition of proteins containing sulfur.
3 Show how nitrogen and sulfur in organic matter are made available by the activity of soil organisms.

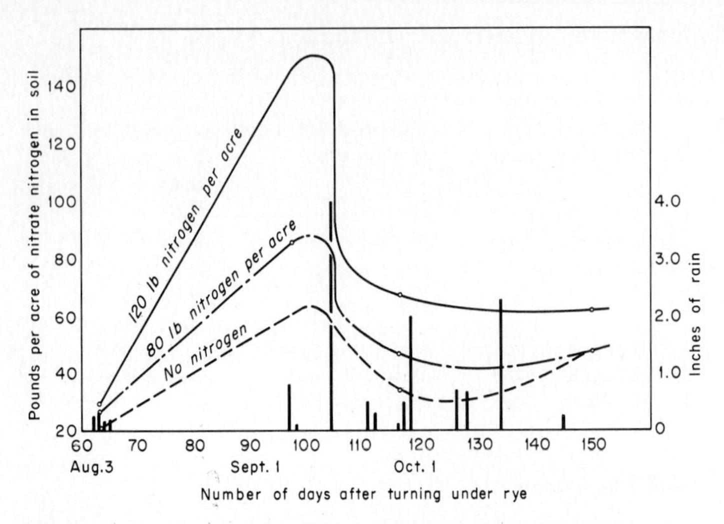

Figure 11-2 The effect of applying varying amounts of available nitrogen to a green-manuring crop of mature rye on the formation of nitrates. Note the effect of the 4-inch (10 centimeter) rain on the leaching of nitrates. *(Adapted from Figure 248,* Virginia Truck Ex. Sta. Bul. *94, 1937.)*

4 In general, organic matter decomposes more rapidly in the southern United States than in the northern United States and more rapidly in greenhouses and hotbeds than outdoors. Explain.

5 Organic matter decomposes more rapidly in warm, moist soils than in cool, wet soils. Give two reasons.

6 The comparatively high rate of decomposition of organic matter in the southern United States should be considered an asset rather than a liability. Explain.

7 What is humus?

8 What is meant by *cation exchange?*

9 In general, plants with high rates of gross photosynthesis combined with normal rates of respiration have a greater capacity to adsorb essential ions on colloidal surfaces than plants with low rates of gross photosynthesis combined with low rates of respiration. Explain.

10 In general, plants with dark green or healthy leaves have a greater capacity to adsorb essential ions on colloidal surfaces than plants with light green or diseased leaves. Explain.

11 In general, plants growing in well-drained soils have a greater capacity to adsorb essential ions than those growing in saturated soils. Explain.

12 How does humus increase the water-holding capacity and aeration of soils?

13 Investigations at the Oklahoma Station have shown that potatoes planted immediately after the plowing under of a mature crop of rye showed symptoms of nitrogen deficiency. Explain.

14 A college professor bought 12 tomato plants growing in cardboard bands for transplanting in his garden. He planted six with the bands and removed the bands from the remainder. The plants set with the bands became stunted and grew poorly, and those set without the bands grew satisfactorily. Explain.

PRODUCTION OF ORGANIC MATTER

Before organic matter is incorporated into the soil, it is produced by green plants. This incorporation takes the form of either a crop or crop residue, or animal or composted manure.

Growing Crops

Crops grown for the specific purpose of adding organic matter to soils are called *soil-improving crops*. These crops have two prime functions: to provide a cover for the soil, called *cover crops* and to add organic matter, called *green-manuring crops*. In certain horticultural enterprises, such a crop may be grown primarily to provide a cover for the soil; for example, in the production of apples or peaches this crop may reduce erosion, reduce dust residues on leaves that may harbor insects, or enhance critical operations such as spraying or dusting. In other enterprises, such as in the growing of potatoes or cabbage, the crop may be grown primarily to add organic matter. The student should remember, however, that both types of crops provide a cover for the soil and add organic matter; each is described according to the main purpose for which it is grown.

Other Effects from Use of Soil-Improving Crops

Soil-improving crops are also used for conserving the supply of essential ions, reducing erosion, checking vegetative growth of crop plants, and retaining snow in cold climates.

Conserving Supplies of Certain Essential Elements The principal ions conserved are nitrate-nitrogen, phosphate, sulfate, calcium, magnesium, and potassium. Nitrate-nitrogen is conserved in the following manner. The crop absorbs the nitrates. In the plant these nitrates are changed to ammonium forms which combine with sugars to form proteins. These proteins are used for making protoplasm and reserve foods. Thus the nitrogen in the absorbed nitrates becomes part of the plant body. When the soil-improving crop is turned under, decomposition sets in and the proteins are changed to nitrate-nitrogen by soil organisms. The conservation of nitrates is illustrated as follows:

$$\underset{\text{(in soil)}}{NO_3-N} \rightarrow \underset{\substack{\text{(plant body)} \\ \text{(conservation)}}}{\text{protein}-N \rightarrow NH_3} \rightarrow \underset{\text{(in soil)}}{NO_3-N}$$

The amount of nitrate-nitrogen conserved is quite marked. Lysimeter experiments at the West Virginia Station have shown that soils of orchards without a cover crop may lose as much as 100 pounds (45.4 kilograms) of nitrate-nitrogen per acre per year and that the same soil with a cover crop may lose only 10 to 20 pounds (4.5 to 9 kilograms) per acre per year.

Calcium, magnesium, and potassium are conserved when nitrates are conserved. The nitrate ion is negatively charged, is not held by soil colloids, and unites with positively charged minerals. In the soil these minerals are calcium, magnesium, and potassium. For every pound of nitrate-nitrogen leached, 0.7 pound (318 grams) of calcium, or 0.4 pound (181 grams) of magnesium, or 1.3 pounds (590 grams) of potassium, or equivalent parts of all three are also leached.

Phosphates and sulfates are conserved in much the same way as nitrates. The soil-improving crop absorbs phosphorus and sulfur in the form of phosphates and sulfates. In the plant, phosphorus and sulfur are used to make certain proteins. When the crop is turned under, the proteins containing phosphorus and sulfur decompose, and the phosphorus and sulfur are changed to phosphates and sulfates. Here again, the amount absorbed by soil-improving crops is quite marked. Investigations at the Virginia Truck Station have shown that a crop of sorghum yielding over 4 tons (3.6 metric tons) of dry matter per acre absorbs about 20 pounds (9 kilograms) of phosphorus. This is more phosphorus than an acre of potatoes or other horticultural crop requires.

Retarding Erosion Soil-improving crops reduce erosion in the following manner. Since the plants are grown close together, their roots ramify practically all the topsoil and a considerable part of the subsoil. In this way the roots serve as a fine, interlacing network which holds the soil together. In other words, they bind the soil. In addition, soil-improving crops reduce the packing effect of heavy rains, check the flow of water, spread it, and filter it from its load of silt. Thus, these soil-improving crops not only increase the fertility of the soil but also hold the soil.

Checking Vegetative Growth of Crop Plants Soil-improving crops are used as cover crops in the production of apples, pears, peaches, grapes, pecans, and citrus fruits. When soil-improving crops are planted during the latter part of the growing season, they check the growth of the crop plants. The cover crop absorbs the nitrates, minerals, and water which may otherwise be used by the crop. This reduction in the essential-element supply, or in the water supply, or in both reduces the rate of cell division. This reduction in the rate of cell division correspondingly reduces the rate of carbohydrate utilization and facilitates carbohydrate accumulation. As stated previously, carbohydrate accumulation is necessary for resistance to adverse conditions of the winter and for growth of roots, flowers, leaves, and twigs the following spring. The planting of cover crops to check vegetative growth of trees in orchards and of vines in vineyards is a standard practice in many parts of the world.

Retaining Snow in Cold Climates In general, a blanket of snow in the orchards or vineyards in the northern part of the United States is distinctly beneficial. The layer of snow keeps the soil comparatively warm, and, in this way, it protects the roots of orchard and vineyard plants from winter injury due to extremely low temperatures. Cover crops vary in their ability to hold snow. For example, in Michigan, amber sorghum and Sudan grass have been found more satisfactory than oats, vetch, rape, soybean, or buckwheat.

Selection of Soil-improving Crops

The selection of a soil-improving crop depends on many factors. Most important are (1) kind of crop (legume or nonlegume), (2) adaptation to soil and climate, and (3) type of horticultural enterprise.

Kind of Crop As previously stated, soil humus consists of lignin and a complex protein. Consequently, crop components high in these materials are more satisfactory as soil-improving crops than those comparatively low. In other words, crops which have comparatively narrow N:C ratios are better builders of soil organic matter than crops with comparatively wide ratios. In general, for the same state of maturity, legumes have narrow ratios and nonlegumes have wide ratios. In addition, the growing of legumes is a means by which nitrogen can be supplied to soils and plants.

Adaptation to Soil and Climate Soil-improving crops differ in their acidity and temperature requirements. Some crops thrive in slightly acid soils only; others thrive in both moderately acid and slightly acid soils. Some thrive in warm weather, others in cool weather. A classification based on soil acidity and temperature requirements follows.

Crops Which Thrive on a Wide Range of Acidity (pH 5.0–7.0)

Cool Crops
- Legumes—crimson clover, crown vetch, Austrian winter pea, Caley pea, and Creole pea
- Nonlegumes—rye, oats, rape, and buckwheat

Warm Crops
- Legumes—cowpea, lupine, soybean, velvet bean, and lespedeza
- Nonlegumes—amber sorghum, Sudan grass, and millet

Crops Which Thrive on a Narrow Range of Acidity (pH 6.5–7.0)

Legumes—alfalfa, sweet clover, red clover, white clover, Ladino clover

Type of Horticultural Enterprise Practically all types of horticultural enterprises use soil-improving crops. For example, many tree fruits, such as apple, pear, peach, and citrus, are produced under the controlled-cover crop system of soil management. This system consists of growing a cover crop for a definite period and keeping its growth under control. With the apple, pear, and peach the cover crop is disked or mowed in the spring and summer and another crop is started the following fall. Disking or mowing the cover crop in the spring and summer eliminates competition for water and essential raw materials. Cover crops used are winter vetch, cowpea, rye, oats, Sudan grass, sorghum, millet, buckwheat, and turnip. With citrus in Florida, the cover crop is grown in the summer and clean cultivation is practiced during the winter. During the summer the rainfall is sufficient for the growth of both the trees and the crop, but during the winter the rainfall usually is sufficient only for the growth of the trees. Cover crops used are beggar weed, and velvet bean. Note the cover crop in the apple orchard in Figure 11-3.

Vegetable crops grown in the northern regions are produced in the summer, and the soil-improving crop is started in the fall. For example, in the market-garden district of Long Island, New York, potatoes, cauliflower, carrots, and other vegetables are planted in the spring and harvested in the fall. Rye or a mixture of rye and vetch is planted immediately after the crop is harvested and is plowed under in the spring. These soil-improving crops are particularly satisfactory in this district, since they grow well in cool weather and are adapted to moderately acid and slightly acid soils. On the other hand, vegetable crops produced in the winter-garden areas in the southern regions are produced in the late fall, winter, and early spring, and the soil-improving crop is grown in the summer. For example, in the Mobile Bay truck-crop district of southern Alabama, potatoes are produced in considerable quantity. The seed pieces are planted in February and the tubers are harvested in May and June. Crotalaria or some other warm-season crop is planted immediately after the potatoes are harvested and allowed to grow until the fall and early winter.

Figure 11-3 A cover crop in an apple orchard. *(Courtesy C. Richard Unrath, North Carolina State University.)*

Production of Manure

Animal Manure Animal manure, just like organic matter, consists of a heterogeneous mass of organic compounds in various states of decomposition. Some of these compounds decompose quickly; others decompose slowly and finally change to humus. Thus, the application of manure supplies essential ions to soil organisms and crop plants. However, it is extremely low in these ions when compared with the commercial fertilizer mixtures on the market. For example, a ton of horse manure of average composition contains about 10 pounds (4.5 kilograms) of nitrogen, 5 pounds (2.3 kilograms) of phosphoric acid, and 10 pounds (4.5 kilograms) of potash. This has the analysis of 0.5 percent nitrogen, 0.25 percent phosphoric acid, and 0.5 percent potash. Moreover, only about one-half of the nitrogen, one-sixth of the phosphoric acid, and one-half of the potash are immediately available to plants. In addition, animal manure is not a balanced fertilizer since it is low in phosphorus.

Before the advent of the automobile industry, many growers of horticultural crops, particularly market gardeners near large cities, obtained large supplies of manure from the many livery stables. In fact, this manure was the only material used to supply the fertilizer requirements of the crops. However, the development of the fertilizer industry has enabled growers to depend less on manure and more on commercial fertilizers. Many experiments with vegetable crops have shown that applications of moderate amounts of manure (10 to 15 tons per acre or 22.5 to 33.7 metrictons per hectare), combined with the use of commercial fertilizers (500 to 1,000 pounds per acre or 560.2 to 1,120.4 kilograms per hectare), have produced greater yields than the use of heavy applications of manure alone (20 to 40 tons per acre or 44.9 to 89.8 per hectare.)

Artificial Manure Artificial or synthetic manure consists of the decomposition of organic matter under more or less controlled conditions. Various types of organic matter may be used, e.g., straw, hay, cornstalks, weeds, lawn clippings, and leaves. The organic matter is treated layer by layer with commercial fertilizers and lime and may or may not be wetted down by artificial applications of water. As previously stated, organic matter is decomposed by fungi and bacteria. Many factors influence their activity. For the making of manure, important factors are (1) nature of the plant material, (2) amount of available nitrogen and other fertilizer materials, (3) moisture supply, and (4) temperature. In general, crop residues high in nitrogen, green material, or warm, moist material will decompose more rapidly than residues low in nitrogen, dry material, or cool, wet material, respectively. Of these factors, the water supply seems to be particularly important. In fact, experiments have indicated that moisture supply seems to be a limiting factor more frequently than temperature. For example, in June of 1933 two lots of straw were prepared for synthetic manure production at the Michigan Experiment Station. One lot, A, was wetted down at frequent intervals throughout the summer, and the other lot, B, was wetted by the rain only. Lot A produced a well-decomposed lot of manure by November of the same year, whereas lot B did not produce a satisfactory manure until July of the following summer. In other words, the annual rainfall of Michigan was sufficient to produce a good grade of synthetic manure in 12 months, but the addition of water produced manure in about 5 months. Thus, regions of comparatively high rainfall should facilitate more rapid decomposition than regions of low rainfall.

Use of Peat

Sphagnum peat has become the successor to manure as a universal source of organic matter. Sold commercially, peat is a brown, spongy, semigranular organic material which when dry is easy to handle. It is comparatively free from disease pests, acid in reaction and has a high water-holding capacity. It comes from bogs where sphagnum moss grows. Year after year sphagnum peat is essentially the same; thus, it is not susceptible to problems caused by variable composition. Peat from sedges, reeds, cattails, and other swamp plants decompose rapidly and do not last long in the soil.

QUESTIONS

1. Soil-improving crops have many uses other than that of supplying organic matter. Explain.
2. Show how cover crops conserve the essential-element supply.
3. How do cover crops reduce soil erosion?
4. Show how cover crops planted in the orchard in the late summer and fall promote "conditioning" of the trees against the low temperatures of the winter.
5. Cool-season cover crops which thrive on a wide range of acidity are well adapted for use in the apple orchard. Explain.
6. In warm-temperate regions, a crop should be growing on a given area of land at all times. Explain.
7. In your opinion, which is more important: the accumulation of organic matter in soil or its efficient utilization? Explain.
8. Many home owners and tenants, particularly in urban areas, burn leaves in the fall. In your opinion, is this a good or a bad practice? Explain.
9. Explain how to make leaf mold. Give all the necessary instructions.
10. A farmer has a large amount of straw at his disposal. Show how he may use this straw for synthetic manure production.
11. The organisms that decompose organic matter are extremely beneficial. Explain.

USE OF COMMERCIAL FERTILIZERS

It is unfortunate that we have allowed the phrase "plant nutrients" to mean those inorganic elements that are essential for plant growth because this causes us to forget the real substances from which the bulk of plants is synthesized.—*A. G. Norman,* Amer. Scientist, *50, 1962.*

As explained in Chapter 7, the green plant is a biochemical factory. Certain raw materials are needed in the manufacture of the foods, fibers, enzymes, hormones, and vitamins. Many soils are deficient in these raw materials; therefore, they are supplemented by using chemical fertilizers.

Commercial fertilizers are frequently called *plant foods*. In reality, this is not true. As previously explained, compounds such as sugars, starch, hemicellulose, fats, and stored proteins, are the actual plant foods; the potential energy stored by them is changed to kinetic energy and liberated by respiration. Commercial fertilizers cannot supply plants with free energy; they do, however contain one or more of the essential elements necessary for plant growth.

Organics versus Commercial Fertilizers

Organics refer to the use of organic matter, e.g., manure, to meet the essential-element requirements of soil and crop plants, whereas *commercial fertilizers* refer to the use of commercial fertilizers to meet these needs. As pointed out in Chapter 7, the crop plant is a manufacturing entity and requires certain ions for the making of the plant tissue and the numerous manufactured compounds. These ions are absorbed and used by the plant regardless of whether they come from organic matter or from commercial fertilizers. For example, the nitrate ion may come from organic matter as a result of its decomposition, or from nitrate of soda, a common ingredient of commercial fertilizers. A similar situation exists with respect to the phosphate ion, the potassium ion, as well as other essential ions. Further, there is no evidence that crop plants have the ability to differentiate the ions from organic matter from those in commercial fertilizers, or that commercial fertilizers have harmful effects on soils and crops, provided they are applied according to the recommendations of the local experiment station or extension service. The main reason that commercial fertilizers are extensively used is that they supply the essential ions much more economically than do the various forms of organic matter.

Principles of Commercial Fertilizer Practice

In the principles of commercial fertilizer practice, there are three important considerations: which commercial fertilizer should be used, when it should be applied, and how it should be applied.

Which Commercial Fertilizer to Use The selection of a suitable fertilizer depends largely on the essential-element level of the soil with respect to the contents of the fertilizer, the essential-element requirements of the crop, and the season of the year.

The Essential-element Level of the Soil The many types of soils in the United States and throughout the world vary in their relative supply of essential elements. Thus, for the same crop grown under practically the same temperature level and under favorable supplies of water and light, each soil may require different ratios. For example, the well-drained sandy loams used for potato production in the Atlantic Coastal Plain of the United States are relatively high in nitrogen in proportion to phosphorus and potassium, and, in general, they require 1-1-1 or 1-1½-1 ratios. A 5-8-7 fertilizer is frequently used in Aroostock County, Maine, and 6-6-5 and 5-7-5 fertilizers are used in the Tidewater district of Virginia and the truck-crop district near Charleston, South Carolina. The well-drained sandy loams given over to potato production in the Middle West are relatively low in nitrogen and potassium in proportion to phosphorus, and, in general, they require 1-3-1, 1-4-1, or 1-4-2 mixtures; 4-12-4, 4-16-8, or 4-16-4 mixtures are used in New York, Michigan, Pennsylvania, and Ohio; and 4-8-4 mixtures are used in southern Mississippi and Louisiana.

The Essential-element Requirements of the Crop Different crops grown on the same soil also differ in fertilizer requirements. As previously explained, vegetable crops grown for their leaves, such as spinach and turnip greens, require relatively large quantities of nitrogen in the fertilizer; whereas crops grown for their fleshy roots, such as sweet potatoes, require relatively small quantities. Tests on well-drained sandy loams in central Georgia have shown that turnip greens require 80 to 100 pounds per acre (89 to 112 kilograms per hectare) of nitrogen, about 80 pounds (36 kilograms) of P_2O_5, and 40 pounds (18 kilograms) of K_2O; whereas sweet potatoes require 30 to 40 pounds (14 to 18 kilograms) of nitrogen and the same quantities of P_2O_5 and K_2O in the mixture.

The Season of the Year Principal factors concerned are temperature and light. As explained previously, with other factors favorable, the temperature of the soil markedly influences the amount of available nitrogen. In general, if the soil has been cold for a considerable period, the natural nitrate-nitrogen supply is likely to be low and artificial applications are necessary. On the other hand, if the soil has been warm for a considerable period, with other factors favorable, the natural nitrate-nitrogen supply is likely to be high. For example, tests at the Louisiana Experiment Station have shown that snap beans started in April on Lintonia silt loam generally require about 35 pounds (16 kilograms) of nitrogen in the mixed fertilizer, whereas the same crop growing on the same soil but started in August generally requires no nitrogen in the fertilizer. Evidently, the relative activity of soil microorganisms is responsible for the different amounts of available nitrogen.

The light supply is of concern in the growing of crops in greenhouses and similar structures. Fall crops in forcing structures are usually started under more favorable light conditions than spring crops. The longer light period and higher light intensity of the early fall promote high rates of gross photosynthesis, and this, combined with normal rates of respiration, provides relatively large quantities of carbohydrates for growth and development. Thus, with moderate quantities of available nitrogen not all the carbohydrates are used for the vegetative phase; some are allowed to accumulate for the initiation of the flower buds and the development of the flowers, fruit, and seed. If the same quantities of available nitrogen were supplied in early spring, practically all the carbohydrates would be utilized in vegetative growth and none would be left for reproductive growth. This explains, partially at least, why tomatoes started in greenhouses in early spring in regions of low light intensity frequently develop relatively few fruits of the first cluster. The low carbohydrate supply and the high available nitrogen supply combine to promote excessive vegetative growth. As a result, very few carbohydrates are available for the setting of the fruit.

When the Commercial Fertilizer Should Be Applied As previously stated, nitrate ions leach readily, whereas phosphorus and potassium ions do not. Therefore, nitrogen may be most efficiently used when applied to a crop just prior to maximum uptake. Thus, commercial fertilizers containing available nitrogen are usually applied just before the nitrogen is needed.

For example, in tests on the fertilization of turnip greens in Arkansas, applications just before planting and after the plants were established were compared. In all cases the commercial fertilizer applied just prior to seeding produced the higher yield. Apparently, natural supplies of nitrogen were insufficient for the rapid growth of the seedling plants. This explains why commercial fertilizers should be applied to woody plants when they are starting growth in the spring; why fertilizers should be applied to vegetable and flower crops a few days before or at the same time the seeds are planted; why additional available nitrogen should be added to crops in the vegetative phase after a series of heavy rains; and why crops grown for their fruits, especially those which contain large quantities of seed, such as tomatoes, require large quantities of nitrogen during the period of fruit growth.

How the Commercial Fertilizer Should Be Applied Any commercial fertilizer may be applied in various ways. Principal methods are broadcast, row, side placement, perforated, and liquid.

In the broadcast method, fertilizer is applied evenly over the entire surface of the soil. It is usually done after the land is plowed and just before it is harrowed, since harrowing mixes the fertilizer with the upper 3 or 4 inches (8 to 10 centimeters) of soil. In general, this method is used to apply the essential elements to crops which are grown close together, such as the small grains and cover crops, and to crops which are grown in narrow rows, such as spinach, carrot, lettuce, onion, lily, and gladiolus.

In the row method, the fertilizer is applied to the bottom of the furrow a week or 10 days before the seed or plants are planted. The fertilizer is either mixed or not mixed with the soil, and usually the land is ridged. This places the fertilizer directly below the plants. At present, this method is used to supply the essential elements to many vegetable and flower crops. A disadvantage of this method is that plant roots growing down and outwards are likely to become injured by the fertilizer salts. This is particularly true on sands and sandy loams.

In side-placement the fertilizer is applied in a continuous band on one or both sides of the row of seed or plants. Investigations at the Virginia Truck Experiment Station have shown that placing the fertilizer on the side and slightly below the level of the seed of many vegetable crops produces greater yields than placing the fertilizer below the seed. There are definite reasons for the greater yields. The fertilizer is placed a short distance from the seed or plants; therefore, relatively large quantities of essential raw materials are available during the early stages of growth. In this way, small quantities of the available forms of phosphorus are changed to the unavailable forms in those soils which fix large quantities of phosphorus. When small quantities of fertilizer are used, the side-placement method is particularly advantageous. In most instances, the most efficient use of a relatively small amount of fertilizer is achieved this way. Figure 11-4 shows the side placement of commercial fertilizers on tomatoes grown for a cannery in New Jersey.

The perforated method is used to apply commercial fertilizers to ornamental trees. This method consists of making small holes about 12 to 18 inches (30 to 46 centimeters) deep around the base of the tree and, at the same time, placing a definite amount of fertilizer in each hole. In this way, the fertilizer is placed close to the roots of the tree and does not accelerate the growth of grasses or other plants under the tree.

In the liquid method soluble fertilizer is applied in solution with water. Three distinct methods are used: by direct application to the soil, by a foliage spray, and with irrigation water.

Figure 11-4 The side dressing of tomatoes grown for the cannery in New Jersey. *(Courtesy, the late Jackson B. Hester, Elkton, Md.)*

By Direct Application to Soils In general, liquid fertilizers applied directly to the soil are used to supply relatively large quantities of essential elements close to the plant roots or to supply a relatively cheap source of nitrogen. Examples are the so-called starter solutions and anhydrous ammonia. Starter solutions frequently used in the transplanting of herbaceous crops are sodium nitrate, ammonium phosphate, monopotassium phosphate, and potassium chloride. Tests conducted at New York, New Jersey, and Florida experiment stations, for example, have shown that starter solutions markedly increase the recovery rate of seedling plants of many vegetable crops. The solution promotes rapid recovery and early growth by providing the plants with an adequate supply of readily available essential ions and by stimulating the rate of root regeneration.

Liquid fertilizers may also be broadcast on the surface and disked in, or applied, as a band in the same way dry-mixed fertilizers are applied. Most liquid fertilizers are popular because of the ease in handling and applying the materials. An example is anhydrous ammonia. The anhydrous ammonia enters the soil as a gas. The ammonia is adsorbed by soil colloids, and under warm, moist conditions it changes rapidly to nitrate-nitrogen. Many experiments have shown that anhydrous ammonia is a profitable source of nitrogen, and, at present, large quantities are used to supply the nitrogen needs of many crops.

By a Foliage Spray In general, foliage sprays are used to correct in a relatively short time a deficiency of some essential element and to supply essential elements which, if applied to soils, would, for some reason, become unavailable to the plants. For example, sprays containing magnesium sulfate have corrected magnesium deficiency of apple trees in England and cantaloupe plants in Maryland; sprays containing certain iron chelates have corrected iron deficiency of citrus trees in central Florida; and sprays containing borax have corrected boron deficiency of rutabagas in southern Ontario. Foliar sprays are used most frequently when a small amount of fertilizer is needed; often trace elements are applied in this way.

With Irrigation Water As previously stated, the essential ions fixed by soils are applied most advantageously at the time of soil preparation. This particularly applies to the phosphate and potassium ions. Ions, however, which are subject to leaching may be deficient at the time the plants need them. In general, these are compounds which dissolve readily in water and which contain the nitrate ion or which change to nitrate in a short time, e.g., sodium nitrate, ammonium nitrate, ammonium sulfate, and urea. Note that the first two compounds contain the nitrate ion and the last two change into nitrate rapidly in warm, moist soils. Since the nitrate ion leaches readily, the irrigation water containing the readily available nitrogen carriers should never penetrate beyond the zone occupied by the root system. Otherwise, large quantities of the expensive nitrogen carriers are likely to be wasted. Two methods are generally used in applying commercial fertilizers with irrigation water. The first consists of placing the fertilizer in solution with water before it is pumped through the irrigation system and the other consists of forcing a concentrated solution of the fertilizer into the irrigation lines at the time the water is being pumped.

CHEMICAL QUICK TESTS

Chemical quick tests have been developed to determine the level of essential-elements required for optimum plant production. There are two types: soil and plant.

Soil tests consist of a quick chemical analysis of essential elements in a small sample of soil. The sample is extracted with a weak acid or with the salt of a weak acid. Aliquots of the extracted solution are treated with chemical reagents. Diphenylamine is used to test for nitrates, ammonium molybdate is used for phorphorus, and sodium cobaltinitrite is used for potassium. The object of the soil tests is to determine the degree of fertility of the soil. If the soil is well supplied with essential elements it gives a good test, if moderately supplied, a fair test, and if poorly supplied, a poor test. Investigations at the Virginia Truck Experiment Station have shown that the soil tests for phosphorus and potassium coincide fairly well with the responses and yield of many vegetable crops grown in the area.

Plant-tissue tests consist of a quick chemical analysis of the sap of the plant. In general, stems or petioles are used. The sap is extracted and the essential elements are determined with reagents, as in the quick soil test. The plant-tissue tests show the essential elements which arc lacking or unavailable in soil and those which are sufficient for optimum yields. Experiments show that the soil and plant-tissue tests have a useful function if representative samples of the soil and plant are used and the tests are done by persons who have a working knowledge of the characteristics of the soil and the needs of the crop under consideration.

Many countries and every state in the United States maintains one or more soil- or plant-testing laboratories. These laboratories provide a complete and extensive evaluation of the commercial fertilizer needs of crop plants.

QUESTIONS

1. In general, readily available forms of nitrogen are required for a rapid development of the vegetative phase, and slowly available forms are required for a moderately rapid development of this phase, particularly when the soils are warm and moist. Explain.
2. Consider complete mixtures—mixture A containing 20 units (5-10-5) and mixture B containing 16 units (4-8-4). On the basis of the current prices for each mixture, determine the cost of each unit in A and B.
3. In commercial fertilizer practices the essential-element level of the soil, the essential-element requirements of the crop, and the season of the year should be considered. Explain.
4. In general, crops in the vegetative phase of growth can use more available nitrogen on sunny days than on cloudy days. Explain.

5 Tomatoes started in January and grown in the greenhouse frequently set very few fruits of the first cluster, whereas the same variety started in August generally sets many fruits of the first cluster. Explain.

6 In general, the time when complete mixtures should be applied depends largely on the time when available nitrogen is needed. Explain.

7 When ammonia forms of nitrogen are applied to soils, they require from one to several weeks to change into the nitrate form. Explain.

8 In general, small quantities of commercial fertilizer, 100 to 300 pounds per acre (112 to 336 kilograms per hectare), should be applied as side dressings or in hills, whereas applications varying from 1,500 to 2,000 pounds per acre (1,680 to 2,240 kilograms per hectare) should be applied broadcast. Explain.

9 In general, foliage sprays may be used to correct a deficiency of certain essential elements in a relatively short time. Explain.

10 Irrigation water containing commercial fertilizers should never penetrate beyond the zone occupied by the plant roots. Explain.

11 What are chemical quick tests? What are their uses?

12 In order for farmers to obtain full returns on their investment in commercial fertilizers, they should control all insect, disease, and weed pests and maintain (for most crops) moderately acid to slightly acid soils. Explain.

13 If tomato fruits contain 0.50 percent N, 0.17 percent P_2O_5, and 0.87 per cent K_2O, calculate the amount of N, P_2O_5, and K_2O removed by a crop yielding 10 tons of fruit per acre (22.5 metric tons per hectare).

14 In general, essential elements removed in harvested products should be systematically replenished. Explain.

SELECTED REFERENCES FOR FURTHER STUDY

Buckman, H. O., and N. C. Brady, 1969. *The nature and properties of soils*. 7th ed. New York: Macmillan. An introduction to soil science answering the questions: What is soil? How is it classified? How does it interact with water, air, fertilizer, lime, microorganisms, and organic matter? How do these interactions further relate to plant growth?

Collings, G. H., 1955. *Commercial fertilizers*. 5th ed. New York: McGraw-Hill. On the manufacture and use of commercial fertilizers, including discussions of sodium nitrate, ammonium nitrate, synthetic and organic nitrogenous fertilizers, use of mineral phosphates, bone phosphate and superphosphate, potash fertilizers, and fertilizers carrying secondary and raw essential elements.

Teuscher, H., and R. Adler, 1960. *The soil and its fertility*. New York: Reinhold. On the interaction between soil and fertilizer for plant metabolism, with sections pertaining to the contents of the soil, its physical properties, soil dynamics (the living soil), major elements and how these behave in the soil, organic and inorganic fertilizers, chemical laws of soil fertility and scientific, and practical procedures of soil and soil fertility management.

Chapter 12

Crop-Plant Growing Structures

He that questioneth nothing, learneth nothing.

Thomas Fuller (1608–1661)

BENEFITS OF USING GROWING STRUCTURES

Many kinds of horticultural plants are grown in various types of plant-growing structures. These structures provide a more favorable environment than is available in the immediate outdoors. For example, the low temperature of the winters in temperate climates—in general from November to March in the northern hemisphere and from May to August in the southern—preclude the growing of crops outdoors. The other extreme, the high temperatures combined with the high light intensity of the summers in warm temperate, subtropical, and tropical climates, is injurious to many valuable food and ornamental plants. Thus, the growing of crops in structures has, among others, the following two advantages: (1) plants can be protected from adverse environmental conditions, and (2) products can be placed on the market either earlier or later than those produced outdoors.

GREENHOUSES, OR GLASSHOUSES

The production of crops in greenhouses, or glasshouses, is sometimes called *plant forcing*. In many communities forcing is an important industry. Three types exist: flower production, vegetable production, and a combination of the two.

Nature of Crop Forcing

The growing of plants in forcing establishments is a most intensive type of agriculture. Plants are grown close together, and crops follow each other in quick succession. The initial capital required and costs of production are higher than for other forms of crop production. However, the water supply and the essential-element supply are under the control of the grower at all times; with the advent of air conditioning, the temperature level is under his control at all times; and he can modify the light supply, particularly with respect to the intensity and to the length of the light and dark periods, according to the needs and requirements of the crop. In other words, the grower of greenhouse crops can regulate the plants' environment to a greater extent than the grower of outdoor crops. For this reason, the grower of greenhouse crops usually produces large yields of high-quality products.

Types of Greenhouses, or Glasshouses

The main types of greenhouses, or glasshouses, are (1) lean-to and (2) even-span. Lean-to houses consist of a single span. This type is built usually on the south or east side of a wall or house for the growing of crops, or on the north side of a wall or house for the rooting of cuttings. Even-span houses consist of two spans equal in width and pitch which may run from north to south or from east to west. Of the two types the even span is the more widely used. This type may be built as a single unit or separate from adjacent units, called *detached houses,* or in two or more units and gutter-connected, called *ridge and furrow houses*. Detached houses provide more ventilation and light exposure, but because of the greater glass area, they lose more heat, an important factor, during the winter. On the other hand, ridge and furrow houses are more difficult to keep in repair and tend to hold excess quantities of snow in northern climates. In general, detached houses are used for the simultaneous production of crops which have different temperature requirements, whereas ridge and furrow houses are limited to the simultaneous production of crops with the same temperature requirements. Figure 12-1 is an airview of the range of the J. W. Davis Company of Terre Haute, Indiana. Note that all the houses are of the even-span type, but that some are detached and others are gutter-connected.

Parts of a Modern Greenhouse, or Glasshouse

The parts of a modern greenhouse, or glasshouse, are the foundation wall, side posts, side glass bars and sash, eave plate, roof glazing bars, columns or purlin

Figure 12-1 The greenhouse range of the J. W. Davis Company, Terre Haute, Indiana. This range, the largest vegetable-forcing establishment in the country, covers 25 acres and is given over to the production of vegetable crops, particularly greenhouse tomatoes and cucumbers. *(Courtesy, W. Keith Owen, J. W. Davis Company.)*

posts, roof ventilating sash, ridge and ridge cap, gable glazing bars and sash, and glass. Study the position of the various parts in Figure 12-2. The foundation wall and side posts support the entire house; the columns and posts support the roof; the eaves carry water into the drainage system; the ventilating sash admits cool air and releases warm air; the glazing bars support the glass; and the glass admits the light. Two grades of glass are available for greenhouses: single strength, weighing 21 ounces per square foot (595 grams per 929 square centimeters) and double strength, weighing 26 ounces per square foot (737 grams per 929 square centimeters). Of these grades the double strength is needed in regions of heavy snowfall. Both the rectangular shape, usually 18 × 24 inches (46 × 61 centimeters) and the square shape, usually 20 × 20 inches (51 × 51 centimeters), are used. Of these the square shape is recommended, since it requires fewer glazing bars to cover a given area and withstands stress and strain to a greater extent than the rectangular shape. For example, actual counts of broken panes of two adjacent houses in New Jersey after a heavy hail storm showed that two out of three panes, or 67 percent, had broken in the 18 × 24 inch (46 × 61 centimeters) shape and only one out of four panes, or 25 percent, had broken in the 20 × 20 inch (51 centimeters) panes.

Framework Material

In general, the framework consists of redwood, aluminum alloys, and steel. Clear-heart, densely grained redwood, or specially constructed aluminum alloys are used for the glazing bars, ventilating sash, and ridges and sills since these

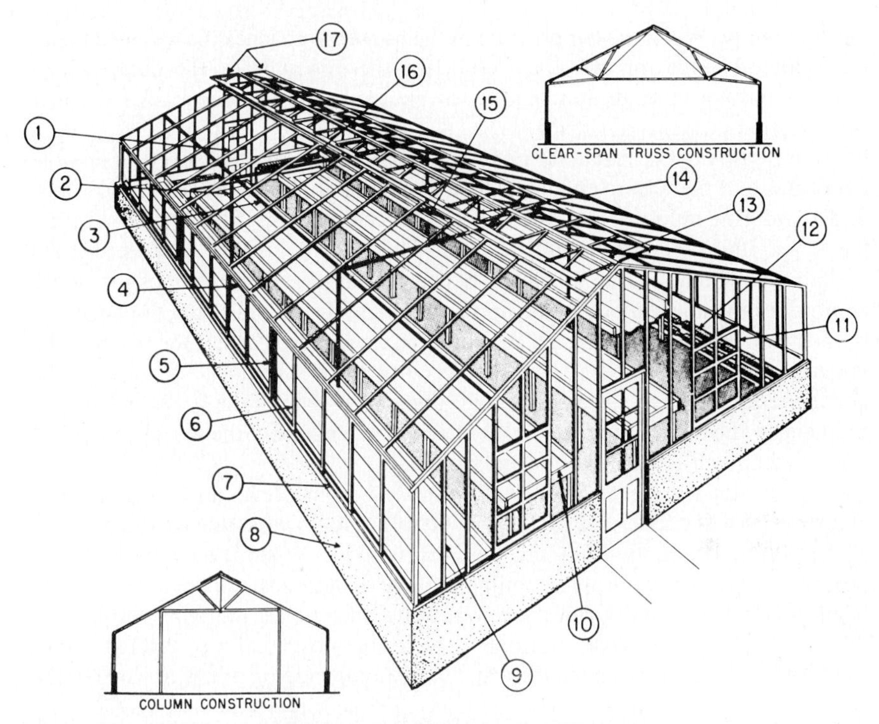

Figure 12-2 Parts of a greenhouse: (1) roof bar, (2) column, or purlin post, (3) roof purlin, (4) eave plate, (5) side post, (6) side sash, (7) wall sill, (8) masonry wall, (9) gable glazing bar, (10) bench, (11) gable casement sash, (12) fin for radiation, (13) ridge and ridge cap, (14) tie rod, (15) strut, (16) crosstie, (17) roof vent sash. *(Courtesy, Alex Laurie et al.,* Commercial Flower Forcing, *6th ed., New York: McGraw-Hill Book Company, Inc.)*

parts are constantly exposed to the weather and to wide fluctuations in temperature. Steel is used for the side posts, inside columns, purlins, and eave plates, since these materials do not excessively shade the plants. Thus, from the standpoint of framework materials two types of houses are used: (1) wood-steel frame and (2) aluminum or steel frame.

Methods of Heating

Greenhouses are heated either by hot water or by steam. In the hot-water system, the boiler is on the same level as the greenhouse, and the hot water is circulated through heating coils by means of an electric pump which is controlled by a mercury tube thermostat. In general, two methods are employed. The first consists in maintaining a high and fairly constant temperature 180 to 220°F (82 to 104°C) at the boiler by means of an aquastat, the thermostat being connected to the pump only. When the temperature of the house approaches the lower limits of the optimum range for any given crop, the thermostat starts the pump which forces hot water through the coils; when the temperature approaches the upper limit of the optimum range, the thermostat shuts off the pump. By this method hot water is always available for circulation into the houses. However, unless the temperature of the water is maintained at a moderately low level (less than 180°F or 82°C) during mild weather, the temperature of the house is likely to go beyond the upper limits of the optimum range. The second involves connecting the thermostat to both the pump and the boiler. When the temperature of the house approaches the lower limits of the optimum range, the thermostat simultaneously starts the pump and then turns on the burner in the boiler. Thus, if the water in the boiler is cold, some time will elapse before warm water begins to circulate in the heating coils; and if the weather is cold, the temperature of the house is likely to drop below the optimum range. With efficient management the disadvantages of each method can be reduced to a minimum.

In the steam system, the boiler is also on the same level as the greenhouse, and steam varying from 212 to 250°F (100 to 121°C), depending on the pressure at the boiler, is admitted through the heating coils. When the temperature approaches the lower limits of the optimum range, a thermostat opens a valve at the head of the house; and when the temperature approaches the upper limits, the thermostat closes the valve. Various types of thermostatically controlled valves can be used. Some types open fully and close completely to admit or shut off the steam; other types open and close gradually. Of these types the latter is the more likely to keep the temperatures within a specified range.

PLASTIC HOUSES

Plastic houses are now used in commercial and home garden production. Examples of plastics are polyethylene, polyethylene UV (meaning ultraviolet-light resistant), polyvinyl, polyvinyl chloride, and reinforced fiber glass panels. Polyethylene is the cheapest per unit area and has the shortest life, and reinforced fiber glass is the most expensive per unit area and has the longest life, with the other types ranging between these two extremes. Of these types, polyethylene is the most widely used. How do crop plants perform under this type of cover as compared to glass? Study the data in Table 12-1. Note that the marketable yield of each of the two varieties was practically the same under each type of cover. Thus, it seems to be largely a matter of economics—the type of cover which gives the grower the greatest profit from his investment. Also note the plastic house presented in Figure 12-3.

Gas Heating of Plastic Houses

The two principal gases used in the heating of plastic houses are natural gas and propane. Natural gas consists mostly of methane (CH_4).

(methane)		(oxygen)		(carbon dioxide)		(water)		
1 CH_4	+	2 O_2	⟶	1 CO_2	+	2 H_2O	+	heat
16		64		44		36		

(propane)	+	(oxygen)		(carbon dioxide)	+	(water)		
1 C_3H_8		5 O_2	⟶	3 CO_2		4 H_2O	+	heat
44		160		132		72		

Note from the equation that the complete combustion of 1 molecular weight of methane requires 2 molecular weights of oxygen. In other words, 1 pound (454 grams) of methane requires 4 pounds (1,814 grams) of oxygen for its complete combustion. Since the air contains 21 percent oxygen, this will require about 19 pounds (8,626 grams) of air. In like manner, 1 molecular weight of propane requires 5 molecular weights of oxygen, and 1 pound (454 grams) of propane requires about 16 pounds (7,256 grams) of air for its complete combustion. Thus, each of these fuels requires abundant supplies of air. Otherwise, incomplete combustion is likely to take place with the formation of unsaturated hydrocarbons, butylene, propylene, and carbon monoxide—compounds which injure the leaves and reduce their rate of photosynthesis, growth, and yield. Note also that with complete combustion of each of these fuels carbon dioxide and water are given off. In fact, 1.0 pound (454 grams) of methane combined with oxygen forms 2.8 pounds (1,268 grams) of CO_2 and 2.3 pounds (1,043 grams) of water, and 1.0 pound (454 grams) of propane combined with oxygen gives off 3.0 pounds (1,362 grams) of CO_2 and 1.6 pounds (726 grams) of water. In general, up to a certain point, increases in both of these compounds are likely to be beneficial. Carbon dioxide concentration may be increased from 300 ppm, the normal concentration, to about 3,000 ppm before some other factor becomes

Table 12-1 The Yield of Two Forcing Tomato Varieties Grown in a Glasshouse and a Plastic House at Michigan State University

Seeds Planted in June, with Plants Producing Fruit from Early October to Mid-December.

Variety	Pounds Marketable fruit per plant	
	Glass	Plastic†
M-O hybrid*	7.7	7.4
W-R7†	6.6	7.0

*Michigan-Ohio Hybrid
†Ohio Wilt Resistant No. 7

Figure 12-3 A plastic house at Mississippi State University. Note how this type of house is ventilated. *(Courtesy, E. L. Moore, Mississippi State University.)*

limiting in the rate of photosynthesis, growth, and yield. Increases in relative humidity (amount of water) with resultant decreases in rate of transpiration are likely to be beneficial up to about 90 or 95 percent. At this percentage, condensation of water vapor within the greenhouse is likely to take place, a situation quite favorable for the spread of certain fungus diseases.

ENVIRONMENTAL DIFFERENCES WITHIN GREENHOUSES AND PLASTIC HOUSES

The environments within greenhouses and plastic houses differ from each other in three ways. (1) Greenhouses permit a greater exchange of air than do plastic houses. This is due to the cracks between the panes of glass and the absence of cracks in the plastic. In other words, plastic houses are relatively tight, and as result, the rate of increase in temperature during sunny weather is likely to be more rapid. (2) Greenhouses have comparatively little drip. This is because the drip grooves in the sash bars carry most of the water to the eaves. Since no provision is made to carry the drip water to the eaves in plastic houses, the dripping of water is likely to interfere with the irrigation schedule of the plants and to increase the relative humidity of the air to the range favorable for the development of certain diseases, for example, leaf mold of tomatoes. (3) Greenhouses permit a greater transmission of light, whereas plastic houses seem to promote a greater diffusion of light. In other words, shadows are more intense in greenhouses. Despite these differences in degree, there seem to be no discernible differences in growth and market yield of plants grown in greenhouses and the same kinds of plants grown in plastic houses. (See Table 12-1).

Air Conditioning of Greenhouses and Plastic Houses

Air conditioning involves reducing extremely high and potentially injurious temperatures during the light period of the summer months. In general, air conditioning (1) lowers the temperature either to within or just above the optimum range, (2) increases the relative humidity, and (3) permits the use of relatively high light intensity. How does the lowering of the temperature, combined with increasing the relative humidity and increasing the light intensity, promote growth and development of crops in greenhouses and plastic houses during the summer months? The lowering of the temperature, combined with increasing the relative humidity, lowers the rate of transpiration. This allows the rate of absorption to keep up with it. As a result, the guard cells maintain turgor, the stomates remain open, carbon dioxide diffuses rapidly into the leaves, and the rate of photosynthesis is high, even in the presence of relatively high light intensity. In addition, the lowering of the temperature lowers the rate of respiration. Thus, abundant carbohydrates are available for growth and, if the plants are managed properly with respect to the vegetative and reproductive phases, high yields of high-quality products are obtained.

At present two methods are in operation: (1) using washed and cooled air and (2) using a high-pressure mist. In general, the use of washed and cooled air requires water-absorbing material to provide a large evaporating surface; a pump, pipes, and a shallow tank or well to circulate water through the water-absorbing material, and exhaust fans to pull the cooled, humidified air through the house. As water evaporates, it absorbs heat. This absorption of heat lowers the air temperature, increases the relative humidity, and permits the use of relatively high light intensity. The extent to which the temperature of the air can be lowered during sunny weather is quite marked. For example, at Mississippi State, Mississippi, during the months of June, July, and August, the temperature of non-air-conditioned houses frequently reaches a maximum of 105 to 120°F (41 to 49°C), whereas the temperature of adjacent air-conditioned houses with very little shade on the roof attains a maximum of 85 to 90°F (29 to 32°C).

On August 24, 1955, an ornamental-plant scientist recorded the air temperature at each of three places on the campus of Texas A. & M. University: (1) within a pad-fan air-conditioned greenhouse, (2) within a comparable non-air-conditioned house, and (3) outside. Note in Figure 12-4 the excessively high temperature of the non-air-conditioned house between 10 AM and 6 PM. During this period the rates of transpiration and respiration were high, and very few carbohydrates were available for growth. According to Laurie et al. the degree of cooling that can be obtained is about 80 percent of the difference between the readings of the outdoor wet-bulb and dry-bulb thermometers. Thus, the greater the difference between the two readings, or the greater the evaporating power of the air, the greater will be the reduction in temperature.

In general, the use of a high-pressure mist requires a supply of water under high pressure, usually not less than 600 pounds per square inch (42 kilograms per square centimeter), nozzle lines to break the water into a fine mist, and water

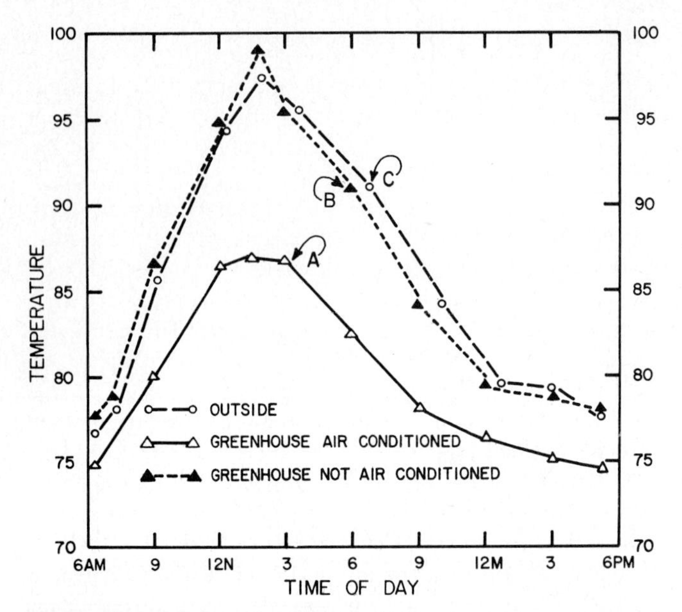

Figure 12-4 Effect of air conditioning by the pad-and-fan method on the temperature and relative humidity within a greenhouse at Texas A. & M. University. *(Courtesy, A. F. DeWerth and R. C. Jaska, Greenhouse Cooling, Texas* Agr. Exp. Sta. Misc. Bul. *163.)*

relatively free of mineral solutes. The fine mist facilitates the evaporation of water. This, in turn, lowers the air temperature, increases the relative humidity, and permits the use of high light intensity. Of the two methods the use of washed and cooled air is the more effective in promoting favorable conditions and is more widely used.

SHADE HOUSES

Shade houses have straight sides and a flat top. Their construction is relatively simple. The frame consists of wood or iron posts, and the cover consists of cloth or a light-transmitting plastic. Principal crops grown in shade houses are chrysanthemums and asters. Under conditions of high temperature and high light intensity, these crops in shade houses usually produce longer stems, larger leaves, and larger and brighter flowers than comparable crops grown outdoors. How is the more satisfactory growth secured? The fundamental process concerned is transpiration and the principal environmental factors concerned are (1) temperature, (2) light intensity, (3) relative humidity, and (4) wind velocity. Experiments at the Ohio Experiment Station have shown that shade slightly lowers maximum daily temperatures and slightly increases the relative humidity. However, it markedly lowers light intensity. Study the data in Table 12-2. Note that the greater the

Table 12-2 Effect of Shade on Reduction of Light Intensity

	In a laboratory		In sunlight outdoors	
Treatment	Ft-c	Reduction, %	Ft-c	Reduction, %
No Shade	45		11,500	
Shade	37	17	7,474	35

Source: Adapted from Alex Laurie, D. C. Kiplinger, and Kennard S. Nelson, *Commercial Flower Forcing,* 6th ed., New York: McGraw-Hill, 1958.

intensity of light, the greater is the reduction. The decrease in light intensity in shade houses during the summer undoubtedly lowers the temperature of the leaves, which, in turn, lowers the rate of transpiration and permits the rate of water absorption to keep up with it. As a result, the guard cells remain turgid, the stomates remain open, and a high rate of photosynthesis takes place throughout the light period. Note the greenhouse roses growing in the shade house in Figure 12-5.

LATHHOUSES

Lathhouses have straight sides and a flat top also. The frame is similar to that of shade houses, but the cover consists of movable lath sash, "snow fence," or nailed down 2 × 2 inch (5 × 5 centimeters) strips placed 2 inches (5 centimeters) apart. During the summers of warm-temperature and sub-tropical regions, lathhouses are required to protect many ornamental plants which are sensitive to high light intensity, such as hydrangaes and azaleas, and are used to grow and maintain stock plants for the foliage industry. Note the lathhouse used for the protection of azaleas in Figure 12-6.

HOTBEDS

Hotbeds consist of three parts: the frame, the cover, and the heating system. In general, the frame is made of concrete blocks, flat sheets of transite, or reinforced polyester plastic. All of these materials are durable, but each has advantages and disadvantages. In general, concrete blocks are effective insulators, but they are relatively costly and obstruct light. Flat sheets of transite are light in weight, easy to handle and construct, but they also obstruct light. Reinforced polyester plastic is also light in weight and easy to handle, though in sharp contrast to concrete and transite, it admits light. Its light-transmitting qualities, however, are likely to decrease with age.

The cover may consist of glass, light-transmitting plastic, canvass, or cloth. The glass panes are usually set in wood frames called *glass sash;* the light-transmitting plastics may be set in wood frames, in panels, or used as continuous sheets; and the canvass or cloth is usually used as continuous sheets. Once again, each has its advantages and disadvantages. In general, glass sash, if handled with care, is durable, permits maximum transmission of light, and is stable in high winds; however, it is expensive, unwieldy, and requires care in handling. The light-transmitting plastics are relatively cheap and easy to handle, but they must be weighed down in windy weather, and they last for a short time only. Canvass or cloth covers are also relatively cheap and easy to handle, but since they obstruct light, their use is limited to protecting plants during cold weather.

In general, hotbeds are heated by three different systems and are classified accordingly.

Figure 12-5 Greenhouse roses growing in a cloth house at the Ohio State University. *(Courtesy, Alex Laurie et al.,* Commercial Flower Forcing, *6th ed. New York: McGraw-Hill Book Company, Inc.)*

Figure 12-6 A lathhouse made of Flexwood supported by iron rods. *(Courtesy, A. C. Oelschig and Sons, Savannah, Georgia.)*

Hot Air

The hot-air-heated bed contains sets of flues which carry the heat and products of combustion from the furnace, usually situated at one end of the bed, to the chimney, usually situated at the other end. The reaction concerned is combustion which, like respiration, liberates heat and other forms of kinetic energy. Note the equation.

Gas or coal or wood + oxygen → CO_2 and other gases + H_2O + heat + other forms of kinetic energy

Flue-heated beds are usually used when growers have plentiful supplies of cheap wood. In general, the direction of the prevailing wind and the length and size of the flue are important considerations. For example, in the production of sweet potato plants in the South, beds 9 × 12 feet (2.7 × 3.7 meters) wide and 50 × 60 feet (15.2 × 18.3 meters), long underlined with two to four rows of 6- or 8-inch (15 to 20-centimeter) tile have produced satisfactory results.

Hot Water

The hot-water-heated bed has pipes under the bed or along the sides of the frame. The size of the pipes, position of the boiler, and slope of the floor are primary considerations in this type of bed. The heaters are usually thermostatically controlled. Thus, they maintain uniform temperatures with little waste of fuel and are economical and efficient. This type of hotbed is very popular among market gardeners in New Jersey.

Electricity

The electric bed is heated by lead-covered resistance coils placed on or under the soil or along the inside walls of the frame or by lamps placed over the bed. This system is automatic, always available, more or less permanent, and reliable. At present, the cost of electricity for the growing of plants is a prime consideration. Thus, many experiments have been made to determine production costs. For example, in 1938 the cost of the electricity was about 2.8 cents per week per sash at the Washington and Pennsylvania Stations and about 1.5 cents per week per sash at the Maryland Station. At the Minnesota Station the combined use of the heating cable and lamps produced better plants than the use of the heating cable alone. This is to be expected, since light is usually the limiting factor in growing plants under glass in the early spring in northern climates. Note the parts of the electrically heated bed presented in Figure 12-7.

COLD FRAMES

Cold frames consist of two parts: the frame and the cover. In fact, the same materials which comprise the frame and the cover of hotbeds are used for cold

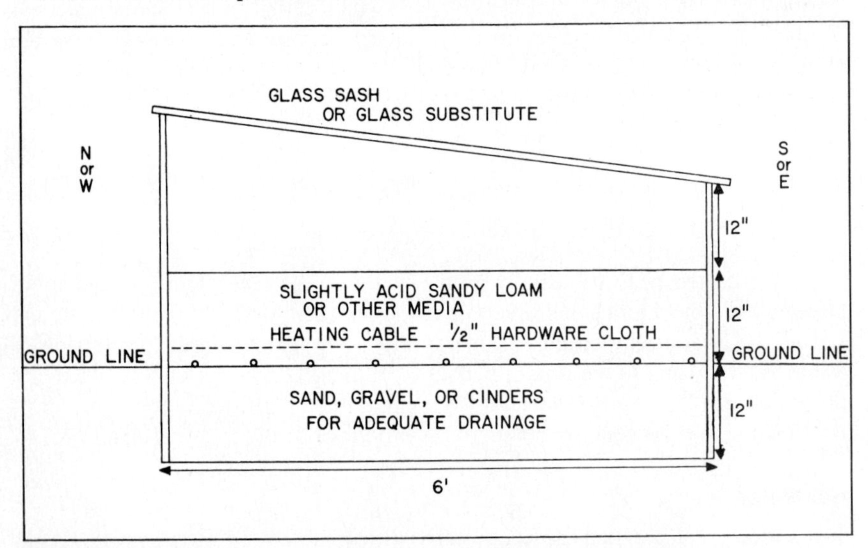

Figure 12-7 Cross section of an electrically heated hotbed. The function of the hardware cloth is to promote a uniform temperature throughout the bedding media. *(Adapted from New York* [*Cornell*] Agr. Ext. Bul. *1043.)*

frames. In other words, cold frames are hotbeds without any man-made heating system. Thus, a hotbed may become a cold frame by simply closing down the heating system. In general, cold frames are used to protect plants from frost, light freezes, hard rains, and heavy wind. In regions characterized by mild weather in both the northern and southern hemispheres, herbaceous crops are started in these structures. Examples are tobacco, tomatoes, and many annual flowering plants. Later, as the weather becomes warm, the covers are removed.

Recent Improvements in Management of Cold Frames and Hotbeds

Recent improvements pertain primarily to the labor involved in ventilating, watering, and servicing operations. To ventilate a large number of contiguous frames in a short time and with a minimum of labor, a worm-lifting gear is attached to an electric motor. Thus, by the mere opening and closing of an electric current, one person can ventilate a large group of contiguous sash in a relatively short time. To water the plants in a large number of contiguous frames, an irrigation line is placed under the eaves. When in operation, this line rotates slowly, and the nozzles break the stream of water into a fine mist. This method prevents the soil from being washed away and facilitates uniform applications of water for all plants. Equally important is the small amount of labor time required, which need not involve more than one person to open a valve in the water line to start the watering operation and to close the valve when the plants have received sufficient water. As another labor-saving device, the frames are placed in a back-to-back arrangement, with a wide distance between each line of double sash. This relatively wide distance is necessary to accommodate the trucks servicing the frames. In this way, the distance between a truck and frame can be reduced to a minimum, thereby reducing the time required to carry materials to and from any given bed.

Protection of Plants in Rows

Protection of plants in rows consists of placing one or two layers of clear plastic on wire loops, anchoring the edges of the plastic with a layer of soil, and making slits in the top of the plastic for ventilation. The hypothesis underlying their use is that the clear plastic produces a greenhouse effect within the enclosure. As the sun shines, the temperature of the soil within the enclosure increases, and some of the heat is given off during the night. This type of protection is particularly advantageous in the early spring, since it helps to keep the temperature within the enclosure close to, if not within, the optimum range for apparent photosynthesis. It also protects plants from the damaging effects of temperatures below 32° F (0°C). In fact, instances are known in which polyethylene clear plastic protected tomato plants from outside temperatures as low as 25°F (−4°C). For example, in northern Kentucky, tomato plants were set in the field on April 5, 1955 and covered with polyethylene plastic. Sometime later in the week the temperature dropped to 25°F (−4°C), but the plants were not damaged. Although the total yield of the protected plants was lower than that of plants which were set later and thus needed no protection, the returns were greater because the protected plants produced ripe fruit earlier, when prices were relatively high. Similar protection has been used with leaf and Bibb lettuce in experiments at the Kentucky Station, Lexington, Kentucky, and with head lettuce at the Truck Crops Research Center, Crystal Springs, Mississippi.

PLANT PROTECTION IN THE FIELD

The kind of protection in the field depends largely on the kind of enterprise. For example, *hedge protection* is used by the growers of slicing cucumbers in the vicinity of Norfolk, Virginia. Cedar hedges on the west, north, and east sides of small tracts (1 to 5 acres, or 0.4 to 2.0 hectares) give the necessary protection from cold winds. *Building protection* is used by market gardeners and florists, particularly in the North, where protection from cold north and northeasterly winds is essential. *Protection by V-shaped troughs* is used by growers of cucumbers in Florida. These troughs are placed on the north side of the rows of cucumbers. In this position protection is given from wind, and the reflection of sun rays from these troughs helps germination and growth of the young plants. If frosts are likely to occur, the troughs are placed over the plants. *Protection by strip cropping* is used by growers of onions in New York and Michigan. Strips of barley or oats protect the young plants from the injurious action of the rather heavy winds. *Southerly or southeasterly slopes* are used by market gardeners, since these slopes warm up earlier in the spring than do northerly or northwesterly slopes.

PLANT CONTAINERS

Plant containers have two rather distinct functions: (1) the growing of plants to full maturity and (2) the growing of plants for transplanting to the field, garden, greenhouse bench, or bed. They are classified as follows: (1) plant containers used to raise plants in groups and (2) plant containers used to raise plants individually.

Group Containers

The principal type of group container is called a *flat*. The flat is essentially a shallow tray. Its width varies from 6 to 24 inches (15 to 59 centimeters), its length from 18 to 36 inches (46 to 92 centimeters), and its depth from 2 to 6 inches (5 to 15 centimeters). Commercial growers usually use flats of standard dimensions—12 × 24 × 3 inches (30 × 59 × 8 centimeters). Their use facilitates standardization of other crop-growing operations, and more efficiently utilizes greenhouse, hotbed, and cold-frame space. In general, durable, nonwarpable wood is used. Recently, metal flats with holes in the bottom for drainage have become available. Note the two general methods of planting seed flats in Figure 12-8.

Individual Containers

Individual plant containers are (1) pots and (2) bands. Pots are round and may or may not have a hole in the bottom for drainage. Bands are square and are open at the bottom. Pots are either porous or nonporous. Porous pots are made of clay or peat fiber, and nonporous pots are made of metal, concrete, rubber, or plastics. Bands are made of wood or paper.

Principal factors concerned in the growing of plants in pots or bands are (1) water supply and (2) nitrogen supply. The water supply pertains particularly to porous and nonporous pots. Investigations at the Massachusetts Experiment Station have shown that, under the same conditions, plants in nonporous pots require less moisture than those in porous pots. Consequently, failures through the use of nonporous pots are likely to be due largely to overwatering, whereas failures through the use of porous pots are likely to be due to excessive drying out or underwatering. If moist soil is constantly maintained in either case, either type of pot should be equally satisfactory.

The nitrogen supply pertains particularly to new clay pots and to paper and peat fiber pots. Tests have shown that new clay pots absorb nitrates. Thus, unless adequate available nitrogen is supplied in the irrigation water, rapidly growing plants are likely to be deficient in nitrogen. Tests also have shown that microorganisms decompose paper pots. Here again, unless adequate available nitrogen is supplied, rapidly growing plants are likely to become deficient in nitrogen. In general, applications of available nitrogen at the rate of 1 ounce of sodium nitrate or ammonium sulfate to 1 gallon (or 7.5 grams to 1 liter) of water at intervals of 1

Figure 12-8 Two standard flats. Left: seed broadcast. Right: seed planted in rows. *(Courtesy, The late K. H. Buckley, Mississippi Agricultural Extension Service.)*

week or 10 days will suffice to supply the nitrogen requirements of the plants and the decomposing bacteria.

Pots are used to raise transplants or to raise plants to full maturity, and bands are used to raise transplants only. Of the many types of pots on the market, clay pots are the oldest and most popular. Many sizes are available, varying from 2, 2¼, 3¼, 4, 5, 6, and 7 inches (5, 6, 8, 10, 13, 15 and 18 centimeters) in diameter. Other types of pots are innovations. The metal type, for example, is quite ornamental and somewhat adapted to the growing of plants in the home or conservatory. Bands are made of either wood veneer or paper. The most popular sizes are 2 × 2 × 3 and 4 × 4 × 5 inches (5 × 5 × 8 and 10 × 10 × 13 centimeters). Figure 12-9 shows various types of pots.

Ornamental Plants and Pot Size

The amount of growth a plant will make and the time it will flower depend largely on the immediate and potential available nitrogen supply. This, in turn, depends on the volume of soil in which the plant is growing. In general, plants

Figure 12-9 Examples of individual containers. Front row: standard clay pots of various sizes. Second row: notched clay pot, veneer band, paper pot for growing, paper pot for shipping, plastic pot, and rose pot. Third row: orchid pot, pan, standard pot showing single hole for drainage. Fourth row: large clay pot, cypress tub. *(Courtesy, E. W. McElwee, University of Florida.)*

growing in small pots produce less vegetative growth and flower earlier than plants growing in large pots. Plants growing in small-sized pots naturally deplete the nitrogen supply earlier than plants growing in large-sized pots. The earlier depletion permits earlier accumulation of carbohydrates and flower-forming hormones, which, in turn, are associated with flower-bud formation, flowering, and fruiting.

Growing Seedling Vegetable Plants

Certain vegetable crops, particularly tomatoes, cabbage, and onions, are raised during the seedling stage in relatively cool weather in the southern United States and shipped as transplants to market gardeners and canners in the northern United States. Because of the necessity for strict control of diseases, the raising of these crops is a highly specialized industry. The seedling plants are grown in open fields of 1 to 100 acres (0.4 to 40.5 hectares) or more. Relatively infertile, well-drained sandy loams and loams are used. In general, the seed is drilled in rows 6 to 12 inches (15 to 30 centimeters) apart. On many plant-growing farms, the amount of seed used is numbered in tons and the number of plants produced is counted in millions.

Two important plant-growing districts are the Rio Grande Valley of Texas and the Tifton-Valdosta of south Georgia. The Rio Grande Valley produces annually some 160 million sweet potato plants, 100 million tomato plants, 400 million cabbage plants, and 900 million onion plants. The Tifton-Valdosta area comprises about five counties and produces approximately 100 million sweet potato plants, 500 million tomato plants, 400 million cabbage plants, and about 20 million onion plants. Small quantities of pepper, eggplant, and cauliflower seedlings are produced also.

QUESTIONS

1 How does the growing of crops in plant-forcing structures differ from the growing of crops outdoors?
2 Higher yields and higher quality generally result when a given crop is planted in greenhouses instead of outdoors. Explain.
3 Compare detached or separate units and ridge and furrow units from the standpoint of ventilation, light exposure, heat loss, and maintenance.
4 State the function of the following parts of a greenhouse: the foundation wall, the columns and posts, the eaves, the ventilating sash, and the glazing bars.
5 In general, how are the optimum temperature ranges maintained by each of the two hot-water heating systems and by the steam heating system?
6 Show how air conditioning of greenhouses and plastic houses during the summer months promotes the production of high yields and high quality.
7 Give an advantage and a disadvantage of light-transmitting plastics for greenhouses.

8 Under conditions of high light intensity and high transpiration, pompon chrysanthemums and asters grown under cloth or light-transmitting plastic produce longer stems, larger leaves, and larger flowers than comparable plants grown outdoors. Explain.
9 How do lathhouses protect plants which are sensitive to high light intensity?
10 Give an advantage and a disadvantage of hot-air-heated, and electrically heated hotbeds.
11 How would you change an electric-heated hotbed into a cold frame?
12 The grower of crops for early market frequently protects his plants. Explain.
13 Explain how to protect tomato plants just set in the home garden. Outline a practical measure.
14 In your opinion how do hedge protection and strip cropping influence the net photosynthesis of the protected crop?
15 How does planting on a southeast slope promote earliness?
16 Flats and pots should be set level on beds and benches. Explain.
17 The soil in a porous pot drys out more quickly than that in a nonporous pot. Explain.
18 Recommend a type of pot which should be used for growing plants in the living room. Give reasons for your choice.
19 Your mother has a vigorously vegetative geranium plant growing in rich soil. She gives the plant optimum supplies of water. She wants the plant to produce flowers at the earliest possible moment. What would you advise? Give reasons.
20 Tomato and cabbage plants grown in the southern United States for shipment to the Middle West are raised in relatively infertile, well-drained sands and sandy loams. Explain.

SELECTED REFERENCES FOR FURTHER STUDY

Cotter, D. J. 1966. Plastics for environmental control in agriculture. *Proc. 7th. Nat. Agr. Plastics Conf.* Lexington, Ky: Department of Horticulture, University of Kentucky. On the growing prominence in the use of plastics in agriculture, both in the United States and in other countries of the world.

Laurie, Alex et al. 1968. *Commercial flower forcing*. 7th ed. New York: McGraw-Hill, chaps. 2, 3. A precise discussion on the location, operation, and management of greenhouses for the commercial production of flower crops.

Meister, R. T. 1965. *Amer. Veg. Grower.* 13 (11). An entire issue report of the greenhouse industry, including new ideas in construction, heating, air conditioning, and reduction of labor and fatigue costs.

Moore, E. L., and T. N. Jones. 1963. Heating plastic houses. *Miss. Agr. Exp. Sta. Bul.* 666. The research experience of a horticulturist and an agricultural engineer on the successful production of tomatoes in plastic houses.

Chapter 13

Crop-Plant Growing Operations

Constructive work is the essence of a man's life.

TRANSPLANTING

Transplanting is a very important practice. Many thousands of deciduous fruit trees, evergreen fruit trees, nut fruit trees, and many millions of ornamental trees and shrubs in this country and throughout the world have at one time or another been transplanted. Moreover, many millions of vegetable and flower plants are transplanted annually. By transplanting, many deciduous fruit orchards, citrus fruit orchards, grape vineyards, and small fruit, vegetable, and flower crop plantations are established; and parks, streets, lanes, highways, factories, business buildings, recreational areas, and homes are made more beautiful.

There are many terms used in transplanting: *potting, repotting,* or *shifting, balling* and *burlapping,* and *setting out*. In general, *potting* is the transplanting of seedlings from a seedbed or flat to a pot. *Repotting,* or *shifting,* is the transplanting of a plant from one pot to another, usually to a larger one. *Balling* and *burlapping* is the transplanting of plants, chiefly evergreens, with a ball of soil supported by burlap or similar material. *Setting out* is the transplanting of plants, usually herbaceous kinds, from pots, flats, or beds to the garden or field.

TRANSPLANTING HERBACEOUS PLANTS

Influence on Plant Behavior

How does transplanting influence the growth of herbaceous plants? What are the fundamental processes concerned? The student should remember that *herbaceous plants always have leaves,* that *most of the water is lost through the leaves,* and that *transplanting destroys part, if not all, of the region of absorption.* Thus, during transplanting operations the amount of water going into a plant is reduced, and this amount is usually less than the amount going out. As a result, a water deficit is likely to take place within the tissues. This deficit produces at least two effects: (1) *a reduction in the size of the cells in the region of elongation and* (2) *a reduction in the rate or cessation of gross photosynthesis.* In the first case the region of enlargement fails to receive sufficient water for normal size of the cells; and in the second, the guard cells lose turgor, the stomates partially or completely close, the rate of diffusion of carbon dioxide into the leaves is reduced, and the manufacture of the primary food substances is accordingly reduced. This, in turn, reduces the amount of growth which would otherwise be made. Therefore, transplanting herbaceous plants reduces their growth and development, and, in general, the extent of the injury or reduction in growth is more or less directly proportional to the severity and duration of the water deficit within the tissues.

Factors Influencing Rate of Recovery

Size and Age of Plant Since size and age are positively associated, these factors may be considered together. In general, the greater the size or age, the lesser the ability of the plant to recover from the check in growth incident to transplanting. Why do large or old plants recover less rapidly? Usually large plants have an extensive root system, and, in transplanting, the younger portion, the tips, is not retained. In this way the region of absorption is reduced considerably. In addition, large plants have more leaves. Thus, the amount of water absorbed in proportion to the amount of water transpired is likely to be materially reduced in the case of the large or old plant. For this reason, whenever it is possible and practical, herbaceous plants should be transplanted in the seedling stage. Growers who transplant large numbers of plants—truck gardeners, market gardeners, florists, and nurserymen—generally transplant seedling plants when they are small. In general, the first transplanting is usually done just before or immediately after the first true leaves begin to develop, and the second transplanting, when necessary, is usually done three to six weeks later.

Age of Absorbing Region When Suberization Takes Place Suberization of the roots involves the laying down of suberin in the walls of the endodermis or cortex or both. Since suberin is impervious to water, suberization of the walls of the endodermis or cortex in the root-hair zone prevents the passage of water from the root hairs to the xylem. Thus, that portion of the root in which suberization has taken place is useless for water absorption. Plants differ in the time they deposit suberin in the walls of the water-absorbing region. For example, in beans, cucumbers, and sweet corn, suberization becomes evident in the three-day-old portion of the root system, whereas in tomatoes and cabbage it becomes evident in the five- to six-week-old portion. Naturally, crops in which suberization takes place relatively early (just back of the tips) are likely to recover less rapidly than those in which suberization takes place relatively late (at a considerable distance from the tips).

Rate of Root Regeneration A direct relation exists between the rate at which the new root system is formed and the rate of recovery. In other words, the faster the new root system is formed, the faster is likely to be the rate of recovery. The rate at which the new root system is formed is largely dependent on a supply of reserve carbohydrates and other foods within the tissues at the time transplanting takes place. Thus, with other factors favorable, plants with large quantities of reserve carbohydrates in their tissues may be expected to recover more rapidly than plants with small quantities.

Classification

On the basis of differential rate of recovery, a classification of herbaceous crops is possible. The three groups are presented in Table 13-1. In general, plants of Group 1 develop new roots at a rapid rate and lay down suberin relatively late in the life of the root system; plants of Group 2 develop new roots at a moderately rapid rate; and plants of Group 3 develop new roots at a slow rate and lay down suberin relatively early in the life of the root system. This explains why plants of Groups

Table 13-1 Classification of Herbaceous Plants Based on Rate of Recovery from Transplanting

Group 1 Crops recovering rapidly	Group 2 Crops recovering slowly	Group 3 Crops recovering very slowly
Cabbage, cornflower, lettuce, petunia, stock, sweet potato, verbena	Aster, Canterbury bell, celery, larkspur, onion, pansy, pepper	Bean, cantaloupe, cucumber, poinsettia, pumpkin, squash, watermelon, sweet corn

2 and 3 require more care in transplanting than those of Group 1. In fact, experience has shown that relatively large plants of Group 3 should be grown in pots or bands if they are to be transplanted.

Hardening off of Herbaceous Plants

As stated previously, the rate of new root formation is dependent on a supply of reserve carbohydrates within the tissues. To promote rapid accumulation of carbohydrates in the tissues, high rates of photosynthesis, combined with low rates of respiration and rates of cell division, are necessary. With high rates of photosynthesis, large quantities of carbohydrates are made per unit time; and with low rates of respiration and cell division, small quantities of carbohydrates are used. As a result, large quantities of carbohydrates are stored in the tissues in a relatively short time. Thus, sunny weather, combined with optimum day temperatures and relatively low night temperatures, would promote a rapid accumulation of carbohydrates within the tissues. The sunny weather, combined with optimum day temperatures, makes possible high rates of photosynthesis; and the low night temperatures maintain low rates of respiration and low rates of cell division. Any deviation from these favorable conditions, as for example cloudy days combined with warm nights, would correspondingly reduce the rate of carbohydrate accumulation within the tissues. This explains why growers transfer plants from greenhouses to hotbeds or cold frames usually during late spring and why they keep plants relatively cool just before the plants are to be set in the field. How long should plants be hardened off to facilitate rapid recovery? In general, about four to six days of favorable conditions as described are sufficient to provide the carbohydrates necessary for the rapid development of the new root system.

With certain plants, this accumulation of carbohydrates permits the making of the hydrophilic colloids—the substances which bind water so that the protoplasm can withstand subfreezing temperatures, intense sunshine, and dry winds. With other plants, the carbohydrates which accumulate do not form the colloids which bind water readily. In general, the cool-season crops possess the ability to bind water, whereas the warm-season crops do not. This explains why the tops of most cool-season crops can be hardened-off to withstand relatively low temperatures, and those of warm-season crops cannot. For example, investigations at the Missouri Station and elsewhere have shown that hardened-off cabbage plants can withstand temperatures as low as 20°F (−6.7°C), and hardened-off tomato plants cannot withstand freezing temperatures. This does not mean, however, that tomato plants should not be hardened off. As stated previously, an advantage of hardening off is the accumulation of carbohydrates within the tissues, and these carbohydrates are used for the development of the new root system.

Transplanting the Weather

As previously explained, low rates of transpiration facilitate rapid recovery from the check in growth incident to transplanting. Thus, environmental factors which induce low rates of this process should be utilized whenever they are available. These environmental factors are relatively low temperature, low light intensity, still air, and high relative humidity. This explains why a light, misty rain is almost ideal for rapid recovery; why cloudy days facilitate recovery to a greater extent than sunny days; why still air is more favorable than rapidly moving air; and why transplanting in late afternoon, particularly in regions of high light intensity, is likely to promote a faster rate of recovery than transplanting in early morning.

Use of Water and Starter Solutions in Transplanting

Most herbaceous crops are transplanted with water. In fact, growers believe that the application of water at transplanting is essential to rapid recovery and full stands. The water settles the soil around the roots, thus eliminating air pockets, and is available for immediate absorption.

Starter solutions are used in the field transplanting of certain kinds of vegetable plants for example, tomatoes, peppers, celery, and sweet potatoes. In general, these solutions contain compounds which supply nitrogen, phosphorus, potassium, and other essential elements. According to tests by the New York, Louisiana, Florida, New Jersey, Texas, and other experiment stations, starter solutions are likely to promote rapid recovery and increase yields, particularly when the plants are grown on relatively infertile soil, when certain elements are deficient, when light applications of commercial fertilizers are made, when the commercial fertilizer is placed some distance away from the seedling plants, or when the soil is very acid.

Transplanting versus Direct Seeding

Direct seeding in the field involves planting seed directly in the field where the crop is to mature and thinning out excess seedlings. The seeding is done by means of precision drills, that is, with machines which plant individual seed at definite intervals, and the thinning is done by hand usually with long-handled hoes. The primary reason for testing this method was to determine its adaptation to once-over, mechanically harvested tomatoes for processing. If feasible, direct seeding could be employed to avoid the check in growth incident to transplanting and to eliminate the cost of raising the transplants. Further, direct seeding is conducive to conducting investigations on the relation of plant populations, or density, that is, the number of plants per unit area, to yield and costs. Extension workers in California in cooperation with growers of tomatoes for processing were the first to test this method. In general, they found that direct seeding, as compared to transplanting, produced high yields consistent with low costs of production more effectively. As a result of this work, practically all of the tomatoes grown for processing in California are direct seeded.

The testing of direct seeding in the tomato processing districts of the humid east—in New York, Indiana, and Pennsylvania—indicates that the method can be successful, provided certain problems specific to humid areas are overcome.

Some of these pertain to the amount and distribution of the rain, the self-crusting and non-self-crusting of the soil, the control of weeds, and the depredations of the flea beetle.

Transplanting Machines

Many herbaceous crops—celery, cabbage, cantaloupe, tomato, capsicum pepper, eggplant, and sweet potato—grown on a commercial scale are transplanted by machine. These machines set plants at definite intervals and depths and release a definite amount of water or starter solution to the roots. In this way, plant setting and watering take place simultaneously. Note the machine transplanting scene in Figure 13-1.

TRANSPLANTING EVERGREEN TREES AND SHRUBS

Influence on Plant Behavior

How does transplanting influence the growth of evergreen trees and shrubs? As with herbaceous plants, the student should remember that *evergreen trees and shrubs always have leaves and that most of the water is lost through the leaves.* Thus, during the transplanting operation the amount of water absorbed is likely to be less than the amount of water transpired. As a result, a water deficit is likely to take place within the tissues with a reduction in the size of the cells in the region

Figure 13-1 Machine transplanting of a herbaceous crop. *(Courtesy, Mechanical Transplanter Company, Holland, Michigan.)*

of elongation and a reduction in turgor pressure and in the rate of gross photosynthesis.

Factors Influencing Rate of Recovery

Size and Age of Plant As with herbaceous plants recovery varies indirectly with size and age. Thus, under the same conditions small evergreen trees and shrubs recover more rapidly than large trees and shrubs. In fact, in the establishment of wood lots and forests certain species of pines are usually transplanted bare-rooted; that is, the plants are removed for transplanting without a mass of soil around the roots. However, evergreen trees and shrubs for landscape use are usually transplanted when they are large. In other words, at the time the plants are transplanted they have a large number of leaves or transpiring surface and the amount of water lost is relatively great. To replace the amount of water lost per unit time a considerable portion of the root system must be retained. This is done by balling and burlapping the roots or by growing the plants in containers. In balling and burlapping, field-grown plants are dug with a large portion of the roots set in a ball of soil which is held together by burlap or similar material. This practice has certain disadvantages: (1) with large plants the severe pruning of the root system removes carbohydrates that have been made; and (2) the operation requires considerable labor. To overcome these disadvantages many nurserymen are now growing ornamental plants in containers. In this way, the root system is not pruned as in balling and burlapping, and it is injured only slightly, if at all in the transplanting operation.

Season of the Year The transplanting of evergreen trees and shrubs is usually done in the fall and winter in warm areas and in early spring in cold areas. In general, during these seasons the rate of transpiration is lower than during the summer, and the deficit of water within the tissues is likely to be mild and of short duration. However, in southern Florida, trees and shrubs are transplanted during the summer. During this season the rainfall is higher than during the winter. Because of the high rainfall, the light intensity is relatively low, the relative humidity is high, and, as a result, the rate of transpiration is correspondingly low.

TRANSPLANTING DECIDUOUS TREES AND SHRUBS

How does transplanting influence the growth of deciduous trees and shrubs? The student should remember that, unlike herbaceous plants and evergreen trees and shrubs, *deciduous trees and shrubs shed their leaves in the fall or early winter and develop a new set of leaves in the spring.* Since most of the water is lost through the leaves and since during the period of recovery transpiration should be low, it follows that deciduous trees and shrubs are transplanted most advantageously when they are without leaves, that is, during the nongrowing season.

During this period transpiration is at a minimum, a water deficit within the plants does not take place, and the root system has an opportunity to become established before stem and leaf growth begins.[1] In general, the time of transplanting during the nongrowing season for any given region depends on the temperature of the soil. Thus, in regions characterized by mild winter temperatures, as for example in the southern United States, transplanting can be done in late fall and early winter, since the soil temperature is sufficiently high for the development of the new root system during the latter part of the winter and air temperature and light intensity increase rapidly in the spring. On the other hand, in regions characterized by severe winter temperatures, for example in the northern United States, transplanting should be done in the early spring only, since soil temperature inhibits root growth during the winter and air temperature and light intensity increase slowly during spring and early summer. However, regardless of the region or climatic condition, if the trees or shrubs are very large, they should be removed with the ball of soil around the base of the plants. (See Fig. 13-2)

Use of Waxes

Investigations have shown that the use of emulsified waxes facilitates the rate of recovery of trees and shrubs in full leaf. These emulsified waxes are

[1]Roots do not have rest periods.

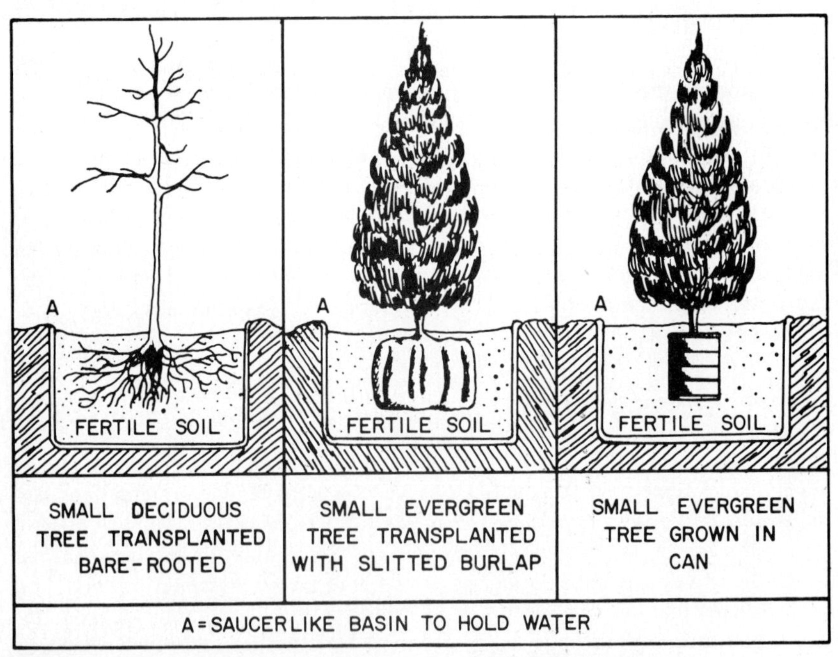

Figure 13-2 The preparation of soil for the transplanting of small trees in the home grounds. (*Adapted from* South Carolina Agr. Ext. Serv. Circ. *430, 1957*).

soluble in water, nontoxic to plant tissues, relatively permeable and colorless, and can be applied to plants in the form of a spray. The principal purpose of the wax is to reduce the rate of transpiration without reducing the rate of photosynthesis or interfering with the rate of respiration. Thus, the outgo of water is reduced, wilting is avoided, photosynthesis takes place, and recovery from transplanting is facilitated. Applications of these emulsions to trees at the time of digging and again after setting facilitate recovery. For example, in 1934 a nursery in Michigan transplanted 22 moderately large elms in full leaf. Of these trees, 11 were sprayed with wax emulsion and 11 were sprayed with water only. Of the 11 trees sprayed with wax emulsion, 10 recovered, and of the 11 trees sprayed with water, only 1 survived the transplanting operation.

QUESTIONS

1. What are the immediate and subsequent effects of transplanting herbaceous plants?
2. Show how transplanting is likely to reduce size of the cells in the region of elongation.
3. Since transplanting herbaceous plants is likely to reduce the rate of photosynthesis, why transplant? Give two reasons.
4. In the same environment a plant with a large leaf surface has a greater outgo of water than a plant with a small leaf surface. Explain.
5. Whenever feasible, herbaceous plants should be transplanted in the seedling stage. Explain.
6. In general, plants with a rapid rate of root regeneration recover more rapidly than plants with a slow rate of root regeneration. Explain.
7. How does hardening off herbaceous plants facilitate recovery from transplanting?
8. Show how hardened-off cabbage plants withstand subfreezing temperatures, whereas hardened-off tomato plants cannot.
9. In your opinion, which method would be more effective in hardening-off herbaceous plants: (1) maintaining night temperatures just below the optimum range and keeping the soil moist or (2) maintaining night temperatures within the optimum range and keeping the soil relatively dry with plants not wilting? Give reasons.
10. Foggy weather or a light, misty rain is ideal transplanting weather for plants with leaves. Explain.
11. Given two comparable or identical lots of herbaceous plants. Lot A is transplanted in the morning; lot B is transplanted in late afternoon. Which lot is likely to recover the more rapidly? Explain.
12. Given two lots of plants transplanted to flats in the greenhouse. Lot A was shaded three or four days; lot B was not shaded. In your opinion which lot would recover the more rapidly from transplanting? Explain.
13. Show how the use of water or starter solution facilitates recovery from transplanting.
14. Large evergreens are usually transplanted with a ball of earth. Explain.
15. The use of containers in growing ornamental plants for transplanting is likely to take the place of balling and burlapping. Explain.
16. In general, the deciduous orchardist transplants his trees when they are without leaves. Explain.

17 In your opinion, which time of the year would be better for transplanting deciduous or evergreen trees or shrubs in your community: (1) the fall or (2) early spring? Explain.

18 Show how water-soluble wax emulsions facilitate recovery of trees and shrubs in full leaf from the check in growth incident to transplanting.

WEEDS

When weeds win all the people lose.

What is a weed? A weed is a plant growing in a place where it is not wanted. It may belong to an economic crop, for example, corn plants in fields of cotton or soybeans; it may have no economic value, for example, henbane in fields of spinach or carrots, or wild onion in lawns. In any case, weeds markedly decrease growth and marketable yields of crop plants. They use some of the finite quantities of carbon dioxide, water, and essential raw materials, particularly the available nitrogen supply; and if the weeds are taller than the crop plants, they use the light. Further, within the temperate climatic zones, many weeds are C_4 plants, whereas most, if not all, temperate crop plants are C_3 plants. As stated in Chapter 1, C_4 plants have a greater rate of net photosynthesis than C_3 plants. Thus, C_4 weeds in fields of C_3 crops have a greater competitive advantage; this may explain, partially at least, why certain weeds markedly reduce marketable yields. However, regardless of whether weeds belong to the C_3 or C_4 group, they cause farmers and gardeners in the United States a loss of about 4 billion dollars per year. Comparable or greater losses are sustained in other parts of the world.

METHODS OF CONTROLLING WEEDS

Cultivation

Cultivation is the intertillage of a crop, and its primary purpose is to control weeds. Study the data in Table 13-2. Note the extremely low yields of the garden beets in the weed plots. The tests were conducted at Cornell University from 1920 to 1925 inclusive. In one set of plots the weeds were allowed to grow, and

Table 13-2 Effect of Weeds on the Yield of Garden Beets Grown at Ithaca, New York, 1920–1925

Number of roots per plot		Average weight of roots, lb	
Cultivated plot	Weed plot	Cultivated plot	Weed plot
381	149	60.6	9.3

in another comparable set they were controlled by cultivation. Experiments with other crops (carrot, cabbage, onion, celery, potato, and tomato) produced similar results. Many crop-plant scientists believe that even a small growth of weeds reduces yields from 20 to 50 percent. Successful growers keep their gardens and fields practically free from weeds.

The question arises: Is cultivation beneficial in the absence of weeds? An important factor is the relative extensiveness of the root systems of crop plants. Note that the marketable yields of the cultivated and scraped plots as presented in Table 13-3 are practically the same for all crops except those of celery and onion. What is the explanation for the similarity of yield of both the cultivated and the scraped plots of cabbage, carrots, and tomatoes, and what is the explanation for the marked increase in yield for the cultivated plots of celery and the moderate increase for the cultivated plots of onions? The investigators studied the root systems of each of these crops. They found that cabbage, carrots, and tomatoes develop extensive root systems, whereas celery and onions develop sparse root systems. Thus, under the conditions of the experiments, only sparsely rooted crops may be expected to respond to cultivation in the absence of weeds. In fact, cultivation in the absence of weeds of extensively rooted crops, particularly during the later stages of growth, is likely to severely prune most of the absorbing roots just below the surface. The severe pruning reduces carbohydrates already made and the area of absorption. This reduces the amount of water going into the plant per unit time, and this, in turn, is likely to lead to a water deficit within the plant, with a resulting decrease in cell enlargement, photosynthesis, and marketable yield.

Another factor is whether soils crust after a rain. Certain soils form a crust after a rain, whereas other soils do not. Theoretically, the crust limits the flow of

Table 13-3 Mean Yields Per Plot for Six Years (1920–1925) of Vegetable Crops on Cultivated and Scraped Soils with Weeds Absent in Both Cases*

		Cultivated		Scraped		Increase in favor of cultivation, %
Kind	Product	Number	Weight, lb†	Number	Weight, lb†	
Celery	Plants	145	145	143	116	24
Onion	Bulbs	269	78	270	72	8
Beet	Roots	381	61	382	58	4
Carrot	Roots	611	87	637	84	3
Cabbage	Heads	50	119	50	119	0
Tomato	Fruit	699	188	714	186	1

*All crops except tomatoes were grown in a gravelly fine loam, and tomatoes were grown in a soil containing a large amount of clay.

†1 lb. = 453.6 grams

Source: Adapted from Tables 1–6, *Cornell Agr. Exp. Sta. Mem.* 107, 1927.

oxygen to and the flow of carbon dioxide away from the roots and the beneficial bacteria. This, in turn, is likely to reduce the growth and the absorbing capacity of the roots and to decrease the available nitrogen supply. For example, experiments in Ohio with Brookston sandy loam, a noncrusting soil high in organic matter, and with Miami sandy loam, a crusting soil low in organic matter, and with corn as the test crop, showed that cultivation of the Brookston soil had no value other than the control of weeds; but cultivation of the Miami soil, in addition to weed control, increased the nitrate and water supply of the soil and the yields of the crop.

Cultivation Tools Cultivators are fitted with various types of attachments, depending on the kind of cultivation to be done. In general, there are two types: (1) scraper and (2) toothed. The toothed types cultivate relatively deep, and the scraper types cultivate relatively shallow. Each type has its specific purpose. Examples of the toothed type are the cultivator steels. These vary in width from 1 to 4 inches (2.5 to 10.2 centimeters). In general, narrow steels are used for fine work, and broad steels for coarse work. Examples of the scraper type are one-sided sweeps, two-sided sweeps, and horizontal blades.

The Cultivation Program

The frequency and depth of cultivation of any particular crop depend largely on the following factors: (1) frequency with which crops of weeds become established and (2) amount of top and root growth of the crop plants.

Weed Establishment The best time to kill weeds is when they are young. During the seedling stage most weeds are easily displaced or covered with soil, and they have not become sufficiently large to seriously compete with the crop for carbon dioxide, light, water, and essential raw materials. Obviously, with other factors equal, the number of cultivations will depend on the frequency with which crops of weeds become established. For this reason, a continuously moist soil usually requires a greater number of cultivations than a continuously dry soil.

Amount of Top and Root Growth A direct relation exists between the amount of top growth a crop is making and the amount and extent of the root system. Thus, during the seedling stage the root system of any given plant is relatively small and inextensive. As the plant continues to grow, the root system continues to grow also, and with most crops it gradually becomes very extensive. Consequently, the first and second cultivations of any given crop may be relatively deep, particularly if the crop has been transplanted. Succeeding cultivations should be relatively shallow and proceed further from the plants to avoid cutting the feeding roots just beneath the surface. Certain authorities believe that if weeds are absent when herbaceous annual crops are half grown, cultivation should cease.

Using Herbicides

Herbicides are chemical compounds which are used to control weeds. They fall into two groups: (1) compounds which are nonselective in action and (2) compounds which are selective.

Nonselective compounds are toxic to all plants, e.g., borax, sodium chloride, and ammonium sulfamate, called Ammate. As would be expected, each of these nonselective compounds has advantages and disadvantages; one compound may be more effective and desirable for use under one set of conditions, and another compound may be more effective and desirable under another.

Selective compounds kill certain kinds of plants and do very little, if any, damage to other kinds. Thus, selective compounds for any given crop should kill all or practically all the weeds in that crop and do no damage to the crop plants. (See Fig. 13-3) These compounds may be placed in two groups: (1) the toxic and (2) the growth regulators. In general, toxic compounds kill the protoplasm by direct chemical action, and the growth regulators induce an abnormal development of certain tissues. Examples of toxic compounds and growth regulators used in the control of weeds, together with the preferred type of application and the weeds found in the crop which are not usually controlled by the chemical, are presented in Table 13-4.

In general, the method of application is based on the time the chemical is applied, with reference to the time the seed is planted. The methods are (1) preplanting incorporated, (2) preemergence, and (3) postemergence. In the preplanting incorporated, the soil is prepared to receive the seed, the chemical is

Figure 13-3 Effect of a petroleum product on control of weeds in carrots. *(Courtesy, W. H. Lachman, University of Massachusetts.)*

Table 13-4 Examples of Selective Herbicides

Type of compound	Common or trade name	Crop	Recommended type of application	Weeds not usually controlled
Dinitro	DNSB	Snap bean Lima bean Garden peas Sweet corn Potato	Preemergence	Late Weeds
Toluidines	Treflan	Snap bean Lima bean Okra Cowpeas	Preplanting Incorporated	Deep germinating dicots
Substituted ureas	Monuron	Asparagus	Before and after harvest	Nutsedge
	Lorox	Carrots	Preemergence or postemergence	Nutsedge
Phthalic	Alanap	Vine Crops	Preemergence	
Amides	Diphenamid	Pepper Tomato	Preemergence or postransplanting	
Triazines	Atrazine	Sweet corn	Preemergence	Creeping grasses
Phenoxyacids	2, 4-D	Sweet corn	Postemergence	Grasses

Source: The Mississippi and Virginia Agricultural Experiment Stations.

applied and mixed with the upper 1 or 2 inches (2.5 or 5.1 centimeters) of soil, and later the seed is planted. Thus, the seed is planted after the chemical has been applied. The main advantage of this method is that the seeds germinate in soil which is practically free from weeds. The main disadvantage is that late-germinating weeds are likely to be a problem. In the preemergence, the soil is prepared to receive the seed, and the chemical is applied to the surface of the soil at the same time or just after the seed is planted. Thus, the seed is planted at practically the same time the chemical is applied. The main advantage of this method is that the chemical is applied when many kinds of weeds are susceptible to injury. The main disadvantage is that heavy rains may leach the chemical to the vicinity of the seed and thus retard or prevent its germination. In the post-emergence, the soil is prepared to receive the seed, the seed is planted immediately afterwards, and the chemical is applied when the crop plants are in the seedling stage. Thus, the seed is planted before the chemical is applied. The main advantage of this method is that possible injury to the germinating seed is avoided. The main disadvantage is that the weeds are likely to be quite resistant to the chemical, since mature tissues are more resistant than immature tissues.

Combinations with Cultivation

Cultivation combined with the use of selective herbicides may be necessary or more desirable for certain crop industries. For example, compact soils may require cultivation even though selective chemicals satisfactorily control the weeds. Study the data in Table 13-5. Note the two weed-control treatments and the marked decrease in yield of carrots due to the compactness of the soil. Evidently, compact soils are unfavorable for the growth and development of carrot roots. As previously stated, compact soils are likely to limit the flow of oxygen to and carbon dioxide away from the absorbing roots and the nitrifying and nitrogen-fixing bacteria. This would seriously interfere with normal respiration and functioning of the root system and the respiration of the bacteria.

Further, combinations with cultivation may be necessary because of the scarcity of hand labor. According to research at the Mississippi Delta Experiment Station, the use of selective herbicides markedly reduced hoe-labor requirements in cotton. But, it did not always result in corresponding reduction in weed-control costs. In many cases the cost of the chemicals and their application offset the savings made in reducing hoe labor.

The use of both nonselective and selective chemicals in the control of weeds is in a state of flux. The U.S. Department of Agriculture and many experiment stations and chemical companies have active research projects. For this reason, the student should consult a local experiment station for the latest information.

Table 13-5 Effect of Two Weed-Control Treatments on the Yield of Carrots, 1948*

Weed-control treatment	Condition of soil at harvest	Yield, lb marketable roots/acre	
		Woodstown sandy loam	Steinburg silt loam
Cultivation and hand weeding	Not compact	11,126	11,658
No cultivation, but hand weeding and oil spray	Compact	7,007	1,204

*A 5-10-10 mixture was applied to each soil at the rate of 1,000 pounds per acre and disked-in immediately after plowing.

Source: Science, Dec. 3, 1948.

MULCHING

Mulching with Organic Materials

Mulching with organic materials consists of placing a layer of organic materials on the surface of the soil.

In general, mulches during the summer particularly in hot climates (1) reduce the rate of evaporation of water from the surface of the soil, thus conserving soil moisture; (2) keep the upper layer of soil from getting extremely hot during the midportion of sunny days, thus keeping soil temperatures close to or within the optimum range for root growth; (3) protect the soil from the bulletlike impact of intense rain, thus reducing soil erosion; (4) eliminate light from the surface of the soil, thus preventing the germination of many kinds of weed seed; and (5) prevent the splashing of soil particles on fruit close to the surface of the soil.

In general, mulches during the winter particularly in cold climates (1) reduce the loss of heat from the surface of the soil, thus keeping the soil comparatively warm; (2) prevent heaving of the soil, thus keeping the root system intact; (3) reduce the absorption of heat in the early spring, thus delaying the growth and, in some plants, the opening of the flowers until the last killing frost; (4) keep the fruit clean, as do summer mulches.

Mulching With Plastics

Mulching with plastics is becoming a standard practice of both commercial and home gardeners throughout the world. In general, the practice consists of laying and anchoring 4- or 5-foot (1.2- to 1.5-meter) sheets of polyethylene plastic on the surface of soil prepared for planting. (See Fig. 13-4) The chemical and physical properties of polyethylene are well adapted to crop production. It is relatively cheap and easy to handle, and more particularly, it is impervious to water and pervious to carbon dioxide and oxygen. In this way, polyethylene does not interfere with the normal respiration of the plant roots and those of beneficial soil organisms, particularly in the diffusion of carbon dioxide away from and the diffusion of oxygen to the respiring cells.

From the standpoint of color, there are two main types: black and clear. Benefits which these types have in common are the ability to (1) markedly reduce the evaporation of water from the surface of the soil, (2) maintain the soil in a loose, friable, and well-aerated condition, (3) eliminate the need for cultivating the covered area, (4) protect the soil from the bulletlike impact of intense rains, and (5) prevent the splashing of soil particles on fruit close to the surface of the soil. In sharp contrast to clear plastic, black plastic prevents the germination and the subsequent growth of weed seeds—a marked advantage. However it increases the temperature of the soil occupied by the plants' roots only slightly, if at all, whereas clear plastic promotes the germination and growth of weeds, and it increases the temperature of the soil occupied by the plants' roots. Thus, unless

Figure 13-4 A film mulch-laying machine in operation. Note the smooth surface of the bed and the way the edges of the film are anchored. *(Courtesy, E. L. Moore, Mississippi State University.)*

a weedicide is applied just before the application of the clear plastic, weeds are likely to become a serious problem. Study the data in Table 13-6. Note the marked increase in both early and total yields for each type of plastic, the differences in carbon dioxide availablity for each group of plants, and the differences and similarities in soil temperature. Why did the black plastic increase yields over bare soil? Why did the clear plastic produce somewhat greater yields than the black plastic? In your own section of the country, would black or clear plastic produce the greater yields or would either type be equally satisfactory? The use of black plastic laid on the surface of the soil and overlaid by clear plastic is in the experimental stage.

Table 13-6 Effect of Clear and Black Plastic on CO_2 Availability, Soil Temperature, and Yield of Muskmelons

	Yield* bus./A		CO_2	Temp.† soil °C at depths (cm)		
Type of mulch	Early	Total	ppm	2.5	15	61
Clear plastic + weedicide	502	666	1297	28	23	18
Black plastic	430	1300	1300	25	20	17
Bare soil	89	325	325	25	21	17

*From Tables 1 and 3, *Cornell Agr. Exp. Sta. Bul.* 1180, 1967.
†From *Conn. Agr. Exp. Sta. Bul.* 634, 1960.

QUESTIONS

1 What is a weed?
2 Weeds markedly decrease yields. Give four reasons.
3 Cabbage has an extensive root system. When the heads begin to form, the roots occupy all the fertile soil. Outline a cultivation program for your particular section of the country.
4 Celery has a sparse root system. The roots occupy soil not more than 6 inches from the basal portion of the plant. Outline a cultivation program for your particular section of the country.
5 In general, rains are light in intensity in the northern United States and heavy in intensity in the southern United States. Would the cultivation program for a crop of tomatoes be the same for each region? Give reasons.
6 Given two types of soil. Soil A is self-mulching; soil B is self-crusting. Which soil requires cultivation after a rain? Explain.
7 In general, the toothed type of cultivator is used when the crop is in the seedling stage. The scraper type is used when the crop is more than half grown. Explain.
8 Given a wet season during the period of head formation of a cabbage crop. Show how deep cultivation would reduce the bursting of the heads.
9 Under what conditions would deep cultivation lower the rate of photosynthesis of a crop, and under what conditions would it have little effect? Give reasons.
10 What are herbicides?
11 State the main difference between nonselective and selective herbicides.
12 How are preplanting incorporated, preemergence, and postemergence treatments made?
13 In general, preemergence treatments should be very effective in weed control. Explain.
14 Compact soils may require cultivation even though the use of selective herbicides satisfactorily controls the weeds. Explain.
15 In general, mulches of black plastic have markedly increased the yield of warm-season vegetable crops in cool climates. Explain.

IRRIGATION

> In the near future farmers will be independent of dry weather, and water for irrigation purposes will be deemed as necessary as food in the cultivation of crops. —*Stephen D. Lee, 1895, first president of Mississippi A. & M. College.*

Natural Supplies of Water

The principal sources of water are rain and snow. The annual amount of rain and snow varies greatly from region to region within any given season and from season to season within any given region. Based on the average annual precipitation, there are approximately three distinguishable regions: the arid, with less than 10 inches (25.4 centimeters), the semiarid, with 10 to 30 inches (25.4 to

76.2 centimeters), and the humid, with 30 to 100 inches (76.2 to 254.0 centimeters) or more. Note the arid, semiarid, and humid regions of the continental United States in Figure 13-5. The arid region has semiarid districts scattered throughout it and comprises the great intermountain area of the West. The semiarid comprises the Great Plains. The humid comprises the eastern one-half and the extreme northwest corner of the country. A similar situation exists in Australia. The region between the East Coast and the Great Dividing Range is humid, whereas the area west of this range is arid and semiarid. Other examples could be cited.

Is the rain from year to year adequate for the growth of crops in each of these regions? Since from 10 to 30 inches, (25.4 to 76.2 centimeters) of water is necessary for the growth and development of most horticultural crops, the rain in the arid and semiarid regions is entirely inadequate. Hence, applications of irrigation water are necessary. In fact, irrigation of the crops of these regions is considered the most important cultural practice, and the value of the land is determined by its accessibility to an abundant source of suitable water. In these regions, laws and regulations governing the use of irrigation water have been established. In general, these laws, usually known as water rights, have been

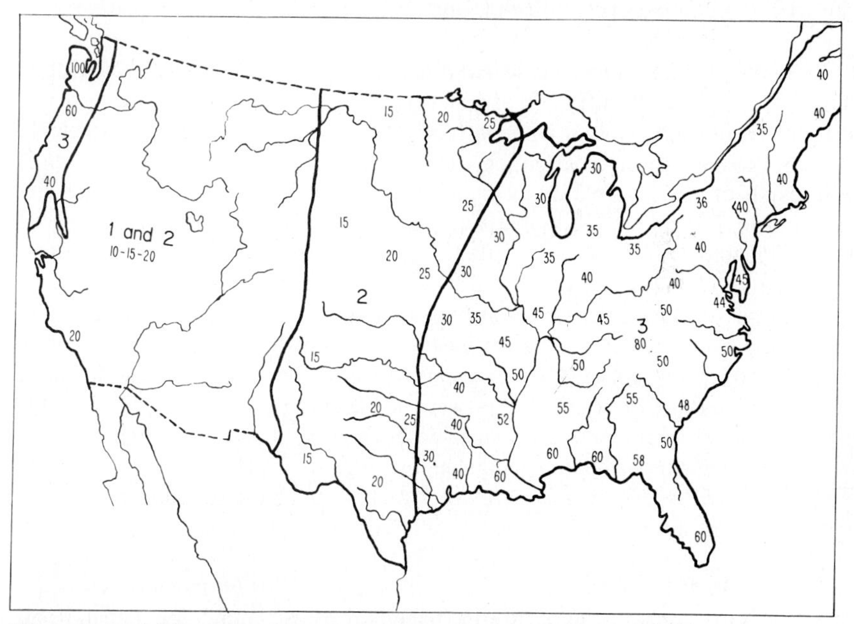

Figure 13-5 Arid, semiarid, and humid regions of the continental United States. 1;2: mostly arid about 10 inches, or 25.4 centimeters, per year, with semiarid districts scattered throughout the area. 2: semiarid (15-30 inches, 37.5–75.0 centimeters, per year). 3: humid (30–120 inches, or 75.0–300 centimeters, per year).

established so that any given individual grower of any given producing district obtains his proportionate share of irrigation water from any given source.

What is the situation in the humid regions? The records show that the rainfall in these regions for the most part is adequate. However, crops growing in these regions frequently suffer from lack of water. In other words, the rains do not always come when they are needed. Every year in some districts there is likely to be a droughty period. This period may last for only one or two weeks; yet it may come at a time when crops require a high moisture content of the soil for satisfactory growth. In general, these droughty periods occur quite frequently. For example, there were 127 droughts in the eastern part of the country from 1919 to 1939. This is an average of six droughts per year. To look at the situation in another way, a study of the annual rainfall at Mississippi State, from 1919 to 1949, a period of 30 years, showed that on the average 42 percent, or 95 days, of the 226-day frost-free growing period did not receive sufficient rain for abundant plant growth. As previously stated, droughts induce water deficits within the plant. This, in turn, decreases the size of the cells in the region of elongation, lowers the rate of net photosynthesis, and decreases yield and quality. As a result, the money spent in buying seed, commercial fertilizers, and other materials and the time spent in preparing the soil, planting the seed, applying the commercial fertilizers, and controlling the weeds are wasted.

Results of Experiments in Humid Areas Since the annual rainfall in humid areas is usually sufficient in quantity, many growers store the water when it is not needed and apply it when needed. What are the increases in marketable yield due to application of irrigation water during periods of insufficient rainfall? The results of nine tests within the humid regions of continental United States are set forth in Table 13-7. Note that supplementary irrigation markedly increased the marketable yield of all of the 10 crop plants. In some cases, only one or two applications of water were necessary for the production of high marketable yields, e.g., in Illinois only one furrow irrigation was made to the tomatoes, in Mississippi only two furrow irrigations were made to the snap beans, and in Michigan only 3.6 inches (9 centimeters) of water were applied to the irrigated crop. Obviously, these applications were made during a critical period in the growth and development of the plants. In general, as a result of numerous tests with supplementary irrigation within humid regions, applications of water are becoming increasingly common for orchards, vegetable-crop and ornamental-crop plantings, lawns, golf courses, parks, and so forth.

Sources of Irrigation Water The principal sources of irrigation water are large springs, streams, lakes, wells, ponds, reservoirs, and municipal water supplies. In the arid and semiarid regions, the source of water is generally above the land to be irrigated and the water is carried over the land by gravity.[2] In the

[2]The western half of the San Joaquin Valley of California and the winter-garden area and Rio Grande Valley of south Texas are exceptions.

Table 13-7 Effect of Irrigation on the Marketable Yield of 10 Crops in the Humid Section of the U.S.

Crop	State in USA	Year	Yield unit	Rainfall only	Rainfall + irrigation	Publication
Apples	Ark.	1964	Tons/A	11.3	16.5	*Ark. Farm Res.* 14 (5), '65
Carrots	Ark.	1965	Tons/A	9.6	19.5	*Ark. Farm Res.* 14 (3), '65
Onions	N.Y.	1952	Bus./A	520	913	*Amer. Veg. Grower* 2 (6), '54
Pickles	Mich.	1957	Bus./A	213	301	*Mich. Quart. Bul.* 40 (4), '58
Potatoes	Mich.	1955	Bus./A	362	609	*Mich. Quart. Bul.* 46 (4), '64
Potatoes	N.Y.	1952	Bus./A	525	713	*Amer. Pot. Jour.* 29, '52
Snapbeans	Miss.	1951	Bus./A	78	239	*Miss. Agr. Exp. Sta. Cir.* 182, '53
Sweet corn	Miss.	1951	Doz. ears/A	399	1013	*Miss. Agr. Exp. Sta. Cir.* 182, '53
Sweet potatoes	Mo.	1955	Bus./A	201	414	*Mkt. Growers Jour.* 84 (12), '53
Tomatoes	Miss.	1948	Boxes/A	90	145	*Miss. Agr. Exp. Sta. Cir.* 182, '53

humid regions, the source of water is usually below the level of the land to be irrigated and pumps are necessary to lift the water. Like growers in the arid regions, growers in this region are interested in the annual precipitation, since this rainfall maintains the lakes, streams, ponds, reservoirs, and surface wells.

Water suitable for irrigation is moderately to slightly acid and free from appreciable quantities of salts, particularly sodium chloride which ionizes, forming sodium and chlorine ions. In districts where the water is quite alkaline, it is acidified to prevent the formation of highly alkaline soil with its resultant deficiency of available nitrogen, phosphorus, and other essential elements. Chemicals generally used are sulfuric acid and phosphoric acid. The Ohio and Michigan Experiment Stations have developed machines for the acidification of water for greenhouse use.

The right to use water in the humid regions is now assuming the same importance as that in the arid and semiarid regions. In fact, many states have made surveys of water resources and are now considering the development of codes embodying water rights and the use of irrigation water by growers of crops.

Methods of Applying Irrigation Water

Three more or less distinct methods by which water is applied to the land and crops are (1) surface irrigation, (2) subirrigation, and (3) spray irrigation. Each method has advantages and disadvantages.

Surface irrigation is the application of water directly over the surface of the land. In general, this method requires a gentle slope, a deep, compact, uniformly textured soil, and plentiful supplies of water. Thus, it is not adaptable on rolling land or on fields which contain shallow topsoils. Two systems of surface irrigation are used: (1) border, or basin, and (2) furrow.

For border irrigation the land is leveled, if necessary, and bordered. The borders vary from 6 to 8 inches (15 to 20 centimeters) in height, generally follow the contours, and run in both directions. This method is commonly used to irrigate orchards in California and spinach in south Texas.

For furrow irrigation the land is leveled or ridged. The furrows vary from 4 to 10 inches (10 to 25 centimeters) in depth and follow the contours of the land. This method is commonly used to irrigate most row crops in the arid and semiarid regions. In fact, the many horticultural crop industries of the West have been made possible by the man's engineering skill in applying water to the land in this region. Note the furrow-irrigated potato field in Figure 13-6.

Subirrigation is the application of water below the surface of the land. In general, this method requires relatively large quantities of water and specific soil strata: (1) an impervious lower layer to hold the water against the force of gravity (clay or hardpan), (2) an open, porous, intermediate layer to serve as a reservoir for water (sand or sandy loam), and (3) a finely textured top layer to facilitate capillary action (fine sandy loam, silt loam, or peat). Subirrigation is used

Figure 13-6 Furrow irrigation of potatoes in Idaho. *(Courtesy, Soil Conservation Service, U.S. Department of Agriculture, Washington, D.C.)*

extensively in certain vegetable districts in Florida and in the celery and onion districts of Michigan and New York. In Florida the water is obtained from artesian wells and is conducted through lines of 3-inch (7.5-centimeter) drain tile placed 18 inches (35.4 centimeters) deep and 24 feet (7.2 meters) apart. In Michigan and New York the soil—a slightly acid, highly decomposed muck—is drained by ditches, and the height of the water table is controlled by a series of dams in the ditches.

A rather unique system of subirrigation is practiced by growers in southern Florida. Two requisites are necessary: (1) an abundant annual rainfall, and (2) a deep, highly decomposed muck, or sapric soil. The source of water is Lake Okeechobee. During the relatively dry winter season, the water is channeled from the lake to grower's fields, first by means of pumps from the lake to large ditches called *canals*, from the canals to relatively small ditches which are connected with and situated at right angles to the canals, and finally through holes in the muck, situated at right angles to the ditches. The holes, called *moles*, are made by pulling a bullet-shaped mass of metal through the soil about 3 feet (0.9 meters) deep with a tractor. The height of the water table for each crop is maintained by pumping water from the canals into the ditches until the desired level is attained. During the relatively wet summer season, the flow of water is reversed again, by means of pumps, from the ditches to the canals and from the canals into the lake.

In the Lake Okeechobee district, the moles are remade at intervals of three or four years, and in West Germany and Great Britain at intervals of 10 to 15 years. Thus, in all seasons the water table is maintained at the desired level with very little labor and at very low cost. In fact, with this method the crop plants seldom suffer from a lack of water during periods of drought or from excess water during periods of heavy rains. From the standpoint of economy, with the welfare of the crop plant also in mind, this method is an excellent way of controlling the water supply to crop plants.

Spray irrigation is the application of water on the surface of the land in the form of spray similar to a gentle rain. This type of irrigation is adapted to all types of soils and to both level and rolling land and generally requires less labor and less water than surface irrigation. The types of spray irrigation systems are classified according to the materials which carry and apply the water. These are (1) the stationary nozzle, (2) the portable sprinkler, (3) the perforated pipe, and (4) the trickle.

The stationary nozzle consists of nonportable parallel lines of galvanized iron pipes. The lines are usually placed 50 feet (15 meters) apart and are supported on low posts 3 to 4 feet (0.9 to 1.2 meters) high, on high posts 10 to 20 feet (3 to 6 meters) high, or on a cable supported by posts 12 to 20 feet (3.6 to 6.0 meters) high. The high-post system is more popular than the low-post, since it allows teams, tractors, and workers to pass under the pipelines and utilizes the land to better advantage. Each line of pipe is equipped with nozzles and an oscillator. The nozzles are of two types: (1) those which deflect and thus break the water into a fine mist (generally used in greenhouses) and (2) those which discharge the water in the form of a small stream (generally used outdoors). When water is running through the system, the oscillator turns the pipe through an arc of 60 to 90°. Thus, pipes equipped with oscillators require less attention and facilitate a more uniform and steady application of water than pipes that are not so equipped. This system of irrigation is used extensively by market gardeners and florists throughout the country. A trip to any of the large market-garden districts in the United States will convince the student of its profitable use and application. Its main disadvantage is the high initial cost.

The *portable sprinkler* consists of feed lines with joints which are quickly assembled and taken apart. The sprinklers are mounted on the pipes and are spaced 20 to 40 feet (6 to 12 meters) apart, depending on the type of nozzle and the water pressure used. The main advantage of the portable pipe system is the relatively low initial cost. This type of irrigation is becoming popular in different regions of the country. Note the portable sprinkler system in operation in Fig. 13-7.

The *perforated sprinkler* consists of the same type of feed lines as that of the portable rotary system, but instead of rotary sprinklers perforated pipes are used. In general, these pipes are laid on the surface of the land with the perforations on the upper side. This system is adaptable to all crops and to all soils which have high infiltration rates.

The *trickle system* consists of large main feeder lines, relatively small laterals, usually ½ inch (1.7 centimeters) in diameter, situated at right angles to the feeder lines, and very small plastic tubes 0.036 inches (0.09 centimeters) in

Figure 13-7 Portable spray irrigation of a celery field in Florida. *(Courtesy, U.S. Department of Agriculture, Washington, D.C.)*

diameter called *emitters*. The main lines carry the water from its source to the laterals, and the laterals, in turn, carry the water to the emitters. (See Fig. 13-8) In orchard irrigation, an emitter is placed under each tree, and in herbaceous row crop irrigation, the emitters are placed 12 to 24 inches (30 to 60 centimeters) apart, depending on the rate of water flow and the infiltration capacity of the soil. In general, the water is applied under low pressure, about 15 pounds per square inch (1055 grams per square centimeter) or less, at a slow rate of 1 to 2 gallons (4 to 8 liters) of water per hour, and on the surface of the soil, directly above the plant's root system. The main advantage of this method is that the soil occupied by the plant's roots is maintained at or just below its field capacity, water deficits within the plant are not likely to take place, and practically no water is wasted. The main disadvantage is that in arid and semiarid districts the presence of excess salts in the soils solution is likely to become a problem, since trickle irrigation does not flush out excess salts as readily as do other types of surface irrigation.

Amount of Water per Application and Time Interval Between Applications

In general, the amount of water that should be applied for any given irrigation depends on (1) the stage of growth of the crop, (2) the depth of the absorbing system, and (3) the field capacity of the soil. Since capillary water moves for short distances only, the amount of water applied at any given time should fill the rooting zone and no more. Thus, crops in the early stages of growth require less quantities per application than crops in the later stages. In like manner, mature crops which develop shallow root systems require less quantities per application than crops which develop moderately deep or deep root systems. For example, most of the roots of mature strawberry, celery, and onion plants are within 6 to

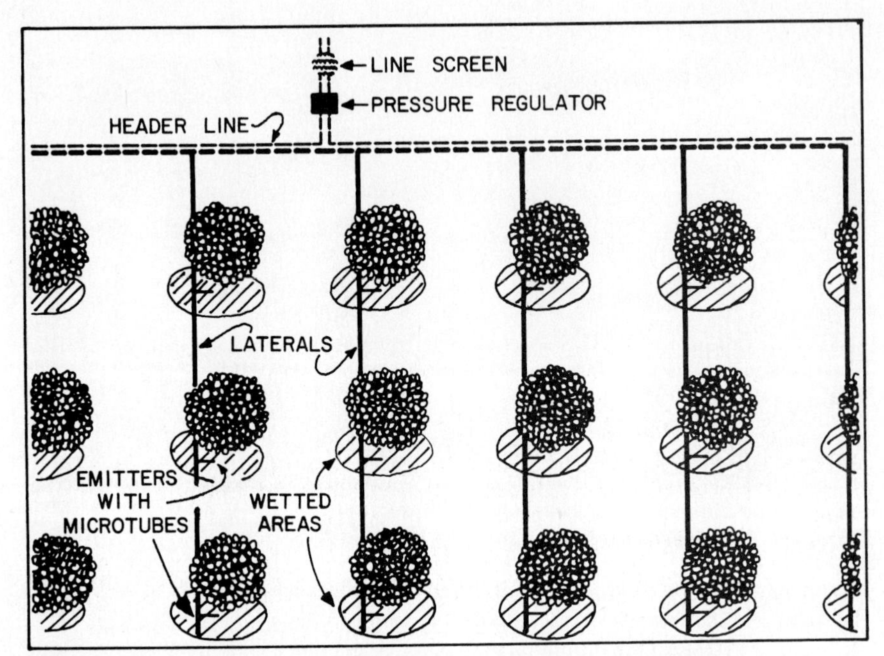

Figure 13-8 Layout of a trickle irrigation system for an orchard. *(Adapted from* Michigan Agr. Exp. Sta. Research Rept. *165, 1972).*

12 inches (15 to 30 centimeters) from the surface. For this reason, nothing is gained by applying more water than is needed to go to these depths. Further, under the same climatic conditions soils with low field capacities, the sands and sandy loams, require less quantities per application than soils with high field capacities, the silt loams and clay loams.

In general, the time interval between applications depends on (1) the area of transpiring surface, (2) the rate of transpiration, (3) the rate of evaporation of water from the soil, and (4) the field capacity of the soil. Thus, under the same conditions plants with a large transpiring surface (large leaf area) need more frequent irrigations than plants with a small transpiring area (small leaf area). In the same type of soil and at the same stage of growth, plants with high rates of transpiration require more frequent applications than plants with low rates. In like manner, soils with high rates of evaporation require shorter intervals between irrigations than soils with low rates. Finally, soils with low field capacities require more frequent applications than soils with high field capacities.

QUESTIONS

1 Increases in yield by irrigation depend largely on the rate of water absorption keeping up with the rate of transpiration. Explain.

2 For the same crop grown in comparable soils more water would be needed in the Southeast than in the Northeast or Middle West. Explain.
3 The mean rainfall in the humid East is sufficient for the growth of horticultural crops, yet these crops frequently produce low yields. Explain.
4 Investigations in apple orchards in the Cumberland Shenandoah Valley of the United States have shown that drought during the period of fruit growth decreases fruit size and that irrigation has increased fruit size. Explain.
5 Growers of vegetable crops for early market frequently irrigate their high-cash crops. Explain.
6 Irrigation systems are part of the necessary equipment of the commercial florist. Explain.
7 What are the principal sources of irrigation water in your community?
8 Distinguish between surface irrigation, subirrigation, spray irrigation, and trickle irrigation.
9 State the three requirements for successful surface irrigation.
10 In furrow irrigation, the slope for sandy loams should be from 10 to 12 inches per 100 feet (25.4 to 30.5 centimeters per 30.5 meters), and for clay loams from 1 to 3 inches per 100 feet (2.5 to 7.6 centimeters per 30.5 meters). Explain.
11 Draw a soil profile of a mineral soil which is particularly adaptable to subirrigation.
12 Compare furrow irrigation and spray irrigation from the standpoint of adaptability to the region east of the Mississippi River.
13 In general, growers who are inexperienced in irrigation should start in a small way and ''grow'' with the practice. Explain.
14 In general, any given farm is an irrigation problem within itself. Explain.
15 From your own observation and experience, describe how the home grounds are supplied with irrigation water.

SELECTED REFERENCES FOR FURTHER STUDY

Dennis, R., 1964. The fascinating story of herbicidal action. *National Agr. Chem. Assoc. News.* 22(4):4–5. A brief but pertinent article on the anatomy and physiology of crop plants and their reaction to herbicides.

Loomis, W. E. 1925. Studies in the transplanting of certain vegetable plants. *N.Y. (Cornell) Agr. Exp. Sta. Mem.* 87. The results of an outstanding investigation which put the transplanting of herbaceous crops on a scientific basis.

Publications by local agriculture experiment stations on irrigation. Many agriculture experiment stations throughout the world have a bulletin or circular on the selection and use of irrigation equipment and the effect of irrigation on yields under local conditions.

Thompson, H. C. 1927. Experimental studies of cultivation of certain vegetable crops. *N.Y. (Cornell) Agr. Exp. Sta. Mem.* 107. The results of the first investigation which placed the cultivation of herbaceous crops on a scientific basis.

Chapter 14

Pruning

They shall beat their swords into plowshares,
and their spears into pruning hooks.

Isaiah 2:4

THE OBJECT OF PRUNING

Pruning is a major horticultural practice. It is necessary for the successful production of tree fruits, grapes, small fruits, nut fruits, and many flower and ornamental plants. In general, pruning involves the removal of parts of the top or root system of plants. Woody tissues are usually removed by means of sharp cutting instruments in order to make clean, nonjagged wounds. Succulent tissues are usually removed by hand.

The object of pruning varies with the viewpoint of the pruner. Commercial orchardists, gardeners, and florists are interested primarily in profits. In other words, they consider such factors as yield, size, color, shape, or quality of flower or fruit only as these factors influence profits. On the other hand, amateur gardeners place costs and profits in a secondary category. They consider such factors as size, color, shape, or quality as important in themselves.

Both the stem and the root systems of plants are pruned. Since pruning the stem and pruning the root influence growth and development differently, they are discussed separately.

PRUNING THE STEM

All plants respond to removal of the top in two definite ways. Pruning the top (1) always reduces the total amount of growth which could otherwise be made and (2) always influences the vegetative-reproductive balance of the plant.

Total Amount of Growth Made

Numerous investigations have shown that pruning the top dwarfs the tree or plant. The removal of twigs of trees or stems of herbaceous plants reduces the amount of carbohydrates and other materials already made and reduces the number of leaves which would have contributed to further food manufacture. In other words, the removal of stems and leaves or of stems containing buds which would later develop into leaves reduces the machinery for food manufacture. This removal, in turn, reduces the total amount of food that can be made in any one season. For example, consider two branches of an apple tree. Each branch has four twigs of equal size and vigor, and each twig has 50 leaf buds which later develop into 50 leaves. Suppose we remove one twig from branch A and none from branch B. Although the individual leaves on branch A will be somewhat larger than those on branch B, with other factors favorable, the 200 leaves on branch B will make more food during the season than the 150 leaves on branch A. Thus, at the end of the growing season branch B will have made a greater amount of growth than branch A. As a second example, consider two tomato plants. Plant A is pruned to a single stem; that is, the laterals on the main axis are "pinched off" as soon as they begin to develop. Plant B is allowed to grow the natural way; that is, no stems are removed. During the fruiting period plant B will have the greater leaf area. This greater leaf area will manufacture a greater quantity of food, and because of this greater quantity of food, plant B will produce the greater number of tomatoes. As a third example, consider two chrysanthemum plants. Plant A is pruned to a single stem, and plant B is allowed to grow in the natural way. Plant A will produce one large flower at the top of the stem, and plant B will produce a small flower at the top of each of several stems. The total weight of the several small flowers produced by plant B will exceed the weight of the single large flower produced by plant A. Clearly, the removal of healthy stems and leaves reduces plant size and yield and the total amount of growth that can be made.

Vegetative-Reproductive Balance

What is the relative effect of pruning the top on the disposition of the carbohydrates and the vegetative-reproductive balance of plants? In general (1) pruning the top reduces the number of growing points of any given plant; (2) this increases the supply of available nitrogen and other essential elements to the growing points which remain; and (3) this, in turn, increases the number of cells which can be made. Pruning the top, therefore, promotes the making of cells and the utilization of carbohydrates. Accordingly, it promotes the vegetative phase and retards the reproductive phase. The stimulation of the vegetative phase and retardation of the reproductive phase may or may not be desirable. Much depends on the vigor of growth any given plant is making. If, for example, orchard trees are young and vigorous, pruning, if necessary, should be extemely light, since heavy pruning of the top delays flower-bud formation. On the other hand, if orchard trees are old and weakly vigorous, severe pruning of the top helps to promote vigor and rejuvention.

Kinds of Top Pruning

There are two kinds of top pruning: (1) heading back and (2) thinning out. In *heading back,* the terminal portion of twigs, canes, or shoots is removed, but the basal portion is not. In *thinning out,* the entire twig, cane, or shoot is removed. In general, heading back stimulates the development of more growing points than a corresponding thinning out. How does heading back stimulate the development of more growing points? The student should remember that terminal buds on

Figure 14-1 Proper method of thinning and heading back limbs of a young peach tree.

shoots secrete growth-inhibiting hormones which are translocated to the lateral buds. These growth-inhibiting hormones prevent the development of the lateral buds. When the terminal bud is removed, as in heading back, the formation and translocation of the growth-inhibiting substances cease and one to several lateral buds develop. Usually, the lateral buds just below a cut develop more, and they, in turn, manufacture the growth-inhibiting hormones which prevent the development of the lateral buds farther down. Thinning out does not have this effect as much as heading back. Thus, heading back induces the compact, dense, or much-branched type of growth, and thinning out induces the open or rangy type of growth.

PRUNING THE STEM OF WOODY PLANTS

Tree Training and Tree Forms

Training is a type of top pruning. It is particularly important in the many tree-fruit industries. Young trees are trained to develop a frame of scaffold limbs sufficiently strong to bear large crops of fruit without breaking. Thus, the development of strong crotches and well-spaced limbs is essential. Generally, this is done by heading back one limb to a greater extent than the other. Furthermore, since even a light pruning of young trees lengthens the time from planting to flowering and fruiting, the practice should be efficiently performed. The various forms to which trees are trained are of three general types: (1) central leader, (2) open center, or vase, and (3) modified leader.

Central Leader The central-leader type of tree has a main branch and a series of well-spaced "subordinate" lateral branches. Its main advantage is the development of strong crotches due to the interlacing of fibers at the junction of the limb and the trunk. Its main disadvantage is shading the interior of the tree. This shading weakens the central leader and thus shortens the life of the tree. Authorities state that trees of some varieties, such as the Wealthy apple, are trained rather easily to this form, but trees of other varieties, such as the Rhode Island Greening apple, are trained to this form with difficulty.

Open Center, or Vase The open-center, or vase, type of tree has no main or central branch but a series of well-spaced "coordinate" lateral branches. These laterals are given the same dominance by cutting them back equally each year. Practically the same number of leaves will develop on each branch. Thus, all the coordinate branches make the same growth each year. The main advantages of this tree form are that light penetration becomes sufficient for the fruiting of inner branches and that a low-headed tree develops, which facilitates pruning, thinning, spraying, and picking operations. Its main disadvantage is that the tree develops weak, crowded crotches which frequently break under severe stress and strain, such as bearing a heavy crop of fruit.

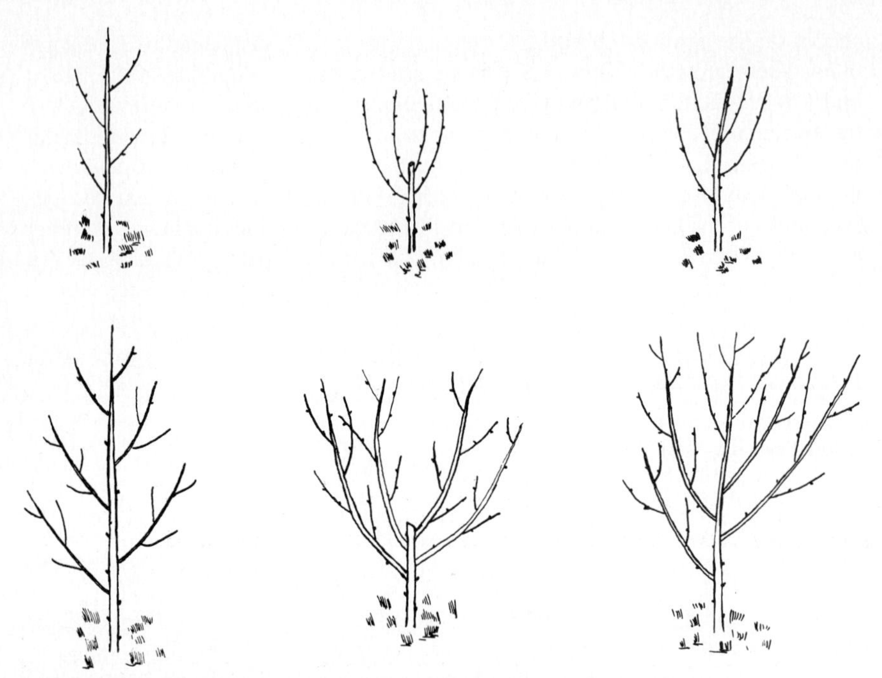

Figure 14-2 Outline of scaffold branches of the tree forms. Left: central leader. Center: open center. Right: modified leader.

Modified Leader The modified-leader type of tree is a happy medium between the central leader and open center. During the period of training the central leader is cut back slightly and not allowed to become dominant. The process of cutting back and selecting laterals is repeated until the proper number and distribution of branches have been obtained. The central leader is then discontinued, and the tree assumes a more or less rounded, open top. In this way the modified-leader type of tree has low and well-spaced limbs, well-distributed fruiting wood and is sufficiently close to the ground to facilitate the many orchard operations. Of the three types the modified leader is the most desirable for many tree fruits because it combines the most important advantages of the other types. Note the relative dominance of the main and lateral branches of each of the three types in Figure 14-2.

Top Pruning Specific Crops

Fruiting Habit of Pome Fruits (Apple and Pears) A working knowledge of the flowering and fruiting habits of crop plants is essential to a proper understanding of pruning requirements. In general, the fruit of the apple and pear is produced from mixed buds which are borne on short lateral branches called fruiting spurs. These spurs develop in the following manner. Certain laterally situated vegetative buds on the second season's wood make a short growth, usually ½ to 3 inches (1 to 7 centimeters) or more, and in late summer they develop a terminally situated mixed bud. (See the second-year wood in Figure 14-3) The following spring each mixed bud develops into a cluster of flowers, makes a short vegetative growth, and develops one or more fruits. (See the third-year wood in Figure 14-3). During the third spring, a laterally situated vegetative bud on any given individual spur makes a short growth, and in late summer it develops a terminally situated mixed bud. This sequence continues throughout the life of any given spur. Thus, an individual spur lays down a mixed bud one year and produces fruit the next.

In the ideal situation, half the spurs lay down flower buds while the other half bear fruit, which makes the tree an annual bearer. However, should a frost, freeze, or some other adverse condition destroy most of the blossoms in any given year, large quantities of carbohydrates accumulate in the tops. As a result, a large number of flower buds are laid down and a heavy crop of fruit is produced. This exhausts the carbohydrate supply, and, as a result, a small number of flower buds are laid down and a light crop of fruit is produced the following year. In this way the so-called alternate, or biennial, habit is established. In general, orchard-crop scientists have tested several methods for breaking the "biennial habit." Thinning the blossoms by using chemical sprays seems to be the most effective. Study the diagram of the fruiting habit of the apple in Figure 14-3.

Vegetative Period of 1 to 4 Years During the vegetative period the trees are trained usually to the modified-leader type. Briefly, training consists of (1) heading back to promote the development of lateral branches and low-headed trees, (2) selecting well-spaced laterals to form the framework and removing the

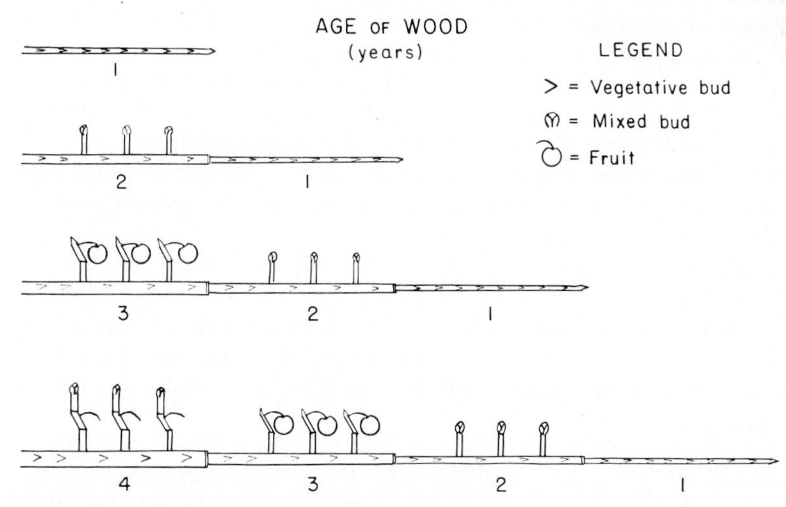

Figure 14-3 Diagram of fruiting habit of the apple.

Figure 14-4 Training of a young spur-type apple tree.

others, (3) disbudding, a special method of pruning, to obtain desired spacing of lateral-scaffold limbs, and (4) pruning the remaining limbs to regulate their growth so that they may occupy a relative proportion of space without interference. In general, one-year-old trees consist of a single stem called a *whip*. This whip is headed back to a height of 42 to 48 inches (about 1 meter) immediately after planting, and laterals are selected about three or four weeks after they have started growth in the spring or during the next winter. In either case, four to six well-spaced lateral shoots are selected and the others are removed. The uppermost lateral usually achieves the greatest growth and is called the *leader*. In general, two-year-old trees have well-developed lateral branches. Immediately after the trees are planted, three to four well-spaced laterals are selected and the remainder are removed. Adequate spacing of the laterals is essential in order to keep them well distributed around the trunk. (See Figure 14-4).

The duration of the vegetative period of apples applies to standard trees. Dwarf trees and spur type start fruiting earlier; therefore, the formative, transition, and bearing periods differ accordingly.

Transitory Period of 5 to 8 Years In the transitory period the trees are changing from an entirely vegetative phase to a period of alternating vegetative and reproductive phases. During this period the plants are pruned lightly, since heavy pruning delays the onset of the fruiting period. In general, this pruning consists in removing undesirable water sprouts and dead or diseased wood and lightly thinning out the top.

Fruiting Period of 8 to 40 Years This period is divided into two stages: (1) early and (2) late. During the early period the young bearing trees are vigorous, and usually only corrective pruning is necessary. Corrective pruning consists of the removal of dead or diseased wood or undesirable parallel branches and the regulation of the growth of the branches to permit the development of strong crotches. As the trees become older the development of small, weak branches become evident. Investigations at the Michigan Experiment Station have shown that a direct relation exists between the diameter of the four-year-old wood and the number, size, and quality of the fruit which it bears. In Michigan orchards, four-year-old wood ½ inch (13 millimeters) or more in diameter produced an average of 10 fruits of good size and color; four-year-old wood from ⅜ to ½ inch (9 to 13 millimeters) in diameter produced an average of seven fruits of medium size and poor color; and four-year-old wood less than ¼ inch (6 millimeters) in diameter produced only three small, poorly colored fruits. Thus, the pruning of mature bearing trees consists largely in the removal of four-year-old wood less than ¼ inch (6 millimeters) in diameter. Most of the thin, unproductive branches are found on the inside of the trees. The apple grower, therefore, in pruning his trees leaves the thick, productive branches and removes the thin, unproductive ones.

Fruiting Habit of Drupe Fruits (Peach, Apricot, Almond, Plum, and Cherry) The fruit of the peach, apricot, and almond is borne on the second season's wood, which is itself developed during the previous season. During the first part of the first season this wood, called *shoots,* grows in length, and during the latter part it develops the laterally situated flower buds. (See the first and second season's wood in Fig. 14-5) Scientists have found a remarkably close

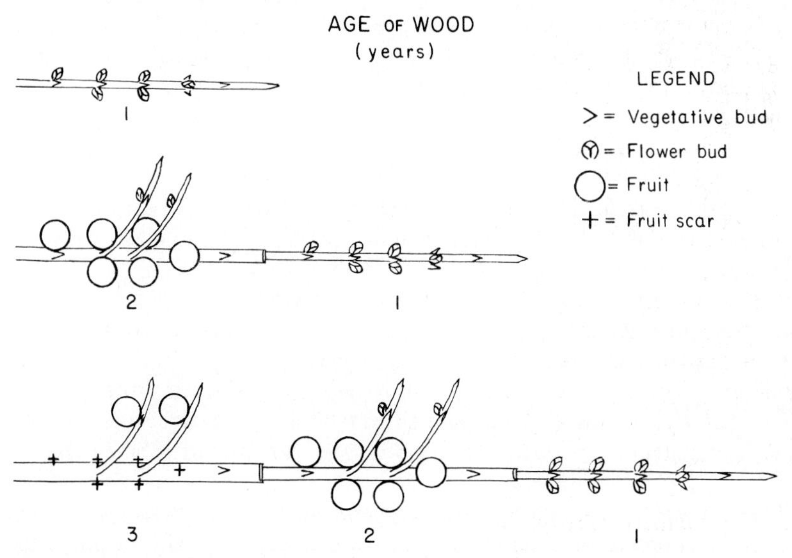

Figure 14-5 Diagram of the fruiting habit of the moderately vigorous peach tree.

relation of vigor of shoot growth to productivity the following year. In fact, the fruiting wood of the peach may be classified as follows: (1) weakly vigorous, (2) moderately vigorous, and (3) excessively vigorous. In general, weakly vigorous shoots make a short growth, 3–6–8 inches (7–15–20 centimeters), and develop relatively few flower buds; moderately vigorous shoots make a moderately long growth, 6–12–18 inches (15–31–46 centimeters), and develop a large number of flower buds; and excessively vigorous shoots make an extremely long growth, 18–24–36 inches (46–61–91 centimeters), and develop relatively few flower buds. Obviously, the moderately vigorous wood is the most productive.

The fruit of the plum and cherry is produced from simple, laterally situated buds also. These buds are laid down on the current season's growth and on spurs. Vegetative buds always terminate shoot and spur growth; therefore, the spurs are usually straight. Like the peach, moderately vigorous growth of the plum and cherry increases the number of nodes, the number of flower buds, and fruitfulness; whereas scant or overvigorous growth decreases the number of flower buds and fruitfulness.

Peach The vegetative period requires two to three years. In general, the peach is trained to the open-center form. In the spring the newly planted trees are headed back to a lateral branch or bud 24 to 36 inches (61 to 91 centimeters) from the ground line, and lateral branches are allowed to grow for two to three years before the scaffold limbs are selected. Thus, during the first two or three years very little pruning is required. The branches which are obviously out of place, broken, or diseased are removed. If any lateral branch becomes overdominant, it is pruned back to an outside branch. This mild pruning during the first two or three years provides for a maximum leaf area and hence a rapid development of the tree. After the tree has grown for two or three years, three to five lateral branches, preferably four, are selected. These should be arranged spirally around the trunk 4 to 6 inches (10 to 15 centimeters) apart. These lateral branches become the main scaffold limbs of the tree.

The secondary laterals in the center of the tree are removed or cut back to a lateral branch. In this way, the open-center type of tree is developed, which exposes the leaves to full sunlight and facilitates spraying, harvesting, and pruning operations. (See two examples in Fig. 14-6)

The transitory period requires three to four years. During the transitory period, moderate pruning is necessary. If any branch becomes dominant, it is cut back to an outside lateral branch. The branches in the center portion are pruned to maintain the open-center type. The objectives are (1) to maintain strong, evenly distributed scaffold limbs, (2) to expose all parts of the tree to adequate sunlight, and (3) to maintain a relatively low-headed tree with well-distributed fruiting wood.

The fruiting period requires 5 to 20 years. Since the moderately vigorous tree is the most productive, the main objective is to establish, if necessary, and maintain the moderately vigorous condition. Thus to establish moderate vigor, weakly vigorous trees will require relatively severe pruning, and excessively

Figure 14-6 Training of the young peach tree. Left, top and bottom: before pruning. Right, top and bottom: after pruning.

vigorous trees will require light, if any, pruning. To maintain moderate vigor two facts should be kept in mind: (1) the moderately vigorous tree develops more flower buds than it can develop into marketable fruit; and (2) the fruiting wood is borne on relatively young wood. Thus, to reduce the number of flower buds and to keep the fruiting wood close to the frame of the tree, heading back is required. In general, this is done by removing one-third to one-half of the terminal portion of the past season's growth or by removing four to five shoots at the terminal of each branch and heading back the remainder of the current season's growth of each branch. Of the two methods the latter requires less time and stimulates the development of shoots close to the frame of the tree. For these reasons it is generally recommended.

Cherry and Plum Like the apple and pear, the many kinds of cherries —sour, sweet, and Dukes—and the many kinds of plums—European, Japanese, and American—are trained to the modified-leader tree form. In general, training involves the selection of three or four laterals and a leader. The laterals should be at least 4 to 6 inches (10 to 15 centimeters) apart and arranged around a main stem to prevent girdling of the trunk. During the transitory period, light pruning, mostly of the thinning-out type, is necessary. With upright-growing varieties attention is given to the development of lateral branches. During the bearing period the amount of thinning out and heading back varies with the species. In general, cherries require the least pruning, since direct light is unnecessary for the development of the fruit, and the crop is seldom heavier than the tree can bear. The European plum requires more pruning than the cherry, but less than the Japanese plum, which requires less than the peach. In any case old trees are pruned more severely than young trees in order to renew the fruiting wood.

Fruiting Habit of the Grape The fruit of the grape comes from mixed buds which are borne on one-year-old shoots called *canes,* which are developed during the previous season. Investigations have shown that fruiting of the grape is positively associated with vigor of the cane. In general, medium-thick (moderately vegetative) canes are more fruitful than thin (weakly vegetative) or thick (vigorously vegetative) canes. For example, in Michigan canes of the Concord variety ¼ inch (6 millimeters) or slightly more in diameter at the fifth or sixth node are more productive than greater or smaller sizes at the same position. Thus, size of cane is an important consideration in productivity. Note the fruiting habit of the grape shown in Figure 14-7.

Pruning the Young Vine Although grapes are pruned to various systems, the pruning requirements of the vine for the first, second, and third years are practically the same for all systems. The main operations are (1) cutting back the most vigorous cane to two or three buds and removing the remainder, (2) setting

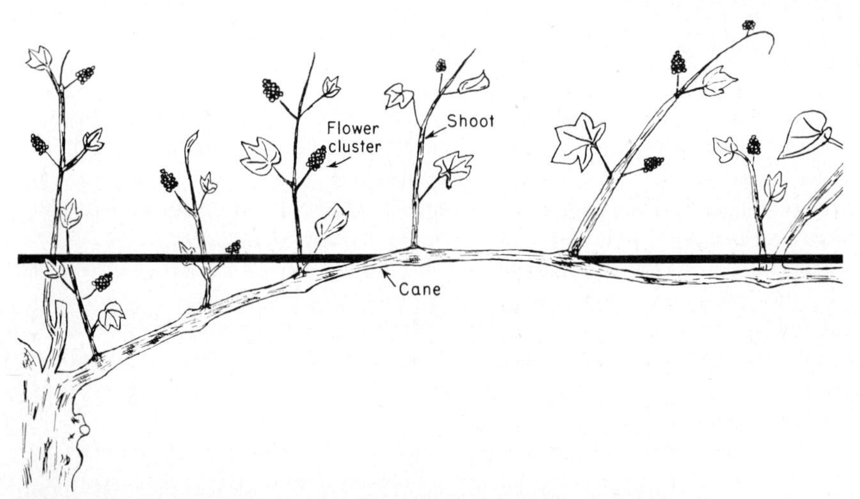

Figure 14-7 Fruiting habit of grape. The fruit-bearing shoots have just started their growth in the spring. Note the flower clusters opposite the leaves. *(Redrawn from H. J. Sefick and J. Harold Clark, Pruning Grapevines,* New Jersey Agr. Exp. Sta. Circ. *423, 1942.)*

Figure 14-8 Pruning the labrusca grape, variety Concord. Left: before pruning. Right: after pruning. *(Courtesy, R. J. Ferree, Clemson University.)*

the plants in the vineyard, (3) tying the canes to a stake to permit close cultivation during the growing season, and (4) cutting back the most vigorous cane to two or three buds and removing the remainder during the dormant season. After the next season the vines are trained to the system desired.

Systems used in training and pruning grapes are the six-arm, the fan, the wire canopy, or overhead, but a common system is the single trunk and arm. In general, this system, or frame, consists of a permanent trunk with lateral branches called *arms*. The length and number of arms depend largely on the size of the individual fruit clusters produced, and this in turn depends on the kind of grape grown.

In the labrusca, or northern, grape the individual fruit clusters are large, and, as a result, only a relatively small number of clusters can develop into marketable fruit and only relatively short arms are required. Four arms are usually used, and this requires a two-wire trellis with the arms, one to the right and the other to the left, on the level of each wire. (See Figure 14-8) In pruning for each arm a moderately vigorous cane is selected for the production of fruit and is headed back to 8 or 10 buds, and, in addition, one or two canes near the arm are selected for renewal purposes and cut back to two or three buds. In this way, the fruiting wood is kept reasonably close to the trunk of the vine.

In the muscadine, or southern, grape the individual fruit clusters are small, and, as a result, a relatively large number of fruit clusters can develop into marketable fruit. Thus, a relatively large number of canes are required, and, in order to support these canes, the arms are extended. (See Figure 14-9). In pruning for each arm all the canes are headed back to two, three, or four buds, depending on the vigor of each individual cane.

Fruiting Habit of Brambleberries (Black, Purple, and Red Raspberry, Blackberry, and Dewberry) Like the grape, the fruit of the brambleberries develop from mixed buds borne on one-year-old canes. The canes develop the

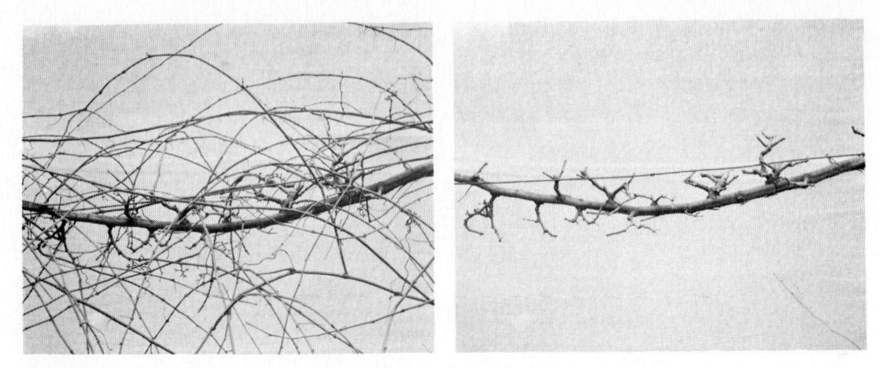

Figure 14-9 Pruning the muscadine grape. Left: before pruning. Right: after pruning. *(Courtesy, H. J. Sefick, Clemson University.)*

first season, fruit, and then die the second season (except the everbearing raspberries and the Himalaya blackberries). Pruning of the canes before they have fruited depends on the variety and method of training.

Black and Purple Raspberry The black and purple raspberry usually set more fruit than the canes can adequately develop, hence they are pruned heavily. The kind and amount of pruning depend on the training system. If the canes are grown with support, the tips of the young shoots of the black varieties are "pinched off" when they are 12 inches (31 centimeters) high if grown in California and when they are from 18 to 24 inches (46 to 61 centimeters) high if grown in the Northeast and South. The tips of the purple varieties are pinched off when they are from 30 to 36 inches (76 to 91 centimeters) high. The pinching off induces the development of lateral branches and prevents the canes from becoming top-heavy. Just before growth starts in the spring the laterals are headed back. The student should keep in mind that the fruit is borne on branches developed from the laterals. Investigations have shown that the number and size of the fruit can be regulated by the number of buds left. In general, if the canes are small, two buds per lateral are sufficient; if they are vigorous, from 8 to 10 buds per lateral can be left. (See Figure 14-10).

Red Raspberry The red raspberry produces tall, erect stems. Pruning, other than removing the canes which have fruited, consists of thinning out small or extra canes. In the Pacific Northwest excess canes are removed in the spring and the remaining canes are cut back in the winter. A Y-trellis is used to support the canes. In the central and eastern states the plants are pruned according to the method of planting. Under the hill system the plants are set 4 × 8 feet (1 × 2 meters), and pruning consists in removing all but about seven of the most vigorous canes per plant. Under the hedge system the plants are set 2½ × 8 feet (0.5 × 2 meters), and pruning consists in removing the small canes and leaving from three to four vigorous canes per linear foot of row.

Blackberry The blackberry is of two types: (1) erect and (2) trailing. The erect varieties produce suckers like the red raspberry and are pruned and trained in much the same manner. In general, the plants are thinned from five to seven canes per hill in hill culture and from two to three canes per linear foot of row in hedge culture. The tips of the young canes may be pinched off when they are from 24 to 30 inches (61 to 76 centimeters) high. The pruning of the laterals is done in the winter. The amount of wood removed depends on the fruiting habit of the variety. Investigations have shown that some varieties produce most of their mixed buds at the base of the laterals and others produce most of them at the tips. In the former case the laterals are pruned to leave three or four buds; with the latter only the tips of the laterals are removed. The trailing varieties are trained to

Figure 14-10 Pruning the black raspberry. Top: before pruning. Bottom: after pruning. *(Courtesy, H. J. Sefick, Clemson University.)*

stakes or on horizontal or vertical trellises. Pruning, other than cutting out canes which have fruited, consists in reducing the number of canes per plant to facilitate the development of berries of large size and the production of vigorous canes for the following year.

Dewberry The dewberry is trained on trellises, on stakes, or is allowed to grow on the surface of the land. All canes, both young and old, are removed after the harvest, and the canes which develop later are used for the next year's crop. These canes are tied to stakes the following spring or are allowed to run along the surface of the land.

Fruiting Habits of Other Small Fruits (Blueberry, Currant, and Gooseberry) Like other small fruits, the blueberry produces fruit on one-year-old wood. Two important considerations are kept in mind in pruning: (1) without pruning most varieties tend to overbear and produce small fruit; and (2) the largest fruit is produced on the most vigorous wood. Consequently, the purpose of pruning is to induce the development of vigorous new wood and to reduce the number of flower buds. This is accomplished by removing weak, slender, and small branches throughout the body of the plant and by cutting back the fruiting branches. The amount of heading back varies with the variety and growing conditions. Rancocus, Concord, and Rubel varieties require little heading back; whereas Cabot, Pioneer, and others require cutting back to about three to six buds per shoot. Usually, little pruning is necessary for the first two years; only the weak, slender wood is removed.

The currant and gooseberry form many stems. With no pruning the plants become dense and unproductive. Since the fruit is borne on one-year-old branches, and on one-year-old spurs on two- and three-year-old branches, and very rarely on one-year-old spurs on older branches, pruning consists largely in removing branches over three years old. Enough one-year-old branches are left to replace those removed. Usually from 8 to 10 branches, well distributed throughout the plants, produce the highest yields. The branches seldom need heading back.

Fruiting Habit of Nut Fruits (Pecans, Walnuts, and Hazelnuts) The pecan, walnut, and hazlenut are monoecious. The pistillate flowers are borne, usually in clusters, on the current season's growth and arise from terminal mixed buds. The staminate flowers are borne in catkins and arise from the base of the leaves on the second season's wood. The individual flowers of both sexes are inconspicuous, and the pistillate flowers are wind-pollinated. Because the pistillate flowers and thus the fruit are borne on terminal branches, except when the crop should be thinned and the trees invigorated, severe heading back would seriously reduce the production of nuts. Consequently, the pruning of vigorous bearing trees consists largely in removing interfering, broken, dead, or diseased branches.

Pecans and hazlenuts are trained to the central-leader type of tree, and walnuts are trained to the modified leader. In this way, higher branches will not excessively shade and thus retard the growth of the lower branches.

Fruiting Habits of Citrus Fruits (Orange, Lemon, and Grapefruit) Like other plants, citrus trees in any one year may produce (1) excessive stem and leaf growth and little fruit, (2) little stem and leaf growth and abundant fruit, and (3) moderate stem and leaf growth and moderately abundant fruit. Obviously, the moderately vegetative and moderately productive condition is the most desirable. Hence, the problem in pruning citrus trees is to maintain a balance between stem and leaf growth on the one hand and flower and fruit production on the other. Since the fruiting wood declines in productivity after the second and third year, the proper balance is maintained by the production of a moderate amount of new wood at a uniform rate each year.

Due to a lack of satisfactory labor, the presence of a disease, or some other destructive factor, growers of certain citrus fruits have modified their production and marketing practices. With orange trees in Florida, they hedge the trees mechanically by reducing the tops and trimming the sides with power-driven saws. In this way, most of the fruit can be harvested while the pickers are standing on the ground.

Research indicates that the removal of dead and weak wood and sprouts that grow below the bud union is the only pruning required by orange trees. Hedging is done by topping the tree and trimming the sides with power-driven saws. This, as compared to the old-time method, which maintained a balance between vegetative and reproductive phases, decreases yields, and reduces tree size, permitting more trees to be planted per acre. Because of high labor cost, it has become necessary to make the trees more accessible to pickers, thus reducing the number of hours required to harvest the fruit.

With grapefruit trees in Florida, very little pruning is necessary except for removal of dead wood for control of the melanose disease. With grapefruit trees in California, a moderate amount of pruning is practiced to encourage fruiting inside the tree.

Fruiting Habits of Figs Figs produce two crops of fruit each year. The first crop, or brebas, is borne near the end of the previous season's wood, and the second crop is borne in the axils of the leaves of the current season's wood. A number of weeks elapse between the ripening of the last fruit of the first crop and the ripening of the first fruit of the second crop.

The type of pruning required depends on the crop grown. For varieties which are grown primarily for the first crop or for both crops, the thinning out of the dead, broken, or diseased branches is all that is necessary. For varieties which are grown mainly for the second crop, such as the Kadota group in California, heading back rather severely is required in order to keep the trees low and thus make picking less expensive.

Ornamental Trees, Shrubs, and Vines Two general types of pruning are required for ornamental trees: (1) removing dead, diseased, or injured branches and (2) keeping the trees within due proportion of the dimensions of the area. Evergreen trees, such as the junipers and arborvitae, must be pruned regularly to keep them within due bounds. Spruces, pines, and similar evergreens seldom require pruning.

In common with the brambleberries, the root system of ornamental shrubs is perennial in nature, whereas the stems live for a relatively short time only, usually one to three or more years. From the standpoint of pruning, these shrubs are divided into three groups: (1) shrubs which grow slowly, (2) shrubs which grow rapidly and bear flowers and fruit on current season's wood, and (3) shrubs which grow rapidly and bear flowers and fruit on one-year-old wood. In group 1 e.g., flowering almond, flowering quince, pearlbush, and viburnum, only the dead stems are removed, and this should be done during the nongrowing season. In Group 2, e.g., althea, butterfly bush, climbing rose, crepe myrtle, coralberry, honeysuckle, hydrangea, spirea, and ligustrum, the two- or three-year-old stems are removed, and this should be done during the nongrowing season also. In Group 3, e.g., barberry, dogwood, deutzia, forsythia, jasmine, spirea, winter honeysuckle, and weigela, one-fifth to one-third of the previous season's wood should be removed, preferably immediately after flowering has taken place. (See Fig. 14-11)

Woody vines, such as English ivy, Virginia creeper, and the bittersweets, usually require little pruning. In general, only dead or injured branches are removed. However, some vines—bittersweet—when comparatively old produce laterals from the upper stems only, thus leaving the base without foliage. To correct this situation the older stems are headed back to the ground. As a result the top is renewed from the younger shoots.

Figure 14-11 Pruning flowering shrubs. Left: before pruning. Right: after pruning. *(Redrawn from F. S. Batson and R. O. Monosmith, Care of Ornamental Trees and Shrubs,* Mississippi Agr. Exp. Sta. Bull *354, 1941.)*

Roses The kind and severity of pruning of the rose depend on the kind of rose grown and the size of flower desired. Some kinds require severe annual pruning, and other kinds require little or no pruning. Roses which require severe pruning are the hybrid perpetuals and the hybrid teas. These kinds are grown for cut flowers. In general, the weak stems are thinned out and the sturdy stems are headed back to varying heights depending on their vigor. In heading back, all cuts are made to an outer bud to promote spreading of the plant. From 6 to 10 buds are usually left on each stem. The student should remember that cutting the flowering stems is essentially a pruning operation. Usually, flower stems are cut so as to leave one to three nodes of the current season's growth on the stem. (See Fig. 14-12)

Roses that require relatively little pruning are the climbing, rambler, and rugosa types. Climbers are pruned immediately after they have finished flowering. Stems that produced flowers are removed to facilitate the development of new shoots which produce flowers the following year. Ramblers require a small amount of heading back and removal of dead and weak wood. Rogusa types require the removal of old, weak, dead, or diseased wood only.

Hedges Ideal hedges make uniform growth within any season and are dense from base to top. To accomplish uniformity of growth the plants must be subjected to a uniform environment—a uniform water supply, essential-element supply, and light supply. This is particularly true of the light supply. Clearly, with other environmental factors in favorable supply, that part of the hedge growing in shade will make less growth than that part growing in sun. To secure density and a compact type of growth from base to top, the frequent formation of lateral shoots throughout the life of the hedge is necessary. Thus, the heading-back type of pruning is used. In general, the plants are headed back to a uniform height when they are set. As the new shoots attain a length of from 6 to 12 inches (15 to 30 centimeters), they are headed back to a length of from 3 to 6 inches (7 to 15 centimeters). This practice is continued until a desired height is reached, after which a severe heading back is practiced. Naturally the number of trimmings

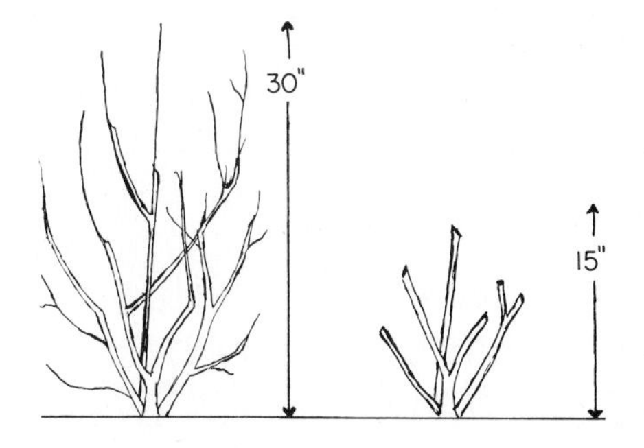

Figure 14-12 Pruning the hybrid tea rose. Left: before pruning. Right: after pruning. The degree of heading back depends on the number of flowering stems desired and on the variety. *(Redrawn from F. S. Batson and R. O. Monosmith, Care of Ornamental Trees and Shrubs,* Mississippi Agr. Exp. Sta. Bul. *354, 1941.)*

in any one season depends on the amount of growth made. Heading back induces the development of lateral buds on the older wood, and, as a result, the body of the hedge is as dense as the top. Figure 14-13 shows satisfactory and unsatisfactory shapes for hedges.

Pruning Tools for Woody Plants

For the pruning of woody plants three kinds of tools are essential: (1) hand pruning shears for small cuts up to ½ inch (1 centimeter) in diameter, (2) lopping shears with 24- to 30-inch (61 to 76 centimeter) handles for cuts between ½ and 1 inch (1.0 and 2.5 centimeters) in diameter, and (3) pruning saws for larger cuts. All pruning tools should be sharp in order to make clean, nonjagged wounds.

Time of Pruning Woody Plants Woody plants are generally pruned between the time of leaf fall and blossoming. Recent investigations have shown that pruning wounds made in early spring heal more quickly and effectively than those made at any other time of year. At this time the wound cork cambium forms readily and produces a layer of corky tissue rather quickly. Furthermore, the temperatures of the spring are more favorable for the activity of the wound cork cambium than those of the winter.

How Cuts Are Made Cuts involving the removal of entire shoots, twigs, and branches are made flush with the adjoining branch or limb. In other words there is no stub or projecting end of the cut branch. Observation shows that cut surfaces of stubs heal less rapidly than the cut surfaces of wounds close to the branch or limb. In fact, unless a shoot forms at the cut end, the surface of the stub rarely heals. Thus, pruning wounds will heal more rapidly if they are made close to the base of removed branches or limbs and parallel to the part from which it was taken.

Treatment of Pruning Wounds Many preparations are available for coating the surface of pruning wounds on woody plants which are 1 inch (2.5 centimeters) or more in diameter. An ideal preparation should (1) facilitate the development of the callus tissue and (2) prevent the invasion of rot-producing organisms. There is no preparation that fully meets these requirements. Some contain disinfectants that retard the rate of callus formation; others retard or

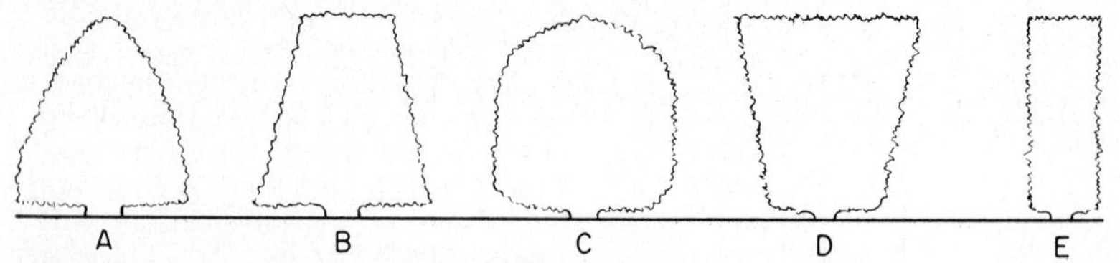

Figure 14-13 Pruning the hedge. Satisfactory and unsatisfactory shapes. A, B: good. C: fair. D, E: poor.

entirely prevent the flow of air into the wound; and others crack and thus allow rot-producing organisms to invade the tissue.

QUESTIONS

1 Pruning a tree or plant reduces its top and root growth. Explain.
2 Given two limbs growing from a common point, having equal size and vigor. Limb A is pruned back severely; limb B is left unpruned. Which will become the dominant limb? Give reasons.
3 Given two lots of tomatoes. Lot A is pruned to a single stem and topped at the sixth cluster; lot B is allowed to grow in the natural way. For a harvesting season of eight weeks which lot would produce the greater total yield? Explain.
4 Show how pruning the top increases the supply of water and essential elements to the remainder of the growing points.
5 Pruning the top promotes vegetative processes and retards reproductive processes. Explain.
6 How does thinning out differ from heading back? Give an advantage and a disadvantage of each.
7 Name the tree forms to which fruit trees are trained.
8 Severe pruning of the top delays flower-bud formation in young deciduous fruit trees. Explain.
9 How is the thin-wood method of pruning applied to mature bearing apple trees?
10 The modified-leader type of apple tree produces stronger crotches and a more substantial frame than the open-center type. Explain.
11 Show the relation of vigor of the tree and number of flower buds formed to yield of peaches.
12 How does the pruning of bearing peach trees differ from that of bearing apple trees?
13 How does the pruning of bearing peach trees differ from that of bearing plum trees?
14 Describe the fruiting habits of grapes and brambleberries.
15 What is the relation of vigor of grape canes to fruitfulness?
16 How does the single-trunk, four-cane kniffin system of training differ from that of the fan system?
17 How does pruning the brambleberries differ from pruning bearing apple trees?
18 How does pruning blackberries and black raspberries differ from that of red raspberries?

PRUNING THE STEM OF HERBACEOUS CROPS

Important herbaceous plants which are pruned are tomatoes in greenhouses, tomatoes outdoors, cucumbers in greenhouses, chrysanthemum, dahlia, zinnia, calendula, and mignonette.

Tomatoes in Greenhouses

Tomatoes grown in greenhouses are always trained and pruned. Individual plants are usually trained to a single stem which is supported by a stake or string. The laterals are pinched off as soon as they arise in the axils of the leaves; thus the thinning-out type of pruning is employed. Pruning in the greenhouse permits close setting, development of fruit above the soil, and facilitates spraying of the plants and picking of the fruit. Since the water and fertilizer supply are under the control of the grower, competition between plants, particularly for water and essential elements, is reduced to a minimum and comparatively large yields per unit area are secured.

Tomatoes Outdoors

Tomatoes are pruned in certain commercial sections for the fresh market and in many home gardens. The plants are trained to one, two, or three stems and are topped at the fourth, fifth, sixth, or succeeding flower cluster. Stakes are usually used for support. The effect of pruning and training tomato plants outdoors on the production of fruit has been investigated by many experiment stations. Advantages and disadvantages of the practice as set forth in *New York (Cornell) Experiment Station Bulletin 580,* 1934, follow.

Advantages

1. Permits close spacing of plants
2. Increases early yield of fruit per acre
3. Keeps fruit off the ground
4. Facilitates spraying plants and harvesting fruit

Disadvantages

1. Increases cost of production
2. Requires stakes and twine for training plants
3. Requires labor for setting stakes and pruning and tying plants
4. Requires labor for removing and storing stakes

Cucumbers in Greenhouses

Like tomatoes, cucumbers grown in greenhouses are always trained and pruned. Individual plants are trained to a single stem and are usually supported by string fastened to a single-wire trellis from 4 to 8 feet (1 to 2 meters) above the level of the land. For about two months after the plants are set, the primary laterals are cut back to the first or second female blossom, and the secondary laterals are pinched out. Thus, both heading back and thinning out are used. As in the case with tomatoes, the pruning and training of cucumbers in greenhouses permit close setting of the plants and the development of fruit above the ground.

Chrysanthemums

Chrysanthemums, particularly the large flowering or standard types, are disbudded. Chrysanthemums produce two types of flower buds: (1) crown buds and (2) terminal buds. Crown buds appear first and are surrounded by leaf buds, and terminal buds appear later and are surrounded by the flower buds. According to authorities at the Ohio Experiment Station, the use of crown or terminal buds depends largely on the variety. On early maturing varieties crown buds produce the most satisfactory type of flower, and on late maturing sorts terminal buds are the more satisfactory. In either case, only one flower is allowed to develop on each plant. In this way a large inflorescence is obtained. Other flower crops which are disbudded are camellia and dahlia.

The stems of many flowering plants are headed back or pinched to produce the development of a large number of stems and flowers per plant. Examples of plants headed back in this way are the pompon varieties of chrysanthemum, calendula, carnation, and mignonette.

PRUNING THE ROOT

Root pruning involves the removal of part of the root system of plants. Root pruning, like top pruning, affects the total amount of growth made and the vegetative-reproductive balance.

Total Amount of Growth Made

Investigations have shown that root pruning dwarfs the tree or plant in much the same way, although not to the same degree, as top pruning. Root pruning removes certain portions of the root system. The roots require carbohydrates for growth, and primary and secondary roots of woody plants store carbohydrates also. Thus, root pruning removes some of the carbohydrates used for the growth and maintenance of the roots, and removes all those stored in the severed roots. To replace the removed roots the plants develop other roots. These new roots require carbohydrates which could be used for other purposes.

Vegetative-Reproductive Balance

What is the relative effect of pruning the root on the disposition of the carbohydrates and the vegetative-reproductive balance? In general, pruning the root reduces the area for the absorption of available nitrogen and other essential elements and water; this, in effect, reduces the amount of available nitrogen and other essential elements and water, which go to each of the growing points in the top; and accordingly, it reduces the number of cells which would otherwise be made. Pruning the root, therefore, decreases cell division and enlargement and the utilization of carbohydrates and promotes the accumulation of carbohydrates. In other words, root pruning favors reproductive processes more than it favors vegetative processes. The student will note that root pruning has the same effect as top pruning on the amount of growth that can be made, but it has the opposite effect of top pruning on the disposition of the carbohydrates and the vegetative-reproductive balance.

Pruning the Root of Woody Plants

Up to the end of the nineteenth century root pruning was practiced and recommended in Europe, particularly in England, for the culture of fruit trees. In

England fruit trees were grown mostly on dwarf stock. A circular trench usually about 18 inches (46 centimeters) from the base of young trees and about 18 inches (46 centimeters) deep was made around the tree and all exposed roots were cut with a sharp knife. For each succeeding year the distance of the trench from the base of the tree was slightly increased so that the roots of any one year were pruned from 2 to 3 inches (5 to 7 centimeters) from the stubs of the previous year. It was generally recommended that the trenching be done in the fall. In this way the trees would show a reduced vegetative growth the following spring and an increase in flower-bud formation in the summer. The student should remember that heavy manuring was practiced also. In fact, the growers generally filled the trench with manure or compost. Since these materials stimulated vegetative growth and since root pruning reduced vegetative growth, the one practice had a nullifying effect on the other. Studies in American orchards during the early part of the twentieth century showed that root pruning markedly reduced the life and the yield of fruit trees. In this country orchards are not heavily manured, and lack of vigor is a more serious problem than excessive vigor. Consequently, root pruning by trenching is no longer recommended and practiced.

QUESTIONS

1 How does the pruning of mature citrus trees differ from that of peach trees? trees?

2 Heavy heading back of pecan branches would greatly reduce the yield. Explain.

3 The heading back of figs grown for the first crop of fruit would seriously decrease the yield. Explain.

4 In what way are the pruning requirements of ornamental shrubs similar to those of the brambleberries?

5 Show how heading back a large spirea bush would reduce its aesthetic and ornamental value.

6 How is density secured in the development of a privet hedge?

7 How does the pruning of hybrid tea roses differ from that of rambler roses?

8 How do woody plants heal their cut surfaces?

9 Pruning wounds made in the early spring just before growth starts heal more quickly than those made in the winter. Explain.

10 Cuts made at the base of removed branches or parallel to the branch from which it was removed heal more effectively than cuts made away from the base. Explain.

11 State the two main requirements of preparations for the treating of wounds. Give reasons.

12 Certain flower crops are disbudded. What is the object of disbudding?

13 The pruning of tomatoes and cucumbers in greenhouse culture is a necessary practice for profitable production. Explain.

14 The removal of healthy leaves from the basal portion of young tomato plants grown in the greenhouse reduces the total yield of the plants. Explain.

15 State two advantages and two disadvantages of pruning and training tomatoes grown outdoors.

16 How does root pruning reduce the total amount of growth that would otherwise be made?

SELECTED REFERENCES FOR FURTHER STUDY

Christopher, E. P. 1954. *The pruning manual.* New York: Macmillan. A discussion of the fundamental principles involved in crop-plant growth as related to pruning, as well as general directions for the pruning of a wide variety of horticultural crops.

Ricks, G. L., and H. P. Gaston, 1935. The "thin wood" method of pruning bearing apple trees. *Mich. Agr. Exp. Sta. Spec. Bul.* 265. Results of research showing the positive relation of light and photosynthesis to the vigor of the wood and to the number, size, color, and market acceptability of the fruit.

Chapter 15

Controlling Pests

Waste is unworthy of a great people.

Ezra Taft Benson

COMPETITION FOR THE PRODUCTS OF PHOTOSYNTHESIS

Three great groups of organisms—all animals, all fungi, and most species of bacteria—are dependent on the products of photosynthesis for their growth and development. Many kinds and species within each group are continually competing with man for these products. For example, man is constantly fighting certain insects which reduce plant yields and certain fungi and bacteria which cause disease. Since these organisms compete with man for the products of photosynthesis, they are called *pests*. This competition is so great that man has organized groups (entomologists, nematologists, and plant pathologists) to study the life cycle and habits of pests. He has developed many materials and many devices for their control. Growers of horticultural plants, like the growers of other crops, must devote a great deal of their time and energy to combat pests in order to obtain profitable crops. These pests are classified as follows: (1) those belonging to the animal kingdom, (2) those belonging to the plant kingdom, and (3) viruses.

PESTS BELONGING TO THE ANIMAL KINGDOM

Pests belonging to the animal kingdom are many kinds of insects, certain kinds of mites, many kinds of nematodes, and a miscellaneous group.

Insects

Insects have three pairs of legs and three body regions. They may be classified according to the way they feed on plants. In general, there are two groups: (1) those with biting mouth parts and (2) those with sucking and/or rasping mouth parts.

Insects with Biting Mouth Parts Insects with biting mouth parts are classified according to the part of the plant on which they feed. In general, there are four more or less distinct groups: (1) those that feed on leaves and/or stems, (2) those that feed on roots, (3) those that bore into stems, and (4) those that feed on fruits, seed, or fleshy storage organs.

Stem and Leaf Eaters How do the stem and leaf eaters reduce photosynthesis? In general, they reduce the chlorophyll content of the leaves. This reduces the amount of light that can be absorbed per unit time, and this, in turn, reduces the amount of initial food substances which can be made. Examples are the caterpillars of certain butterflies and moths, cabbageworm, tomato worm, celery worm, cutworm, apple-tree tent caterpillar, and webworm; certain beetles and their larvae, such as the Japanese beetle, blister beetle, asparagus beetle, common bean beetle, Colorado potato beetle, and rose chafer; grasshoppers, both adults and nymphs; and leaf miners, such as spinach leaf miner, holly leaf miner, and arborvitae leaf miner.

Root Feeders How do root feeders reduce photosynthesis? In general, they eat the younger portions of the root system and reduce the area for the absorption of water. This reduces the amount of water which can be absorbed per unit time, while transpiration is not reduced. As a result, as previously explained, a water deficit occurs within the plant, the guard cells lose turgor, the stomates close, carbon dioxide cannot diffuse rapidly into the leaves, and photosynthesis is accordingly reduced. Examples of root feeders are the larvae of the cucumber beetle and the white-fringed beetle, and strawberry rootworm and white grubs —the larvae of the May or June beetle.

Stem Borers Stem borers may be placed in two groups: (1) herbaceous stem borers and (2) woody stem borers. How do herbaceous stem borers reduce photosynthesis? In general, they bore into the stems and eat the xylem (and other tissues). This stops the flow of water into the leaves above the damaged area, while transpiration continues. As a result, the guard cells collapse, the stomates close, carbon dioxide cannot diffuse into the leaves, and photosynthesis slows down or entirely stops. Examples of herbaceous stem borers are the squash vine borer and the corn borer.

How do the woody stem borers reduce photosynthesis? In general, they puncture or sever the secondary phloem. This reduces the flow of manufactured substances to the roots, and as a result these substances accumulate in the tops; and, in accordance with the law of mass action, photosynthesis slows down. In other words, the effect of the injury caused by woody borers is practically the same as that caused by girdling. Examples of woody stem borers are peach tree borer, apple tree borer, and raspberry cane borer.

Feeders on Fleshy Fruits, Seed, and Storage Organs These are usually the larvae of moths and beetles. These larvae eat large quantities of food and make the products unfit for human consumption. In a sense these insects have successfully competed with man for their food, e.g., sweet potato weevil, bean weevil, pea weevil, tomato fruitworm, corn earworm, plum curculio, and Oriental fruit moth worm. Some of the more common insects with biting mouth parts are shown in Figure 15-1.

Insects with Sucking or Rasping Mouth Parts How do insects with sucking or rasping mouth parts reduce photosynthesis? In general, they pierce the epidermis, suck the tiny chloroplasts, soluble foods, and vitamins from the leaves, and make the leaves incapable of making chlorophyll. This reduces the amount of light which would otherwise be absorbed and the amount of initial food substances which would otherwise be made. Examples of insects with sucking mouth parts are the many kinds of aphids, e.g., apple aphid, garden pea aphid, cabbage aphid, and rose aphid; the bugs, e.g., squash bug, harlequin cabbage bug, and tarnished plant bug; thrips, e.g., onion thrips and citrus thrips; leafhoppers, e.g., potato leafhopper and aster leafhopper; and scales and mealybugs, e.g., oystershell scale, San Jose scale, white peach scale, grape mealybug, and taxus mealybug. Some of the more common insects with sucking or rasping mouth parts are shown in Figure 15-2. Figure 15-3 shows aphid injury on leaves of sweet potato.

Mites

Mites have four pairs of legs and two body regions. They comprise the spiders, scorpions, daddy longlegs, and ticks. Of these kinds the most serious pests on plants are (1) the foliage mites and (2) the bulb mites. How do the foliage mites reduce photosynthesis? In general, they pierce the foliage, usually the underside of the leaves, with two pronglike projections, and cause the leaves to turn yellow

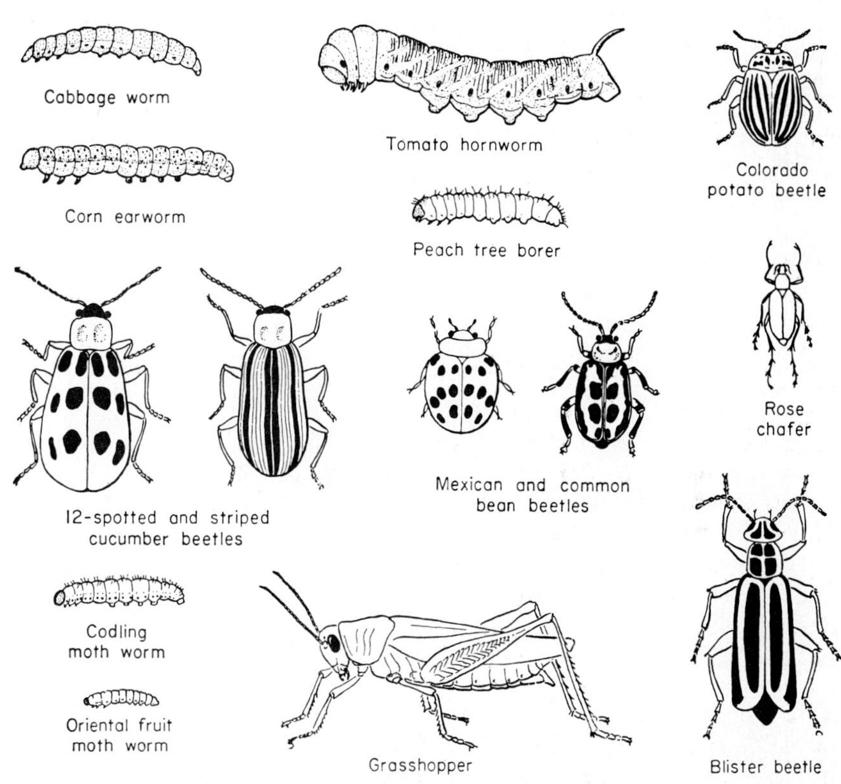

Figure 15-1 Examples of insects with biting mouth parts.

at first and then brown. Thus, foliage mites reduce the amount of light which would otherwise be absorbed, and the amount of foods which would otherwise be made. Examples of foliage mites are the common red spider, cyclamen, or strawberry mite, and citrus mite.

Bulb mites, as the name suggests, feed on bulbs and similar organs and in a sense they have successfully competed with man for their food. Examples of storage organs susceptible to infestation are bulbs of lily and onion, corms of gladiolus and crocus, and rhizomes of asparagus and peony.

Parasitic Nematodes

Parasitic nematodes are minute, mostly dioecious, unsegmented worms. In general, they suck the juices from plant tissues by means of a hollow tube called the *stylet*. Parasitic nematodes may be placed in two groups: (1) those which attack the roots and (2) those which attack the foliage. Nematodes which attack the root system are the root knot, the root lesion, and the root tip. How do these kinds reduce photosynthesis? Tissues invaded by the root-knot nematode form galls,

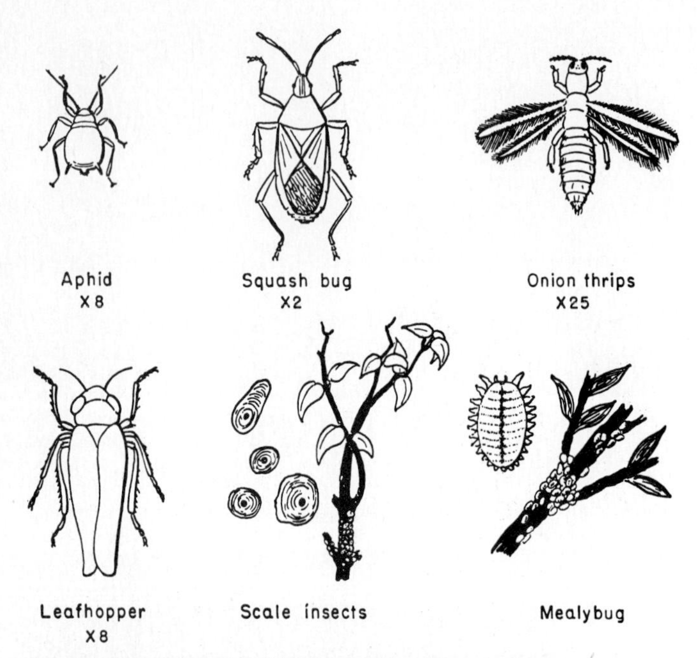

Figure 15-2 Examples of insects with sucking mouth parts.

Figure 15-3 Insect with sucking mouth parts, and predator insect feeding on sucking insect. *(Courtesy, Lewis Riley, Clemson University.)*

but the cells are not killed; and tissues invaded by the root-lesion and root-tip nematodes do not form galls, but the cells are killed. In each of these three cases the water-absorbing capacity of the root system is reduced, and under conditions of high transpiration wilting takes place; the guard cells lose turgor, the stomates close, carbon dioxide cannot diffuse into the leaves, and photosynthesis is accordingly reduced. Most horticultural crops are susceptible to parasitic nematodes. In fact, only a few varieties are resistant; these belong to the following kinds: apricot, avocado, citrus, date, peach, plum, and snap bean. Note the formation of galls on the roots of tomato in Figure 15-4.

Nematodes which attack the foliage are known as *foliar nematodes*. How do these foliar nematodes reduce photosynthesis? In general, they move from leaf to leaf by films of water, enter the stomates, and feed on the tissue that makes the initial food substances. Thus, they reduce the chlorophyll content, and this, in turn, reduces the amount of light that would otherwise be absorbed and the amount of foods which would otherwise be made. Horticultural crops susceptible to the foliar nematode are potato, chrysanthemum, African violet, and begonia.

Miscellaneous Group

The miscellaneous group includes rodents and certain birds. Rodents comprise ground moles, field mice, and rabbits. The *ground mole* frequently damages lawns and crops grown in the garden and occasionally damages crops grown in

Figure 15-4 Nematodes on roots of tomato. Left: nematode infested. Right: healthy. *(Courtesy, Lewis Riley, Clemson University.)*

the field. These animals, searching for white grubs and other insects in the soil, make tunnels and thus loosen the soil and seriously disturb the root system. *Field mice* and *pine mice* are frequently troublesome during the planting season, since they eat the seed; this often necessitates replanting. These mice and *rabbits* frequently damage trees in the orchard by eating the bark at the base of the tree.

PESTS BELONGING TO THE PLANT KINGDOM

Plant pests are minute plants. They have no chlorophyll, hence they cannot manufacture their own food. They must live, as all animals must live, on the food that green plants have made. The student should remember that not all minute plants are harmful to crop plants. Some are distinctly beneficial, such as the fungi which decompose organic matter and the bacteria which change ammonia to nitrates.

Many plant pests live on horticultural crops. They attack the plant in various ways. Some attack crop plants in the seedling stage and thus reduce the stand; others attack the leaves and thus reduce the plant's capacity to make food; others attack the stem and disrupt the food, water, and mineral transportation systems; and others attack the marketable product. In all instances profitable yields per plant or per unit area are decreased. Plant pests are classified as follows: (1) fungi and (2) bacteria.

Fungi

Fungi consist of many cells. The body may be divided into two parts: (1) the fine threadlike strings of cells collectively called *mycelia* and (2) the fruiting bodies. The mycelia are the vegetative part and the fruiting bodies are the reproductive part. Fungi can multiply and spread by fragments of the mycelia or by germination of the spores. Under favorable conditions each type of structure is capable of growing into a new individual. The fungi which attack horticultural crops constitute a large group. Their mode of living and manner of feeding on the host plant vary greatly. They are classified as follows: (1) fungi with practically all their mycelia and all their fruiting bodies on the surface of the plant; (2) fungi with all their mycelia within the plant and fruiting bodies that break through the outer cover; (3) fungi with their entire bodies within the plant.

Fungi with practically all their mycelia on the surface of the host plant are external feeders. An example is the powdery mildews. On infected plants the mycelia can be seen as fine welts of grayish threads among which are interspersed small brown or black fruiting bodies. The fungus attacks the lower side of the leaves more than the upper side and thrives in warm humid conditions. Fungi with mycelia within the tissues of the plant and fruiting bodies that break through the outer cover are the lesion producers. They are divided into four groups: (1) anthracnoses, (2) downy mildews, (3) leaf spots, and (4) rusts. In general, the lesions are small at first and water-soaked. They gradually enlarge and assume a characteristic color, gray, brown, or black, depending on the color of the fruiting bodies of the fungus. Frequently the fungus grows so rapidly that the spots grow together (coalesce). The lesion producers attack many horticultural crops. Examples are presented in Table 15-1 and in Figures 15-5 and 15-6.

Fungi with their bodies within the plant are the fusarium wilts. These minute plants live on organic matter in soils, enter the host plant body through the region of the root-hair zone, and attack the xylem. The xylem becomes plugged, turns brown, and the host plant wilts. Examples are fusarium wilt of aster, cotton, protepea, sweet potato, tomato, and watermelon.

Bacteria

Bacteria are microscopic, one-celled, and reproduce by cell division. They vary in size and shape. Some are small and round; others are large and rod-shaped. In general, the bacteria which cause disease on plants are short rods. They possess the faculty to live on dead and living tissue, but they can enter plants only through natural openings, such as stomates, hydathodes, and nectaries, and through wounds. There are three more or less distinct groups: (1) those which

Table 15-1 Types of Fungus and Bacterial and Virus Diseases

Group	Disease	Crops
Fungi	Powdery mildews	Grain, legume, cucurbit, chrysanthemum, rose, zinnia
	Anthracnoses	Bean, protepea, cucumber, cantaloupe, watermelon
	Downy mildews	Grass, grain, cantaloupe, cucumber, lettuce, onion, spinach, tobacco, violet, pansy, grape
	Leaf spots	Apple scab, cherry leaf spot, rose black spot, black rot of sweet potato
	Fursarium wilts	Cotton, sweet potato, tomato, watermelon
Bacteria	Local infections	Fireblight of pear, bacterial blight of snap bean, soft rot of carrots
	Bacterial wilts	Sweet corn, carnation, ring rot of potato
	Galls	Crown gall
Viruses	Mosaics	Potato, tomato, sweet potato, protepea, raspberry
	Yellows	Peach, aster

Figure 15-5 Leaf spot of strawberry. *(Courtesy, John T. Presley, U.S. Department of Agriculture.)*

cause local infection, (2) those which induce wilting, and (3) those which induce galls. Bacteria which cause local infection are rather specific in their action. Some produce lesions on stems and leaves, others produce cankers on woody stems, and others produce soft rots. Like the fusarium wilts, bacteria which induce wilting invade the xylem and thus disrupt the flow of water from the roots

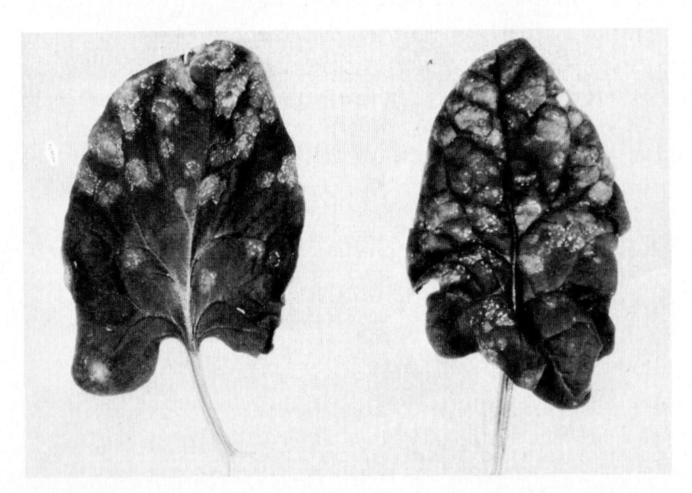

Figure 15-6 White rust of spinach. Left: upper side. Right: lower side. *(Courtesy, The late S. S. Ivanoff, Mississippi Agricultural Experiment Station.)*

to the leaves, and bacteria which induce galls cause a rapid rate of cell division. Examples of each of the three types of bacterial disease are shown in Table 15-1.

Viruses

In general, viruses are large, complex molecules. The central, or inner, part consists of either ribonucleic acid, called RNA, or deoxyribonucleic acid called DNA, and the other part consists of a large number of different proteins. For example, a virus infects the tomato plant which causes a disease called mosaic. An individual unit of this virus has a single molecule of RNA to which is attached about 2,100 different proteins. In some plant viruses, the proteins are arranged on the RNA in the form of a helix and appear under the electron microscope as short rods, whereas in others, the proteins are arranged in numerous planes and appear as polyhedrons.

Essentially, plant viruses cause disease by disrupting the biochemistry of the crop plant. Immediately after a virus particle invades a living cell, the particle becomes unstable and its RNA and protein parts separate. The RNA of the virus displaces the DNA or RNA of the crop plant. The RNA duplicates itself by mitosis and sets the pattern for the synthesis of its own characteristic proteins. These proteins are made from the amino acids and other substances made by the plant. The new RNA molecule and the new set of proteins combine in the formation of a new virus particle. Thus, since plant viruses use amino acids made by the crop plant, they are parasites.

Under favorable conditions the formation of new virus particles is quite rapid, and the derangement of the biochemistry and physiology of the crop plant is correspondingly rapid. In many crops, there is a reduction in the chlorophyll content of the leaves and a corresponding reduction in the absorption of light, in the rate of photosynthesis, growth, and yield. In some crops, the reduction in chlorophyll is manifested by the appearance of mosaic, yellow areas at irregular places on the leaf blade, and in others by the yellows, the loss of chlorophyll throughout the entire leaf blade. Examples of virus diseases are presented in Table 15-1 and Figures 15-7 and 15-8.

QUESTIONS

1. All animals, most fungi, and most bacteria are dependent on green plants for their existence. Explain.
2. Horticulturally speaking, what is a pest?
3. Man always has been seriously concerned about his food supply. Explain.
4. Distinguish between insects, mites, and nematodes.
5. How do the leaf eaters reduce photosynthesis, growth, and yield?
6. How do the root feeders reduce photosynthesis, growth, and yield?
7. How do woody stem borers reduce photosynthesis, growth, and yield?
8. How do foliar mites reduce photosynthesis, growth, and yield?

9 How do root nematodes reduce photosynthesis, growth, and yield?
10 How do foliar nematodes reduce photosynthesis, growth, and yield?
11 Given a tomato plant wilting from 10 A.M. to 4 P.M. only. Examination of the roots shows galls or swellings. State the pest.
12 Distinguish between fungi, bacteria, and viruses.
13 How do lesion-producing fungi and bacteria reduce photosynthesis, growth, and yield?
14 How do the fusarium wilts lower the rate of photosynthesis, growth, and yield?
15 How do virus diseases reduce photosynthesis, growth, and yield?

METHODS OF CONTROL

In order that any pest may grow and develop, three conditions must be fulfilled: (1) the pest responsible for the damage must be present; (2) the environment must be favorable for the spread of the pest; and (3) the crop must be susceptible. Pests, therefore, may be controlled by practicing any one or more of the following methods: (1) eliminating or eradicating the causal organism, (2) using natural enemies of pests, (3) using chemical compounds, (4) using two or more of the above methods in combination, and (5) growing resistant varieties.

Figure 15-7 Tomato mosaic. Left: mosaic-infected plant. Middle: healthy leaf. Right: mosaic-infected leaf. *(Courtesy, John T. Presley, U.S. Department of Agriculture.)*

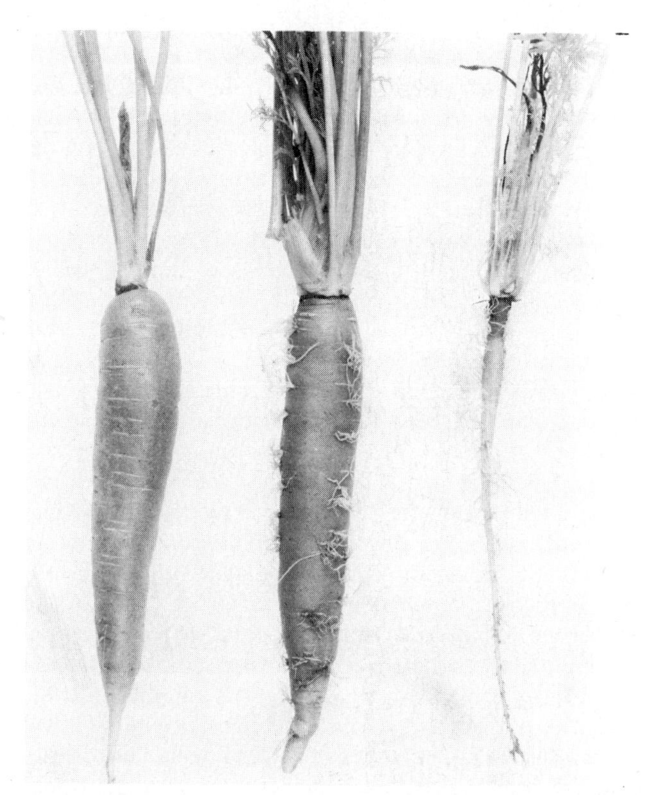

Figure 15-8 Aster yellows on carrot. Left: healthy plant. Middle: moderately infected plant. Right: severely infected plant. *(Courtesy, the late S. S. Ivanoff, Mississippi Agricultural Experiment Station.)*

Eliminating the Causal Organism

Eliminating the casual organism involves (1) removal of the organism from seed or plants, (2) crop rotation, and (3) crop sanitation.

Removal of the Organism Removal of the organism may be secured with a specific lot of seed, a plant or tree, or a large group of plants. As an example, consider the control of black rot of cabbage. The causal organism is a bacterium which lives on the coat of the seed and on organic matter in the soil. Soaking the seed in hot water at a temperature of 122°F (50°C) for 25 minutes followed by rinsing in cold water eliminates the bacterium from the coat of the seed. Unless the bacterium is present in the soil after the seed or seedlings have been planted, the crop will be entirely free of the disease. As a second example, consider the control of citrus canker. Citrus canker was at one time a serious disease of oranges, lemons, and related fruits in Florida. This disease was controlled permanently by promptly removing and burning all trees that showed the disease. As a result, citrus trees in Florida now are free from canker because the causal organism no longer is present. As a third example, take the production of certified potato seed. Many varieties of potatoes are susceptible to many kinds of virus diseases, e.g., mild mosaic, rugose mosaic, leaf roll, and spindle tuber. These viruses impair the capacity of the plant to make chlorophyll. As a result, the rate of photosynthesis is low. These viruses are present in the sap of infected plants and are carried from plant to plant by insects. Control of the virus diseases, therefore, consists of controlling the insect carriers, combined with prompt removal of plants showing symptoms of any of the virus diseases. Certified seed of the potato are tubers which are practically free from several viruses or mosaic diseases. During the growing season the plants which are raised for certification are inspected frequently, and all infected plants are removed and destroyed. The removal of infected plants prevents the disease from spreading to healthy plants and permits the raising of a crop of tubers which is practically free from the several virus diseases.

Crop Rotation Crop rotation consists of growing a group of crops in a definite sequence on the same land. In other words, a definite period is allowed to elapse before the crop is planted again. In this way, the insect or disease organism is starved out. In general, crop rotation is most effective in controlling pests which live in the soil for a short time only. It has little or no effect on organisms which live in the soil for a period of 10 to 25 years. For example, the spores of clubroot of cabbage can live in the soil for two years only. Consequently, if the land is kept free of crops and weeds of the cabbage family for two years after a cabbage crop has been grown, the land will be free of this disease. On the other hand, a short rotation is ineffective in the control of fusarium wilt of watermelons since the wilt fungus lives in the same soil for a period of 20 to 25 years.

Crop Sanitation Crop sanitation consists of (1) promptly removing insect-infested and diseased plants and (2) plowing under refuse immediately after the harvest. Both methods of sanitation assist in controlling pests. Removing insect-infested and diseased plants assists in reducing and may, in many cases, entirely prevent the dissemination of the pest. It is particularly essential in controlling pests of crops which are grown close together and in quick succession, as in the raising of plants in greenhouses, hotbeds, and cold frames. Plowing under plant remains reduces the multiplication of fungi and bacteria in the plant and soil. It is generally practiced to control pests of crops grown in the field.

Using Natural Enemies

Predators and parasites which are natural enemies of a given pest can be used advantageously with some crops. Two examples of this kind of pest control follow. In 1869, cottony-cushion scale, a serious pest on citrus trees, was accidentally introduced into California on a species of Austrialian acacia. Neither of its two natural enemies, the vedalia beetle and a parasitic fly, was introduced with it. As a result, the scale spread rapidly throughout citrus orchards, many crops were lost, and the orchards depreciated in value. In 1888, both enemies were introduced, and they soon became established in citrus orchards. However, the vedalia beetle was found to be the more effective in the controlling of the scale. Within a year, orchards in which the beetles were first released were practically free of the scale, and within two years the cottony-cushion scale was no longer a serious problem in California.

The tomato hornworm eats the leaves of tomato plants. Healthy, rapidly growing worms reduce the photosynthetic surface in a short time, and accordingly they reduce the yield and quality of the fruit. In the southeastern United States these worms rarely do serious damage. In a short time after they appear, many of them become heavily parasitized. The female parasitic wasp lays large quantities of eggs in the body of the caterpillar, and the numerous larvae rapidly consume the flesh and vital organs of the worm. As a result, the worm dies or is at least rendered inactive quickly, and the application of an appropriate stomach poison is generally unnecessary.

In general, a biological balance exists between certain insect pests and their enemies. With many horticultural industries the maintenance of this balance assumes great practical significance. In fact, it may entirely eliminate the need for other methods of control. For example, the use of chemicals is seldom necessary for the control of insects and mites in avocado orchards of southern California. The natural and imported predators and parasites keep the potentially serious insects and mites under control, e.g., the looper, the long-tailed mealybug, the black scale, the soft brown scale, the six-spotted mite, and the avocado brown mite. Numerous experiments have shown that when the predators and parasites are excluded from the trees these insects and mites multiply rapidly and reduce the photosynthetic capacity of the leaves and the vigor and yield of the trees. Thus, the main advantage in using natural enemies—the so-called biological method—is that the use of chemical compounds and other methods are unnecessary. However, the main disadvantage is that the pests are not entirely eliminated.

Using Biological Insecticides and Viruses

Biological insecticides are sometimes called *microbial insecticides.* The one most commonly used is derived from a bacterium called *Bacillus thuringiensis* and is available under the trade names Dipel, Bitriol, and Thuricide. These materials are very effective against certain caterpillars, such as loopers, cabbage worms, and cutworms. They leave no harmful residue on the plant and are completely safe to humans.

There are certain viruses to which insects are susceptible. The use of these viruses is a new approach to insect control, making them one of the potential replacements for many chemical insecticides.

Using Chemical Compounds

A large number and a wide variety of chemical compounds are used in the control of pests. These compounds are classified according to their principal use as follows: (1) stomach and contact poisons, (2) contact poisons, (3) systemics, (4) fumigants, (5) seed and seed-stock treatment materials, (6) fungicides and bactericides, and (7) antibiotics.

Stomach and Contact Poisons Stomach and contact poisons are used for the control of insects with either biting or sucking or rasping mouth parts. For insects with biting mouth parts applications should be made just before the insect begins to feed, and the poison should remain on or adhere to the tissues until the insect has fed. For many insects with sucking or rasping mouth parts applications should be made just as the insect begins to feed, and the poison should adhere to the body of the insect until the insect has been killed, e.g., rotenone, pyrethrum derivatives, malathion, methoxychlor, and parathion. A particular advantage of the pyrethrum products is that they are nonpoisonous to man. However, because of their high price and short residual life, their use has been limited.

Contact Poisons As previously stated, contact poisons are used to control insects with sucking or rasping mouth parts. In general, the contact insecticides are specific in action, e.g., oil emulsions for the control of scale and mealybugs, nicotine sulfate for the control of many kinds of aphids, and Aramite for the control of phosphate-resistant red spider mites.

Systemics Systemics are chemical compounds which enter the conducting systems of plants. They may be applied as a foliage spray or to the soil. In the first case, they enter the plant through the openings of the leaves at the end of large veins, and, in the second case, they enter through the region of water absorption.

Most systemics are highly toxic to man, and care should be exercised in their use. Examples of systemics are Phorate, Di-Syston, Furadan, and Demeton. As with all pesticides, as a result of continuing research, many new products are appearing on the market each year.

Fumigants Fumigants are chemical compounds which change from either the solid or liquid form to the vapor or gaseous form. These gases stop the respiration of insects and nematodes and enter the cells of fungi and bacteria. In general, they may be placed in two groups: (1) compounds applied to soil and (2) compounds applied to crops growing in greenhouses or similar structures. Compounds applied to soil are used mainly for the control of injurious nematodes and for the control of the peach tree borer. Examples of nematocides are chloropicrin, methyl bromide, Nemagon, Vapam, and VC-13. Compounds used for the control of the peach tree borer are paradichlorobenzine and ethylene dichloride. Compounds applied to crops growing in greenhouses or similar structures are used mainly for the control of insects with sucking mouth parts and the red spider mite, e.g., the aerosols of dithio compounds, parathion, and malathion.

Seed and Seed-stock Treatment Materials Seed treatment materials are chemical compounds which are applied to the surface of seeds, tubers, bulbs, and fleshy roots. In general, they are used to protect the young plants against certain types of fungi or bacteria which would otherwise damage or destroy them. They are applied in carefully measured doses which kill the cells of the bacterium or fungus and at the same time do no damage to the seeding plants, e.g., Captan, Chloronil, and Diclone.

Fungicides and/or Bactericides In general, these compounds are used to prevent or inhibit the growth of germinating spores or mycelium of fungi or the cells of certain bacteria on the surface of plants. In other words, these compounds prevent any given fungus or bacterium from invading the tissues and getting inside the plant. Once these organisms have successfully invaded the tissues the use of these compounds is ineffectual. Consequently, correct timing of applications is essential. In general, these compounds should be applied just before rather than after a rain for crops outdoors, since they will have better opportunity to kill germinating spores. Examples of fungicides and bactericides are copper sulfate and fresh hydrated lime in a suspension with water, commonly called Bordeaux mixture; quick lime and sulfur in solution with water; colloidal, or wettable, sulfur; sulfur dust; and the dithiocarbamates—Dithane, Fermate, Manate, Parzate, Captan, Mildex, and Cyprex.

Antibiotics Antibiotics are definite chemical compounds which are made by a given organism for its protection against disease-producing organisms. For example, the fungus *Penicillium notatum* makes and secretes the antibiotic penicillin for protection against its bacterial enemies. In other words, the penicillin kills the invading bacteria. For this reason, plant scientists believed that antibiotics may be effective in the control of certain bacterial diseases of plants. Experiments have shown that this is the case, and in some instances the antibiotics are systemic in action; that is, when they are applied in solution with water, they are absorbed and become part of the sap of the plant. In this way they are not washed off by irrigation water or rain, and they protect the internal tissues and destroy the disease-producing organism which may be present. Examples of antibiotics used to control certain plant diseases are streptomycin, Streptomycin-Terramycin, and Actidione. Figures 15-9 and 15-10 show two types of sprayers in operation: one for trees in the orchard and the other for large areas of herbaceous crops.

Precautions in Using Pesticides Since many pesticides are highly poisonous to man, they should be handled with care and should be used only for the purpose for which they are recommended. Thus, the grower would do well to

Figure 15-9 Spraying a peach orchard in South Carolina. *(Courtesy, Lewis Riley, Clemson University.)*

(1) read and follow the directions on the label or container to the letter; (2) keep pesticides in their original containers, preferably in a room for their storage only, and under lock and key; (3) destroy the containers as soon as they are empty; (4) avoid breathing the gas of volatile chemicals and use a mask if necessary; and (5) become familiar with the tolerance requirements of each pesticide.

Tolerance Requirements As previously stated, certain compounds are nonpoisonous to man. Thus, the amount of residue of these compounds on the product for human consumption is of no consequence. However, other compounds are poisonous to man even in small quantities. Thus, the amount of

Figure 15-10 Spraying a bean field in South Carolina. *(Courtesy, Lewis Riley, Clemson University.)*

residue of the compounds on the product is of considerable consequence. Naturally, the amount which has no injurious effect on the consumer must be determined, and this amount is called the *tolerance requirement*. In other words, it is the maximum amount of pesticide as determined by scientists that may remain as a residue on the product with no injurious effect on the consumer. In general, the amount is expressed in parts per million (ppm), and the amount varies with the poisonous nature of the pesticide, its persistence, and the plant part to which it is applied.

Disadvantages in Using Pesticides Disadvantages in using pesticides are as follows. (1) Insecticides are likely to kill beneficial insects, for example, the honeybee and predators and parasites of certain pests. The killing of honeybees may result in inadequate pollination of the crop, and the killing of predators and parasites may upset the biological balance between the insect and its natural enemies. This is likely to result in a rapid infestation of insects which are rare and of no economic importance before the insecticide is applied. (2) All pesticides add to the cost of production. (3) Many pesticides have unfavorable effects on the fundamental processes, particularly on photosynthesis and transpiration. For example, tests have shown that Bordeaux mixture, lime sulfur, and certain oil emulsions markedly reduce the rate of photosynthesis of apple leaves and that Bordeaux mixture markedly increases the rate of transpiration of leaves of many herbaceous plants, e.g., tomatoes, cucumbers, cantaloupes, and carrots. All these effects are undesirable and have led to the development of less injurious materials and to an intensification in the breeding of resistant varieties.

Integrated Control

Integrated control involves the use of chemicals, beneficial insects (parasites and predators), and other nonchemical controls, such as certain cultural practices or partially resistant varieties. Using two or more of these methods in combination to control the pests of a given crop is called *integrated control*. Although this method is relatively new, it is being used currently to control pests in citrus and apple orchards. For example, in apple orchards predatory mites and predatory insects control destructive mites. Insecticides are applied when they are least destructive to the predatory mite population and to the pollinating insects. Weeds in or near orchards are controlled to prevent rapid buildup of a particular pest population. Thus, three different methods—the biological, the chemical, and the cultural—are integrated in the control of pests of a specific crop.

Growing Resistant Varieties

Growing resistant varieties is a very practical method of pest control. The buying and applying of chemicals are unnecessary, the rate of photosynthesis is not reduced, and the rate of transpiration is not increased. Thus, great economies are

Table 15-2 Resistant Varieties of Certain Crops

Crop	Variety	Resistant to:
Cucumber	Poinsett	Downy mildew
Pear	Baldwin	Fireblight
Peach	Nemaguard	Nematode (certain strains)
Potato	Katahdin, Kennebec	Mild mosaic
Onion	White Persian	Thrips
Squash	African	Vine borer
Tomato	Manalucie	Fusarium wilt
China aster	Tilford	Fusarium wilt
Gardenia thunbergii		Nematodes
Snapdragon	Outdoor varieties	Rust

effected and high marketable yields are obtained. Within the past 30 years considerable emphasis has been placed on the development of resistant varieties. Examples of horticultural crops are presented in Table 15-2.

QUESTIONS

1. State the three conditions necessary for the spread of insects and diseases.
2. What is meant by elimination of the causal organism?
3. Crops in greenhouses are usually markedly free from disease, whereas the same crops grown outdoors frequently show disease infection. Explain.
4. What are certified seed tubers of the potato?
5. In the production of certified seed tubers of the potato, control of insects is essential. Explain.
6. How does crop sanitation control pests?
7. What are the limitations of crop rotation in pest control?
8. How do natural predators and parasites control pests?
9. Give an advantage and a disadvantage of the so-called biological method.
10. Stomach poisons are used for insects with biting mouth parts. Explain.
11. Contact poisons are used for insects with sucking or rasping mouth parts. Explain.
12. What is the main difference in time of application of stomach poisons and contact poisons?
13. What are systemics? What is the chief limitation in their use?
14. How do fumigants kill pests?
15. State the purpose of using seed and seed-stock treatment materials.
16. What is meant by correct timing in the application of fungicides and bactericides?
17. What are antibiotics? How do they protect plants from certain fungi and bacteria?
18. Growers should become familiar with precautions in using pesticides. Explain.
19. What is meant by the tolerance requirements of a given pesticide?
20. Reduction in photosynthesis is a distinct disadvantage in the use of Bordeaux mixture, lime sulfur, and oil emulsions. Explain.

21 In the development of new insecticides and fungicides, attention should be given to their effect on beneficial insects, on photosynthesis, respiration, and transpiration of the crop, and on the bacteria and structure of the soil. Explain.

22 With other factors favorable, pest-resistant varieties should be used whenever they are available. Give three reasons.

SELECTED REFERENCES FOR FURTHER STUDY

Christensen, R. P. 1966. Man's historic struggle for food. *U.S. Dept. of Agr. Yearbook* pp. 2-15. A forceful reminder that the growth of our population has always been in direct relation to the food supply, and that our food supply, in turn, is dependent on increasing crop-plant yields, either per plant or per unit area.

Whitten, J. L. 1966. *That we may live*. Princeton, N.J.: D. Van Nostrand. The results of a Congressional inquiry on the effect of pesticides on the behavior of the crop plant, the environment, and the well-being of the human family, as compiled by a U.S. Representative.

Chapter 16

Marketing

In the world as a whole the job of creating abundance has hardly begun.

Henry R. Luce.

The marketing of horticulture products involves many practices and operations. Sequentially, the major practices are harvesting and preliminary grading, removal of field and greenhouse heat, curing, storing, processing, shipping, and selling the product.

FUNDAMENTAL PROCESSES

While a product is attached to the parent plant, it receives food, hormones, and vitamins made by photosynthesis, and water supplied by the root system. Since the product is a living entity, it requires a source of free energy. This energy comes from the sugars decomposed in respiration. Thus, when no food, hormones or vitamins are being manufactured and no water is being absorbed, the fundamental processes concerned in the marketing of horticulture products are respiration and transpiration. In general, each of these processes should be maintained at as low a rate as possible. Note in the following equation that stored food is hydrolyzed into soluble food, for example, starch into glucose, that the soluble food in the presence of abundant oxygen is oxidized to carbon dioxide and water, and that with the assistance of the ADP-ATP system, useful energy is liberated.

$$\text{Stored food} + H_2O \xrightarrow{\text{(hydrolyzing enzymes)}} \text{Soluble food} + O_2 \xrightarrow{\text{(oxidizing enzymes)}} CO_2 + H_2O + ATP + \text{Useful energy} + \text{Heat}$$

$$ATP \rightleftharpoons ADP + P$$

Note also that water is produced by respiration and given off by transpiration. In general, when the amount of water lost is equal to the amount made, little if any shrinkage takes place. When the amount lost is greater than the amount made, shrinkage takes place; the degree of shrinkage is roughly proportional to the amount of water lost. Thus, since the rate of respiration should be low, the rate of transpiration should be low also.

CAREFUL VERSUS CARELESS HANDLING

In many marketing operations the handling of living entities is necessary. This should be done as carefully as possible to keep the number and severity of bruises per unit product to a minimum. Healthy fruit maintained at relatively low rates of respiration and transpiration retains its market attractiveness and is not as susceptible to rot-producing organisms. On the other hand, careless or rough handling has opposite effects. Study the data presented in Table 16-1. Note that in all five orchard operations, careful handling kept the number of bruises at a low level. Investigations with other varieties of apples and with varieties of peaches, potatoes, and sweet potatoes have also shown that careful handling keeps the number and severity of bruises at a minimum.

HARVESTING AND PRELIMINARY GRADING

Harvesting involves removing the product from the parent plant, and preliminary grading involves discarding unmarketable specimens. Both operations constitute the first step in the journey of the product from the producer to the consumer and may be performed (1) by hand, or (2) by machine.

Harvesting and preliminary grading by hand consists in manually removing the product from the parent plant and in manually discarding unmarketable or

Table 16-1 Effect of Careful and Careless Handling on Two Varieties of Apples

Orchard operation	Number of bruises per 100 apples of each variety			
	Careful handling		Careless handling	
	McIntosh	Jonathon	McIntosh	Jonathon
Picking the fruit	20	3	214	40
Placing apples in crates	75	3	366	80
Transporting filled crates	32	3	90	26
Placing fruit on grader	97	7	333	20
Filling bushel basket	78	2	203	57

Source: Tables 7 and 9. How to reduce apple bruising, *Mich. Agr. Exp. Sta. Spec. Bul.* 374, 1951. Data based on 5-bushel sample of each variety.

unwanted specimens. Naturally, this method is essential for the harvesting of fragile, highly perishable products, such as all cut flower crops and nursery plants, and all other products which are profitable when harvested by hand, such as bananas, coffee, tea and cacao. Harvesting and preliminary grading by machine involves mechanically removing the product from the parent plant and grading the product either by hand and/or by machine.

During World War II and for about 20 years thereafter, there was a continuous migration of farm workers to urban centers, resulting in an increase in the wages of farm workers. In fact, in 1945 there were about 18.8 million farm workers in the United States, earning an average wage of $3 per day. However, in 1967 there were only 7.5 million workers, they were earning an average wage of $10 per day.[1] Thus, labor on the farm became scarce and expensive. To take the place of the nonexistent farm labor, growers turned to machines. As a result, many types of harvesting machines were developed. In general, the creation of these machines required the combined skill and ingenuity of two groups of plant scientists: agricultural engineers, specialists in the design, development, and operation of farm machines, and fruit and vegetable crop horticulturists, specialists in the design and development of varieties and cultural practices adpated to mechanization. Five types of machine harvesters are briefly described as follows.

Shake-Catch-and-Collect System

The shake-catch-and-collect system was designed for the harvesting of deciduous tree fruits, grapes, and blueberries. Examples of the kinds of fruit they bear are apples for drying, peaches for canning, prunes for drying, cherries and blueberries used in pies, and grapes used in making juice or wine. In general, the system

[1]From Agricultural Statistics for 1945 and 1967, U.S. Department of Agriculture, Washington, D.C.

requires a vibrator to shake the fruit from the plant, and an underlying and overall frame to catch and collect the fruit into pallets or field boxes. Various types of vibrators and catching frames have been developed in order to meet the specific needs of a particular crop. Figure 16-1 shows a shake-catch-and-collect harvester developed for the harvesting of apples and peaches.

Pickup-and-Collect System

The pickup-and-collect system, which picks up fruits from the surface of the land, was designed for the machine harvesting of walnuts, almonds, pecans, filberts, and tung. As described in Chapter 17, these fruits possess a hard ovary wall, or pericarp. Consequently, bruising, so common with soft fruits, is not a problem. In general, the vibrator, or shaker, of either the cable or boom type is used for the harvesting of walnuts and almonds. They may or may not be used for pecans, and they are not used for filberts, since with both of these kinds the fruits naturally fall to the ground when they are ripe. In all cases, the fruits are picked up from the surface of the land by machines specially designed for this purpose. In general, these machines consist of two rollers: a front roller elevated about ¾ to 1 inch (1.91 to 2.54 centimeters) above ground level and run in a clockwise direction, and a rear roller elevated ¼ to ½ inch (0.64 to 1.27 centimeters) above ground level and run in a counterclockwise direction. These rollers working in unison pick the fruit off the ground and elevate it to a collecting compartment on the machine. The main advantage of this method is that the fruit can be harvested with little labor. The main disadvantage is that the machine requires a smooth,

Figure 16-1 A mechanical shaker-catch harvester developed for the harvesting of apples or peaches. *(Courtesy, E. T. Sims, Jr., Clemson University.)*

clean surface from which to pick up the fruits since old leaves, pieces of twigs, clods, small rocks, and uneven land are likely to reduce the efficiency of the pickup operation.

Once-Over Harvest

Once-over harvesters were designed for the harvesting of vegetable crops grown for canning and pickling. Examples of these crops are peas, snap beans, tomatoes for canning, and cucumbers for pickling. The type of machine used varies with the crop. With peas, the vines are cut at their bases, the tops are swarthed, and the pods (fruits) are separated from the plants and shelled in a machine called a *viner*. With pole snap beans, rotary tynes, or fingers, attached to a reel or chain work downwards from the top to the bottom of the plants as the machine moves forward. These fingerlike tynes strip the pods from the plants and place them on a moving conveyor belt which carries the pods to pallets or boxes. With tomatoes and cucumbers the basal stems are cut and the tops are carried to a compartment of the machine which shakes or squeezes the fruits from the vines. All over ripe and otherwise undesirable fruits are removed by hand. With all of these crops, the requirements of the variety and its culture and the needs of the machine have been considered collectively. Adapted varieties have relatively small tops, their fruit matures in a short time, and they can be spaced at short distances, generally from between 10,000 and 20,000 plants, to 80,000 and 100,000 plants per acre. In addition, careful attention is paid to the chemical control of weeds and to applications of commercial fertilizers and irrigation water.

Multiple-Over Harvest

The multiple-over harvester was primarily designed for harvesting head lettuce. It consists of a sensor, or detector, a control unit, a cutting assembly, a series of recovery belts, and an elevator. The sensor actually selects the marketable heads, based on the size of the head and air pressure the head can withstand. If the size and air pressure are within the marketable range, the sensor sends an electric signal to the control unit which activates the cutting assembly and the recovery belts. The cutting assembly cuts the stem at the base of the plant, the recovery belts carry the heads to a vertically inclined elevator, which, in turn, carries the heads to a bulk container. Recently, a sensor based on the use of gamma radiation has been developed. When the intensity of radiation drops below a certain level, the sensor sends an electric signal to the control unit which is then activated as previously described. Thus, with either type of sensor, the need for "stoop" labor in the harvesting of head lettuce is practically eliminated.

Digger-Grader System

Digger-grader combines were designed for harvesting crops grown for their storage organs, e.g., tubers of potato, bulbs of onion, and fleshy roots of carrot, table beet, and sweet potato. In all of these crops, the tops require specific treatment before the storage organs are harvested. Potato tops are killed by the application of vine killing chemicals, the so-called defoliants; onion, carrot and beet tops are cut or twisted off in the first step of the harvest operation; and sweet potato vines are cut just before the roots are removed. The digger-grader consists of a V-shaped digging blade which lifts the roots out of the soil, and a vertically inclined conveyor belt which carries the roots to a horizontal conveyor belt from which the roots are graded by hand. Figure 16-2 shows a digger-grader harvester designed for harvesting sweet potatoes.

THE SCIENCE AND TECHNOLOGY OF THE FRESH-MARKET ROUTE

In the fresh-market, or living-product route, the product remains a living entity throughout its journey to the consumer. As such, it requires a source of free energy and a source of water. As pointed out previously, both free energy and water are supplied from sugars within the product, and for long storage life, the rates of respiration and transpiration should be relatively low. The question arises, How do growers and shippers attain the desired low rates of these processes?

By removing the field and greenhouse heat, the internal temperature of the product immediately after it has been harvested is lowered. This is done in various ways depending on the nature of the product. Some products are placed in rooms maintained at relatively low temperature, e.g., cut flowers of rose, carnation, and chrysanthemum; others are submerged in water maintained at a low temperature, e.g., ears of sweet corn, tart cherries, and clingstone peaches;

Figure 16-2 A digger-grader harvester designed for the harvesting of sweet potatoes. Note how the soil cushions the flesh roots as they travel up the elevator. *(Courtesy, Johnson Manufacturing Company, Pendleton, North Carolina).*

others are placed in refrigerated cars or trucks, e.g., head lettuce, cantaloupes, and watermelons; and others are placed directly in rooms maintained within the optimum storage temperature. (See Table 16-2.)

Curing is a method which involves subjecting the product to a relatively high temperature and a high relative humidity for a period of 5 to 10 days, depending on the nature of the product. The purpose of curing is to facilitate the development of periderm of both bruised and intact tissue. As stated in Chapter 2, the periderm consists of three distinct tissues: (1) the phellogen, the cork cambium; (2) the phellum, the layers of corky dead cells on the outside; and (3) the phelloderm, the layer of parenchyma cells on the inside. In other words, all of these tissues are collectively called periderm. As stated previously, the periderm lowers the rates of respiration and transpiration and prevents the entrance of soft-rot producing organisms. Products which require curing are the tubers of the potato, the bulbs of the onion, and the fleshy roots of the sweet potato.

Storing is another method which consists of placing the product in an environment where the life processes, respiration and transpiration, are maintained at a minimum consistent with the market quality of the product.

Respiration

Since products in storage are alive, their living cells respire to secure energy. The equation for respiration follows:

$$\text{Stored foods} + H_2O \rightarrow \text{soluble foods} + \text{oxygen} \xrightarrow{\text{(enzymes)}} CO_2 + H_2O + \text{heat and useful energy}$$

A comparatively large number of factors influence the rate of respiration of products in storage. These factors are divided into two groups: (1) plant and (2) environmental.

Plant Factors Scientists have shown that the soluble sugars, particularly glucose, are the chief sugars used in respiration. Hence, with other factors favorable, the greater the concentration of soluble sugars within the living tissues, the greater is the rate of respiration. The effect of storage temperature on potato tubers shows how soluble sugars may become the principal factor affecting respiration. At 32°F (0°C), the potato tubers become sweet; that is, comparatively large quantities of sugar are formed at the expense of starch. Because of the relatively greater concentration of sugars in the tubers stored at 32°F (0°C) than in those stored at 40°F (4.4°C), the tubers stored at 32°F (0°C) will respire at a greater rate than those stored at 40°F (4.4°C) when they are placed in higher temperatures, as is usual in marketing operations.

Because living cells require a constant supply of energy and because respiration liberates the necessary energy, the rate of respiration will be directly proportional to the number of living cells. Thus, 2 bushels of apples will require more energy and will produce more carbon dioxide and liberate more heat than 1 bushel. For the same reason, a cabbage or lettuce head will respire more rapidly in the same environment than a potato tuber or sweet potato root. In fact, the greater the number of living cells in proportion to the dead cells of a product, the greater will be the amount of carbon dioxide and water given off.

In general, the rate of respiration varies directly with the water content of the product. For example, at the Oklahoma Experiment Station, various kinds of vegetable products were stored at 3°C (37°F) for a period of 26 hours. The succulent vegetables lost 10 to 13 percent of their dry weight, whereas moderately succulent vegetables lost only from 1 to 2 percent of their dry weight. Obviously, succulent products respire more rapidly, weight for weight, than relatively nonsucculent products. Thus, lettuce heads respire more rapidly than potatoes or sweet potatoes or even peppers. This means that immature fruit respires more rapidly than mature fruit.

Environmental Factors In general, with other factors in favorable supply and with the temperature from 10 to 35°C (50 to 95°F), the van't Hoff rule applies; that is, the higher the temperature from 10 to 35°C (50 to 95°F), the greater the rate of respiration. In storage the respiration rate should proceed at a minimum. Note the effect of temperature on the respiration rate of the three fruits and three vegetables shown in Figure 16-3. The curves show that the rate of respiration varied directly with the temperature. In other words, the higher the temperature, within 0 and 35°C (32 and 95°F), the higher was the respiration rate. Numerous investigations have shown that practically the same situation

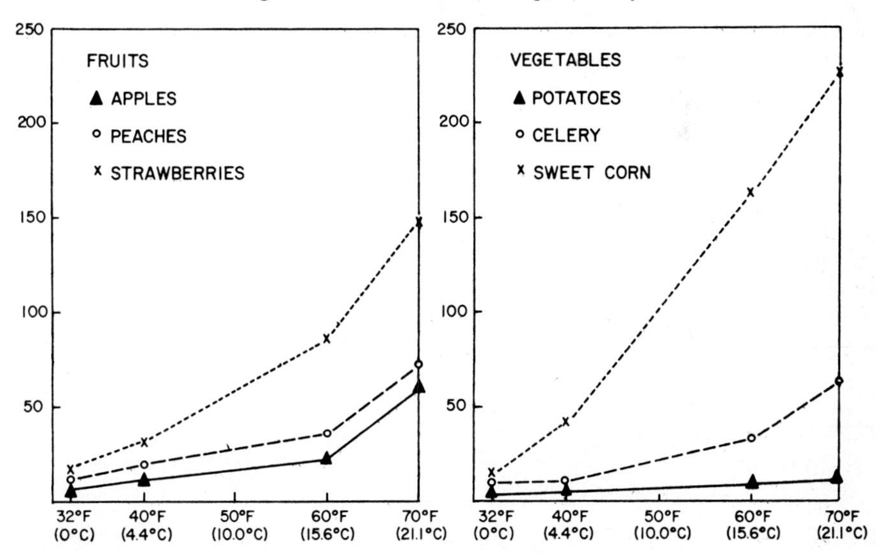

Figure 16-3 Rate of respiration of three fruits and three vegetables. *(Courtesy, R. E. Hardenburg, U.S. Department of Agriculture, Beltsville, Maryland).*

exists with other products. From these studies optimum storage temperatures of many products have been determined. Examples are shown in Table 16-2.

A study of the equation for respiration shows that oxygen is absorbed and carbon dioxide is given off. Thus, if living products are placed in a gastight room, the oxygen supply of the air will gradually decrease and the carbon dioxide will gradually increase. This decrease in concentration of oxygen and increase in concentration of carbon dioxide decreases the rate of respiration in accordance with the law of mass action. As the products of the reaction (carbon dioxide) pile up, the speed of the reaction (respiration) slows down. This principle is now used in the storage of certain varieties of apples, and the practice is known as *controlled-atmosphere storage*. In general, the apples are stored in gastight rooms maintained at 40 to 42°F (4 to 6°C) the oxygen and carbon dioxide are kept at 3 and 5 percent, respectively; and gases given off by the apples, such as ethylene, are removed by means of activated carbon. However, for all products in storage the concentration of oxygen should always be sufficient for the complete combustion of the sugars. If inadequate quantities are available, compounds are formed which are injurious to the tissues and protoplasm. Consider the following equations:

(respiration with adequate oxygen)

1. $C_6H_{12}O_6 + 6O_2 \rightarrow 6CO_2 + 6H_2O$

(respiration with inadequate oxygen)

2. $C_6H_{12}O_6 + O_2 \rightarrow \underset{\text{(alcohol)}}{C_2H_5OH} + \underset{\text{(acetic acid)}}{CH_3COOH} + 2CO_2 + H_2O$

Note that 6 molecules of oxygen are required for the complete combustion of 1 molecule of glucose and that if only 1 molecule of oxygen is present, alcohol and acetic acid are formed. Alcohol is injurious to the tissues and induces death

Table 16-2 Storage Temperatures for Certain Horticulture Products

31 to 33°F (−1 to 1°C)	31 to 40°F (−1 to 4°C)	40 to 45°F (4 to 7°C)	33 to 35°F (1 to 2°C)	55 to 65°F (13 to 18°C)
Pear, apricot, peach, plum, nuts, grape, small fruits, fig, cabbage, lettuce, celery, onion, carrot, beet, garden pea	Apple	Potato	Cut flowers (except) orchid and gladiolus	Capsicum pepper, squash, pumpkin, sweet potato

Source: U.S. Dept. Agr. Handbook 66, 1954.

of the protoplasm. Other intermediate compounds induce death also. For example, with inadequate oxygen potato tubers form tyrosine which is responsible for blackheart, certain varieties of the apple form aldehydes which induce storage scald, and cabbage and celery form compounds which induce speckling and pitting of the petioles and veins. Thus, every storage room should be supplied with sufficient oxygen for normal respiration.

The equation for respiration shows that heat is liberated. In the storage of certain products the liberation of heat is a vital factor, particularly with products which have a high respiration rate. As an example consider celery. Investigations have shown that the temperature of plants inside the storage crates is always higher than that on the outside. This high temperature is undoubtedly due to the lesser flow of air around the stalks in the center of the crate. Consequently, since relatively large amounts of heat are given off in storage, adequate quantities of air are necessary to carry the heat away. For this reason storage in bulk is generally less satisfactory than storage in various types of containers. This is particularly true for products which have a high respiration rate.

Transpiration

As with respiration, both plant and environmental factors influence the rate of transpiration.

Plant Factors Plant products differ in the degree of differentiation of their tissues and hence they differ in the rate of transpiration under the same conditions. In general, nondifferentiated, very succulent tissues contain more water than highly differentiated, nonsucculent tissues. Under the same conditions tissues with a high water content lose water more rapidly than tissues with a low water content. Examples of highly succulent tissues are asparagus spears, young spinach leaves, and young turnip greens. Examples of highly differentiated products are mature cabbage heads, onions, and celery.

Two kinds of tissues constitute the outer cover of plant products. These are the epidermis and the periderm. The student will recall that the epidermis consists of a single layer of living cells. On the outer walls a layer of wax may or may not be present. Since a layer of wax on the epidermis retards transpiration, plant products with a cutinized epidermis will shrink less rapidly in storage than those with a noncutinized epidermis. Examples of products which have an epidermis are asparagus spears, spinach, and other vegetables grown for their leaves.

Experiments have shown that plant products with a well-developed and noninjured periderm lose water less rapidly and keep longer in storage than those products with a poorly developed or badly injured or bruised periderm. In fact, long storage life of many products depends on the maintenance of a healthy and intact periderm. Plant products which possess a periderm as the outer cover are apples, pears, citrus fruits, root crop vegetables, sweet potatoes, potatoes, and gladiolus corms.

Environmental Factors The student will recall how temperature influences the rate of transpiration of a growing plant. Since products in storage transpire, the influence of temperature on their rate of transpiration is much the same as that on growing plants. In other words, the rate of transpiration of products in storage is roughly proportional to the temperature. Thus, a comparatively high storage temperature induces a greater rate of transpiration and greater shrinkage than a comparatively low temperature. Since a high rate of transpiration and excessive shrinkage are undesirable, comparatively low temperatures in storage houses are maintained.

The term *relative humidity* refers to the amount of water vapor in the air compared with the amount of water vapor when the air is fully saturated for any given temperature. Thus, when the relative humidity is 50 percent the air contains only one-half as much moisture as it would contain if it were saturated for any particular temperature. The student will recall the effect of relative humidity on the rate of transpiration of growing plants. As might be expected, its effect on the transpiration of products in storage is much the same. Thus, with other factors of the storage environment favorable, the rate of transpiration is inversely proportional to the relative humidity. In other words, a low relative humidity induces a high rate of transpiration and a high relative humidity induces a low rate of transpiration. Since a high rate of transpiration induces excessive shrinkage and since excessive shrinkage is undesirable, a comparatively high relative humidity is maintained in storage houses.

In most crops, experiments have shown that the relative humidity should vary between 70 and 85 percent. If the humidity is lower than 70 percent, shrinkage is likely to become excessive. On the other hand, if the humidity is greater than 85 to 90 percent, environmental conditions become favorable for the condensation of moisture on the surface of the stored products and for the development of storage rots.

Horticultural products classified according to their humidity requirements follow: fruits, 80 to 85 percent; beets, carrots, parsnips, salsify, turnips, radishes, celery, cabbages, potatoes, and sweet potatoes, 75 to 85 percent; onions, beans, squash, and pumpkins, 70 to 75 percent; cut flowers, 80 percent; flowering bulbs, 60 to 70 percent.

Types of Storage Houses

Storage houses may be classified as (1) refrigerated and (2) ventilated, according to the type of temperature and relative humidity control used.

Refrigerated storage houses have insulated walls and doors and a thermostatically controlled cooling system. How are optimum temperature ranges and relative humidity ranges maintained in this type of storage? In general, when the temperature approaches the upper limit of the optimum temperature range, the thermostat starts the refrigerating machinery. Substances which are gases at ordinary temperature and pressure are compressed and released through pipes or coils in the storage room. As the compounds expand, they take up heat—the heat of respiration of the living products and the heat which has passed through the insulated walls and doors. (The student will recall that free energy is needed to change a substance from the liquid to the vapor state). When the temperature approaches the lower limit of the optimum range, the thermostat stops the machinery. In this way, the temperature is maintained within narrow limits, and because of the inverse relation of temperature to relative humidity, the relative humidity is maintained with narrow limits also.

Ventilated storage houses have insulated walls and doors, but instead of a refrigeration system, they have side and ceiling vents, a dirt floor, and a thermostatically or manually controlled heating system. How are optimum temperature and relative humidity ranges maintained in this type of storage? On the one hand, the weather must be taken into consideration. During mild periods, the temperature of the storage house rises. Before it reaches the upper part of the optimum range, the side vents are kept open. This is usually during the night or 24-hour period when the temperature of the outside air is lower than that of the inside air. The vents are usually kept closed during the day, or that part of the 24-hour period when the temperature of the outside air is higher than that of the inside air. The ceiling vents are kept open at all times in order to get rid of heat—the heat of respiration of the living products and that which has passed through the insulated walls and doors. On the other hand, during cold weather the temperature of the storage house declines and all of the vents are kept closed—the side vents to keep out the cold air and the ceiling vents to keep in the heat of respiration. In addition, there is a thermostat which turns on the heating system, if the system is thermostatically controlled, or starts the fires, if the system is manually controlled.

Grading

The individual fruits, vegetables or flowers of any given variety vary greatly in size, shape, color, and freedom from blemishes. However, the consumer demands uniformity—not variability. To establish uniformity in market quality, most horticultural products are divided into three or four commercial classes, or grades. For example, the commercial classes for peaches are U.S. Fancy, U.S. No. 1, and U.S. No. 2. The U.S. Fancy grade requires that the size be at least 1⅞ inch in diameter, that the red pigment cover at least 50 percent of the surface of highly colored varieties and 25 percent of the surface of poorly colored varieties, and that all the fruit shall be practically free of blemishes. The commercial classes for sweet potatoes are U.S. Extra No. 1, U.S. No. 1, U.S. Commercial, and U.S. No. 2. The U.S. Extra No. 1 requires that the length of the roots be between 3 and 9 inches, that the width be between 1¾ and 3¼ inches, and that all of the roots be practically free from diseases, insect damage, cracks, freezing injury, and internal breakdown.

Packing

Packing consists of placing specimens of any given product in an appropriate container so that the individual units remain in place and present an attractive appearance. Packages have the following functions: they protect the product

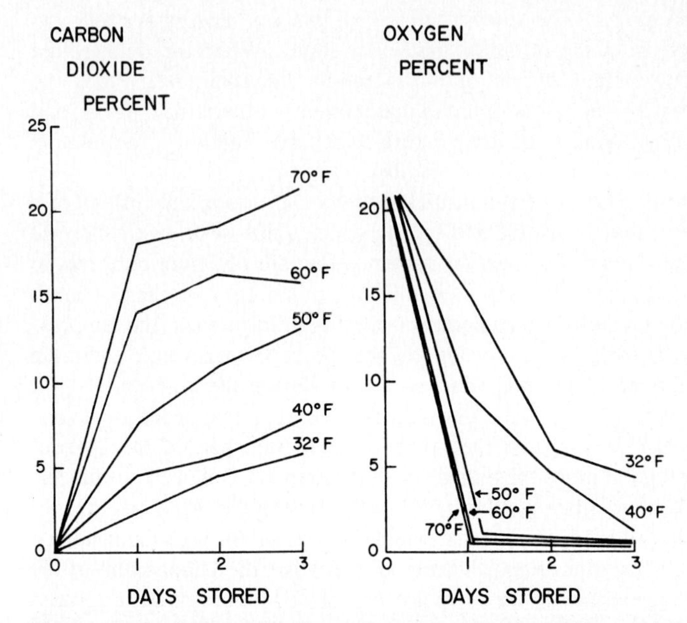

Figure 16-4 Carbon dioxide and oxygen levels within sealed polyethylene bags containing green snap beans. *(Courtesy, R. E. Hardenburg, U.S. Department of Agriculture, Beltsville, Maryland).*

from excessive bruising; they provide ventilation for oxidative respiration; and they serve as a measure of the contents and a means by which the product may carry identification marks, shipping directions, or advertising matter. In general, there are two sizes—relatively large master units for transporting the product to the wholesale market or warehouse and small consumer units for displaying the product in the retail store. Examples of master units are flat-bottomed bushel baskets, wood crates, and corrugated paper boxes treated to withstand hydrocooling, icing, or similar wet-package treatments. Examples of consumer units are plastic, film bags, and trays overwrapped with plastic film.

Packaging in consumer-sized units consists of placing a small number of individual specimens in a transparent plastic film bag, generally made of Saran, pliofilm, or polyethylene. At first their use was unsatisfactory, since no provision was made for ventilating the products while they were displayed under the warm temperature of the retail store. A rapid buildup of carbon dioxide and a rapid depletion of oxygen occurred within the package. Study the curves in Figure 16-4. Note that the pods of the green snap beans in the nonperforated bags had practically no oxygen for their oxidative respiration. As a result, the anaeorobic, or fermentative, type of respiration took place, causing the formation of compounds associated with poor quality and off-flavor. Investigations have shown that, in general, the optimum concentrations are 10 to 20 percent carbon dioxide, 3 to 5 percent oxygen, and 90 to 95 percent water vapor. To maintain these concentrations, ventilation holes in the films are necessary. According to the authorities, consumer-sized packages with 2, 3, or 4 ⅛- or ¼-inch holes per package or numerous pinlike holes will let in sufficient oxygen and let out sufficient carbon dioxide to prevent the formation of compounds associated with off-flavor. This explains why green-mature tomatoes and hard-ripe peaches never ripen satisfactorily in nonventilated film bags and why vegetable products with high respiration rates, e.g., broccoli, spinach, and cauliflower, develop off-flavors under nonventilated conditions. Figure 16-5 shows the outside view of a modern peach packing house.

THE SCIENCE AND TECHNOLOGY OF THE NONLIVING PRODUCT ROUTE

In the nonliving-product route, the individual units are living entities only for a short part of their journey to the consumer—usually from the time they are severed from the parent plant until they arrive at the processing plant. During this period the entities require a source of free energy and a source of water, and, as with the products of the fresh-market route, free energy and water are made available by respiration. Thus, while the products are living entities, the rates of respiration and transpiration are maintained at a minimum. However, soon after the products arrive at the processing plant their enzyme systems are destroyed by one of several methods of preservation. The most common methods are canning, drying, pickling and fermenting, freezing, and using sugar concentrates.

Canning

Essentially, canning consists of subjecting the plant product to high temperatures, usually from 200 to 240°F (93 to 116°C) for 5 to 90 minutes, depending on

Figure 16-5 Outside view of a modern peach packing house. *(Courtesy, D. O. Ezell, Clemson University).*

the kind of product, degree of maturity, and size of can used. The fundamental process concerned is respiration. The high temperature inactivates the enzymes. When the enzymes of respiration are destroyed the process ceases. With the cessation of respiration the sugars, proteins, minerals, and vitamins are conserved. Naturally, the shorter the time between harvesting and processing, the lesser will be the utilization of starches and sugars in respiration.

The canning of fruits and vegetables is an important horticultural industry. In fact, with many crops the acreage given over to canning is greater than that given over to growing the crop for market in the fresh state. Examples are peaches, tomatoes, sweet corn, and lima beans. Figure 16-6 shows a modern canning plant near Griffin, Georgia.

Drying, or Dehydration

As the name suggests drying consists of removing sufficient water from the product to prevent it from spoiling in storage. As with other forms of food preservation the principal process concerned is respiration. Removing most of but not all the water inactivates the enzymes for respiration. In this way the product is conserved. In general, horticultural products are dried in two ways: (1) under natural conditions and (2) artificially. In either case the drying is done by means of warm, relatively dry air. To prevent spoilage of the product the rate of drying is relatively rapid.

Drying under Natural Conditions This practice is sometimes called *sun-drying*. Principal environmental factors concerned are (1) degree and amount of sunshine and (2) temperature and dryness of the air. In general, the greater the intensity and duration of sunlight, the higher the temperature; and the drier the air, the greater the rate of drying. Thus, warm-dry climates characterized by abundant sunshine will be more favorable to sun-drying than warm-moist and

Figure 16-6 A modern canning plant located near Griffin, Georgia. *(Courtesy, H. L. Cochran, Pomona Products Company).*

cold-dry climates. Of the commercial crop-growing regions in the continental United States the climate of certain intermountain areas of California is particularly favorable for the production and sun-drying of many horticultural crops. In this region large quantities of raisins, prunes, apricots, peaches, figs, and pears are produced. The winters are sufficiently cold to bring the trees out of the rest period; the springs and early summers are sufficiently warm to permit the production of large quantities of high-quality fruits; and the late summers and falls are sufficiently warm and dry to permit rapid drying of the fruit. In these intermountain valleys most of the rains occur during the winter and spring; the summers are practically dry.

Drying Artificially Drying artificially consists of passing heated air over the product. The principal factors concerned are (1) temperature and dryness of the air and (2) velocity of the air. Heating the air is necessary to increase its drying capacity. For example, within the temperature range for drying, an increase from 20 to 30°F (−7 to −1°C) will double the water-holding or drying capacity of the air. Thus, the higher the temperature, without impairing the quality of the product, the greater will be the rate of drying and the lesser will be the reduction in quality. In general, the optimum temperature for drying depends on the type of product and its initial water content. For example, the optimum temperature range for apples and pears varies from 130 to 165°F (54 to 74°C) and for beans and peas from 115 to 140°F (46 to 60°C). For any one product the lower limit of the temperature range is used when drying begins, and the temperature is gradually increased to the upper limits as drying proceeds. A type of drying is called *vacuum drying*. As the name suggests, the atmospheric pressure surrounding the product is reduced. This, in turn, promotes a rapid loss of water from the tissues and a reduced rate of oxidation of valuable constituents, e.g. Vitamin C. Products vacuum dried are peaches and apricots.

Pickling and Fermenting

Essentially, pickling and fermenting consist of placing the plant product in a salt brine of definite concentration. The salt brine stops or limits the oxidative phase of respiration, allows the fermentative phase to continue, inhibits the growth of spoilage microorganisms, and promotes the growth of *Lactobacillus*, the lactic acid which produces bacteria. The combination of salt and acid preserves the product and results in its being pickled. The final product from the brine tanks may be flavored in various ways, e.g., dill, sweet, sour, or relish. A modification of pickling is the addition of salt to shredded cabbage under conditions favorable for the growth of *Lactobacillus* and the production of lactic acid. The familiar name for this product is *sauerkraut*.

Horticultural crops which are grown for pickling are cucumbers, cabbage, onions, and beets. Special varieties of these crops have been developed for the purpose of pickling. Michigan and North Carolina are foremost in the production of cucumbers for pickling; New York and Wisconsin are foremost in the production of cabbage used for sauerkraut. Numerous pickling and salting stations are located in these states. Figure 16-7 shows a group of vats used in the processing and curing of cucumbers for pickling.

Freezing

Essentially, freezing involves exposing fruits and vegetables to temperatures ranging from 0 to −50°F (−18 to −46°C). It is generally accomplished by placing the food in a blast of cold air or directly on refrigeration plates. The low temperatures stop enzyme activity and, consequently, the rate of respiration. Vegetables are usually precooked or blanched for a very short time before freezing in order to quickly inactivate the enzymes which are present. Freezing and storing foods at temperatures of 0°F (−18°C) or lower retain the freshness, color, flavor, vitamin content, and nutritive value of fresh fruits and vegetables to a high degree.

An important factor for success in freezing horticultural products is the growing of suitable varieties. Investigations have shown that not all kinds of fruits and vegetables are adaptable to freezing and that certain varieties within a given kind are unsuitable. Examples of products adaptable to freezing are strawberries, raspberries, blueberries, asparagus, green broccoli, and certain varieties of peaches, plums, grapes, snap beans, lima beans, peas, and sweet corn.

Figure 16-7 Vats for the processing of pickling cucumbers. *(Courtesy, Mount Olive Pickle Company, Mount Olive, North Carolina, and Tom Byrd, North Carolina State University).*

Some foods are partially dried before they are frozen. This process is known as dehydrofreezing, and since the water content of the product is greatly reduced, it greatly aids in reducing the weight of food which is shipped. An example of such a product is slices of apple fruits.

In the process of freeze-drying, the food is dehydrated to complete dryness at temperatures below the freezing point of the food. This process retains the original freshness of the product, and, when properly packaged, the food may be stored at room temperature. Tomatoes and sweet potatoes are examples of freeze-dried vegetables.

Using Soluble Sugar Concentrates

This method consists of placing the product in an aqueous solution of sugar at a specific concentration and sealing the container. Both low and high concentrations of syrup may be used. The use of low concentrations is a recent development and has been found to be successful with peaches. The procedure consists of placing fresh slices in glass jars; covering the slices with a dilute concentration of syrup; adding small quantities of sodium bisulfite, sodium benzoate, citric acid, and ascorbic acid; sealing the jars; hydrocooling the jars and product to about 50°F (10°C); and storing the product at 28 to 30°F (−2.2 to −1.1°C).

High concentrations of sugar have been employed for a long time. In general, this method consists of placing fleshy fruits with a pH of 7 or less in a highly concentrated sugar solution, gently boiling the mixture to a relatively low water content, placing the concentrate in an appropriate container, and sealing the container so that it is airtight. The type of preserve which results depends on the nature of the original product, e.g. jellies are made from juices of fruit, and marmalade from citrus fruit and pieces of rind.

TRANSPORTING PRODUCTS

Market Routes or Channels

How do products go from the producer to the consumer? Note that the three principal routes, presented in Figure 16-8, are for fresh fruits and vegetables, processed products, and flower and ornamentals, respectively. With fresh fruits and vegetables, the route may be short and direct, such as to a roadside market. It may be relatively long and indirect; in other words, the product may pass through the hands of one or more middlemen, e.g., the shipper, the wholesaler, or the retailer, before it reaches the consumer. With processed products, the route is moderately long and may go directly from the processing plant to a retail store such as a supermarket, or indirectly from the processing plant to a wholesaler or retailer. With flowers and ornamental plants, the channel is relatively short and direct, e.g., from the producer to a florist shop or a garden center.

MARKETING CHANNELS OR ROUTES

I. The Fresh-Living Product Route

Route	Grower of:				
1	Fruits Vegetables	"We grow them, you pick them" method			Customer
2	Fruits Vegetables	Grading and packing by grower	Roadside market		Customer
3	Fruits Vegetables	Grading and packing Individual grower Grower Co-operative Shipper	Warehouse Wholesaler Supermarket for storing, grading, repacking	Fruit and vegetable stands Hucksters Restaurants Supermarket	Customer
4	Flowers Ornamentals	Wholesale florist or nursery	Retail flower shop or garden center		Customer

II. The Non-Living Product Route

5	Fruits Vegetables	Processing plant	Warehouse Wholesaler Supermarket	Restaurants Independent stores Supermarket	Customer

Figure 16-8 The principal market routes from the producer to the consumer.

Transportation Environment

The optimum transportation environment depends on the nature of the product. In general, fresh fruits or vegetables are shipped within or just below their optimum temperature and relative humidity ranges for storage. As in storage, so in shipping, the life activities of these products are maintained at a minimum. Otherwise, there would be a rapid decrease in valuable foods and vitamins and a corresponding decrease in quality, flavor, and palatability. Canned, pickled, and preserved products, however, are shipped in uniformly cool (above freezing temperature) dry air, since alternate freezing and thawing reduces their flavor and quality, and moist air induces the rusting of tin containers. In sharp contrast, frozen products are shipped at subfreezing temperatures, since these temperatures prevent the growth and destructive activity of spoilage organisms. Conveyances equipped for this purpose are the refrigeration railway car and the refrigerated truck. Finally, flowers and ornamentals are shipped in vehicles maintained at temperatures just below or within the optimum temperature range. In this way, respiration and transpiration are not likely to be excessive.

THE MODERN SUPERMARKET

The modern supermarket is a colossal retail store—a place for the display of consumer-sized packages of individual units of an item or a small number of the same items. The variety of commodities offered for sale almost staggers the imagination. Magazines, stationery, soft goods, hardware, cigars and cigarettes, candy, baked goods, TV dinners are available, as well as the usual grocery items, meat, dairy, poultry products, and both fresh (living) and processed (dead) fruits and vegetables. In fact, the number of items offered for sale in a supermarket may vary from 1,000 to 5,000 or more, depending on the size and needs of the community the supermarket serves.

The supermarket came into being as a result of the growth of urban areas. Suburban areas, in turn, developed adjacent to or near cities and large manufacturing establishments. They are characterized by one-, two-, even three- car family homes, hard surface roads, and at least one shopping or service center which usually contains at least one supermarket. Thus, supermarkets are usually located within a short distance from most homes and are readily accessible.

The principals in the operation of supermarkets are the manager of the store and the home maker. The duties of the manager are many and varied. In general, he makes surveys among the home makers of the community as to their needs and requirements; estimates the needs of the store at least one week in advance so that he can place orders to meet these needs from the chain-store warehouse, the wholesaler, or the retailer; maintains a continuous supply of any given item on the display shelves or storage cabinets; keeps fresh fruits and vegetables at their optimum temperature and relative humidity; keeps records of the rate of sale of any given item; and maintains an attractive and pleasant environment throughout the store. The home makers select the items they want from the display shelves or storage cabinets; place them in the pushcarts; pay for them at the checkout counter; and transport them, usually in the family car, from the store to home. Thus, the manager provides a service for which the customer pays, resulting, hopefully, in a high degree of satisfaction attained by everyone.

What are the distinguishing features of the present day supermarket? What differentiates these retail stores from the old-time grocery stores? Doubtless, the distinguishing features are the self-service selection of individual items and the self-delivery of these items on the part of the home maker. These features have resulted in lower operating costs. According to research by the U.S. Department of Agriculture, an individual employee, in a grocery store during cracker-barrel days sold an average of 2,500 dollars worth of goods each year, whereas an individual employee in a supermarket in 1954 sold an average of 30,000 dollars worth of goods. On the assumption that a dollar in 1954 has one-half the value of a dollar when grocery stores predominated, the efficiency of employees of supermarkets, in terms of the value of goods sold, has increased six times over that of their grocery-store counterparts. Thus, the self-service and self-delivery features have significantly decreased the cost of retail marketing, and the resultant savings have accrued to the consumer.

QUESTIONS

1 In the marketing of fresh fruits and vegetables and cut flowers, the rates of respiration and transpiration should be maintained at a minimum. Explain.
2 Fresh fruit and vegetables and cut flowers in storage and in market operations always decrease in dry weight. Explain.
3 The temperature of products in storage is usually higher than that of the air in the storage room. Explain.
4 Why do products with an intact periderm transpire less and have a longer storage life than those with a broken periderm?
5 With other factors favorable, should a product be harvested in the morning? In the afternoon? Or would either period be equally satisfactory? Explain.
6 Give reasons for the development of new types of harvest machines in the years following World War II.
7 Mechanical harvesting of a crop may necessitate marked changes in its cultural requirements. Explain.
8 How do the shake-and-collect harvesters operate? The pickup-and-collect, the once-over, and the digger-combine harvesters?
9 How is the law of mass action applied to the controlled-atmosphere storage of certain varieties of apples?
10 The amount of oxygen in storage rooms should always be sufficient for the complete combustion of the sugars. Explain.
11 Perforated plastic bags are necessary for the marketing of fresh fruits and vegetables. Explain.
12 Most horticulture products are stored at a relatively high humidity. Explain.
13 State the functions of the ceiling vents and the side vents in ventilated storage houses.
14 In average weather, with the temperature in a sweet potato storage house gradually rising and the top vents continually open, when should the side vents be open during a 24-hour period? Explain.
15 If a cold storage is not available, will tree fruits keep better for a short period on the tree or off the tree? Give reasons.
16 Explain why the warm-dry climate of the Pacific Coast region of the United States is particularly favorable for the sun-drying of fruit and vegetable products.
17 What season of the year, if any, is most favorable for sun-drying in your particular community? Give reasons.
18 How do canning, drying, pickling and fermenting, freezing, and using high sugar concentrates preserve horticultural products?

19 How do U.S. grades and standards facilitate the marketing of horticulture products?

20 What is a supermarket? How do supermarkets facilitate the buying, displaying, and selling of horticulture products?

SELECTED REFERENCES FOR FURTHER STUDY

Cargill, B. F., and G. E. Rossmiller. 1969. *Fruit and vegetable harvest mechanization, technological implications.* East Lansing, Mich.: Rural Manpower Center, Michigan State University. A valuable and up-to-date presentation of papers by agricultural engineers, horticulturists, and agriculture economists concerning the problems of harvesting horticulture products by machine.

Hardenburg, R. E. 1971. Effect of in-package environment on keeping quality of fruits and vegetables. *Hort Sci.* 6 (3):198–201. A discussion of the results of research which examined the effect of prepackaging fruits and vegetables in consumer-sized transparent films.

Smock, R. M., 1949. Controlled atmosphere storage of apples. *N.Y. (Cornell) Agr. Ext. Serv. Bul.* 759. A discussion of the principles involved in the storage of fruits of the apple, with emphasis on the control of carbon dioxide and oxygen in the storage room.

Wright, R. C. et al. 1954. The commercial storage of fruits and vegetables and florist and nursery stocks. *U.S. Dept. Agr. Handbook 66.* A discussion of recommended conditions for the storage of all kinds and varieties of horticulture products, including 216 citations to the literature.

Part Three

Principal Horticultural Crops

Chapter 17

The Tree Fruits

The finest gift a man can give to his age and time is the gift of a constructive and creative life.

TYPES OF TREE FRUITS

Tree fruits are divided as follows: deciduous and evergreen. In general, deciduous trees shed their leaves in the fall, whereas evergreen trees shed most of their leaves in the spring when the young leaves are expanding. Thus, deciduous trees make foods during limited periods of the year only and evergreen trees make foods the entire year. Principal deciduous-tree fruits are the pomes and drupes, and the principal evergreen-tree fruits are orange, lemon, grapefruit, lime, banana, and olive.

DECIDUOUS-TREE FRUITS

Pome Fruits (Apple and Pear)

Root System The root system usually consists of a relatively short tapering taproot and several large, spreading lateral roots which branch into a network of smaller, threadlike roots. The larger roots are woody and serve mainly for anchorage, transporation, and storage; the smaller roots are nonwoody and serve as the absorbing system. The depth and spread of the roots depend largely on the texture and moisture of the soil. Where the soil is very loose and the soil moisture is deficient, the roots often penetrate many feet into the soil. In general, however, the major portion of the feeding roots is within the upper 15 inches (36 centimeters) of the soil, and the root spread is several feet beyond the tips of the scaffold branches. Thus, deep plowing or tillage will destroy many of the absorbing roots near the surface.

Stems and Leaves The woody stem develops from alternate lateral and terminal vegetative buds. These lateral and terminal buds produce the twigs, each with a leaf and vegetative bud at each node. As the top of the tree develops, the twigs become small branches, and finally limb or scaffold branches—the main framework of the tree. The shape of the top of mature trees varies with the species and with the variety. In general, apple trees are more or less globular, and pear trees are shaped like an inverted cone. However, certain apple varieties (Henry Clay and Delicious) have the upright habit of growth, whereas others (Stayman) have the spreading type.

The leaves are simple, alternate, and toothed or lobed. They vary in size, shape, color, thickness, pubescence, and texture. By these differences the several species of apple and pear and certain varieties within a given species can be identified. For example, the leaves of the Baldwin variety are broad and distinctly saucer-shaped; those of the Delicious are moderately broad; and those of the Jonathan are small and narrow at base and apex. In general, the leaves of the pear are usually thicker, greener, less pubescent, more finely toothed, and more ovate than those of the apple.

Flowers and Fruit The flowers are perfect, with a five-lobed calyx, five moderately large separate petals, numerous distinct stamens, and a five-celled, five-styled ovary. Five or more blossoms are borne from the mixed flower bud on the end of each spur. The fruit varies in shape, color, texture, size, time of maturity, and other characteristics. In general, apples are spherical, with cavities at the basal (stem) end and apical (blossom) end, and the skin is green, yellow, or red or may develop two or all three of these pigments. The flesh is white or yellow and free of grit cells. On the other hand, the pear is pyriform with less distinct basal and apical cavities, and the color of the skin varies from dull yellow in some varieties to dark reddish-brown in others. The flesh is usually white or creamy white and contains numerous grit cells.

Economic Importance and Principal Commercial Districts The commercial production of apples in the continental United States is a major horticultural enterprise. In fact, this country leads all others in the production of the fruit. Principal centers of production are south and east of large bodies of water or east of mountain ranges. These afford protection from severe cold. Important commercial districts are (1) the Shenandoah-Cumberland located between the Blue Ridge and Cumberland Mountains, (2) the western New York on the south shore of Lake Ontario, (3) the Hudson and Champlain Valleys in southern New York, (4) the New England extending from southern Maine to Connecticut, (5) the eastern shore of Delaware and Maryland, (6) the western Michigan on the east shore of Lake Michigan, (7) the Ozark in southwestern Missouri, (8) the Wenatchee and the Yakima Valley in Washington, (9) the Hood River Valley in Oregon, (10) the Watsonville and the Sebastopol in central California, (11) the Payette in Idaho, and (12) the Grand Valley in Colorado. The United States and Canadian crops combined amount to more than 20 percent of the world's production of apples.

Leading European apple-producing countries are France, Italy, and West Germany. Production in France has rapidly increased as a result of the cultivation of new areas, made possible by the introduction of irrigation. The Netherlands have modernized apple production by replacing old orchards with large numbers of dwarf trees per unit area. In the United Kingdom, total production of apples has increased, in spite of a reduction in area, as a result of employing modern systems of orchard planting and management.

The importance of apple production for export is increasing throughout the world. Leading producers in Eastern Europe are Hungary and Poland. In the Ukraine, Moldavia, and the Kuban areas of the USSR, considerable new orchard plantings are being made. Leading apple-producing countries in the Southern

Figure 17-1 Airview of apple orchards in South Carolina. *(Courtesy, Lewis Riley, Clemson University.)*

Hemisphere are Argentina, Australia, South Africa, and New Zealand. In the Far East, apples are grown commercially in Japan, China, and Korea.

Varietal Groups or Types In all apple-producing areas of the world, there are early nonstorage types, for example, Lodi and July red; mid-season storage types, for example, Red Delicious, Golden Delicious, and McIntosh, and late storage types, for example, Rome Beauty and Winesap.

Spur-type strains are available in many of the leading varieites. In spur-type trees, vegetative growth is controlled by a genetic factor in the scion itself. Many spurs, but relatively few lateral shoots, are produced, so that spur-type trees are considerably smaller and produce fruit earlier than standard trees.

In the continental United States pears are grown much less extensively than apples. Fireblight, a serious bacterial disease, is one of the chief limiting factors in commercial production. Important commercial districts are located in California, Michigan, Washington, and Oregon.

Western Europe is the world's major pear-producing area. Italy is the most significant producer followed by France, the Netherlands, Germany, the United Kingdom, and Belgium.

Use of Growth Regulators In general, growth regulators have an important place in the production of apples. In fact, they are used in many phases in the growth of the trees and the development of the fruit.

Growth of the Spurs and Shape of the Fruit Cytokinins when applied to apple trees promote the growth of the spurs and the development of lateral branches; and applications of cytokinins and gibberellins to the flower clusters of Red Delicious a few days after full bloom have been known to increase the length-diameter ratio of the fruit, enhance the growth of the lobes at the calyx end, and improve their overall appearance.

Overcoming Biennial Bearing Biennial bearing is a term applied to deciduous fruit trees which bear exceedingly heavy crops of fruit one year and very light crops the next. The year of heavy crops is called the *on* year, and that of light crops is called the *off* year. Experimental tests have shown that hand thinning of the blossoms during the on year promotes the production of moderately heavy crops each year. However, hand thinning is time-consuming and expensive, workers are not always available, and with certain varieties, it is ineffective. Fortunately, apple-tree scientists have found that NAA, naphthaleneacetic acid, is very effective in reducing the number of blossoms during the on year, and its application is a common practice.

Preventing or Reducing Preharvest Drop of Fruit With certain varieties of the apple, the abscission layer forms before the fruit is at the proper stage for picking. As a result, the fruit drops from the trees and becomes badly bruised. This lowers market value and reduces storage life. Investigations have shown that certain growth regulators retard the formation of the abscission layer and enable the fruit to remain on the tree. Of the numerous chemicals which have been tried, alphanaphthaleneacetic acid (NAA) and its metallic salts, alphanaphthaleneacetamide (NAd), 2-, 4-, 5-trichlorophenoxypropionic acid (2-, 4-, 5-TP), and succinic acid, 2-, 2-dimethylhydrazide (SADH) have been found to be particularly effective. In fact, these compounds are now used in most commercial orchards. As would be expected, many factors influence their effectiveness. Some of these factors are the variety, the concentration of the chemical, the time and number of applications, and the temperature of the weather. In general, when preharvest dropping of the fruit is a problem, a single application just before dropping begins provides the necessary protection.

QUESTIONS

1. What is the essential difference between deciduous trees and evergreen trees?
2. The depth of root penetration of an apple or pear tree depends largely on the texture and internal drainage of the subsoil. Explain.
3. Most apple-growing districts of continental United States are in the Northeast, Middle West, and on the upper Pacific Coast. Explain.
4. A fruit spur usually bears in alternate years, whereas the tree as a whole bears annually. Explain.
5. How do frosts during the blossoming period induce biennial bearing of apple varieties?
6. How do certain growth regulators prevent the preharvest drop of fruit of certain varieties of apple?
7. Sunny days and cool nights during the fruit-ripening period are particularly favorable for the production of highly colored fruit and cloudy days and warm nights are particularly unfavorable. Explain.

Drupe Fruits (Peach, Nectarine, Plum, Apricot, Cherry, and Almond)

Root System Mature trees of the drupe fruits have several large woody roots which branch into many small roots and numerous absorbing roots. In general, the root system is less extensive and less deep than that of the apple and pear. For these reasons the trees suffer more quickly from drought or from weed competition. However, the root spread is roughly proportional to the size of the tree. As with pome trees, the lateral spread of the roots is usually a foot or more beyond the spread of the branches.

Stems and Leaves The stem system consists of (1) main or scaffold stems called *limbs* and (2) smaller branches and twigs from which arise shoots containing leaves, flowers, and fruit. In general, the buds are produced on current-season wood. Buds are (1) lateral or (2) terminal. Lateral buds are (1) vegetative or (2) flower, whereas terminal buds are always vegetative. Both terminal and lateral vegetative buds give rise to (1) shoot growth (long) or (2) spur growth (short). The peach and almond produce shoots mostly, and apricots, plums, and cherries develop both shoots and spurs. The leaves are alternate, simple, long in proportion to width, short-petioled, and finely toothed. The teeth and petiole often contain glands. As is common with apples and pears, the leaves unfold about the time of blossoming and absciss after growth ceases in the late summer or fall.

Flowers and Fruit The flowers of the drupe fruit crops have characteristics in common and characteristics in distinction. Characteristics in common are the perfect type of sex expression, regular arrangement of flower parts, with a five-lobed, bell-shaped, or tubular calyx, a five-petaled corolla, numerous stamens attached to the rim of the calyx, and a solitary pistil attached at the bottom of the calyx. Characteristics in distinction are the number of flowers per node, the color of the petals, and length of the flower stalk. In the peach and apricot, the flowers are in singles or pairs, the petals are white, pink, or red according to the variety, and the flower stalk is very short. In the plum and cherry, the flowers occur in groups, the petals are white, and the flower stalk is long.

The fruit develops from ovary tissue only and varies greatly in size, shape, pubescence, color of the skin, and color and texture of the flesh. Plum fruits are moderately large, globular, and have a smooth skin and smooth pit. Apricot fruits are similar to peach fruits in size and shape and have a fine pubescence and smooth pit. Cherry fruits are comparatively small, smooth, long-stalked and have flesh which is either sweet or sour to the taste and small, smooth pits. Fruit of the sour cherry is about ½ inch (1 centimeter) in diameter and pale red or dark red; and fruit of the sweet cherry is about ¾ inch (1.5 centimeters) in diameter with light yellow, red, or dark red skin, and firm flesh. Almond fruits are much compressed. The mesocarp, instead of being fleshy as in the other drupe fruits, becomes leathery and tough and at maturity separates from the stone. The kernel or seed of the almond is the edible portion.

Varietal Groups or Types In general, peach fruits are comparatively large, subglobular, slightly grooved on one side, and have a pubescent skin, firm greenish, white, or yellow flesh which either separates from or clings to the stony endocarp. Varieties which produce fruit in which the fleshy mesocarp and the stony endocarp separate on ripening are called *freestones,* and varieties which produce fruit in which the fleshy endocarp and stony exocarp remain attached are called *clingstones*.

Types of plums are the European, which consists of dessert plums, and prunes, the Oriental, and the American. Types of cherries are sour cherries, which are commercially produced for canned and frozen products, and sweet cherries, which are produced for fresh market.

Economic Importance and Principal Commercial Districts In the continental United States, a large number of drupe-fruit industries have developed. *Peaches* are produced chiefly on the Pacific Coast and in the Southeast, Northeast, and Middle West. The leading peach-producing states are California, South

Figure 17-2 Airview of peach orchards, supply store, and packing shed in South Carolina. *(Courtesy, Soil Conservation Service, U.S. Department of Agriculture.)*

Carolina, Georgia, North Carolina, and Pennsylvania. In California, yellow-fleshed clingstone varieties are grown mainly for canning, and yellow-fleshed freestones are grown for drying. In states other than California, yellow-fleshed freestones are grown for immediate consumption and for canning. Italy is Europe's leading peach-producing country. France, Chile, and Japan also produce substantial quantities of peaches.

Plums are grown chiefly on the Pacific Coast. In this region, the principal varieties belong to the prune group—a type of plum high in sugar content and adapted to natural dehydration. Leading plum-producing countries are Germany, Yugoslavia, and Rumania.

Apricots and *nectarines* are grown chiefly in California and southern and central Europe. Turkey, Iran, Australia, Argentina, Africa, and Chile also have commercial plantings of apricots.

Cherries are produced in two regions—the Pacific Coast and the Great Lakes—with sweet cherries grown mostly in the former and sour cherries grown mostly in the latter. Italy, France, Germany, and Turkey are the leading cherry-producing countries.

Almonds, like apricots, are grown chiefly in California. Sweet, edible almonds consist of two groups: (1) hard shell and (2) soft shell. Of these the soft shell is the more important.

Use of Growth Regulators Research workers in South Carolina report that various plant-growth regulators are effective in altering peach metabolic processes sufficiently to allow preplanned manipulation of maturation and ripening rates, harvest dates, and development or suppression of specific quality attributes. For example, SADH accelerates maturity, promotes the development of red and yellow pigments, and aids in adapting the fruit to a single once-over harvest. Ethephon also accelerates maturity of the fruit.

QUESTIONS

1. How does a peach fruit differ from an apple fruit?
2. In general, drupe fruit trees are less drought resistant than pome fruit trees. Explain.
3. How does the fruit of the peach differ from that of the apricot?
4. How does the fruit of the almond differ from that of the peach?
5. Which is the leading region in peach production? Give reasons for your answer.
6. The vigor of shoot growth of the peach tree is associated with the number and kind of buds formed. Explain.
7. A student goes into a peach orchard and examines two trees, A and B. The shoots of tree A have many fruit buds in twos at the node. Those of tree B are mostly single. Which tree will produce more fruit? Explain.

EVERGREEN-TREE FRUITS

Citrus Fruits (Orange, Lemon, Grapefruit, and Lime)

Citrus trees are broad-leaved evergreens. The trees vary in height from 10 to 15 feet, or 3 to 4.5 meters, (lime), 10 to 20 feet, or 3 to 6 meters, (lemon), 25 to 40 feet, or 7.5 to 12 meters, (orange), and 30 to 50 feet, or 9 to 15 meters, (grapefruit). They differ little in habit of growth. When young, the trees are upright and spreading; as they grow older they become somewhat pyramidal in shape. In general, citrus trees are long-lived and begin bearing at four to six years of age.

Root System Citrus trees are propagated by budding. This helps to ensure the production of uniform trees and the development of satisfactory root systems. The principal stocks are (1) sour orange, (2) rough lemon, (3) sweet orange, and (4) trifoliate. Of these the sour orange stock was the most widely used for many years. However, since this stock is susceptible to a disease known as *quick decline* (tristeza disease), resistant stocks are now used in districts where this disease is prevalent, i.e., trifoliate Troyer citrange stocks in California and rough lemon, sweet orange, and Cleopatra stocks in Florida. Rough lemon stocks are more adaptable to sandy soils, and trifoliate stocks are more adaptable to the well-drained flatwood soils, especially where cold resistance is needed.

Stems and Leaves The trunks are usually short, and their diameter depends on the kind of crop and age of the tree. In general, most trees are headed "low." At planting, the young trees are headed back 24 to 30 inches (60 to 75 centimeters) to induce low heading. This produces a dense to fairly open head, with the lower branches drooping almost to the ground, thus protecting the trunk from cold weather. The wood is very strong and elastic. For example, heavily loaded orange and grapefruit limbs, though their normal position is from 8 to 12 feet (2 to 4 meters) above the ground, often bend at fairly narrow arcs without breaking.

The leaves are rather rough and leathery in texture, light green on the under side, and deep glossy green on the upper. They vary in shape and size according to the kind of fruit. In general, the larger the fruit, the larger the leaf; thus the leaves of the lime are small and ovate, those of the lemon are a little larger than the lime and also ovate in shape, those of the sweet orange are medium, fairly broad, and long ovate, and those of the grapefruit are large, long, and broad. The petioles are broadly winged (grapefruit), narrowly winged (sweet orange and lime), or not winged (lemon), a characteristic which serves as a means of identifying the various kinds.

Flowers and Fruit. The flowers occur singly or in groups consisting of small axillary or terminal cymes. Individual flowers have four or five white or purplish petals (lemon petals are white inside and purple outside), numerous stamens, usually 20 to 40, and a single 7- to 15-celled pistil. Some kinds are extremely fragrant (sweet orange and grapefruit), which can often be detected for a considerable distance from the orchards. The size of the flower varies with the kind of plant. For example, flowers of the grapefruit are large; those of the sweet orange are moderately large; and those of the tangerine (mandarin) are small.

From February to April, depending upon the area in which the orchards are located, the main bloom of the year takes place. In areas that are sufficiently warm, lime and lemon flowers appear each month, but most of the flowers open in late winter or early spring.

The fruit is derived from ovary tissue only and consists essentially of a leathery rind or skin (the outer portion) and of soft juicy flesh (the inner and edible portion). The rind is derived from the outer ovary wall and has numerous oil sacs embedded in its tissue. It varies in depth (thin to thick) and in color (light yellow to deep orange-red), depending on the kind of fruit. The rind serves as a cover and provides protection for the inner ovary wall which is divided into five or more segments by means of thin, radial, grayish membranes. Each segment is packed with club- or spindle-shaped sacs, called *vesicles,* which contain soluble sugars, various organic acids, and more particularly comparatively large quantities of ascorbic acid or vitamin C. Most varieties of orange and grapefruit are self-fruitful, but citrus flowers are very attractive to several kinds of insects, including honeybees, and when two or more varieties are available these insects cross-pollinate many flowers. A few, the Washington Navel orange and Tahiti lime, for example, produce parthenocarpic fruit.

Economic Importance and Principal Commercial Districts Citrus fruits are grown in many parts of the world. Important commercial industries [illegible] located in Spain, Italy, Palestine, India, China, Australia, and North and South

America. The industry is foremost in value and importance in the continental United States and is located in the southern regions of the country. Outstanding industries exist in Orange, Los Angeles, San Bernardino, Tulare, Riverside, Ventura, and San Diego counties of southern California, in the Rio Grande Valley of south Texas, in central and southern Florida, and the Plaquemines district below New Orleans, Louisiana. Most of the oranges are produced in California and Florida; most of the grapefruit in Florida and Texas; most of the lemons in California and limes in southern Florida.

Uses of Growth Regulators A number of growth regulators have been found to promote root growth, to reduce preharvest dropping of fruits, and to have other influences on the tree, its flower or its fruit.

In extensive trials in California and Florida, growth-regulator sprays before harvest reduced the preharvest drop of several varieties of oranges. The sprays given long enough before maturity tended to increase thickness of stem and rind, and usually increased size of fruit.

Figure 17-3 Airview of citrus orchards in Florida. *(Courtesy, A. H. Snell, New Port Richey, Florida.)*

QUESTIONS

1 Name the important citrus fruits.
2 How do the size and habit of growth of citrus trees compare with those of apple and peach trees?
3 The various kinds of citrus are not grown on their own roots. Give two reasons.
4 Compare the life of citrus leaves with those of the apple.
5 Some citrus flowers have a strong, sweet scent. Compare this with deciduous fruit flowers.
6 Some citrus fruits are always seedless. Explain.
7 During the days of sailing ships, fresh vegetables and deciduous-tree fruits were not part of the diet of the crews of the ships, but limes, lemons, or oranges usually were found on all ships. Explain.
8 Florida and California grow most of the citrus fruits in the United States. Explain.

SELECTED REFERENCES FOR FURTHER STUDY

Batchelor, L. D., and H. J. Webber. 1948. *The citrus industry*. Vol. 2. Berkeley and Los Angeles, Calif.: University of California Press. The result of the combined efforts of 16 authorities on various phases of citrus production, with the principal topics being nursery methods, selection and use of rootstocks, planting, cultivating, fertilization, irrigation, pruning, frost protection, and pest control.

Chandler, W. H. 1957. *Deciduous orchards*. Philadelphia: Lea and Fibiger. A discussion, by an outstanding and widely recognized authority, on important aspects of deciduous-tree fruit culture: (1) the specific characteristics of the trees of each kind, (2) the anatomic, physiologic, and biochemical processes in the initiation, setting, and development of the fruit, and (3) the effect of the environment and orchard practices on these processes.

Childers, N. F. 1961. *Modern fruit science*. New Brunswick, N.J.: Horticultural Publications, Rutgers University. Information on the growth, development, and cultural requirements of deciduous-tree and small fruits grown in the cool-temperate and warm-temperate zones of the world, as presented by an outstanding teacher and investigator in the field of pomology.

Chapter 18

Nut Fruits, Tropical and Subtropical Fruits, Tung, and Beverage Crops

The life and soul of science is its practical application.

Lord Kelvin, 1883

NUT FRUITS

The most important nut fruits of the world fall into two groups: (1) the evergreen tropical nonhardy and (2) the deciduous semihardy.

Evergreen Nonhardy

Principal evergreen nonhardy crops are coconut, Brazil nut, and cashew. These crops require very warm growing seasons and extremely mild nongrowing, or dormant, seasons. At present coconuts are grown widely in the tropical regions of the world, Brazil nuts are confined largely to the valleys of the Amazon River and its tributaries, and cashew nuts are grown mainly in the tropics of India.

Deciduous Semihardy (Pecan, Walnut, Filbert)

Root System Although little experimental evidence is available on the nature and extent of the root system, authorities believe that the taproot forms a system of laterals which branch to form the absorbing system. In general, the degree of ramification is proportional to the size of the tops. For example, trees of the Persian walnut produce relatively small tops and a correspondingly sparse root system. On the other hand, trees of the black walnut and pecan develop a large top and a correspondingly well-developed and extensive root system. As with other tree crops, the depth of penetration depends largely on the looseness, drainage, and aeration of the subsoil.

Stems and Leaves In general, the size and shape of the top vary with the kind of plant. For example, California walnut trees are small and round-headed; Persian walnut trees are medium-sized and round-headed; and pecan trees are tall and globular-headed. As with many other tree crops, the stem system consists of a trunk, primary, secondary, and tertiary scaffold branches, and small branches or shoots. On the twigs and young branches the bark is smooth, and on the large branches and trunk it is rough and shaggy. The leaves are alternate and pinnately compound, and vary in number, size, and shape with the kind of crop. For example, pecan leaves have a large number of small, pointed leaflets, and walnut leaves have a small number of large, oval leaflets.

Flowers and Fruit The deciduous nut crops are monoecious. In other words, the pistils and stamens are borne in separate flowers and both types of flowers are borne on the same tree. The pistillate inflorescence, called a *cluster,* develops from terminal buds on the past season's growth, and the staminate inflorescence, called a *catkin*, develops from buds at the base of one-year-old wood. Individual flowers are small and inconspicuous. The stamens produce large quantities of fine, fluffy pollen and the pistils present a comparatively large stigmatic surface. Air currents, the agent of pollen transfer, carry the pollen from the stamens to the stigmas.

Individual fruits are dry, indehiscent, one-seeded ovaries. According to the kind of plant, they vary in size from ½ to 1½ inches (1 to 4 centimeters) in diameter and length from 1 to 2 inches (2 to 5 centimeters), and in shape from round to oblong. The edible kernels are enclosed in a hard pericarp or ovary wall and contain from 65 to 70 percent fat, from 10 to 16 percent protein, and about 14 percent carbohydrates. They are good sources of the B vitamins.

Economic Importance and Principal Commercial Districts The Persian walnut grows wild in the Carpathian Mountains in Eastern Europe; throughout Turkey, Iraq, Iran, Afghanistan, and southern Russia; and in the foothills of the Himalaya Mountains. In the United States Persian walnuts are produced mostly in the Pacific Coast region, with California the leading state; and pecans are

Figure 18-1 Pecan orchards in southern Mississippi.

produced mostly in the southern regions, with Texas the leading state. Other important states are Georgia, Alabama, Florida, Mississippi, and Louisiana.

QUESTIONS

1 How do the pecan and Persian walnut differ in the size of the tree?
2 The deciduous nut crops produce enormous quantities of pollen, and the stigmatic surface is quite large. Explain.
3 Although pecan trees grow along the southern boundary of the Middle West, they rarely produce nuts. Explain.
4 Shading of the leaves of nut trees results in the development of poorly filled nuts of low oil content. Explain.
5 On heavily loaded filbert trees, nuts in singles or in twos are usually better filled than those in threes or fours. Explain.

TROPICAL AND SUBTROPICAL FRUITS

The Date

Plant Characteristics The date is a member of the palm family —monocotyledonous evergreen trees which produce leaves from terminal buds only. The date attains a height of 100 feet (30.5 meters). Its wood is fibrous and spongy. The root system is adventitious, arises at the base of the stem, and consists of a large number of lateral roots which give rise to numerous absorbing roots. Roots penetrate to 20 feet (6 meters) or more. The leaves which arise from a terminal bud are large, 9 to 20 feet (2.7 to 6 meters) long, and parallel-veined.

The blade consists of 100 to 260 leaflets called *pinnae*. At the base of each pinna several stiff hairs are borne. The date, like the persimmon, is dioecious. The inflorescence, both male and female, arises in the axils of the leaves, is large, much-branched, and contains thousands of individual flowers. Scientists have found that the pollen has a marked effect on the size, shape, and time of maturity of the fruit. After pollination, two of the three carpels in each female flower drop and the remaining carpel grows to maturity. If pollination does not take place, all three carpels or fruits remain on the tree but do not develop into high-quality fruits. Hence, pollination of all three carpels is necessary for the proper development of the fruit. The fruit is a one-seeded, oblong berry; the skin color varies from yellow to purplish-black; and the flesh is dry, semidry, or moist. The water content varies from 22 to 51 percent, and the sugar content varies from 71 to 86 percent. Thus, dates are very nutritious.

Fruiting Habit Since the main axis grows like corn, the only pruning necessary is removing old leaves as the terminal bud elongates. Pruning is done by cutting the leaf petiole about 8 to 12 inches (20 to 30 centimeters) from the trunk at the rate of 12 to 20 leaves per year and by removing all suckers from the base of the trees during the early period of growth. In Arizona, growers usually remove the shoots between April 15 and July 1. In general, growers are very careful to avoid wounding of the trunk. Wounds of this type of tree, a monocot, fail to heal. The date is essentially a high-temperature plant. Temperatures as low as 15 to 18°F (−9.4 to −7.8°C) severely injure the leaves, but apparently fail to harm the terminal bud.

Uses The date, believed to be a native of northern Africa or Arabia, has been in cultivation for over 4,000 years. It is grown in arid tropical and subtropical regions where sufficient water is available for the trees. In the United States commercial crops are grown in the Coachella Valley of southern California and Arizona. In this country only the fruit is used, as dessert either in the fresh or dry state. In many other countries (North Africa) both the fruits and trees are used. The fruit constitutes the chief article of diet of the Arabian people; the sap is fermented for the making of alcoholic liquors and vinegar; the trunk is made into fence posts; and the leaves are made into ropes, baskets, and crates.

The Avocado

Plant Characteristics The avocado is a dicotyledonous evergreen. Trees attain a height of 60 feet (18 meters). They differ in habit of growth. Some are low and spreading; others are tall and upright. The roots are shallowly situated and hence occupy the upper portion of the soil. The leaves are thick, leathery, and bright green. They vary in shape and in length—3 to 16 inches (7 to 35 centimeters). The flowers, which are borne in clusters at the tips of branches, are smooth and possess two series of petallike structures called *perianth lobes*.

Scientists have discovered a remarkable synchronization in the opening and closing of the flowers and in pollen shedding and pistil receptivity. Certain varieties open their flowers in the morning, at which time the pistil is receptive, close again in midday, and open again the following afternoon, at which time the pollen is shed. Other varieties open their flowers in midafternoon, at which time the pistil is receptive, close again in late afternoon, and open again the following morning or morning of the second day, at which time the pollen is shed. Thus, when the pistils of the one variety are receptive, the pollen of the other variety is shedding; conversely, when the pistils of the other variety are receptive, the pollen of the one is shedding. In this way, the various varieties are successfully pollinated. Within any variety marked changes in the weather will greatly change the normal schedule.

The fruit is a one-seeded, fleshy berry and varies from round to oval to pyriform or necked. Both seed and flesh vary greatly in size. The skin is yellowish-green, purplish-green to purplish-blue, thin and membranous in some varieties and thick and woody in others. The flesh, which separates readily from the skin, is light green to yellow and has a buttery consistency. It contains from 10 to 20 percent fat and about 80 percent water and is a good source of vitamin C. Thus, the fruit is very nutritious.

Importance and Uses The avocado is a native of tropical America. In the United States it is grown in southern Florida and in southern California only. Avocados are used as salads, in ice-cream making, as a base for milk shakes, and in the manufacture of mayonnaise. There are three horticultural races: (1) West Indian, (2) Guatemalan, and (3) Mexican. The West Indian is the least resistant to cold and is grown principally in Florida; the Guatemalan is moderately resistant to cold and is grown mostly in California; and the Mexican is the most resistant to cold and is grown in Florida and California.

The Persimmon

Plant Characteristics There are two kinds: (1) Japanese and (2) native. With the Japanese, the tree is upright spreading, has a round, open top, and attains a height of 40 feet (12 meters). With the native type, the tree is upright, has a pointed top, and attains a height of 50 feet (15 meters). In both types, the trees develop a taproot and several laterals which penetrate the soil rather deeply; the leaves are simple, thick, large, leathery and glossy green on the upper surface and light green on the lower surface; and the sex expression is dioecious. In other words there are male and female trees. Both sexes are borne in the axils of the leaves of the current season's wood. The female flowers are solitary, and the male flowers occur in clusters of three or four. The fruit is a round to oblong conicle, few-seeded and nonseeded, fleshy, juicy ovary. The skin is light orange or bright red. The flesh is yellow to cinamon brown, soft and pasty in some varieties and stringy in others. A marked characteristic of the flesh is its astringency, which is due to tannin. When the fruits are immature, the walls of the cells which contain tannin are soluble in the mouth; hence, the tannin is released and the mouth puckers. When the fruits are mature, the walls of the cells containing tannin are insoluble; hence, the tannin is not released and the mouth does not pucker. Scientists have found that the astringency of the fruits can be removed by exposing them to ethylene. Ripe persimmons contain about 66 percent water and 32 percent sugars and are a good source of vitamin C.

Fruiting Habit The young trees are usually headed back to about 36 inches (72 centimeters) when they are set and trained to four or five scaffold branches in much the same way as for peaches. The fruit is borne on the current season's wood, usually in alternate years. If annual crops are desired, some pruning is necessary. The Japanese prune off all shoots which have borne fruit in any particular year. The Chinese prune off the twigs bearing fruit at the harvest with long poles or sticks.

Importance and Uses The Japanese persimmon originated in China. It has been grown in both China and Japan for hundreds of years. In the United States, the trees thrive well in the Southeast and Southwest and in California. The fruit is used mostly as a dessert.

The Fig

Plant Characteristics The fig is a deciduous bush[1] or tree. The root system is extensive. The laterals occupy the upper layer of soil and extend considerable distances, usually 50 feet (15 meters) or more. The branches or trunks are short. The bark is gray, and the twigs have well-developed pith. The leaves are simple, alternate, three- to seven-lobed, thick, leathery, light green, and hairy (pubescent). The flowers are borne on the interior of a pear-shaped peduncle. This peduncle ripens and thus becomes the edible portion. When ripe, the flesh is coarse, pink, red, purple, or violet, and contains about 80 percent water, 1.5 percent protein, and 15 percent sugar.

Fruiting Habit There are two main types: (1) the Adriatic, or common, which grows in the Southeast, and (2) the Smyrna, which grows in California. The Adriatic fig is parthenocarpic; hence, it is seedless and does not require pollination. The Smyrna fig is nonparthenocarpic and hence requires pollination. The pollen used on Smyrna figs is secured from wild figs called *caprifigs.* A small wasp called *blastophaga* carries the pollen.

The fig is prolific. In the United States, two crops are usually borne and a third crop is not uncommon. The first main crop is borne on previous season's wood. The second and third crops are borne on current season's wood. As stated

[1]The fig is trained to several main branches in the southeastern United States and is called a *bush*.

previously, figs are trained to the bush or tree form. Of these the bush form is preferred in the Southeast, since the tops are often killed by subfreezing temperatures.

Importance and Uses The fig is now widely distributed through tropical and subtropical countries. In the United States, it is grown in the southeast region and in California. In the Southeast, many homes have one or two fig bushes growing in the yard. Figs are used in the fresh state or in various forms of preservation—canned, dried, preserved, candied, and pickled.

TUNG

Tung trees are grown for their seeds, which contain large quantities of oil. This oil has excellent drying properties. Thus, it is used extensively in the making of paints, varnish, linoleum, oilcloth, and printer's ink.

Plant Characteristics

The trees are deciduous and monoecious. They attain a height of 12 to 20 feet (4 to 6 meters) and have a tendency to produce whorled branches. The leaves are comparatively large, simple, heart-shaped or three-lobed, and dark green. The inflorescence is a cluster. This cluster arises on the terminals of twigs of the past season's growth and contains from one to three pistillate flowers and a relatively large number of staminate flowers. The individual flowers are showy, open just before the unfolding of the leaves, and are white or tinged with pink in the throat. Since no pollination problems have been reported, dichogamy or incompatibility is apparently not present. The mature fruit is round or globular, 2 to 3 inches (5 to 7 centimeters) in transverse diameter, has a dark brown outer cover, and holds from four to seven relatively large, firm, brown seeds which contain the oil. Figure 18-2 is a close-up of a tree in the Wade orchard, Lucedale, Mississippi. Note the leaf and fruit characteristics and the age of wood on which the fruit is borne.

Economic Importance and Principal Commercial Districts

Tung is a native of central China. The trees were first grown in the United States in 1904. Since that time, ecologic studies and trial plantings in the warm sections of the United States showed that the trees thrive best in a belt from 50 to 100 miles (80 to 161 kilometers) wide extending from northern Florida westward to southern Louisiana and eastern Texas. At present Mississippi leads in commercial plantings with over 10 million trees, followed by Florida, Louisiana, Alabama, and Georgia in descending order named.

Figure 18-2 Close-up of tung tree. Note that the fruit is borne on the young wood. *(Courtesy, G. F. Potter, U.S. Department of Agriculture.)*

BEVERAGE CROPS

The horticulture beverage crops are coffee, tea, and cacao. The chemical compounds concerned are two closely related alkaloids: caffeine ($C_8H_{10}N_4O_2$) in coffee and tea, and theobromine ($C_8N_4O_2$) in cocoa and chocolate. In general, these compounds taken in moderate quantities mildly stimulate the metabolism of the human body with no depressive or otherwise deleterious effects. For this reason, the beverages made from these crops are often called the *benign stimulants*.

These crops have at least four characteristics in common: they are woody, dicotyledonous evergreens; they grow into small trees; they require a high temperature level combined with high relative humidity; and their stomates are adversely affected by high light intensity. For example, studies on the behavior of the stomates of arabian coffee show that if the leaves are exposed to full sun of the tropics, the stomates are open during the early part of the morning, closed during the middle of the day, and open during the late hours of the afternoon.

However, if the leaves are exposed to partial shade, the stomates remain open during the entire light period. This explains why companion cropping is quite prevalent in the tropics, and why a tall-growing crop which can withstand high light intensity is grown with the beverage crops which can not withstand high light intensity.

Coffee

Plant Characteristics Coffee belongs to the family *Rubiaceae* and to the genus *Coffea.* Within this genus, there are several species which are grown for the making of coffee. Of these several species, the specie *arabica* is foremost in popularity and production. In the tropics, mature plants of arabica are essentially small trees. The root system consists of a short 12- to 18-inch (30- to 46-centimeter) taproot and a series of primary and secondary laterals. These laterals, in turn, give rise to the slender, whitish feeder roots. According to authorities, the area for the absorption of water and essential elements of any given plant is enormous. The stem system consists of two types: the vertical and the horizontal, or reproductive, which bear the flowers, fruit, and seed. A unique feature of arabica is that flowers and fruit in all stages of development exist at the same time on the stem. The leaves are simple, dark green, glossy, and alternate on the stem. The flowers are borne in clusters at the nodes, and an individual flower has five whitish petals, five stamens, and a single-bilobed, mostly self-pollinated pistil. The fruit is a drupe, about the size and color of a cranberry. When ripe, the fruit contains two seeds commonly called *beans.*

Economic Importance and Principal Producing Countries Since the production of coffee beans is limited to the tropics and since large quantities of coffee are consumed by countries outside the tropics, huge quantities of coffee beans are imported each year. In other words, the coffee bean is a significant commodity in world trade. People producing the beans sell or exchange them for goods they cannot produce themselves economically.

From the standpoint of the amount of coffee consumed, producing countries belong to one of four groups: (1) Brazil, 40 percent, (2) Columbia, 20 percent, (2) countries other than Brazil and Columbia, 20 percent, and (4) tropical countries of Africa and Asia, 20 percent. The student will note that countries of Central and South America produce 80 percent of the world's coffee.

Tea

Plant Characteristics Tea is a member of the family *Theaceae;* its scientific name is *Camellia sinensis* L.; and it is a close relative of and its characteristics are similar to the ornamental plant *Camellia japonica* L. The root system is extensive with a well-developed taproot and a much-branched system of primary and secondary laterals and feeder roots. The taproot and laterals store starch, an important feature with regard to the pruning of the stems and the plucking of the leaves; and the feeder roots become suberized a short distance from the root cap, an important feature with regard to the transplanting operations. Several investigators have reported the presence of endotrophic mycorrhiza within the tissues of the feeder roots. The leaves are simple obovate or acuminate, and alternate on the stem with stomates on the lower surface only. The flowers are borne on short pedicels which arise from the axils of the leaves. An individual flower is perfect with five to seven sepals, five to seven white, waxy-appearing petals, numerous stamens, and a single pistil with a three-lobed stigma. The fruit is a pod which dehisces by splitting from the apex into three parts or valves.

Economic Importance and Principal Producing Districts In common with the production and consumption of coffee, the production of tea is limited to the tropics, though tea is consumed in practically every country in the world. Thus, as with coffee, tea is a significant product in world trade. Principal producing districts are the Darjeeling, Douars, Assam and Surman Valley of North India; the Nilgeris and Anamallais of South India and Ceylon, and the Sumatra and Java of Indonesia. Small quantities are produced in northern Taiwan, Uganda, Kenya, Tanganyika, Nyasaland, and southern Rhodesia.

Cacao

Plant Characteristics Cacao belongs to the family *Stericulia,* and its scientific name is *Theobroma cacao* L. Mature trees attain a height of 15 to 25 feet (5 to 8 meters) and a spread of 20 to 26 feet (6 to 8 meters). The roots consist of a taproot and an extensive system of lateral branches. The bark of the trees is thin and greyish. The leaves are simple, entire, with a ⅜ phylotaxy, and are dark green on the upper surface. The number of stomates per unit area appears to be associated with the degree of light intensity, since studies indicate that a greater number per unit area occurs at 50 percent shade than at 25 percent shade or no shade. The flowers are perfect with five pink or white sepals, five pink or white petals, five stamens, five staminodes, and a single superior pistil. Although some types are self-incompatible, all cacao trees are cross-compatible, and a high rate of cross pollination takes place. The fruit is a moderately large, five-carpelate, central-placentated pod. Each carpel contains a white or purple seed invested in a layer of mucilagenous material. A unique feature of the cacao tree is that the flowers (and fruit) are borne on the trunk and main branches, a phenomenon called *cauliferous flowering* (and fruiting).

Economic Importance and Principal Producing Countries As with coffee and tea, the production of cacao for the making of cocoa and chocolate is limited to certain equatorial regions of the world, even though large quantities are consumed annually by people outside the tropical regions. Thus, cacao is also a significant commodity in world trade. Principal producing districts lie between 0 and 15° North and between 0 and 20° South. These are the coastal areas of

Central and South America, and the coastal areas between Liberia and Cameroon. Limited quantities are also produced in Ceylon, Sumatra, Java, New Guinea and New Caledonia.

QUESTIONS

1. Describe a date tree.
2. A wound in a date tree does not heal as readily as a wound in an apple tree. Explain.
3. The three races of avocados differ in the degree of low temperature they can withstand. Which race will do best in Florida? In California?
4. Describe the opening and closing of avocado flowers.
5. Immature fruit of the Japanese persimmon is astringent; mature fruit is not astringent. Explain.
6. Compare pruning the Japanese persimmon with pruning the peach.
7. The fig is difficult to graft, and pruning wounds often fail to heal. Explain.
8. The Adriatic fig is not propagated by seed. Explain.
9. How are Smyrna figs pollinated?
10. Although tung is monoecious, considerable self-pollination takes place. Explain.

SELECTED REFERENCES FOR FURTHER STUDY

Kilby, W. W. 1969. History and literature of the domestic tung industry. *Miss. Agr. Exp. Sta. Tech. Bul.* 56. A thorough discussion of all phases of tung crop production by a horticulturist who has spent his entire research career on investigations with tung.

Ochsee, J. J., et al. 1961. *Tropical and subtropical agriculture.* Vol. 2. New York: Macmillan, chap. 10. An excellent chapter for the student who wishes to become acquainted with the plant characteristics and economic importance of each of the three beverages, with the principal topics being areas of production, botany, varieties, breeding and selection, climate and soil requirements, culture, control of pests, harvesting and processing, as well as a list of selected references at the end of the chapter.

Wellman, F. L. 1961. *Coffee, botany, culture and utilization.* New York: Interscience. A discussion by a world authority on all aspects of coffee production and its consumption in a well-illustrated researched reference.

Woodroof, J. G. 1967. *Tree nuts, production, processing, products.* Vol. 2. Westport, Conn.: Avi Publishing. A discussion of the production, processing and utilization of important nut fruits: pecans, pistachio nuts, and walnuts.

Chapter 19

Small Fruits

By their fruits ye shall know them.

Matthew 7:20

GRAPES

The most important kinds of grapes belong to three species of the same genus: *Vitis vinifera, Vitis labrusca,* and *Vitis rotundifolia. V. vinifera* is native to southwestern Asia and has been cultivated for more than 4,000 years. In Europe and Asia, it is known as the *European* or *Old World* grape, and in the United States as the *California* grape. *V. labrusca* and *V. rotundifolia* are native to North America and are not grown extensively in other countries.

Plant Characteristics

Root System The root system of mature vines is relatively extensive but the depth and spread vary with the species and variety. In well-aerated soils, the roots extend from 8 to 12 feet (2.4 to 3.7 meters) in the horizontal direction and from 5 to 8 feet (1.5 to 2.4 meters) in the downward direction. As with pomes and drupes, the primary and secondary roots are woody and function largely for support, translocation, and storage. The much-branched smaller roots function primarily for the absorption and conduction of water and essential ions.

Stems and Leaves The stems are climbing, woody vines which possess tendrils, enlarged nodes, well-developed conducting tissues, and a large pith. The buds are classified as (1) terminal and (2) lateral. The terminal buds are always vegetative, and the lateral buds are either mixed or vegetative. The mixed buds give rise to flower-bearing canes, and the vegetative buds give rise to nonflowering canes. The grape bears its fruit on lateral shoots of the current season's growth from mixed buds on last season's canes. These stems, until mature and woody, are called *shoots;* later, when mature, they are called *canes*. The leaves are simple. They vary in size, shape, color, and number of lobes, according to the species and variety.

Flowers and Fruit Flowers are borne in clusters, or panicles, on the current season's shoots. Individual flowers are hypogynous and perfect with five united, greenish-white petals, five reflexed or upright stamens, and a single rudimentary, or functional, pistil. The upright or reflexed position of the stamens and the functional condition of the pistil are important horticultural characteristics. Varieties with upright stamens produce functional pollen, and varieties with reflexed stamens produce abnormal, impotent pollen. Nonfunctioning, impotent pollen cannot germinate and is generally useless in fruit production. The mature fruit is a moderately large, juicy, round to elongated ovary which contains relatively large quantities of soluble sugars, notably glucose or grape sugar.

Reaction to Phylloxera *Phylloxera* is the name of a root louse. It seriously damages and eventually destroys the roots of susceptible varieties. Varieties of vinifera are susceptible, whereas varieties of labrusca are resistant. This explains why the top system of all varieties of vinifera is grafted on the root system of varieties of labrusca and is an interesting case of resistance and susceptability of two closely related species.

Sex Expression and Fruiting Habits In general, the sex expression of vinifera is perfect, the individual berry is very large, the number of berries per cluster varies from 100 to 200, and the skin of the fruit adheres to the pulp. The sex expression of labrusca is also perfect, but the individual berry is moderately large, the number of berries per cluster varies from 40 to 100, and the skin separates from the pulp. The sex expression of rotundifolia is either perfect or dioecious, according to the variety, the size of the individual berry is small, the number of berries per cluster varies from 5 to 20, and the skin adheres to the pulp. This explains why plants with functional pistils only are interplanted with plants with functional stamens or with both functional stamens and pistils. Note the flowering shoots and cluster of fruit of labrusca in Figure 19-1.

Figure 19-1 Flower-bearing shoot of *V. labrusca*. Note on the left the flower clusters at the 3rd, 4th, and 5th nodes of the new shoot and on the right, at the 2nd, 3rd and 4th. *(Courtesy, Virginia Agricultural Experiment Station.)*

Species Used for Root Systems In general, varieties of *V. vinifera* are grown on rootstocks of related American species. These root systems are needed to provide resistance to phylloxera and, in some districts, to nematodes. Most rootstocks used are hybrids of two or more American species. The principal species used are *V. aestivalis, V. berlandieri, V. champini, V. cordifolia, V. monticola, V. riparia,* and *V. rupestris*. Of these species the more desirable and most frequently used are *V. champini, V. riparia, V. rupestris,* and *V. berlandieri*.

In the United States ,varieties of *V. labrusca* and varieties of *V. rotundifolia* are usually grown on their own roots. However experiments have shown that rootstocks of either *V. riparia* or *V. champini* if grafted on the stems of *V. labrusca* impart vigor and productivity to the vines, particularly to varieties of labrusca which are poor in vigor and low in productivity when grown on their own roots.

Use of Growth Regulators

Numerous tests have shown that treating the basal portion of stem cuttings with indolebutyric acid (IBA) and napthaleneacetic acid (NAA) markedly and consistently promotes the development of roots. In fact, this response has been so pronounced and consistent that the use of NAA and IBA in the rooting of stem cuttings is a standard practice.

Figure 19-2 *V. berlandieri V. riparia* cross, resistant to phylloxera and moderately resistant to nematodes, used as rootstock by varieties of *V. vinifera*. *(Courtesy, A. An Jugar, Wine Experiment Station, Provide de Valentia, Spain.)*

In increasing the size of the individual berry and the individual cluster of Thompson Seedless, a valuable variety of Vinifera grown for table use, experiments with gibberellic acid (GA_3) have shown that this compound increases the size of the individual berry and the size of the individual cluster. As a result, treated clusters are larger, more uniform, and more attractive than nontreated clusters.

Tests have shown that benzothiazole-2-oxyacetic acid (BOA) strikingly delays the maturity of berries on a cluster of either seeded or seedless varieties. This delay in maturity is directly proportional to variations in concentration from 5 to 40 ppm. Although few growers are interested in delaying the maturity of the clusters, BOA may be useful in staggering harvest dates.

Plant scientists in Australia have discovered that dipping the clusters in a solution of 2-chloroethylphosphonic acid (ethephon, CEPA) at 500 ppm at the beginning of ripening period advanced the rate of ripening by 6 to 14 days, depending on the variety.

Black Corinth Seedless is a long-established variety from which currants are made. Girdling the canes to increase the set of the fruit and the size of the berry has been a common practice for many years. In recent experiments, however, gibberellic acid (GA_3) has been found to be more effective than girdling. As a result, GA_3 is now being used in practically all commercial vineyards for Black Corinth Seedless in California. The application of a concentration of 10 ppm induces a satisfactory set of small- to medium-sized, rounded berries, which are desired by the bakeries.

Economic Importance and Principal Producing Countries

Of all of the fruit crops grown in the world, grapes rank first and foremost. Of the three species, the economic value of vinifera is greatest. About 52 million metric tons are produced annually. Leading producing countries are Italy (10 million tons), France (8 million tons), Spain (4 million tons), and the United States (2 million tons). Somewhat lesser quantities are produced in Portugal, Greece, Rumania, and Hungary. Within each of these countries, the commercial producing districts are characterized by mild winters and dry summers, such as the mild-winter–dry-summer climates adjacent to the northern coast of the Mediterranean and the lower coast of California. By far, the greatest quantities of vinifera are produced in the Fresno, Tulare, and San Joaquin counties of the Central Valley.

In sharp contrast to vinifera, varieties of labrusca and rotundifolia are grown in the United States only. Labrusca varieties are grown in the northeastern part, with principal producing districts along the south shores of Lake Erie and Lake Ontario, in the Finger Lakes district of New York, and the Paw Paw district of southwestern Michigan. Rotundifolia varieties are grown in the southeastern United States, with scattered acreages in the upper Coastal Plain of North Carolina, in South Carolina, Georgia, Alabama, and Mississippi.

BRAMBLEBERRIES

The bramble fruits include the red, purple, and black raspberry, the erect and trailing blackberry, or dewberry.

Plant Characteristics

The root system of the brambleberries is moderately extensive. As with grapes, the primary and secondary roots are woody and are used for support, transportation, and storage. The stems possess numerous terminally curved spines called *prickles. Stems are erect (for example, the red and black raspberry) or decumbent (the dewberry). As a rule these stems, called canes,* have a life cycle of two years. During the first year they attain their growth and differentiate their flower buds, and during the second year these flower buds develop into fruit-bearing shoots.[1] When the fruit has matured, the canes die and are removed by pruning. The leaves are alternate and contain from three to five leaflets. The flowers are

[1]However, the everbearing varieties produce terminal clusters of flowers and fruit the first season.

borne in clusters. Individual flowers are comparatively large and perfect in all kinds of varieties, except the Pacific Coast dewberry, which develops both perfect and pistillate plants. The receptacle is convex and bears numerous stamens and pistils; and each pistil develops into a small drupe or drupelet. In the raspberry the cluster of drupelets of each fruit separates from the receptacle at maturity, forming the familiar thimblelike mass of fruit. In the blackberry, dewberry, and loganberry the drupelets remain attached to the receptacle at maturity. Figures 19-3 and 19-4 show examples of the brambles.

Figure 19-3 Typical clusters of erect and trailing brambles. A: trailing, Dewberry. B: erect, Blackberry. C: flower cluster of thornless blackberry. D: fruit cluster and leaf of thorny blackberry. *(Courtesy, U.S. Department of Agriculture, Office of Information, Washington, D.C. for A, B, and Department of Horticulture, Clemson University for C, D.)*

Figure 19-4 Two improved varieties of red raspberry. Left: Cherokee. Right: Pocahontas. *(Courtesy, George Oberle, Virginia Polytechnic Institute and State University.)*

Economic Importance and Principal Producing Countries

In general, the bramble fruits are grown in home and commercial gardens in low and moderately high elevations of the temperate zones, and in high elevations of the tropical and subtropical zones. Considerable quantities are grown in Europe, Japan, Lebanon, Mexico, and the United States. In the United States, the major commercial districts are as follows: (1) red raspberry—the Pacific Coast, the western sections of Washington and Oregon, the Great Lakes, southern Michigan, western New York, and eastern Minnesota; (2) black raspberry—the Great Lakes, southwestern Michigan, western New York, northeastern Ohio, and northwestern Pennsylvania; (3) blackberry and dewberry—the South, Southwest, and Middle West, Texas, Oklahoma, and Kentucky; and (4) dewberry—the Pacific Coast, western Oregon, and Washington.

STRAWBERRIES

Root System The root system of the strawberry is relatively shallow and moderately extensive. Studies at the Nebraska Experiment Station have shown that most of the roots extend horizontally and vertically for a distance of about 12 inches (30 centimeters). A few roots were found between the 1- and 2-foot (30 and 60-centimeter) levels. Since the range of the root system is limited and shallow, strawberry plants are mulched in regions of comparatively high transpiration.

Stems and Leaves The stems are short and thick and are called *crowns*. They bear three kinds of buds: (1) those which develop into short, thick stems, or crowns, (2) those which develop into long, slender stems called *runners,* and (3) those which develop into flowers. New plants are formed from runners. They have long internodes and form a new plant at the second node and every other node thereafter. Gardeners take advantage of this method of asexual reproduction in the establishment of new plantations. The leaves arise in rosettes around the short crownlike stem and are long-petioled and trifoliate.

Flowers and Fruit The flowers occur in groups or clusters. Mature individual flowers are relatively large with five or more green sepals, five or more white petals, and numerous stamens and numerous pistils distributed over a fleshy receptacle. The mature fruit is the fleshy receptacle to which is attached a large number of small seedlike fruits called *achenes*.

Varieties

The present-day strawberry has developed from at least three species of the same genus, *Fragaria chiloensis, Fragaria virginiana* and *Fragaria vesca*. *F. chiloensis* and *F. virginiana* are native to the New World, have 56 somatic chromosomes, and produce relatively large fruit; whereas *F. vesca* is native to the Old World, has 14 somatic chromosomes, and produces small fruit. According to authorities, these three species and probably one or two others hybridized under natural conditions, producing the parents of modern varieties. These varieties vary in adaptation to the climate and in length of physiologic dormant period. For example, the variety Klondike has a short rest period, is relatively nonhardy, produces large quantities of fruit during cool weather, and endures relatively hot summers. Thus, the Klondike is adapted to the warm, humid summers and mild winters of the warm-temperate zones. On the other hand, the variety Premier has a long rest period, can withstand mild subfreezing temperatures, and cannot endure hot summers. Thus, the Premier is better adapted to the cool-temperate zones. Note the fruit clusters of strawberry of Figure 19-5.

Use of Growth Regulators

Experiments have been conducted to examine (1) the production of runners and runner plants, (2) the rate of ripening of the fruit, and (3) the resistance of flower buds to cold. Plants sprayed with gibberellin at the rate of 550 ppm produced a markedly increased number of runners and runner plants, and those sprayed with the same substance at 10 ppm at one-week intervals showed an increased rate in the ripening of the fruit. Finally, plants sprayed in October with a solution containing SADH at 5,000 ppm protected their flower buds from injury due to cold. A group of treated plants produced 238 marketable fruits, whereas a comparable group produced 139.

Figure 19-5 Fruit cluster of strawberry. Top: two flowers have set fruit; several are sterile. Bottom: all flowers have set fruit. *(Courtesy, U.S. Department of Agriculture, Office of Information, Washington, D.C.)*

Economic Importance and Principal Producing Countries

The strawberry is the most widely adapted and most widely grown of the small fruits and is one of the most popular fruits in the world. Strawberries are grown from Florida to Alaska, from New England to California—in every state in the United States. They are grown throughout Europe, in Canada, in South America, and in many other countries. Important commercial producing centers include Europe (479,000 metric tons), the United States (220,300 metric tons), Lebanon (130,000 metric tons), Japan (128,000 metric tons), and Mexico (85,000 metric tons).

In the United States, a major commercial district exists in California, which currently produces more than 50 percent of the commercial crop. Other important districts exist in Washington, Oregon, Michigan, Tennessee, Arkansas, Louisiana, New York, New Jersey, and Florida. The wide variance in climates within these regions and the wide adaptability of varieties to meet specific needs, permit growing, harvesting, and marketing of the fruit during the greater portion of the year.

CURRANTS AND GOOSEBERRIES

Plant Characteristics

In general, the root system is not extensive. Individual roots are small and fibrous, and they have a tendency to grow from the stem in two whorls. The stems are relatively numerous since they arise from many points of the root system. In gooseberries, the stems possess spines, or prickles, and in currants, they are smooth. In fruit-bearing plants, the crops are borne on one-year-old wood or on one-year-old spurs borne on two- or three-year-old wood. When the stems become four or five years old, they cease to bear fruit. Thus, productivity is maintained by removing the old stems. The leaves are alternate and palmately lobed. The inflorescence is a raceme. Individual flowers are perfect and epigynous with four or five calyx lobes, four or five petals, four or five stamens, and a single, inferior pistil. Cross-pollination, chiefly by insects, is the rule and, with rare exceptions, all varieties are self-fertile. As with pomes, the mature fruit consists of ovary-wall and receptacle tissue. In currants, ripe fruit is black, red, or white, according to the variety.

Specie Types

Currants and gooseberries belong to the genus *Ribes*. The red- and white-fruited forms of currants belong to the species *sativum,* and the black-fruited form belongs to the species *nigrum.* With gooseberries, varieties of American origin belong to the species *hirtillum,* and those of the European belong to the species *grossularia*.

Economic Importance and Principal Producing Districts

Currants and gooseberries are grown in home and market gardens of 15 countries of northern Europe, where they have attained a high degree of popularity. However, their production in the United States is discouraged and restricted because the plants are the alternate hosts of the white pine blister rust—a serious disease of the white pine. In areas where the white pine is not important commercially, limited quantities are grown for home and market gardens. The annual area varies between 4,000 and 5,000 acres (1,620 and 2,025 hectares).

BLUEBERRIES[2]

Plant Characteristics

There are many kinds of blueberries. Six important commercial kinds of species are listed in Table 19-1. Note the differences in plant height, habitat, and regional adaption. In general, the root system of blueberries is shallowly or deeply situated, according to the species and habitat, finely divided, and devoid of root hairs; the woody stems develop from underground rootstocks forming suckers; the flowers are relatively small, perfect, sympetalous, and usually occur in clusters; and the fruit is a relatively small, juicy ovary which, when ripe, varies from black to light blue. Figure 19-6 presents (1) a plant of the highbush, (2) a plant of the lowbush, and (3) the fruiting habit of the blueberry.

CRANBERRIES

Plant Characteristics

The cranberry has a root system similar to that of the blueberry, which is of the same genus, but the roots are finer and fibrous. The roots are devoid of root hairs, among which mycorrhizal fungi are ramified. The roots possess the ability to live submerged in water for months provided oxygen does not drop below a

[2]The material on blueberries is contributed by Dr. W. T. Brightwell, The Coastal Plain Experiment Station, Tifton, Georgia.

Table 19-1 Important Species of *Vaccinium* (Blueberries)

Species	Common name	Plant height, ft	Habitat	Regional adaptation
V. australe	Highbush	10–15	Moist places	Atlantic Coast Plain; Middle West
V. ashei	Rabbiteye	4–8	Well drained	Southeast
V. pallidum	Dryland	1–3	Dry, well drained	Appalachian Mts.; northwest Arkansas
V. angustifolium	Lowbush	½–1½	Dry, well drained	Northeast; Canada
V. ovatum	Evergreen	20	Dry, well drained	Northwest
V. membranaceum	Mountain	3	Dry, well drained	Northwest

Figure 19-6 Top left: a highbush blueberry plant. Bottom left: a low-bush blueberry plant. Right: rabbiteye blueberry twigs. *(Courtesy, James E. Moulton, Michigan State University, for photos at left, and Department of Horticulture, Clemson University, for photo at right.)*

certain level. The stems are woody and are either trailing or upright. In general, the upright, or vertical, stems vary in length from 6 inches to 12 inches (15 to 30 centimeters), and those which are about 6 inches (15 centimeters) are the most productive, hence the most desirable. The flowers are relatively small, perfect with a single compound pistil, and the fruit is a small, juicy ovary which when ripe contains the pigment anthocyanin. Figure 19-7 presents cranberry shoots and inflorescence, cranberries on shoots, and a cranberry bog.

Use of Growth Regulators

The clarity and intensity of the red pigment has always been an important quality factor. Plant scientists of the New Jersey Experiment Station found that an application of ethephon (CEPA) at 400 ppm just before the harvest increased the intensity of the red pigment. However, applications of succinic acid-2,2-dimethylhydrazide (SADH) caused poor coloration of the fruit.

Economic Importance and Principal Producing Districts

Practically all of the world's commercial cranberries are grown in the northern United States and Nova Scotia. The crop amounts to between 850,000 to 900,000 barrels annually with a value of between 20 and 25 million dollars. The industry is highly specialized. Most of the crop is grown in Massachusetts, New Jersey, Wisconsin, Washington, Oregon, and Nova Scotia.

Figure 19-7 Top left: cranberry shoots and inflorescence. Top right: cranberries on young shoots. Bottom: view of cranberry bog and storage building. *(Courtesy, Irving E. Demoranville, University of Massachusetts.)*

QUESTIONS

1 Differentiate between the root system of the strawberry and the root system of the grape.
2 How do the flowers of muscadine grapes differ in sex expression from those of vinifera?
3 How does the flower and fruit of the grape differ from those of the raspberry?
4 Gibberellins have facilitated the production of the seedless grape, including the black currant grape, in California. Explain.
5 The use of N, N-dimethylaminosuccinamic acid (Alar-85) has resulted in a remarkable increase in the yield of Concord grapes. Explain. State other effects of growth regulators on grapes.
6 Although the strawberry is grown in all states of the United States, an individual commercial variety has only limited adaptation. Explain.
7 Currants and gooseberries are grown in the region of the Great Lakes and on the upper Pacific Coast. Give two reasons.
8 State two ways in which the roots of blueberry plants differ from the roots of apple and peach trees.
9 Cranberries are grown on bogs high in organic matter and sand; blueberries are well adapted to similar soils with a high water table. Explain.

SELECTED REFERENCES FOR FURTHER STUDY

Eck, P., and N. E. Childers. 1966. *Blueberry culture,* Rutgers, N. J.: Rutgers University Press. A broad, well-presented discussion of general information on blueberry culture, including the botany, breeding of varieties, and production of this special crop.

Flint, W. P., and C. L. Metcalf. 1932. *Insects, man's chief competitors.* Baltimore: Williams and Wilkins, chap. 6. A brief and interesting account of how the grape phylloxera reduced thousands of prosperous people to poverty and how these people became prosperous again.

Winkler, W. A., 1962. *General viticulture.* Berkeley: University of California Press. A clear and concise text covering general viticulture, including the application of research findings to viticulture, which, although based largely on research and experience in California, are generally applicable to all arid and semiarid countries of the world.

Chapter 20

Vegetable Crops Grown for Their Stems or Leaves

When tillage begins, other arts follow. The farmers, therefore, are the founders of civilization.

Daniel Webster

THE PERENNIALS

Asparagus

Plant Characteristics Asparagus has two kinds of roots: (1) fleshy (storage) and (2) absorbing. The fleshy roots are about the size of an ordinary lead pencil. In general, they grow in a lateral direction from 8 to 14 inches (20 to 36 centimeters) per year for about three to four years. Their principal storage constituent is sucrose. The absorbing roots arise from the young portions of the storage roots. Thus, the roots of mature asparagus plants have a spread of 3 to 5 feet (0.9 to 1.5 meters) or more.

Asparagus has two kinds of stems also: (1) underground (rhizomes) and (2) aerial (spears and stalks). The rhizomes are short, thick, and stubby. They grow upward at the rate of about 2 inches (5 centimeters) per year and form buds which develop into the aerial stems.

The stems which are cut for market are called *spears,* and those which grow after the cutting season are called *stalks*. In general, spears are cut when they are 6 to 8 inches (15 to 20 centimeters) long for a period of six to eight weeks. The leaves are small, triangular, scalelike, vestigial structures. They do not manufacture food. The stems which contain chlorophyll are the real food manufacturing organs. Thus, asparagus stems have been modified for the manufacture of the initial food substances.

Asparagus is a dioecious plant. Investigations have shown that male and female plants differ in earliness, yield, and size of spears. In general, male plants produce spears smaller in size, greater in quantity, and earlier in the spring than female plants. The flowers of both sexes are small, numerous, and axillary. Various species of small bees carry the pollen.

The fruit is a small, spherical berry, green when immature and red when ripe. The seeds are relatively large, black-coated, and angular. The principal reserve carbohydrate is hemicellulose. Note the spears on the female plant in Figure 20-1.

Figure 20-1 Asparagus spears, large and straight. Typical spears of female plants growing in loose, friable, stone-free soil. *(Courtesy, Lewis P. Watson, Raleigh, North Carolina.)*

Use of Growth Regulators The *shelf life of asparagus* refers to the period the spears are on display in the retail store. During this period the rate of destruction of chlorophyll, or degreening, is relatively rapid. Experiments have shown that the immersion of fresh spears in a solution of 6-benzylamino purine (BA) at a concentration of 25 ppm for 10 minutes reduces the rate of destruction of chlorophyll for as long as 10 days. Experiments with succinic acid 2–2 dimethylhydrazine (SADH, Alar) also reduces the rate of degreening. In your opinion, how do these chemicals prolong the shelf life of the spears?

Economic Importance and Principal Producing Districts Asparagus is strictly a cool-season crop. Therefore, it thrives best in the cool-temperate zone of the world, and at high elevations in the other three zones. Under these cool-temperate conditions, it is grown in home and market gardens for consumption in its fresh state and in commercial districts for canning. Commercial districts are located in Taiwan, Japan, the Netherlands, and Spain, and in California, New Jersey, and Illinois. In the United States, about 123,000 acres (49,718 hectares) were given over to the production of spears in 1967.

Rhubarb

Plant Characteristics The roots of rhubarb consist of extensive absorbing roots. The stems are of two kinds: (1) rhizome and (2) aerial. The rhizomes are fleshy and woody and bear buds at the crown. The aerial stems attain a height of 3 to 6 feet (0.9 to 1.2 meters) and produce a spike with large numbers of small, greenish-white, perfect flowers. The leaves arise in the form of a rosette from the buds on the crown. The blade is simple, large, and contains oxalic acid, a compound toxic to humans. The petiole, or leafstalk, is the edible portion and is long, fleshy, and well developed. The color of the petiole varies from green (due to chlorophyll) to dark red (due to anthocyanin). Varieties vary in thickness and color of petioles; for example, McDonald, Ruby, Valentine, and Strawberry develop the thick, moderately attractive red petioles, whereas Victoria and Linnaeus produce thick, somewhat green petioles.

Use of Growth Regulators As with stems in general, the crowns and rhizomes have a physiologic dormant period. Ordinarily an exposure to cold between 20 and 40°F (−7 and 4°C) for about one month is necessary. Limited tests indicate that applications of gibberellin reduce the amount of cold which is necessary to break physiologic dormancy and increase the yield of the petioles when grown in forcing establishments.

Economic Importance and Principal Producing Districts As with asparagus, rhubarb is strictly a cool-temperate plant. Thus, its growth and development is limited to the cool-temperate zones of the world—to Canada, the British Isles, and other cool-temperate areas in Europe, Asia, Africa, South

America, Australia, and New Zealand. In the United States, it thrives in the region north of the Potomac and Ohio Rivers and in the mountains of Virginia and North Carolina. Forcing establishments exist in the vicinity of Detroit, Michigan, Boston, Massachusetts, and Chicago, Illinois. Note the rhubarb plant, the rhubarb petiole, and the bunch of rhubarb petioles ready for the market in Figure 20-2.

Globe Artichoke

Plant Characteristics The root system is extensive. The stem is also either (1) rhizome or (2) aerial. The rhizomes are the underground, fleshy, storage organs. The aerial stems are relatively long and bear long, deeply lobed, thistlelike leaves. The flower head, the edible part, consists of a mass of fleshy bracts which, when harvested young, are tender and edible. They contain inulin, a form of starch.

Use of Growth Regulators Experiments indicate that gibberellin applied to plants in the fall in the United States at a concentration of 25 to 50 ppm accelerates the development of the heads.

Economic Importance and Principal Producing Districts The globe artichoke is of only minor importance commercially, and its growth is limited to districts characterized by mild temperatures. Production districts exist along the Mediterranean areas of Europe and Africa and along the coast of the Monterey, Santa Cruz, and San Mateo counties of California.

Figure 20-2 Left: rhubarb plant. Note the abundant long-fleshy red petioles. Right: rhubarb leaf, a long, straight petiole, and a bunch of harvested petioles ready for market.

QUESTIONS

1 Large-sized one-year-old asparagus crowns are preferred. Explain.
2 The crowns are set from 8 to 10 inches (20 to 25 centimeters) deep. Explain.
3 Why is the asparagus plant allowed to grow two to three years before it is cut?
4 What is the relation of the size of the spear to the carbohydrate content of the storage roots? Give reasons.
5 The sugars in asparagus roots gradually decrease throughout the cutting season, rapidly decrease during the period of stalk growth, and markedly increase after most of the stalks have fully developed. Explain.
6 A severe infestation of asparagus beetles during the latter part of the growing season will result in low yields the following year, whereas this is not true at the beginning of the growing season. Explain.
7 The cutting of spears usually extends from six to eight weeks. Why is cutting stopped at the end of this period?
8 The life of the average commercial planting of asparagus is about 15 years; yet some home garden plantings maintain production from 20 to 30 years. Explain.
9 In the northern United States, harvesting of rhubarb does not take place until the plants are two years old. Explain.
10 Harvesting longer than six to eight weeks is likely to thoroughly exhaust the carbohydrate supply in the rhizomes and thus excessively devitalize the plant. Explain.
11 On commercial plantations, seedstalks are usually cut off when they begin to develop. Explain.
12 In what ways is the growing of rhubarb similar to the growing of asparagus? In what ways is it different?
13 Rhubarb grows satisfactorily in the mountains of Virginia and North Carolina; it does not thrive in the Gulf Coast section of the United States. Explain.
14 In what respects is the growing of the globe artichoke similar to that of asparagus and rhubarb? In what respects is it different?

THE BIENNIALS

The biennals include plants such as—Brussel sprouts, cauliflower, broccoli, kale, and kohlrabi. During the first year, these plants develop the edible part, a distinct storage organ or organs, and during the second season after a period of exposure to cold, they develop flowering stems, fruit, and seed. Other biennials are celery and chard.

Plant Characteristics of Cabbage and Closely Related Crops

The plants develop a much-branched, highly ramified root system. Investigations have shown that the roots of half-grown cabbage plants extend laterally for a distance of from 2 to 4 feet (0.6 to 1.2 meters). These roots are much-branched, especially in the upper 4 inches (10.6 centimeters) of soil. Consequently, deep cultivation of the cabbage crops during the late stages of growth is likely to cut the feeding roots just beneath the surface.

The vegetative stems are comparatively short, and the leaves are simple, large, well developed, and fleshy. Those with a storage organ contain large quantities of starch which gradually change to sugar. The flowering stems arise from the axils of the leaves of the storage organ and are from 2 to 4 feet (0.6 to 1.2 meters) high.

The inflorescence is a terminal raceme. Individual flowers are complete and regular with four sepals, four white or light yellow petals, six stamens, and a two-celled pistil. The flowers are mostly insect-pollinated and varieties within each crop cross readily. The fruit is a long, slender pod called a *silique*. The seed are quite similar in appearance and germinate readily under favorable conditions.

Cabbage

Cabbage develops a short stem and a large terminal bud called the *head*. The head is the principal food storage organ and is the part used for human consumption. It varies in size, shape, color, edible quality, and adaptability for storage, according to the variety.

Varietal Types Three general types are recognized: (1) early, which may be pointed or round and small-headed; (2) mid-season, which may be round, flat, or savoyed medium- to large-headed; and (3) late, which may be round, or medium- to large-headed.

Use of Growth Regulators If young cabbage plants are exposed to suboptimum temperatures between 40 and 50°F (4 and 10°C), they will produce flowering stems and seed without producing heads, a condition known as *bolting*. Naturally, the grower of heads cannot sell bolted plants, and unfortunately, bolting occurs frequently in producing districts characterized by mild-winter temperatures. Certain vegetable-crop scientists found that although chlorophenoxypropionic acid (CIPP) and dichlorophenosyacetic acid (2,4-D) prevented the formation of flower buds, they did not delay their formation for a period of two weeks.

Economic Importance and Principal Producing Districts Of the cabbage group, cabbage is the most widely grown and the most commonly known. In fact, considerable areas are given over to the production of this crop on all five continents. For example, according to the *Production Yearbook* of the United Nations, in 1970 20 European countries grew cabbage on 879,676 acres (356,000 hectares), approximately 18 Asian countries grew cabbage on 558,446 acres (226,000 hectares), and several countries of Africa, South America, and Central America grew cabbage on considerable acreages.[1] In the United States,

[1]United Nations, *Production Yearbook,* 24:197, 1971.

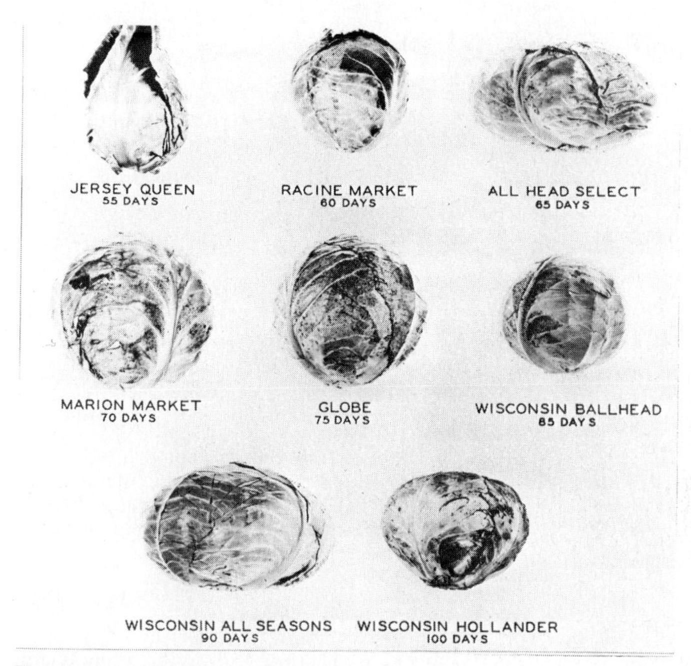

Figure 20-3 Types of cabbage heads. *(Courtesy, U.S. Department of Agriculture, Office of Information, Washington, D.C.)*

cabbage is widely grown in home gardens, in market gardens for local markets, in the winter-garden areas of the Southeast and Southwest for long-distance shipment, in the Northeast and Midwest for storage, sauerkraut manufacture, and pickling.

Brussels Sprouts

Brussels sprouts develop an elongated, unbranched, erect stem, and a large number of small, lateral vegetative buds. In reality, these buds are small heads and comprise the storage organs. The food-manufacturing leaves are comparatively large and are approximately as broad as they are long.

There are two types, or forms: (1) tall, such as Half Dwarf Improved, grown mostly in California, and (2) dwarf, such as Catskill and Long Island Improved. Note the characteristics of the plant presented in Figure 20-4.

Market gardeners and home gardeners raise Brussels sprouts quite extensively in northern France, Belgium, Holland, and the British Isles. In the United States, they are not widely grown in home gardens, and commercial areas are confined to specialized districts. These districts indicate the mild-temperature requirements for Brussels sprouts. They are located on Long Island, in upstate New York, in Oregon, and in the Moon Bay Region just below San Francisco. Note that these areas are adjacent to large bodies of water.

Figure 20-4 Brussels sprouts. Note the small headlike cabbages on the upright stem. *(Courtesy, Joseph Harris Seed Company, Moreton Farm, Rochester, New York.)*

Cauliflower

Cauliflower is grown for its so-called curd which is formed by the stem system with short internodes. No flower primordia are present in the flowerlike stem terminal, or curd. The leaves of cauliflower are eliptical, longer than those of cabbage, and are used to cover and protect the curd from light and wind which causes it to become brown.

There are two types: (1) the quick-growing Snowball, developed in Holland and Denmark, and (2) the slow-growing, late, or winter type from the Mediterranean region.

In general, cauliflower is popular and grown in uniformly cool and humid areas of the world—in Europe, Asia, Africa, North America, Australia, and New Zealand. In the United States, commercial districts are located along the coast of Oregon and California, on Long Island, in upstate New York, and in southwestern Michigan. Note the long leaves and the white curdlike head of the plant in Figure 20-5.

Green Broccoli

Green broccoli is grown for its green curdlike flower clusters with its young tender leaves and stems. One large head is formed terminally and several smaller ones are formed laterally. Of these, the terminal heads are considered to be the more desirable. Figure 20-6 shows typical terminal flower clusters. Green broccoli is more often an annual than a biennial since the plant frequently produces flowers and seed the first year.

Figure 20-5 Cauliflower. Note the white curdlike head and the long leaves which are usually tied together over the head to blanch it. *(Courtesy, Ferry Morse Seed Co., Inc., Mountain View, California.)*

Green broccoli is grown in many countries of the world, in home and market gardens in the cool-temperate zone during the spring and summer and in the warm-temperate zone during the winter and early spring. In the United States, the principal commercial areas are Texas and Arizona in winter; California in early spring and New York, New Jersey, Oregon, and California in the fall. Broccoli for processing (mainly freezing) is grown in California. About 40,000 acres (16,147 hectares) annually are given over to the commercial production of the crop.

Figure 20-6 Green broccoli. Note the compact terminal head. *(Courtesy, Joseph Harris Seed Company, Moreton Farm, Rochester, New York.)*

Collards

Collards develop large fleshy leaves which are used for greens. They are widely grown in home gardens in the southern United States, and limited quantities are grown for local markets.

Kale

Like collards, kale develops large green leaves which are used as greens. Kale varieties vary in plant height (dwarf and tall), shade of green of the foliage (grass green and gray green), and indentation of the leaves (indented and smooth). Note the fine indentation of the leaves of the plant in Figure 20-7.

Kohlrabi

Kohlrabi develops a short, fleshy, turniplike stem. The enlarged stem is the storage organ and contains starches and sugars. The leaves are comparatively small and oval. Note the storage stem and the position of the leaves of the kohlrabi plants in Figure 20-8.

Land Cress

Land cress is a hardy biennial mustard. The species is identified by the four to seven lobes on the base of the leaves and by the pleasant sweetish taste of the leaves, the edible part. The stems are short and inconspicuous and the leaves develop in the form of a rosette.

Figure 20-7 A typical plant of Vates kale. This savoyed variety is adapted to and grown in the eastern United States. *(Courtesy, Joseph Harris Seed Company, Moreton Farm, Rochester, New York)*

Figure 20-8 Kohlrabi plants. Notice the leaves arising from the enlarged, fleshy stems. *(Courtesy, Joseph Harris Seed Company, Moreton Farm, Rochester, New York.)*

Use of Growth Regulators Experiments have shown that spraying succinic acid-2,2-dimethylhydrazide (Alar) just before the advent of cold weather delays the development of the inflorescence for a period of 10 to 15 days. In this way, the harvest period can be extended before bolting takes place.

Importance and Principal Producing Districts Land cress is of minor importance in both home and commercial gardens. Wild plants grow throughout

Figure 20-9 Land cress growing in Roanoke County, Virginia.

the eastern United States and southern Canada and are harvested for greens. However, small commercial areas exist in Virginia, Tennessee, North Carolina, and Georgia. Bitter Crest, or Golden Rocket, a related species, is grown for human consumption in Europe.

QUESTIONS

1. Distinguish between cabbage, cauliflower, green broccoli, Brussels sprouts, collards, kale, kohlrabi, and land cress.
2. Cabbage is grown in late fall, winter, and early spring in the southern regions of the United States, in summer and fall in the northern regions. Explain.
3. In the southern United States, young plants are frequently exposed to long periods of low temperatures and often go to seed without forming a head. Explain.
4. Cabbage is normally a biennial. Under what conditions does it become an annual?
5. Brussels sprouts are grown extensively in the British Isles and not extensively in the United States. Explain.
6. In the southern United States, the collard is quite popular in home gardens. Explain.
7. List the members of the cabbage group which are easily grown; those which are not easily grown.

Celery

Plant Characteristics The root system of celery is nonextensive. The taproot is destroyed by necessary transplanting and, as a result, a moderate number of laterals develop. Studies of the root system show that the lateral roots of full-grown plants extend for a comparatively short distance only. For example, at Cornell University, most of the roots of plants growing in sandy loam were found within a radius of 6 inches (15 centimeters) from the base of the plants, and most of these roots were within 2 to 3 inches (5 to 8 centimeters) from the surface. The comparatively sparse root system explains why the plants can be set close together—from 6 to 8 inches (15 to 20 centimeters); why celery cannot readily compete with weeds; and why the topsoil should be abundantly supplied with water and essential elements.

During the first year the plants produce a short, stubby stem usually from 3 to 6 inches (8 to 15 centimeters) long and a large number of thick-petioled, pinnately compound leaves. The petioles are conspicuously ribbed, very broad at the base, and contain comparatively large quantities of starch and related substances. During the second year, the much-branched flowering stem arises from the axils of the fleshy petioles and on maturity attains a height of 2 to 3 feet (0.6 to 0.9 meters).

The inflorescence is an umbel. Individual flowers are small, perfect, and are mostly self-pollinated. Mature fruit, called *seed,* are small and dry and have corky ribs. For satisfactory germination, the seed require shallow planting and a uniform supply of moisture.

Varietal Types There are two types of celery: (1) the yellow-petioled and (2) the green-petioled. In general, the yellow-petioled varieties mature their petioles in a relatively short time and are grown for the early market; the green-petioled varieties mature their petioles in a somewhat longer time, and are grown for late market and storage. Examples of the former are Golden Plume and Cornell; examples of the latter are Utah and Giant Pascal. Of these two types, the green-petioled varieties are the more popular and the more widely grown.

Use of Growth Regulators The *shelf life of celery* refers to the period during which the trimmed plants are on display in the consumer market. Naturally, the fundamental processes concerned are respiration and transpiration. For a long shelf life, the rate of each of these processes should be relatively low. Experiments have shown that the momentary immersion of trimmed plants in a solution containing 6-benzylaminopurine (BA) at a concentration of 10 ppm and stored at 4°C (39°F) prolonged shelf life for a period of 40 days; whereas the nontreated plants stored at the same temperature were market unacceptable at the end of the same period.

Economic Importance and Principal Producing Districts Celery is grown in many countries within the cool-temperate zone during the spring, summer, and fall; within the warm-temperate zone during the winter and spring; and within subtropical zones in high elevations during the summer. Commercial districts exist in England, Canada, and Mexico. In the United States, important districts are the Lake Okeechobee in Florida and the Delta in Los Angeles, California, constituting 19,000 acres (7,689 hectares) in winter and early spring; the Los Angeles in California, the Kalamazoo and Muskegon in the Lower Peninsula of Michigan, the Williamson in New York, and the Ohio, constituting about 4,500 acres (1,821 hectares in summer); the California constituting 6,000 acres (2,428 hectares) in late fall. Each year about 32,600 acres (13,193 hectares) are given over to the production of the crop.

Chard, or Swiss Chard

Chard is a beet developed for its fleshy petioles and leaves. During the first year the plants develop a fleshy taproot, a short stem, and a large number of well-developed simple leaves. The outer leaves mature first. As they are harvested, new leaves develop. In this way, the plant produces leaves throughout a relatively long growing season.

QUESTIONS

1 On mineral soils, celery responds to continuous cultivation throughout the growing period. Explain.
2 Celery is grown in late fall, winter, and early spring in Florida; in summer and fall in New York and Michigan; and the entire year in Los Angeles county of California. Explain.

3 Celery thrives best and the petioles attain their highest quality in cool weather. Explain.
4 Celery grown on upland soil in the humid East usually requires irrigation. Explain.
5 Celery thrives well on well-drained, slightly acid mucks with a controlled water table. Explain.
6 The climate of Great Britain is adaptable to the raising of high-quality celery. Explain.
7 Celery cannot readily compete with weeds. Explain.
8 What is the purpose of blanching celery?
9 If commercial celery is grown in your state, describe briefly the type of soil used, the system of soil management, the variety, method of plant production, disease-control program, and growing and shipping season.
10 How do applications of BA increase the shelf life of celery?

THE ANNUALS

Lettuce

Plant Characteristics The root system of mature plants is moderately extensive. In upland soil, the taproot extends to the 4- or 5-foot (1.2- to 1.5-meter) level. Branches of the first order extend laterally to a distance of 6 to 8 inches (15 to 20 centimeters) and then turn downward. Branches of the second order are the most numerous. They usually fill the upper 10 to 12 inches (25 to 30 centimeters) of soil. A direct relationship exists between the density of the root system and the compactness of the soil. In compact soils, the root system is more dense and more shallow than in loose soil.

During the vegetative stage, the stem is short, usually from 4 to 6 inches (10 to 15 centimeters) long. Around it, the leaves arise in a rosette. They vary in size (large and small), shape (spatular and circular), color (light green to dark green), and crispness. During the reproductive stage, the stems elongate and branch, and each of the various branches forms a terminal inflorescence.

The inflorescence is a panicle. Individual flowers are perfect, with five stamens and a one-celled ovary. They are usually self-pollinated. The fruit, called *seed,* are very small; each contains a single embryo and is planted shallowly.

Varietal Types There are two general types: (1) the solid heading and (2) the loose, or open, heading. The solid heading may be arranged in three subgroups: (1) the crisp with crinkled leaves, (2) the butter with smooth leaves, and (3) the cos with upright leaves. The loose, or open heading, may be divided into two subgroups: (1) the crinkled and light green, and (2) the smooth and dark green. Of these types, the crisp, solid headed is the most widely grown.

Economic Importance and Principal Producing Districts In common with other cool-season vegetable crops, lettuce is grown in home gardens, market gardens, and greenhouse establishments in countries within the cool-temperate

Figure 20-10 Varietal types of lettuce. Top left: nonheaded leaf. Top right: loose-headed, green. Bottom: firm, crisp headed. *(Courtesy, Ferry Morse Seed Co., Inc., Mountain View, California.)*

zone and within the warm temperate zone at high elevations. Commercial districts exist in Europe, Asia, Africa, America, New Zealand, and Australia. In the United States, principal producing regions are California, Arizona, Texas, and Florida in winter; California, Arizona, and New Mexico in early spring; New Jersey, Connecticut, and Massachusetts in late spring; California, New York, and Colorado in summer; California, New Mexico, and Texas in early fall; and Arizona in late fall.

Chinese Cabbage

Chinese cabbage is closely related to and resembles the cabbage crops in its root system, stems, and leaves. The leaves are the edible portion and, during the latter stages of growth, they form loose, upright heads which are fine in texture, crisp, succulent, and used in salads or as greens. (See Fig. 20-11).

Spinach

Plant Characteristics Spinach has a distinct, well-developed taproot. The laterals are relatively few and short. Most of the absorbing roots arise directly from the taproot.

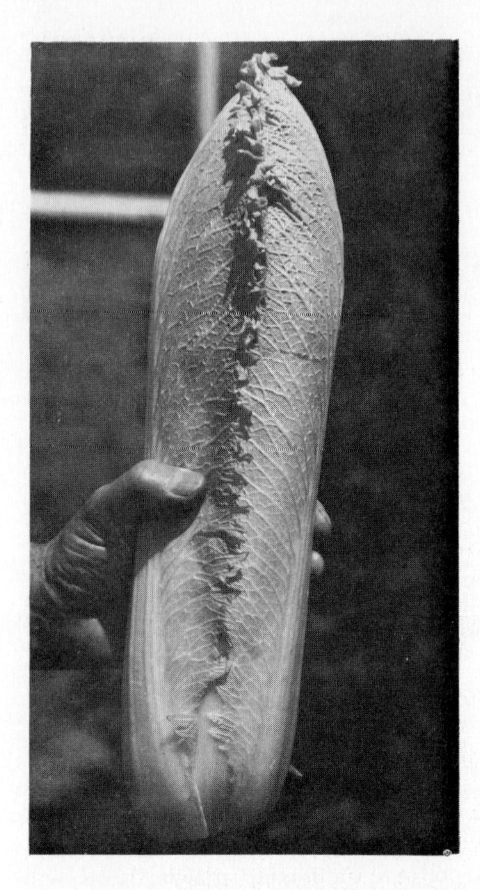

Figure 20-11 A head of Chinese cabbage. *(Courtesy, Associated Seed Growers, New Haven, Connecticut.)*

The stem is short and platelike and bears a rosette of well-petioled, simple, moderately large leaves. The leaves vary in shape (spear-shaped to round) and character of the surface (smooth, slightly undulated or savoyed). The savoyed surface is due to extensive growth of the parenchyma tissue between the veins. The leaves are exceedingly dark green. They are an excellent source of carotene (provitamin A), ascorbic acid, and minerals, as well as being a good source of thiamin and riboflavin.

Spinach is dioecious or monoecious according to the variety. In dioecious varieties the males are of two types: (1) extreme and (2) vegetative. The extreme males have few, if any, well-developed leaves toward the tip of the seedstalk, and the vegetative males have well-developed leaves toward the tip. Since spinach is grown for its leaves in home and commercial gardens, vegetative males are more desirable as breeding parents than extreme males. Female plants are always vegetative. Thus, highly vegetative strains of the dioecious varieties have been developed by roguing out the extreme males in seed-producing fields. In monoecious varieties the plants are highly vegetative.

The flowers are without petals, small, and relatively inconspicuous. Pollen grains are small, produced in abundant quantities, and are carried from one plant to another by the wind. The fruit, called *seed,* have the ability to germinate at rather low temperatures and require a continuous supply of water. Poor germination frequently occurs in dry soil in the summer.

Varietal Types Varieties of spinach are classified into two main groups: (1) varieties which develop savoyed or crumbled leaves and (2) varieties which develop smooth leaves. *Savoying* refers to the development of the tissue between the veins. The tissue is more highly developed in savoyed leaves than it is in smooth. Each of these main groups are divided into two subgroups: (1) varieties which develop flowers and seed during short days and long nights, the so-called quick-seeding varieties, and (2) varieties which develop flowers and seed during somewhat longer days and shorter nights, the so-called long-standing varieties. Examples of the former are Virginia Savoy and Hollandia, and examples of the latter are Long Standing Bloomsdale and Nobel.

Economic Importance and Principal Producing Districts Two environmental factors affect the growth and distribution of spinach. These are the temperature level and the length of the light period. Spinach is primarily a cool-season, long-day–short-night plant. This explains why spinach is grown for its leaves in locations characterized by cool temperatures combined with short days and why it is grown for its seed in locations characterized by cool temperatures combined with long days. Thus, spinach grown for its leaves is limited to the

Figure 20-12 Field of spinach growing in the eastern United States. *(Courtesy, V. R. Boswell, U.S. Department of Agriculture, Washington, D.C.)*

cool-temperature zones during late winter and early spring and to the warm-temperate zones during the winter and early spring. In the United States, spinach is the most important vegetable crop grown as greens. Principal commercial areas are in the vicinity of Baltimore, Maryland, Norfolk, Virginia, and Crystal Springs, Texas, and in central New Jersey and southern California. The Virginia and New Jersey districts supply markets during the fall and spring; the Texas district supplies markets in the winter and early spring; and the California district supplies the Pacific Coast markets during the winter. The principal districts which are given over to the raising of spinach for canning are located in California and in southern Texas.

QUESTIONS

1 In Florida, lettuce is grown during the late fall, winter, and early spring. Explain.
2 In New York, lettuce is grown in the summer, and in the Imperial Valley of California, it is grown in the winter and early spring. Explain.
3 In general, growth should be rapid during the early stages of plant development and moderately rapid during the later stages. Explain.
4 Comparatively cool, sunny days and cool, crisp nights favor the formation of firm heads. Explain.
5 Lettuce heads stored in the refrigerator remain more sweet and crisp than those stored in a warm room. Explain.
6 Name the two types of lettuce.
7 Spinach in home and commercial gardens is grown during the fall, winter, and early spring. Give two reasons.
8 Spinach grows poorly on soil greater than pH 7.0 and less than pH 6.0. Explain.
9 Investigations have shown that spinach requires abundant moisture and essential elements, particularly nitrate-nitrogen. Explain.
10 Spinach maturing in the late fall is sweeter than that maturing in late spring. Explain.
11 Since 1910 spinach has become very popular in the United States. Can you think of any reasons for this?

Chapter 21

Vegetable Crops Grown for Their Fleshy Storage Organs

Whether the task be great or small, do it well or not at all.

Lord Chesterfield

THE ONION CROPS

The onion crops consist of onion, shallot, leek, garlic, and chive. These crops have at least four characteristics in common: they possess a characteristic odor and flavor, store carbohydrates in the basal portion of the leaves, develop a nonextensive root system, and produce perfect, insect-pollinated flowers. The characteristic odor and flavor vary with the crop and with the variety within the crop. For example, garlic is more pungent than onions, and the American varieties of onions are more pungent than the European varieties. The basal portion of the leaf is thick and fleshy and constitutes the edible portion. These basal leaves with the short, platelike stem are called *bulbs* and contain comparatively large quantities of inulin (a type of starch), moderate amounts of sugars, and moderate amounts of ascorbic acid or vitamin C. The nonextensiveness of the root system, with the major portion of the absorbing system of mature plants within a radius of 6 inches (15 centimeters) from the stem, requires the absolute control of weeds and the selection of soils with high water-holding capacity. The leaves are simple and present a rather small photosynthetic surface. The inflorescence is an umbel; individual flowers possess six stamens and a simple pistil; and they are chiefly pollinated by various species of bees.

Onion

Distinguishing Characteristics The onion develops *distinct bulbs*. According to the variety, these bulbs vary in size (small, medium, and large), color (white, yellow, or red), shape (flattened, round, or globular), texture (fine or coarse), and pungency. The plant is normally a biennial. The fleshy bulbs develop during the first season, and seedstalks develop during the following season. The leaves develop from a short, flattened stem at the base of the bulb. They consist of two parts: (1) sheath and (2) blade. The sheaths are fleshy and surround the younger leaves within. The blades are green, pointed, and hollow.

Varietal Types Onions are classified according to the degree of mildness of the bulb. Three distinct types are recognized: (1) mild, (2) semimild, and (3) pungent, or strong. In general, the mild type develops larger, finer-textured bulbs than the pungent, or strong, type. The commercial varieties of onions are rather specific in their optimum growing requirements. For example, the long-storage, pungent varieties grown in the Middle West and Northeast do poorly in the Southwest and Pacific Coast, and the nonstorage, mild varieties grown in the Southwest are ill-suited to the Middle West and Northeast.

Use of Growth Regulators A serious problem in the storing and marketing of onion bulbs is the production of sprouts. These sprouts use food and water which would otherwise remain in the bulbs. Thus, sprouting decreases the market value of the bulbs. Researchers are trying to find a growth regulator which would prevent or inhibit the development of growing points and thereby enhance the storage and market life of the bulbs. Of the numerous chemicals which have been tested, maleic hydrazide (MH) seems to be the most promising, and of the several stages in the development of the bulb, applications two weeks before the bulbs are harvested, or when 50 percent of the plants in the field have toppled tops, are the most effective. Applications before this critical stage have failed entirely to prevent sprouting; applications after this stage caused the production of "puffy," unmarketable bulbs.

Economic Importance and Principal Producing Districts The leading commercial onion producing countries are the United States, Japan, Rumania, Italy, Mexico, and Turkey. More than 100,000 acres (40,500 hectares) of onions are grown commercially in the United States annually. The leading commercial producing states are California, New York, Oregon, Colorado, Arizona,

Figure 21-1 Varietal types of onions. Top left: Granex, short day, mild, yellow, short storage life. Top right: Early Supreme, short day, mild, white hybrid, short storage life. Bottom left: Amigo, medium to long day, mild, yellow, high yield. Bottom right: Southport Red Globe, long day, pungent, deep red scales and flesh, long storage life. *(Courtesy, Dessert Seed Co., El Centro, California.)*

Michigan, New Mexico, and Texas. Onions are widely grown in many market-garden areas throughout the United States and the world. They are very popular in home gardens in many countries.

Shallot, Leek, Garlic, and Chive

The shallot develops *several small bulblets, or cloves, held together at the base;* the leek develops *thick, mild, fleshy leaf sheaths;* the garlic develops *a group of small bulbs, or cloves, enclosed in a membranelike skin;* and chive develops *small, distinct bulbs*. Commercial districts of shallots exist in southern Louisiana, and commercial districts of garlic are located in central California. The climate and cultural requirements are similar to those of the onion.

QUESTIONS

1 In general, the onion group of crops requires continuous cultivation, cannot compete readily with weeds, and can be planted at close distances. Explain.

2 In ancient times masters of sailing vessels frequently took on a supply of onions before making long voyages. Explain.
3 Distinguish the three main types of onions.
4 In the southern regions of continental United States onions are grown in the winter and early spring; in northern regions they are grown in the summer. Explain.
5 Onions are particularly adapted to loose, friable loams and to slightly acid mucks. Explain.
6 In general, the larger the plants grow before they begin to form bulbs, the larger will be the yield. Explain.
7 In what regions of the United States are most of the pungent varieties of onions grown? In what region are most of the mild types grown?
8 How does maleic hydrazide prevent sprouting of onions in storage?
9 Name and distinguish the members of the onion group.

THE TUBER CROPS

Potato[1]

Root System The root system of asexually propagated crops is fibrous and adventitious. In other words, the roots arise from the nodes of the stem situated in the soil. In adult plants the root system is moderately extensive. Although a few roots extend from 3 to 4 feet (0.9 to 1.0 meter), both vertically and laterally, most of the roots are from 6 inches to 2 feet (15 to 61 centimeters) long. These roots are situated in the topsoil, with the greater density in the upper 3 to 4 inches (8 to 10 centimeters). Hence, if cultivation is necessary during the later stages of growth, it should be shallow to avoid cutting the absorbing roots just beneath the surface.

Stems and Leaves The stems are of two types: (1) aerial and (2) underground. The aerial stems are angular, green or greenish-purple, depending on the variety, and bear in a spiral arrangement pinnately compound leaves. Under humid conditions the leaves are broad and flat, and under arid conditions they are narrow and cupped.

The underground stems consist of stolens and tubers. The stolens are about the size of a lead pencil and extend laterally for a distance of 1 to 4 inches (3 to 10 centimeters). The tubers arise at the end of the stolens and are short, thick, and fleshy. They develop scalelike leaves called *eyebrows* which subtend buds called *eyes*. These buds are undeveloped branches. Each eye contains both terminal and lateral branches, and each potato has both terminal and lateral eyes. The terminal eyes develop sprouts before the lateral eyes, and the terminal sprouts in each eye develop before the lateral sprouts. However, cutting of the tubers, an important practice, destroys the dominance of the terminal eyes, and removing the first sprouts destroys the terminal dominance within each eye.

[1]In the southern United States, the term *potato* usually refers to the sweet potato, and the potato is called the *Irish potato*.

The anatomy of immature and mature tubers differs considerably. The immature tubers consist of an epidermis, a wide band of cortex, pericycle, vascular bundles, and pith. As the tuber develops, the epidermis is replaced by a phellum —the layer of corklike cells. The cortex becomes a narrow band just beneath the periderm, and the vascular bundles extend to the eyes. The pith becomes greatly enlarged and constitutes the major portion of the tuber. The function of the periderm is to keep the all-important water within the tuber and to resist the attacks of rot-producing organisms. The lenticels permit exchange of carbon dioxide and oxygen, and the cortex and pith are abundantly filled with grains of starch. In fact, most of the dry matter of the tuber consists of starch.

Flowers, Fruit, and Seed The flowers are borne in clusters terminating the stem. Individual flowers are perfect and either white, yellow, purple, or striped, according to the variety. Flowering is usually more profuse in regions characterized by low summer temperatures than in regions characterized by high summer temperatures. The fruit, or seedball, is round, small—½ to 1 inch (1.3-2.5 centimeters) in diameter—and contains from 100 to 300 seeds. Seed are used to develop new types and varieties.

Varietal Types Varieties grown in the United States are placed in two groups: (1) early maturing and (2) late maturing. Early-maturing varieties develop tubers in a relatively short time, and late-maturing varieties require a relatively long time. Because of extensive breeding programs, particularly those involving the U.S. Department of Agriculture in cooperation with many state experiment stations, new varieties are constantly being developed. For the most part, these new varieties combine disease resistance with high yields of high-quality tubers.

Use of Growth Regulators In the southern United States, plantings of the fall crop are frequently made by using tubers of the preceding spring crop. At this time, the tubers are either wholly or partially in their physiologic dormant, or rest, period. In either case, poor, irregular stands are obtained. Potato scientists have been endeavoring to shorten this period of rest by subjecting the tubers to various combinations of temperature, relative humidity, and exposure to currents of air. They have found that dipping the tubers in a solution containing 1.2 percent chlorohydrin and storing them in an airtight place breaks the physiologic dormant period in one day, and that dipping the tubers in solution of gibberellin at 0.5 to 1.0 ppm also shortens the rest period.

When the tubers are out of their physiologic dormant period, the buds begin to develop into stems, called *sprouts*. If the storage temperature is much above the optimum range, the sprouts develop rapidly, which detracts from the market value of the tubers. Experiments have shown that if the growth retardant, maleic hydrazide, is sprayed on the foliage when the tubers are almost fully formed, the tubers remain practically free from sprouts while in storage. Evi-

Figure 21-2 Examples of U.S. varieties of potatoes. A: Cobbler, early, good quality, deep eyes, fair yields. B: Cherokee, medium early, good quality, shallow eyes, and resistant to scab. On opposite page. C: Superior, mid-season, excellent for chips. D: Kennebec, late, general purpose, excellent for cooking, satisfactory for chips, widely adopted nationwide. E: Russet-Burbank, late, excellent for baking and for making French fries. *(Courtesy, Raymon Webb, U.S. Department of Agriculture, Beltsville, Maryland.)*

dently, the maleic hydrazide is translocated, moving from the tops to the buds of the tubers where it exerts its regarding effect.

Economic Importance and Principal Producing Districts The potato is one of the major food plants of the world. Annual production is between 5 and 6 billion hundred-weights. The leading producers are Russia, Poland, West Germany, the United States, France, and the United Kingdom.

In the United States, two rather distinct commercial industries exist: (1) the early-crop and (2) the main-crop. Major early-crop producing districts are Mobile Bay, Alabama, Charleston-Beaufort, South Carolina, Eastern Shore, Virginia, Long Island, New York, and Sacramento Delta, California. Major main-crop producing districts are Aroostook County, Maine, western New York, eastern Pennsylvania, northern Ohio, the central part of the lower peninsula of Michigan, the Red River Valley of Minnesota and North Dakota; and the high altitude areas of Idaho, Washington, and Oregon.

Jerusalem Artichoke

The Jerusalem artichoke is a tuber-forming sunflower. The root system is extensive. The stems consist of two types: (1) underground (tubers) and (2) aerial. The tuber stores a type of starch called *inulin* and develops in much the same way as that of the potato. In the United States, the crop is better adapted to the climate of the northern two-thirds of the country than to the southern one-third.

Dasheen, or Taro

The dasheen, or taro, produces corms which store starch. These starchy corms are the edible portion of this widely grown tropical plant. It is closely related to

Figure 21-3 Jerusalem artichoke. In general, the Jerusalem artichoke produces high yields of rough tubers as shown, sometimes more than 40 tubers on one plant. *(Courtesy, U.S. Department of Agriculture, Office of Information, Washington, D.C.)*

and resembles the common ornamental caladium, commonly called *elephant's ear*. Dasheen is cultivated extensively in tropical America, in South China, on the tropical islands of Japan, as well as on other tropical islands of the world. Its use and culture are more highly developed in Hawaii than elsewhere.

QUESTIONS

1. In general, the potato requires shallow cultivation, particularly during the period of tuber formation. Explain.
2. Potatoes harvested in the immature state skin easily. Explain.
3. Investigations have shown that the potatoes with a well-developed periderm shrink less and rot less in storage than those with a poorly developed periderm. Explain.
4. The climate of northern Maine, the Delta district of California, West Germany, Great Britain, and Ireland is particularly favorable for the production of high yields of potatoes. Explain.
5. How does chlorohydrin break the physiologic dormant period of potato tubers?
6. How does maleic hydrazide prevent or retard the sprouting of potatoes in storage?
7. Give reasons why the Jerusalem artichoke generally produces higher yields in the northern two-thirds of the United States than in the southern one-third.

THE ROOT CROPS

The vegetable crops known as *root crops* include carrot, parsnip, salsify, beet, radish, turnip, rutabaga, and horseradish. These crops develop enlarged storage organs called *roots*. They contain fairly large quantities of starch and the size, shape, color of skin, and flesh vary greatly with the crop.

The root crops are herbaceous biennials with the exception of the radish which is either annual or biennial. They develop an extensive absorbing system and long-petioled leaves. Their growth and development consists of two distinct stages: (1) the production of the enlarged storage organ and (2) the production of flowers, fruit, and seed. In home and commercial gardens, these crops are grown for the enlarged roots only.

Carrot

A transverse section of the fleshy root shows two distinct regions: (1) the outer and (2) the inner. The outer tissues consist of a thin periderm and a relatively wide band of storage tissue. The periderm reduces transpiration to a minimum and resists the attacks of invading organisms, and the storage tissue of mature roots stores relatively large quantities of starch and carotene (the precursor of vitamin A) and moderate quantities of sugar, thiamin, and riboflavin. The inner core consists of xylem and pith. High-quality carrots contain a relatively small

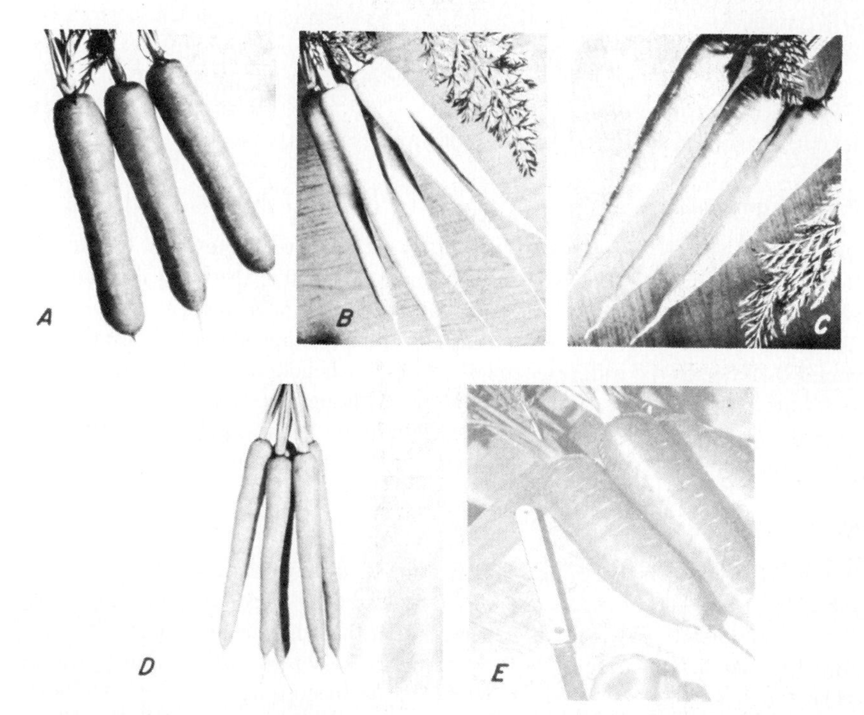

Figure 21-4 Types and varieties of carrots. A: Nantes. B: Imperator. C: Danvers. D: Gold Pak. E: Chantenay, Red Cored. *(Courtesy, Ferry Morse Seed Co., Inc., Mountain View, California.)*

inner core. Scientists have shown that the outer tissues contain more carotene than the inner.

Stems, Leaves, Flowers, and Fruit The stem is short and platelike during the first growing season and long and erect during the second. The leaves are decompound and the inflorescence is a conspicuous compound umbel. Individual flowers are small, perfect, and white and are largely insect-pollinated. The fruits are called *seed* and are small, dry, and indehiscent. Each individual fruit contains one seed. The embryo germinates slowly and requires a fine friable seedbed and uniform supplies of moisture.

Varieties are classified according to shape and length of the root. Shape refers to the tip of the root, whether blunt or pointed, and length is considered with reference to the diameter. Principal classes are moderately long and blunt; long and blunt; long and pointed; and very long and pointed. Within recent years, scientists have developed highly colored strains of certain varieties which are adpated to specific producing districts in the country.

Economic Importance and Principal Producing Districts Because the enlarged root contains large quantities of beta carotene, it is an excellent source of Vitamin A. As a result, the demand for carrots is quite high. In general, the enlarged roots are grown in home and market gardens in cool-temperate zones, and during the cool portion of the season, in warm-temperate zones. In the United States, commercial districts exist in California, Texas, New York, New Jersey, Arizona, Michigan, and Wisconsin.

Parsnip and Salsify

Parsnip and salsify develop enlarged roots which have white flesh but do not have distinct zones. The absorbing system is extensive; the stem is short and platelike during the first stage of growth and tall and branched during the second. The inflorescence of parsnip is an umbel, and the fruits, called *seed,* are flat and roundish; whereas the inflorescence of salsify is a head and the fruits, also called *seed,* are relatively long and oval-pointed. Although these crops are of minor importance commercially, they are grown in home and market gardens in many countries within the cool-temperate areas of the world.

Beet

The enlarged root of the garden beet is quite distinct from that of the other root crops. A cross section shows alternate circular bands of storage and conducting tissues. Frequently, the wide bands are darker than the narrow bands. The contrast in color between these alternate bands is known as *zoning*. Zoning varies greatly between varieties, within varieties, and according to the environment. Distinct zoning, however, is undesirable. The leaves are simple, oblate and long and are arranged on a short stem called the *crown*. They vary from dark purple to

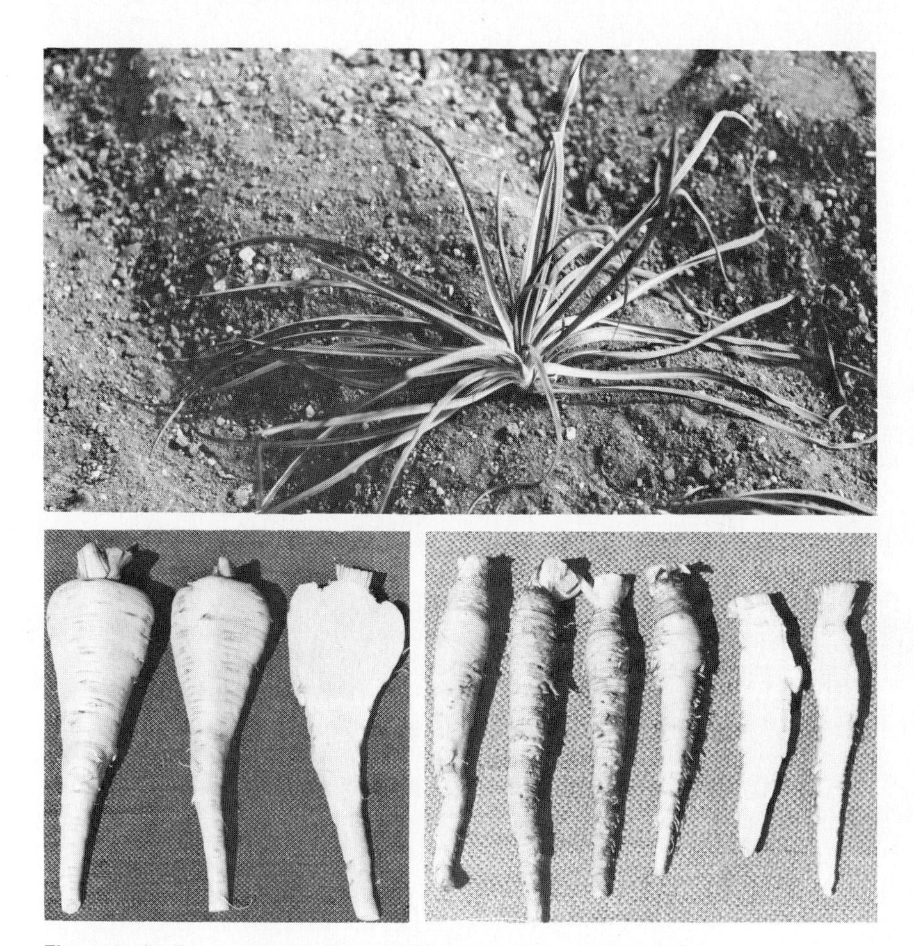

Figure 21-5 Top: salsify top. Bottom left: parsnip. Bottom right: salsify roots. (*Courtesy, U.S. Department of Agriculture, Washington, D.C., for top photo and W. P. Judkins, Virginia Polytechnic Institute and State University, for bottom photos.*)

light green. The fruit cluster, or seedball, usually consists of several ovaries, each of which contains a single seed. Since most seedballs contain more than one seed, the thinning of the seedlings is necessary.

Varietal Types Varietal types consist of the red-fleshed and the golden-fleshed. Of these, the red-fleshed type is the more common and the more widely grown. Varieties and strains of this type vary greatly in the degree of zoning —the alternate light and dark bands in the flesh. The ideal type of red-fleshed beet is a solid dark red, free from zoning. The golden-fleshed beet is now also available. Ideal types are free from zoning and are considered highly desirable by certain consumers.

Figure 21-6 Types and varieties of garden beet. Top left: Crosbys Egyptian, early and flat. Top right: Detroit Dark Red, early and round. Bottom: plant ready to harvest. *(Courtesy, U.S. Department of Agriculture, Office of Information, Washington, D.C.)*

Economic Importance and Principal Producing Districts Garden beets are grown in home and market gardens in countries within the temperate zones of the world. In the United States, the winter-garden areas in the southern part of the country supply the markets during the winter and spring, and commercial districts in the northern part supply the markets in the summer and fall and produce the enlarged roots used for canning.

Radish

The enlarged roots of the many varieties and types vary greatly in color, shape, size, season of maturity, and texture of the flesh. The leaves are simple and are arranged in the form of a rosette on a short stem. Eventually the stem elongates and produces flowers, fruit, and seed.

Varietal Types Varieties are generally classified according to the time the roots require to attain maturity. Three groups exist: (1) spring, (2) summer, and (3) winter. In general, spring varieties grow quickly and their roots mature in a relatively short time (25 to 30 days); summer varieties grow less quickly and their roots mature in a relatively long time (45 to 50 days); winter varieties grow even more slowly and produce large roots which have a long storage life under favorable conditions.

Economic Importance and Principal Producing Districts Radishes are grown in most home gardens, particularly in early spring. The crop is important commercially in Japan and several other countries. In the United States, most market gardeners and greenhouse growers produce small quantities to supply local markets, and truck-crop growers in a few localities in the southern part of the country produce radishes for long-distance shipment.

Figure 21-7 Types and varieties of radish. Left: French Breakfast. Center: White Icicle. Right: Crimson Globe.

Turnip and Rutabaga

Although turnips and rutabagas are closely related botanically, they are quite distinct from each other. On the one hand, the turnip has green, hairy leaves, an indistinct crown, and a relatively small, fleshy root, which is free of secondary roots frequently arising from the sides of the relatively enlarged fleshy root. On the other hand the rutabaga has smooth, nonhairy leaves, a distinct crown, and secondary roots arising from the sides of the enlarged root.

Varietal Types Two groups of turnips exist: (1) white and (2) yellow. Of these, the white-fleshed is the more widely grown. Within this group, certain varieties are grown primarily for their tops which are used for greens, and other varieties are grown primarily for their fleshy roots. All rutabaga varieties have yellow flesh.

Economic Importance and Principal Producing Districts Turnips and rutabagas are popular and grown widely in home and market gardens throughout the world, particularly in the cool-temperate zones.

Horseradish

Horseradish is grown for its thick, fleshy white roots which have a pungent flavor. The fleshy roots are ground, preserved in vinegar, and used as a condi-

Figure 21-8 Left: rutabaga. Right: turnip. Note differences in the size of leaves, roots, and the lateral roots *(Courtesy, U.S. Department of Agriculture, Office of Information, Washington, D.C.)*

ment. The pungent compounds are soluble in water but are volatile; hence, ground horseradish root is kept in sealed containers. The plants form a rosette of long-petioled, narrow, dark green leaves.

The principal commercial district in the United States is located near St. Louis, Missouri. Other commercial areas exist in New York and New Jersey.

Sweet Potato

Plant Characteristics The root system consists of (1) an extensive absorbing system and (2) fleshy roots. The fleshy roots are small at first; later the base of most of them becomes thickened. A few become much thickened and form the edible roots. These enlarged roots have a typical root structure. The young root has an epidermis, a relatively thick cortex, a pericycle, endodermis, and radial bundles. As the root enlarges, a phellum with lenticels takes the place of the epidermis. Cambium arises between the phloem and secondary xylem and secondary phloem is produced in scattered strands.

The function of the phellum, as in the case of the potato, is to keep the all-important water within the fleshy root and to resist the attack of rot-producing organisms. Its formation is influenced by temperature and relative humidity. Investigations have shown that high temperatures, 80 to 85°F (26.7 to 29.4°C) combined with high humidity, 80 to 85 percent, greatly facilitate phellum formation. Since, at the harvest, the phellum is not fully developed and is easily bruised by handling, the primary purpose of curing is to heal the bruised surfaces and to thicken the skin.

The fleshy root is primarily a storage organ. It contains large quantities of starch which gradually change to sugar. According to the variety, color of the skin varies from creamy white to dark red, and color of the flesh varies from white to salmon pink. Thus, some varieties possess the ability to make carotene, whereas others do not.

The sweet potato produces a relatively large top which consists of a main axis and primary laterals. The stems vary in length, depending on the variety and the environment. For example, the so-called vineless sorts produce stems from 2 to 4 feet (0.6 to 1.2 meters) in length, and the so-called viny sorts produce stems from 6 to 20 feet (1.8 to 6.1 meters) in length. For the same variety soils with abundant essential elements, particularly nitrogen, combined with optimum moisture and favorable temperature, produce plants with longer stems than soils with moderately abundant nitrogen and low water.

The leaves are long-petioled, simple, cordate, and slightly or deeply lobed depending on the variety. The veins are prominent on the lower surface and usually show the same degree of pigmentation as the stems.

The flowers are complete, axillary, and resemble those of the morning glory. They occur singly or in clusters on stout peduncles. Color of the corolla varies from white to light purple. In general, there are five stamens clustered around a single superior pistil. The fruit is a round, hairy or nonhairy pod which

Figure 21-9 Types and varieties of sweet potato. Top: Porto Rico, moist-fleshed, long storage life. Bottom left: Goldmar, moist-fleshed, short storage life. Bottom right: Big Stem Jersey, dry-fleshed, long storage life. *(Courtesy, Lewis P. Watson, Raleigh, North Carolina, for top photo, L. E. Scott and J. C. Bouwkamp, University of Maryland, for bottom left photo, and U.S. Department of Agriculture, Office of Information, Washington, D.C., for bottom right photo.)*

contains from one to five seeds. The mature seed are black, angular, and have a hard coat. This coat greatly delays germination of the seedling. Recent tests have shown that soaking the seed in concentrated sulfuric acid followed by washing scarifies the seed coat and ensures prompt germination.

Varietal Types Varietal types are based on the characteristics of the cooked flesh: (1) varieties with firm, dry, mealy flesh when cooked and (2) varieties with soft, moist, sugary flesh when cooked. Big Stem Jersey, Orlis, and Nancy Hall are examples of the former, and Centennial, Goldrush and Porto Rico are examples of the latter.

Use of Growth Regulators Certain growth regulators have been investigated to determine their effect on flower induction, plant production, and sprout prevention. In the tests on flower induction, applications of 2,3,5 tri-iodobenzoic acid (TIBA), ammonium salt of 2,4 dichlorophenoxyacetic acid (2,4-D), and naphthaleneacetic acid (NAA) failed to induce flowering in Porto Rico and Orlis, but applications of 2,4-D at 500 ppm induced flowering in Porto Rico, Goldrush, and Yellow Jersey. In the tests on plant production, applications of 2,4-D markedly increased plant production of Unit 1 Porto Rico, applications of thiourea increased plant production in Kandee and Goldrush, and applications of ethylene chlorohydrin increased plant production of Texas Porto Rico and several other varieties. In the tests on the prevention of sprouts in storage, applications of maleic hydrazide (MH) had no visible effect on the foliage and neither increased nor decreased sprouting. Applications of the methyl ester of apha naphthaleneaceticacid (MENA) injured the foliage for a short period only, did not lower the yield nor the keeping ability of the roots, and reduced sprouting.

Economic Importance and Principal Producing Districts The sweet potato is a major crop in many countries in the warm-temperate, subtropical, and tropical zones. These countries include Nigeria, Taiwan, Africa, Japan, and Brazil. In the United States, commercial districts are the San Joaquin and Coachella of California, the Gilmer of Texas, the Sunset and Oak Ridge of Louisiana, the Vardaman of Mississippi, the Mobile Bay of Alabama, the Orangeburg and Horry of South Carolina, the Wilson of North Carolina, the Norfolk-Eastern Shore of Virginia, the Salisbury of Maryland, and the Swedesboro of New Jersey.

QUESTIONS

1 Name three common characteristics of the vegetable root crops.
2 Young carrots contain less carotene than old carrots. Explain.
3 The development of a large top during the early stages of carrot growth is necessary for the development of highly colored roots. Explain.

4 In Louisiana, two crops of carrots are grown each year. The fall crop matures in December and January, and the spring crop matures in May and June. Usually the roots of the fall crop are more highly colored than those of the spring crop. Explain.

5 In general, beets harvested in the summer have poorer color than those harvested in the fall. Explain fully.

6 Usually, the spring varieties of radishes maturing in hot, dry weather tend to become pithy. Explain.

7 Parsnip roots become sweet after exposure to subfreezing temperatures. Explain.

8 Distinguish between turnips and rutabagas.

9 Turnip greens are more widely adapted to growing in southern home gardens than is spinach. Give two reasons.

10 Distinguish between the moist-fleshed and the dry-fleshed types of sweet potatoes.

11 Highly colored roots are more desirable for human consumption than poorly colored roots. Explain.

12 Only roots from high-producing hills should be saved for "seed." Explain.

13 Soils containing abundant available nitrogen make for abundant vine growth and low yields of poorly colored roots. Explain.

14 Abundant rainfall during the later stages of growth is likely to crack sweet potatoes. Explain from the standpoint of income and outgo of water.

15 What is the prime purpose of curing? Give reasons for your answer.

16 Sweet potatoes cure best at temperatures varying from 80 to 85°F (27 to 30°C) with high humidity. Explain.

17 The sweet potato is a short-day—long-night plant. How would you induce flowering under long-day—short-night conditions?

Chapter 22

Vegetable Crops Grown for Their Fruit or Seed

Always do a little bit more than is expected of you, and you will never fail.

THE SOLANACEOUS CROPS

The Solanaceous crops are tomatoes, capsicum peppers, and eggplants. These plants belong to the nightshade family and have similar climate, soil, and cultural requirements.

Tomato

Plant Characteristics The young seedlings develop a taproot and a subordinate system of lateral branches. When plants are transplanted the taproot is destroyed, the laterals become thick and well developed, and adventitious roots arise from the stem located below the surface of the land. In adult plants, the lateral and adventitious roots extend horizontally from 3 to 5 feet (0.9 to 1.5 meters). Thus, the tomato develops an extensive root system.

The plant develops a main stem and a system of lateral branches. In all commercial varieties the main stem is erect for the first 1 to 2 feet (0.31 to 0.62

meters) of growth when it becomes decumbent. In some varieties the stems extend for a small number of nodes only—the so-called determinate sorts; in others they elongate throughout the growing season—the so-called indeterminate sorts. The leaves are alternate, compound, relatively large, well developed, with rather broad leaflets in some varieties, and rather long, narrow leaflets in others. They possess glandular hairs, which when disrupted liberate the odor and stain characteristic of the tomato plant.

The flowers are borne in clusters on the main axis and on lateral branches. The number of clusters varies from 4 to 100 or more depending on the type and variety. Individual flowers contain a green calyx, a sulfur yellow corolla, five or more stamens, and a single superior pistil. They are mostly self-pollinated. The ripe fruit is a comparatively large, juicy, and fleshy ovary. According to the variety it varies in size (4 to 12 ounces, or 113 to 340 grams), shape (oblate, globular, or flattened), color (yellow, pink, or red), cell number (5 to 25), and arrangement of cells (regular or irregular). The juice contains moderate quantities of soluble sugars, several organic acids and mineral salts, and relatively large quantities of vitamin C. The seed are imbedded in a jellylike mass of tissue containing large quantities of phosphorus. They are relatively small and are covered with a mass of fine hairs. Under favorable conditions the seed germinates in a short time, usually from 5 to 10 days.

Varietal Types As previously stated, the determinate varieties extend their stems for a small number of nodes only. As a result, plants of this type are low in stature, compact in growth, and develop their fruit within a relatively short period. This, in turn, makes the mechanical harvesting of the fruit feasible and explains why the determinate varieties are grown for canning. In sharp contrast, the indeterminate varieties extend their stems for a large number of nodes, in fact, as long as growing conditions are favorable. As a result, plants of this type are relatively high in stature, open or rangy in growth, and develop their fruit during a relatively long period. In fact, under a favorable environment the plants grow and develop fruit for a period of two years or more. This explains why the indeterminate varieties are grown in places where the production of fruit over a long period is essential, as in home gardens, market gardens and greenhouses. Both determinate and indeterminate types produce fruit which vary in size and shape. Note the various shapes shown in Figure 22-1.

Use of Growth Regulators In the United States, millions of tomato plants are grown each year for shipment to and transplanting within the canning districts of Indiana, Ohio, Maryland, and adjacent states. When the plants are grown in warm-moist soil supplied with liberal quantities of available nitrogen and under crowded conditions in the row, they develop long, drawn out, slender stems. Such plants cannot withstand the period of shipment and fail to recover from the check in growth incident to transplanting. Experimental tests with succinic acid 2,2-dimethylhydrazide (SADH, Alar) applied to seedlings in the field show that

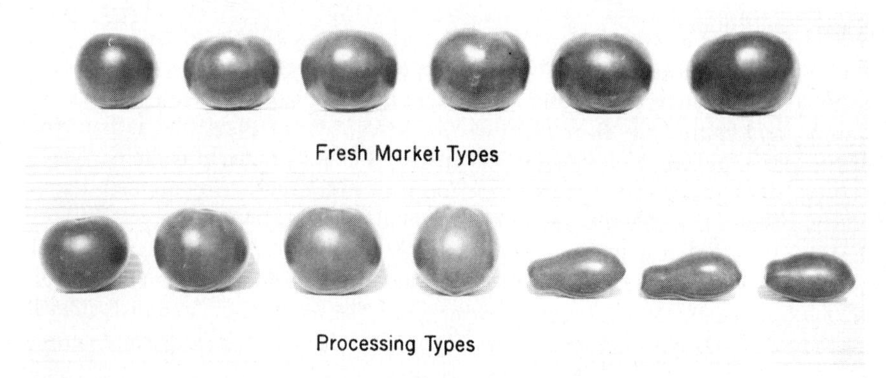

Figure 22-1 Fresh-market and processing types of tomato fruits. *(Courtesy, U.S. Department of Agriculture, Office of Information, Washington, D.C.)*

this regulator promotes the development of plants with short internodes and thick stems, and this, in turn, facilitates rapid recovery from transplanting.

The optimum temperature range for the growth of the pollen tube down the style is between 59 and 65°F (15 to 18°C). Frequently, open blossoms of the first cluster are subjected to temperatures slightly below this range. As a result, the pollen fails to grow down the style in time to release the sperm to unite with the egg, and the ovary fails to develop. For reasons not fully understood, applications of naphthaleneacetic acid (NAA), gibberellic acid (GA_3), and 2 chlorethylphosphonic acid (ethephon) have markedly increased the setting of fruit of the first cluster. In fact, applications of any one of these compounds is becoming a standard practice in home gardens, market gardens, and greenhouses throughout the world.

Economic Importance and Principal Producing Districts In general, tomatoes are grown intensively in the temperate and/or subtropical zones of Europe, Asia, Africa, South and Central America, Mexico, the Philippines, and Australia. In the United States, tomatoes are grown in most home gardens for family consumption, in most market-gardens and in many greenhouse establishments to supply the needs of local markets, in many winter-garden areas for long-distance shipment, and in special districts for canning. Principal market-garden areas are located near large cities such as New York, Philadelphia, Boston, Detroit, Chicago, and Los Angeles. Principal winter-garden areas are the east Florida near Miami, the west Florida, the Rio Grande Valley and the Jacksonville of Texas, the Imperial Valley and the Los Angeles of California, the Humboldt of west Tennessee, and the Swedesboro of New Jersey. Principal canning districts are located in southern Indiana, the Sacramento and the Santa Clara Valleys of California, the eastern shore of Maryland, southern New Jersey, eastern Virginia, and southeastern Missouri.

Capsicum Pepper

Plant Characteristics The root system is moderately extensive. The main stem is erect, woody at the base, and much-branched. The leaves are flat, shiny, simple, and entire. The flowers, fruit, and seed occur singly in the axils of the leaves. The flowers are both self- and cross-pollinated and have white or purple petals, five stamens, and a single superior pistil. The fruit is a moderately large, fleshy ovary, dark green when immature, and red or yellow when mature, according to the variety. The outer wall is fleshy and thick, and the inner walls bear placenta, which, in turn, bear seed. Both green-mature and mature fruits are high in carotene, the B vitamins, and ascorbic acid. The seed are flat and disk-shaped and require a fairly high temperature, 70 to 75°F (21 to 24°C) for prompt germination.

Varietal Types Three types of capsicum peppers are grown: (1) sweet, (2) pimiento, and (3) hot, or pungent. Sweet peppers are relatively large and are eaten fresh or cooked. Pimiento peppers are heart-shaped and are used in cheese and in seasoning. Hot peppers are moderately small and are used for canning and the making of paprika and chili sauce. Figure 22-2 shows three types of peppers.

Economic Importance and Principal Producing Districts Peppers are grown in many warm- and cool-temperate areas of the world: in South and Central America, Mexico, the United States, Europe, and Japan. In the United States, commercial districts exist in Florida, Texas, and New Jersey. Peppers for canning are raised in Georgia, Texas, and California; peppers for paprika are raised in Louisiana; and peppers for chili sauce are raised in California and New Mexico.

Eggplant

Plant Characteristics Eggplants develop a moderately extensive root system. As with capsicum peppers the main stem of mature plants is woody at the base and is much-branched. The leaves are simple, alternate, large, somewhat angled or lobed. The flowers occur singly or in clusters opposite the leaves. They are moderately large, perfect, violet or purple, and mostly self-pollinated. The fruit is a fleshy ovary varying in size (2 to 6 inches or 5 to 15 centimeters in diameter), shape (long, ovate, or pyriform), and color (purple, light purple, yellowish, striped, or white), according to the variety. Figure 22-3 shows fruit of Black Beauty, a variety adapted to practically all crop-producing areas of the United States.

Varietal Types Three distinct types are grown: (1) varieties which develop large, dark purple fruit; (2) varieties which develop small, light purple fruit; and (3) varieties which develop long, slender, dark purple fruit, 2 to 3 inches (5 to 6 centimeters) in diameter and 6 to 8 inches (15 to 20 centimeters) in length.

Figure 22-2 Varietal types of pepper. Top left: bell sweet. Top right: pimiento. Bottom: hot-pungent. *(Courtesy: U.S. Department of Agriculture, Office of Information, Washington, D.C., for top left and bottom photos, and W. H. Greenleaf, Auburn University, for top right photo.)*

QUESTIONS

1 Tomatoes growing on light sandy loams produce smaller yields than those growing on clay loams. Explain.
2 Important canning states are Indiana, New Jersey, Maryland, and California. Explain.
3 Why are tomato yields relatively low in the extreme South and the extreme North?
4 How do growth regulators condition seedling plants to recover rapidly from the check in growth due to transplanting?

Figure 22-3 Black Beauty eggplant. *(Courtesy, Associated Seed Growers, New Haven, Connecticut.)*

5 Pruning tomatoes reduces yield per plant, usually increases fruit size, decreases blossom end rot, limits the extension of the root system, and permits close planting. Explain fully.
6 What are the similarities and differences between the leaves of the tomato and the leaves of the capsicum pepper? The leaves of the eggplant?
7 The tomato is more popular than the eggplant or capsicum pepper. Give two reasons.

THE VINE CROPS

The vine crops are cantaloupes, cucumbers, pumpkins, squash, and watermelons. These crops belong to the cucumber family, and they have similar climate, soil, essential-element, and cultural requirements and are attacked by the same insects and diseases.

Plant Characteristics in Common

Root System The vine crops develop extensive and moderately deep to deep root systems, and as pointed out in Chapter 13, they deposit suberin in the walls of the region of absorption relatively early. Thus, cultivation when necessary should be shallow to avoid cutting the absorbing roots just below the surface, and if the plants are to be transplanted, they should be grown in containers and handled carefully to reduce injury of the root system to a minimum.

Stems and Leaves The stem system consists of a main axis and a series of primary and secondary laterals. In adult plants the branches are long and trailing.

Consequently, each plant requires a large area of land and planting distances are quite wide. To supply the plant with large quantities of water, the extensive root system is necessary. In addition, the many leaves present a large transpiring surface. Thus, under conditions of rapid transpiration the amount of water lost is large. The leaves are alternate, simple, long-petioled, and palmately veined. Tendrils are formed opposite the leaves. These tendrils twine around objects and help to anchor the vines to the surface of the land.

Flowers, Fruit, and Seed The vine crops are monoecious and andromonoecious. In monoecious crops, staminate and pistillate flowers occur on the same plant. In contrast, in andromonoecious crops, staminate and perfect flowers occur on the same plant. In all cases the flowers are axillary and develop a moderately large, yellow corolla. These crops are wholly insect-pollinated, the honeybee being the principal carrier. The student will recall the discussion on the effect of temperature and wet weather on the activity of honeybees in the orchard. In like manner, if bees are inactive in cucumber, cantaloupe, squash, pumpkin, and watermelon fields, yields are likely to be low.

The fruit is an enlarged fleshy structure consisting of ovary, the inner portion, and receptacle, the rind. The seed is relatively large, elliptical, with a hard coat. Under favorable conditions the seed germinates in from two to five days.

Cantaloupe—Muskmelon

In general, there are two types: (1) the netted and (2) the winter. Netted melons have a netted skin, shallow sutures and ribs, loose-textured flesh, and keep for a short time in storage. Winter melons, however, have a smooth or ridged skin, firm-textured flesh, and keep for a long time in storage. Of these types the netted is the more important commercially. Figure 22-4 shows fruits of each of these types.

Use of Growth Regulators The plants of certain varieties of muskmelon are entirely gynoecious, that is, their flowers contain functional pistils only. In order for these plants to produce fruits with viable seed, they are interplanted with a variety which produces functional stamens. Experiments have shown that foliar applications of 5-methyl-7-chloro-4-ethoxycarbonmethoxy-2,1,3-benzothiazole induce the development of perfect flowers and that hand pollination of these flowers results in the development of viable seed. Thus, investigators have discovered a means by which gynoecious lines can be maintained for the production of F_1 seed.

The dipping of fruits in 2-4 chlorophylthiotriethylamide hydroxide (CPTA) at the one-half or full slip stage of maturity at 500 or 1,000 ppm induced the development of more pink in the flesh. Injecting the fruits with CPTA produced

Figure 22-4 Types and varieties of muskmelon—cantaloupe. Top: closed cavity, netted, var. Planters Jumbo, Hale. Bottom left: open cavity, netted, var. Netted Gem. Bottom center: smooth var. Honey Dew, fall. Bottom right: ribbed casaba, var. Golden Beauty, fall or winter. *(Courtesy, Otis Tilley Seed Co., Salisbury, Maryland for top photo, and Department of Horticulture, Virginia Polytechnic Institute and State University, for bottom photos.)*

similar results. Researchers believe that the greater intensity of pink is associated with greater lycopene formation.

Economic Importance and Principal Producing Districts The cantaloupe is the most valuable vegetable crop belonging to the cucumber family. It is grown commercially to a limited extent in several countries of Europe, Asia, Africa, South and Central America. The United States leads in commercial production. Important commercial districts are the early-west coast of Mexico, the Imperial Valley of California, and the Rio Grande Valley of Texas.

Cucumber

The stems are long and trailing, the leaves are simple, alternate, and angular. The staminate flowers occur in clusters, and the pistillate flowers occur singly or occasionally in groups of two or more. The fruits are elongated or cylindrical, and vary in size (long, moderately long, and short), color of the rind (light to dark green), and color of the spines (white or black).

In the United States, two varietal types and two separate corresponding industries exist: (1) the slicing and (2) the pickling. Varieties for slicing produce long, cylindrical, dark green fruits with crisp flesh. They are grown in many home gardens throughout the country; in greenhouses until they are ready for market, or in greenhouses or hotbeds to the transplanting stage and then in open fields, in the northern part of the country; and entirely in open fields in the southern part. Principal commercial districts in the southern part of the country are central Florida, southern Texas, southeastern South Carolina, eastern North Carolina, and the Norfolk section of Virginia.

Cucumbers for pickling are grown entirely in fields. The fruits are usually graded into two or three sizes and delivered to the salting station for preservation in brine. Principal producing districts are located in the southern peninsula of Michigan, in Wisconsin, Indiana, Ohio, eastern North Carolina, and southern Mississippi. Districts producing lesser quantities are located in Virginia, New York, Illinois, and California. Note the differences in size and color between fruits of the two types presented in Figure 22-5.

Pumpkin and Squash

Pumpkins belong to two botanical species: *Cucurbita pepo* and *Cucurbita moschata*. Varieties of *C. pepo* are divided into two groups: (1) those which develop short, erect stems and mature their fruits in a relatively short time, and (2) those which develop long, 6- to 20-foot (1.8- to 6.1-meter) trailing stems and mature their fruits in a relatively long time. Both groups have prickly, harsh-textured, deeply notched leaves and a five-sided longitudinally grooved fruit stem. Varieties of *C. moschata* develop long, trailing stems, soft-textured leaves, and a five-sided indistinctly grooved fruit stem. In addition, these varieties develop

Figure 22-5 Types of cucumber. Top: slicing. Bottom: pickling. Note the differences in size and color. *(Courtesy, Associated Seed Growers, New Haven, Connecticut.)*

white spots at the junction of the veins of the leaves. Varieties of both types are usually monoecious. The staminate flowers have long, slender stalks, and the pistillate flowers have short, thick stalks. Both are axillary and occur singly. The fruiting habit is like that of the cantaloupe. As with the cantaloupe, there are periods of fruit setting, which are alternated with periods of female flower abortion. The fruits are fleshy and vary greatly in size, shape, and color. Carbohydrates are stored in comparatively large quantities. The yellow-fleshed varieties contain carotene.

Varietal Types Pumpkins are divided into two types: (1) bush and (2) trailing. The bush-type fruits are usually harvested when immature and are called *squash*. The trailing-type fruits are harvested when fully mature. Mature fruits are high in sugars and keep well in storage.

Squash belongs to one species only, *Cucurbita maxima*. The plants have long, trailing stems and smooth, soft-textured leaves. The fruit stalk (petiole) is round, soft, and spongy. The fruits are large and fleshy, and contain large quantities of carbohydrates.

Economic Importance and Principal Producing Districts Pumpkins and squash are grown in many countries within the temperate and subtropical zones of Europe, Asia, Africa, North and South America, and Australia. In the United States, they are grown in home gardens throughout the country, in market gardens for the fresh market, and in truck-crop districts of California for canning and drying. Note the shape and size of the fruits of *C. pepo* and *C. moschata* in Figure 22-6.

Watermelon

The stems are angular in cross section, and the leaves are divided into three or four lobes. The flowers occur singly in the axils of the leaves and usually open at sunrise and close on the afternoon of the same day. According to the variety, the fruits vary greatly in size (5 to 40 pounds, or 2.3 to 18.2 kilograms), shape (round,

Figure 22-6 Top: summer squash, Yellow Straight Neck (left) and Zucchini (right). Bottom: fall and winter squash, Blue Hubbard, Buttercup, Quality, Butternut, Green Delicious, Royal Acorn, Golden Hubbard, Table Queen (left to right). *(Courtesy, Joseph Harris Seed Company, Inc., Moreton Farm, Rochester, New York.)*

oval, oblong, or cylindrical), color or rind (light grey, blackish green, mottled, or striped), color of flesh (dark red, light red, yellow), and sweetness of the flesh.

Use of Growth Regulators For many years, vegetable crop scientists have endeavored to develop a seedless watermelon. They have attained this objective by using a drug called *colchicine*. Applications of a dilute solution (0.2 percent) to the buds of diploid varieties ($2n = 22$) has induced the formation of tetraploids ($2n = 44$), and crossing of the tetraploids with the diploid has resulted in the formation of sterile (seedless) triploids ($2n = 33$). Note the various types of fruits presented in Figure 22-7.

Economic Importance and Principal Producing Districts Watermelons, like other cucurbits, are popular and widely grown in most of the warm-temperate and subtropical countries of the world. They are grown in Africa, their native home, and in many other countries which have warm growing seasons long enough for the fruits to mature (from 110 to 130 days).

The United States leads in commercial importance. The principal commercial districts are located in the Southeast and Southwest—southern Georgia,

Figure 22-7 Types of watermelon fruits. Top left: large, round, dark green, var. Florida Giant. Top right: large, oblong, dark skin, var. Tom Watson. Bottom left: medium to large, oblong, striped light and dark green, var. Garrison. Bottom right: small, rounded, light green, ice-box type, var. Sugar Baby. *(Courtesy, Ferry Morse Seed Co., Inc., Mountain View, California.)*

southeast Texas, northern Florida, southeastern South Carolina, southern Alabama, and eastern North Carolina.

QUESTIONS

1. Distinguish between netted melons (cantaloupes) and winter melons.
2. If the weather is cold and wet during the blossoming season, very little pollination is likely to take place. Explain.
3. Given two cucumber plants. On plant A, two fruits are allowed to remain on the vine until they are ripe; on plant B, all fruits are picked while immature. Which plant produces the greater number of fruits? Give reasons.
4. In general, abundant water and moderate supplies of fertilizers are necessary throughout the fruiting period of cucumber. Explain.
5. Watermelon fruits are not harvested until they are ripe. Explain.
6. Differentiate between pumpkins and squashes.
7. How are seedless watermelons produced?

Sweet Corn

Plant Characteristics The root system consists of two parts: (1) the absorbing and (2) the buttress. The absorbing roots are adventitious and very extensive. Investigations have shown that the roots of mature plants have a lateral spread of 4 to 5 feet (1.2 to 1.5 meters) and a vertical spread of 6 to 8 feet (1.9 to 2.4 meters) depending on soil conditions. The buttress roots arise from the first or second nodes above the soil in the form of a whorl. They proceed outward and downward and penetrate the soil, thus providing for additional anchorage and wind resistance and additional absorbing roots.

The stem system consists of primary stems and secondary stems. The secondary stems are frequently called *suckers*. Under field conditions the primary stem bears the staminate and pistillate inflorescences, and the secondary stems may or may not bear flowers. An individual leaf consists of three distinct parts: (1) sheath, (2) ligule, and (3) blade. The sheath is the basal portion of the leaf which is wrapped around the stem; the ligule is attached to the top of the sheath and fits tightly around the stem; and the blade is comparatively long, parallel-veined, and pointed.

Sweet corn is monoecious. The staminate inflorescence contains a large number of small flowers, terminates the primary stem, and is called the *tassel*. Each staminate flower has three stamens and a rudimentary pistil. The pistillate inflorescence also contains a large number of flowers, terminates a short lateral, and when mature is called the *ear*. The internodes of each lateral are very short. Thus, the sheaths of the leaves overlap and form the husk of the ear. Each pistillate flower has a single pistil and rudimentary stamens. The ovary is the immature kernel, and the style is long and branched near the tip. These styles are

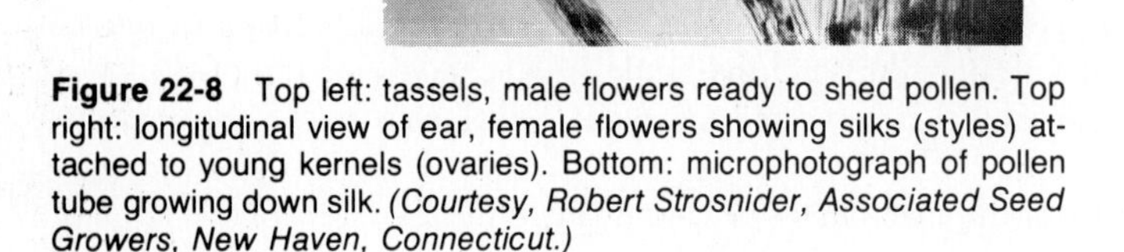

Figure 22-8 Top left: tassels, male flowers ready to shed pollen. Top right: longitudinal view of ear, female flowers showing silks (styles) attached to young kernels (ovaries). Bottom: microphotograph of pollen tube growing down silk. *(Courtesy, Robert Strosnider, Associated Seed Growers, New Haven, Connecticut.)*

called the *silk* and protrude through the top of the husk to catch pollen grains which are carried by gravity or wind. Corn is most readily cross-pollinated.

Effect of Pollen on the Quality of the Kernels Investigations have shown that the pollen of any one variety may or may not affect the quality of the kernels of another variety. For example, if the pollen of a sweet corn variety grows down the style of an adjacent sweet corn variety, no noticeable changes in the quality of the kernel takes place. However, if the pollen of a field corn variety grows down the style of a sweet corn variety, a marked reduction in the quality takes place. For this reason, sweet corn should not be planted adjacent to field corn which blooms at the same time. Also, when varieties are being tested, those varieties which bloom at the same time should be isolated or planted at least 100 feet (30.5 meters) from any other variety.

Varietal Types Three general types are grown: (1) normal sweet corn, (2) extra sweet—either the sugary or the shrunken mutant which forms no starch, and (3) certain varieties and/or strains of semisweet field corn. Varieties within each type vary in size of ears, size, color, and depth of kernel, sugar content, toughness of pericarp (outer skin of kernel), and length of time the kernels remain in high-quality condition. Many excellent varieties are now available.

Economic Importance and Principal Producing Districts Sweet corn is native to and popular in northern South America, Central America, and Mexico. In the United States, it is grown in practically all home gardens, in market-garden areas for local markets, in commercial areas for long-distance shipment and for canning and freezing. The principal states supplying long-distance markets are Tennessee, Alabama, Florida, Georgia, and Mississippi. The principal states which grow large quantities for canning and freezing are Minnesota, Indiana, Ohio, and Maryland.

Okra

Plant Characteristics The root system is extensive, and the stem consists of an erect axis and primary laterals which become woody with age. The leaves

Figure 22-9 A typical plant of okra, var. Clemson Spineless, showing leaves, flower, bud, and pods. *(Courtesy, Lewis P. Watson, Raleigh, North Carolina.)*

are ovate and lobed. The flowers are axillary, large, showy, and contain both stamens and pistils. The fruits are elongated, relatively large pods. They vary in color, degree of ribbing, pubescence, and presence of spines, according to the variety, and are consumed in the immature state. When mature, the pods are hard and woody. The seed are relatively large and dark in color.

Economic Importance and Principal Producing Districts Okra is grown in the southeastern United States to a greater extent than in other regions of the country. In this region, because of its adaptation to high temperatures and high light intensities, it is a standard home-garden vegetable. Limited quantities are grown for canning. Statistics on the extent of its culture in other countries are not readily available. Note the typical plant of Clemson Spineless presented in Figure 22-9.

QUESTIONS

1 From the standpoint of human nutrition, the yellow varieties of sweet corn are superior to the white. Explain.
2 On sandy loams in New York, tests have shown that removal of suckers (a type of pruning) has decreased sweet corn yields. Explain.
3 In the home garden, sweet corn should be planted in blocks—not in a single row. Explain.
4 A home gardener wants to know how to harvest sweet corn so that it will remain sweet. Outline two methods.
5 Sweet corn should never be planted adjacent to field corn. Explain.
6 Harvesting of the immature pods of okra should be thorough and systematic. Explain.

THE VEGETABLE LEGUMES

The vegetable crop legumes are snap bean, lima bean, protepea, certain varieties of soybean, and garden pea. All are warm-season crops with the exception of the garden pea, which is distinctly a cool-season crop.

Plant Characteristics in Common

Root System The root system consists of a taproot and extensively branched laterals. These laterals may extend horizontally from 1 to 5 feet (.3 to 1.5 meters). Thus, the root system is extensive and widespread. Like other members of the legume family, vegetable legumes have roots which support the growth and development of nitrogen-fixing bacteria called *Rhizobium*. These bacteria possess the ability to use nitrogen from the air. This nitrogen is combined with sugars by the bacteria in the formation of amino acids which are used by the bacteria and the bean plant. Unless the soil contains abundant quantities of an adapted strain of these bacteria, inoculation with pure cultures is generally advisable. Tests in Washington and Wisconsin, where large quantities of garden peas are grown, have shown that good strains of bacteria have increased yields growing on land which had previously been planted to peas. Growers often test the effect of inoculation by comparing the yield of treated and untreated strips planted side by side in the same field.

Flowers, Fruit, and Seed The inflorescence is a raceme in the case of bean and protepea and consists of single or two flowers in the case of garden pea. Individual flowers are moderately large and showy and possess the characteristics of the legume family. The large upper petal is called the *standard*, the two lateral petals are called the *wings*, and the two lower petals are called the *keel*. The keel encloses 10 stamens and a single pistil. Both self- and cross-pollination take place. Honeybees and bumblebees possess the ability to enter the flowers and induce cross-pollination. The fruit is a one-celled, elongated pod, called a *legume*. The pod varies in shape (flat or round, straight or curved); length, 2 to 8 inches (5 to 20 centimeters) or more; color (yellow, yellowish-green, green, and dark green); and number of seeds (3 to 10 or more), depending on the kind of crop and variety. The seeds are relatively large, consist mostly of two well-developed seed leaves—the cotyledons—and are highly nutritious since they contain large quantities of starch, proteins, and relatively large quantities of thiamin, riboflavin, calcium, and iron.

Snap Bean

Snap beans are of two types: (1) the dwarf and (2) the pole. Dwarf varieties produce short, erect, much-branched, determinate stems; and pole varieties develop long, twined, rarely branched indeterminate stems. The leaves of both

Figure 22-10 Bush type snap beans at harvest stage. Left: round podded. Right: flat podded. *(Courtesy, Joseph Harris Seed Co., Inc., Moreton Farm, Rochester, New York, for photo at left, and Ferry Morse Seed Co., Mountain View, California, for photo at right.)*

Figure 22-11 Bush type lima beans. Left: small-seeded, var. Henderson Bush. Right: large-seeded, var. Fordhook 242. *(Courtesy, Ferry Morse Seed Co., Mountain View, California.)*

types consist of three leaflets. These leaflets are either broad and ovate, or long and narrow, and vary from light green to dark green, according to the variety. The pods vary in shape (flat, oval, or round in cross section and straight or curved in longitudinal section) and color (yellow, light green, or dark green), according to the variety.

In the United States, important producing districts for long-distance shipment are located in central Florida—March to June (spring crop) and August to November (fall crop), near Charleston, South Carolina—March to June (spring crop) and September to November (fall crop), near Norfolk, Virginia (April to June, in central Mississippi (May to June), and in southern Louisiana (April to June). Important canning areas exist in New York, Maryland, and California. In other temperate countries, snap beans are largely grown in home and market gardens.

Lima Bean

As with snap beans, there are dwarf and pole types. However, unlike snap beans, there are two groups within each type: (1) small-seeded and (2) large-seeded. Plants of the small-seeded type have erect stems, glabrous, nonhairy leaves, and small, numerous pods; whereas plants of the large-seeded types have large, thick leaves, and large, relatively few pods.

Lima beans are grown in home gardens, in market gardens for local markets, in certain trucking regions in the Southeast for long-distance shipment, in certain areas of the Middle West and the Northeast for canning and freezing, and on the coast of California for dehydration—in this region, both the small-seeded and the large-seeded types are raised. About 400,000 hundred-pound bags of the small-seeded type and 800,000 hundred-pound bags of the large-seeded type are produced annually. In fact, the United States leads all other countries in commercial production.

Protepea

Protepeas are beans because they possess the flower and trifoliate leaves which are characteristic of beans. In general, the plants have short, much-branched, determinate stems. A unique feature is their ability to develop flowers and fruit, the pods, just above the upper level of the foliage. Note the relative position of the pods shown in Figure 22-12.

Protepeas are widely grown in home and market gardens in the southeastern United States for consumption as a fresh vegetable and in commercial districts of Texas and California for canning and drying. In other countries, production is of minor importance.

Edible Soybean

Field varieties of soybeans are grown for feed, for oil, and for the manufacture of plastics, whereas horticulture varieties are grown for human consumption. The stems are erect and become woody with age. The three leaflets are hairy and broad. Individual flowers are relatively small and white or purple, and the pods are small and contain from two to three seeds. Although soybeans are grown

Figure 22-12 Protepea, dwarf type. Note the long slender pods above the foliage. *(Courtesy, U.S. Department of Agriculture, Office of Information, Washington, D.C.)*

extensively in China for human consumption, at present the crop is not widely grown for this purpose in the United States. Recent investigations have shown marked varietal differences in yield, flavor, and palatability of the cooked seed.

Garden Pea[1]

Garden peas differ from beans in that the stems are hollow and the leaves are pinnately compound, with one, two, or three pairs of leaflets, a branched terminal tendril, and large stipules. The flowers are borne singly or in pairs on long stalks; the pods are nonconstricted; and the seed is round and either smooth or wrinkled when dry. Note the types of peas shown in Figure 22-13.

Garden peas are grown in home and market gardens of countries which have a uniformly cool climate in order to obtain the maturation of at least one crop of pods. The United States leads in commercial production. Producing districts for long-distance shipment include the winter-garden areas in California, Arizona,

[1]In the southern United States, garden peas are usually referred to as *English peas*.

Figure 22-13 Varietal types of peas. Top left: smooth-seeded, dwarf, var. Frosty, early. Bottom left: wrinkle-seeded, dwarf, var. Progress. Right: tall, wrinkle-seeded, var. Alderman. *(Courtesy, Joseph Harris Seed Co., Inc., Moreton Farm, Rochester, New York.)*

the coastal plain section of South and North Carolina, the eastern shore of Virginia, and Long Island, New York. Producing districts for canning and freezing are located in Wisconsin, Minnesota, New York, Washington, Montana, Oregon, Maryland, Illinois and Michigan.

QUESTIONS

1. Botanically speaking, what is the fruit of the bean and garden pea?
2. In general, beans and peas require smaller applications of nitrogen than most other vegetable crops. Explain.
3. State the two types of snap beans.
4. In the United States, two crops of dwarf snap beans are raised annually in the South; only one crop is raised in the North. Explain.
5. Beans should be planted shallowly and in loose, friable soil. Explain.
6. Usually beans are extremely vegetative and nonfruitful when grown in highly fertile soil. Explain.
7. In general, the light green-leaved varieties produce lesser yields than the dark green sorts. Explain.
8. Pole varieties of snap beans are grown more extensively in the southern regions than in the northern regions. Explain.
9. In general, the commercial fertilizer bill for growing snap beans is less than for many other vegetable crops. Explain.
10. What types of beans are grown in your community? Are they grown for long distance market, local market, or the cannery? Give reasons.
11. Name the two types of lima beans.
12. During hot, dry weather the Henderson Bush lima develops a deeper root system and produces higher yields than the Fordhook. What is the relation of a deep root system to the yielding ability of the crop under conditions of high transpiration?
13. Gardeners use brush for tall peas and stakes for pole beans. Explain.
14. The climate of England provides an almost ideal temperature for the cultivation of garden peas. Explain.
15. In the continental United States, peas for early market are raised in the extreme South; whereas peas for canning are raised in Wisconsin and on the eastern shore of Maryland. Explain fully.

CULTIVATED MUSHROOMS

Cultivated mushrooms are varieties of the common meadow mushroom (*Agaricus compestris* L.) and are saprophytic fungi; that is, they live on dead organic matter and contain no chlorophyll. The plant body may be divided into two distinct parts: (1) the vegetative portion, analogous to the roots, stems, and leaves of green plants, and (2) the reproductive part, analogous to the flowers, fruit, and seed. The vegetative part consists of fine, white threads, each about 0.001 to 0.002 inch (0.025 to 0.050 millimeter) in thickness, collectively called *mycelia*. The reproductive part consists of the edible portion and has three distinct parts: (1) the convex cap, or pileus, (2) the stalk, or stipe, and (3) the annulus. Beneath the cap are large numbers of radiating plates or spore-bearing tissue, called *gills*. The stalk supports the cap, and the annulus protects the gills when the fruit is small. As the cap grows, the annulus is stretched and finally ruptures, giving the appearance of a collarlike structure around the stipe. The function of the mycelia is to absorb water and food nutrients from the compost in which they grow. The function of the fruiting body is to produce spores which are microscopic in size but are analogous to the seed of higher plants.

Mushrooms are grown within the cool- and warm-temperate zones of many countries. In the United States, the center of commercial production is in the vicinity of Lancaster, Pennsylvania.

HERBS

Herbs are flavoring agents and, like spices, are used in cooking to season, enrich, or otherwise alter the flavor and odor of certain foods to make them more pleasing to taste. Different parts of the plants, such as the leaves, fragrant seeds, fruits, buds, bark, and roots, have been used for this purpose since ancient times. Most of the spices, black pepper, nutmeg, cinnamon, cloves, and allspice, for example, are derived from tropical plants. Savory herbs are aromatic plants, the various parts of which possess pleasing odors and tastes. These plants grow in different parts of the world and have long been considered essential in food preparation both in the home and in public eating places in practically all countries.

Herbs comprise a miscellaneous group. Some are annuals, some are biennials, and others are perennials. Some are grown for their leaves, some for their flower parts, and others for their seed. In general, a few plants of any one kind will supply the needs of an individual family.

Herbs grown for their leaves
- Perennials—balm, catnip, lavender, peppermint, spearmint, rosemary, sage, thyme, and chive
- Biennials—parsley and celery
- Annuals—summer savory, sweet marjoram, and sweet basil

Herb grown for its flower heads
- Annual—dill

Herbs grown for their seeds
- Biennial—caraway
- Annuals—anise, coriander, and sesame
- Annual or perennial—fennel

Herbs grown for their fruits or pods
- Annual—pungent peppers, hot (cayenne, tobasco, and chili) and mild (paprika, pimiento)

QUESTIONS

1 In what respects are mushrooms similar to higher plants? In what respects are they different?
2 The mushroom is a fungus and a saprophyte. Explain.
3 Why are herbs popular? Give four reasons.
4 In the home garden, would you place the annual and perennial herbs in the same location? Give reasons for your answer.

SELECTED REFERENCES FOR FURTHER STUDY

Edmond, J. B., and G. R. Ammerman. 1971. *Sweet potatoes: production, processing, and marketing*. Westport, Conn.: Avi. A thorough, well-organized, clearly presented monograph which gives experimental evidence which supports the current principles and practices of sweet potato production, processing, and marketing.

Jones, H. A., and K. L. Mann. 1963. *Onions and their allies: botany, cultivation, and utilization*. New York: Interscience. A monograph on cultivated alliums grown throughout the world.

Knott, J. E. 1971. *Vegetable production in southeast Asia*. Laguna: University of Philippines. A discussion of the characteristics of the vegetable crops grown in southeast Asia, particularly the Philippines with its aim being to acquaint the student with the principles and practices which underlie vegetable production in the tropics.

Thompson, H. C., and W. C. Kelly. 1957. *Vegetable crops*. 5th ed. New York: McGraw-Hill. An excellent text and reference on vegetable crops grown in the temperate and subtropical zones.

Chapter 23

Floriculture

Habits and customs differ, but all people have the love of flowers in common.

Chinese proverb

COMMERCIAL FLORICULTURE

Most people like flowers. In general, flowers are used to express human sentiments and to beautify the environment. Since in this age of specialization many people have neither the time nor the facilities nor the "know-how" required for raising flowers, they buy flowers. Most flowers are sold for use at weddings, funerals, social functions, and as gifts.

The initiation and development of commercial production coincided with the growth of cities and the social activities of the people. According to Laurie and Kiplinger, the growing of flowers for sale in the United States began in the vicinity of Philadelphia during the early part of the nineteenth century.[1] At first,

[1]According to the first official census, taken in 1790, the three largest cities ranked as follows: Philadelphia with 42,444, Boston with 33,131, and New York with 18,038.

only outdoor gardens and fields were used. Later, as the demand for "out-of-season" flowers increased, greenhouses were erected. Still later, as happened in the growing fruit and vegetable crop industries, the services of the middleman became necessary. Thus, at present, there are wholesale florists who are primarily production specialists or commission merchants, retail florists who are primarily salesmen, and retail growers who primarily grow and sell their own products.

Principal factors concerned in the growth of commercial floriculture are (1) the formation of the Society of American Florists, (2) the development of the Florists' Transworld Delivery (FTD), and (3) the work of the Land Grant College System and the U.S. Department of Agriculture. The Society of American Florists initiated the unique and effective advertising slogan "Say It with Flowers." This slogan appeals to the public; it is catchy and all inclusive. The Florists' Transworld Delivery Association provides facilities for the buying of flowers in any part of the United States and in many other countries of the world. For example, suppose a student wants to buy a bouquet of roses for a friend, and he wants to have the bouquet presented to his friend as she starts from home to the campus for a visit. He would go to, or phone, the local florist; the local florist would transmit the order, usually by wire, to a florist in the friend's hometown who would make up and deliver the bouquet to her. The service rendered by FTD ensures quick and prompt delivery of fresh, high-quality products. As such, it has been of great benefit both to the producer and to the consumer. The work of the Land Grant College and University System and the U.S. Department of Agriculture involves training of students in floriculture, solving problems of the many ornamental industries, and disseminating research results to producers and consumers. The training of students in the solution of problems by the scientific method have assisted greatly in placing the many ornamental crop industries on a comparatively firm foundation.

There are four cultural facets of the floricultural industry: (1) the production of cut flowers (see Figures 23-1 and 23-3), (2) the production of potted plants (see Figure 23-2), (3) the growing and forcing of flowering bulbs and corms in Figure 23-6, and (4) the production of bedding plants (see Figure 23-8). Each industry is highly specialized with definite requirements, facilities, and technical knowledge.

PRODUCTION OF CUT FLOWERS

Cut flowers are produced in greenhouses, in shade houses, and outdoors. Greenhouses are used for both warm- and cool-season crops. Carnation, chrysanthemum, orchid, rose, and snapdragon are the major greenhouse crops. The principal environmental factors—water, temperature, light, and essential elements—are either entirely or partially controlled. Water is supplied by specialized irrigation systems and devices. Night temperature is either thermostatically or manually controlled. Light intensity during the late spring and summer is reduced by the use of shading materials and increased in the late fall and

Figure 23-1 Cut-flower production of single stem, standard chrysanthemum. *(Courtesy, Joe W. Love, North Carolina State University.)*

winter by the removal of these shading materials. The length of the light period is reduced by the use of black cloth or increased by the use of electric lights. Essential elements are provided as solutions of moderately refined salts and in various slow-release forms. These slow-release forms include slowly soluble materials, plastic incorporated fertilizers, various urea aldehydes, and fritted salts.

Figure 23-2 Chrysanthemum pot plants grown and flowered in an arch-design, fiber glass greenhouse, utilizing automatic black shade cloth treatment to provide short light and long dark periods. *(Courtesy, Joe W. Love, North Carolina State University.)*

Cut flowers are grown in either ground beds or in raised 4-foot (1.2-meter) wide benches. (See Fig. 23-3). The benches may be either watertight or non-watertight. Watertight benches require a V-shaped bottom and a layer of tile or gravel at the bottom of the V. Crops are irrigated by maintaining a constant level of water in the bottom of the bench, by running water through tile lines, or by the use of overhead pipes. Nonwatertight benches are either V-shaped or flat and require a layer of sand or gravel at the bottom. With this type of bench, the crops are watered by means of overhead irrigation systems. Usually water is applied through various automatic systems rather than by hand. Asbestos rock, concrete, concrete slabs, tile, or wood (cypress and redwood) can be used to construct benches. Asbestos rock is the most satisfactory because of its strength and durability and its resistance to heat, moisture, oxidation, bending, and buckling.

Cloth houses are used whenever outdoor temperatures are favorable for the growth of crops. In general, their use is limited to the frost-free growing period, since, unlike greenhouses, these structures provide no protection against sub-freezing temperatures. As discussed in Chapter 12, their use is particularly advantageous during the summer and fall of regions characterized by high temperatures and high light intensity. The covers slightly lower the temperature of the air, but they markedly lower the light intensity. This, in turn, markedly lowers the temperature of the leaves, which lowers the rate of transpiration and allows

Figure 23-3 Production of carnations in ground beds and raised beds. Note wire support and frame. *(Courtesy, Joe W. Love, North Carolina State University.)*

the rate of absorption to keep up with it. As a result, the guard cells remain turgid, and the stomates remain open. Thus, with other factors favorable high rates of photosynthesis take place throughout the entire light period. At present the principal crops grown within these structures are chrysanthemums and asters. Note the greenhouse roses growing in the cloth house in Figure 12-5.

Outdoor culture requires the utilization of temperatures which are favorable for the growth of the crop in question. In other words, the crops are grown at temperatures within their optimum temperature range. The principal crops are chrysanthemum, aster, and gladiolus. Chrysanthemums and asters are grown principally in the vicinity of Los Angeles, California, and gladiolus, in many parts of the country. Producing districts exist in central Florida, southern Alabama, the coastal plain section of North Carolina and South Carolina, northern Indiana, northern Ohio, and southern Michigan. In this way a succession of crops is produced, with Florida, the earliest producing district, supplying the markets from October to May; and Michigan, the latest producing district, supplying the markets from August to September. Thus, the flower spikes are made available in many markets throughout the entire year.

Use of Growth Regulators

Growth regulators are used on a limited scale in the production of cut flowers. Foliar applications of B-Nine (succinic acid 2,2-dimethylhydrazide) are used to restrict the growth of the length of the stem of standard chrysanthemums.

Keeping Quality

The fundamental processes concerned are (1) water absorption and transpiration and (2) respiration. With water absorption and transpiration the principal plant factors are (1) the relative area of absorption and (2) the water-holding capacity of the tissues. The principal environmental factors are (1) temperature, (2) relative humidity, and (3) wind velocity. Since cut flowers absorb water only through the stem, the area of absorption is exceedingly small compared with the area of transpiration. Thus, the utilization of environmental factors which reduce the rate of transpiration will accordingly reduce water deficits and prolong the life of the flowers. With respiration the principal plant factor concerned is the amount of sugars available, and the principal environmental factors are (1) temperature and (2) the use of certain chemicals. Since sugars are used in respiration, the greater the actual and potential sugar supply at the time of cutting and the lower the rate of respiration after cutting, the longer will be the life of the flowers. Thus, high sugar content in the tissues should be promoted and low temperatures utilized whenever feasible. In regard to the use of chemical compounds, certain substances have been found that substantially lower the respiration rate and greatly retard the activity of bacteria and fungi that rot the stems. These substances are placed in the water in which the stems are emerged. Two preparations are (1) a solution consisting of hydrazine sulfate, manganese sulfate, and sugar and (2) a solution consisting of potassium, aluminum sulfate, sodium hypochlorite, ferric oxide, and sugar. Commerical preparations are Floralife and Bloomlife. Retail florists frequently place a tablet of these materials in each pack of flowers sold.

Water uptake also is reduced by the secretion of exudate in the vascular channels. This is overcome in the case of poinsettias by placing the stem base in boiling water to stop the flow of this exudate. Water uptake is also increased with some plants such as stocks by crushing the stem end to expose more absorptive surface.

Prepackaging

Prepackaging consists of packing an average-sized consumer unit in an appropriate package. The processes concerned are (1) respiration, (2) transpiration, and (3) rate of cell division. Naturally, for long storage life and long keeping quality, the flowers should have a low rate of respiration, a low rate of transpiration, and a low rate of cell division. How does the package maintain a low rate of these processes? To maintain a low rate of these processes the package should fulfill the following requirements: it should have a small volume, and it should be nonwater absorbent, gasproof, and sufficiently strong to withstand handling. The relatively small volume, combined with "misting" of the flowers and the nonwater-absorbent quality of the material, maintains a high relative humidity. Thus, a low rate of transpiration is maintained. The relatively small volume, combined with the gasproof quality of the cellophane wrap, maintains a certain concentration of carbon dioxide—usually from 5 to 15 percent. Thus, a low rate of respiration is maintained. The fact that the stems are not placed in water permits a slight water deficit within the tissues. Thus, a low rate of cell division is maintained.

PRODUCTION OF POTTED PLANTS

The growing of plants in pots or similar containers differs from the growing of plants in the greenhouse bed or bench, the garden, field, or orchard. The volume of soil in the container is exceedingly small; the root system is greatly restricted; the natural essential-element supply is limited; and the necessity for frequent watering is conducive to the leaching of nitrates and possibly other essential ions. Thus, soils for growing plants in containers are reinforced with heavy applications of highly decomposed organic matter or a mixture of both of these materials. These finely divided forms of organic matter increase the capacity of the mixture to hold available water and exchangeable essential cations and promote drainage and aeration.

The root medium for pot plants is usually amended with dolomitic limestone to obtain the optimum pH level, and with superphosphate. The lime provides calcium and magnesium, and the superphosphate provides phosphorus, calcium, and sulfur. Although it is possible to incorporate sufficient quantities of these two essential materials at the time of soil preparation to last a year, nitrogen and potassium are applied according to a routine fertilization schedule since they are readily leached from the soil and thus are not stored in the soil for an extended period of time. Seedlings are transplanted to small pots of 1 to 2¼ inches (2.5 to 5.7 centimeters) and then to larger or "finish" pots. Rooted cuttings, however, may be transplanted directly to the final pot. The nitrogen plus potassium, or complete mixture, is applied usually in the liquid form when the soil is moist. If fertilizer is applied when the soil is dry, a rapid uptake of solutes from the soil solution which will retard plant growth takes place.

Use of Growth Regulators in Potted Plants

Growth regulators are extensively used in the production of pot plants. These are most frequently used to produce more compact plants that will blend better into home decor. Chemicals used for this purpose have the desirable side effect of reducing internodal length without reducing the number of leaves.

In 1950, the growth retardant Amo 1618 (2-isopropyl-4-dimethylamino-5-methylphenyl-1-piperidinecarboxylatemethylchloride) was developed. The compound was very effective for dwarfing plants but too expensive for commercial use. Later a related material, Phosphon (2-4-dichlorobenzytributylphosphoniumchloride), appeared on the market and is now used for retarding the height of chrysanthemums and lilies. This material is applied as a solution to the soil. Cycocel (2-chloroethyltrimethylammoniumchloride), a more recent introduction, can be used as a foliar spray or a soil drench for controlling the height of poinsettias and azaleas. B-Nine (succinic acid 2,2-dimethylhydrazide) used as a foliar spray effectively controls the height of chrysanthemums, bedding plants, poinsettia, azalea, and hydrangea. The most recent growth retardant developed is A-Rest [a-cyclopropyl-a-(4-methoxyphenyl)-5-pyrimedinemethanol]. This chemical controls plant height and is active at a low concentration.

Another growth regulator category is composed of the "pinching" or "pruning" chemicals. The product Emgard 2077 (a combination of emulsifier and methylnonanoate) and the recently developed product Off-Shoot-O (a combination of emulsifier and methyloctanoate plus methyldecanoate primarily) are applied together as a foliar spray for azaleas and result in the destruction of the shoot terminals, an equivalent to pruning these plants. The saving in labor is considerable since this crop is generally pruned five times before marketing. Chrysanthemums and many woody ornamentals also can be chemically pruned by this method.

An intriguing growth regulator is Ethrel (2-chloroethylphosphonic acid) which induces ethylene release within the plant. The wide range of ethylene effects (including ripening of fruits, dropping of leaves, and promoting of branching) are found to result from foliar and root applications of this chemical on various plant species. The beneficial responses are flowering induction of bromeliad, height reduction of poinsettia and kalanchoe, increased branching of azalea, and enhanced root formation of carnation, gardenia, kalanchoe, and chrysanthemum, when used in combination with IBA (indolebutyric acid) as a quick dip for cuttings.

Pot Plants Grown for Sale

Ornamental pot plants grown for sale may be divided into two groups: (1) those grown primarily for the beauty of the flowers and secondarily for their foliage and (2) those grown for their attractive foliage only. Principal kinds of the first group are African violet, azalea, chrysanthemum, begonia, cineraria, cyclamen, geranium, hydrangea, calceolaria, and poinsettia. Principal kinds of the second group are *Asparagus plumosus, Asparagus sprengerii,* caladium, fern (Boston and maidenhair), philodendron, pothos, rubber plants, sansevieria, cacti, dieffenbachia, and succulents. Millions of excellent plants are produced each year.

GROWING AND FORCING FLOWERING BULBS AND CORMS

In general, bulbs and corms are developed for the storage of relatively large quantities of reserve carbohydrates—usually hemicellulose and/or starch. These reserve substances are changed to sugars and other similar compounds for the development of the roots, stems, leaves, and flowers. Naturally, with other factors favorable, the greater the amount of carbohydrates which are stored, the greater will be the production of roots, foliage, and flowers. Major crops are gladiolus and lily; minor crops are iris, narcissus, hyacinth, and tulip. In the commercial production of these crops two more or less distinct industries have developed: (1) the production of the storage organs with the simultaneous development of foliage and flowers as shown in Figure 23-6 and (2) the forcing of the storage organs in greenhouses.

Production

Although the details of cultivation vary with the crop, the main objective for all crops is the rapid development of large storage organs with reserve food. Thus, as in the development of onions, potatoes, root crops, and sweet potatoes, the sequence of events is as follows: the development of absorbing roots and a large photosynthetic surface during the first part of the growing season, and the development of the storage organ during the latter part. For example, in the production of gladiolus corms in southwestern Michigan cormels (small corms) are planted 4 to 6 inches (10 to 15 centimeters) deep and 3 to 6 inches (7 to 15 centimeters) apart in rows 2, 2½, or 3 feet (61, 76, or 91 centimeters) apart in well-drained, moderately acid sandy loams. At first, the cormel develops the absorbing root system and then it develops one, two, three, or four stems. When

Figure 23-4 Potted azaleas scheduled for Christmas blooming. Plants were treated with a sequence of six weeks of long days, six weeks of short days, and six weeks of artificial precooling, prior to placement in the greenhouse for forcing. Note buds ready to flower. *(Courtesy, Joe W. Love, North Carolina State University.)*

Figure 23-5 Production of poinsettia pot plants in a Dutch-type greenhouse that are scheduled for Christmas sale. Note the care instruction labels attached to each plant. *(Courtesy, Joe W. Love, North Carolina State University.)*

Figure 23-6 Production of narcissus bulbs in the field. *(Courtesy, L. A. Burnett, Soil Conservation Service, U.S. Department of Agriculture.)*

these stems are 6 to 8 inches (15 to 20 centimeters) high, their basal portion starts to enlarge, and when the stems are 18 to 24 inches (46 to 61 centimeters) high, the young corms are fully grown. At the same time, cormels arise from adventitious buds between the old and new corms. In this way, a comparatively large number of corms are produced. The new crop of corms is harvested when the leaves turn yellow or just before the first frost.

In general, commercial districts may be divided into two groups: (1) the southern district in which flower production is the main consideration and (2) the northern district in which corm production is the chief consideration. Outstanding southern districts are located in central Florida and the Mobile Bay area of Alabama. An important northern district is the Holland-Zeeland area of southwestern Michigan.

Forcing

The main objective is to produce attractive foliage and brilliant flowers for sale at the right time. To obtain this objective certain practices are common to all flowering bulbs and corms, and certain practices are specific. Common practices are (1) the use of large, well-developed, disease-free bulbs or corms, (2) the development of the absorbing system and most of the stems and leaves during the first part of the forcing period, and (3) the development of stems, leaves, and flowers during the latter part. Specific practices are a storage period of four weeks at 32 to 35°F (0 to 1.7°C) for the Easter lily and a conditioning period of two weeks at 80 to 85°F (26.7 to 29.4°C) in moist media for the gladiolus. The specific treatment for the Easter lily is necessary to conserve the carbohydrate supply, which is directly related to the number of flower buds produced; whereas the conditioning period of the gladiolus is necessary to develop a large number of cells within the growing points before production begins.

Of the various kinds of flowering plants that are forced from storage organs, the Easter lily is the most important commercially because of the demand for it as a potted plant at Easter. Since the Easter lily should look its best on only one or two days, preferably on Easter Sunday, a very precise synchronization of treatments and practices has been worked out from the time the bulb is harvested until Easter Sunday occurs. Height is controlled by the use of Cycocel or A-Rest. (See Fig. 23-7)

The storage organs of iris, hydrangea, narcissus, and tulip are also forced. The proper size of the bulbs or corms, the preconditioning treatment before forcing, e.g., by the use of chemicals, the type of forcing container and media, and the level of the forcing temperature for Christmas or for the winter and spring trade are specific and exacting. These requirements have been developed by growers and certain experiment stations.

PRODUCTION OF BEDDING PLANTS

Ornamental bedding plants are grown in beds of the home grounds and in public places or in porch boxes, window boxes, or urns. In general, the greatest demand takes place just before or immediately after the last frost in the spring, and operations of the commercial florist in the raising of these plants are timed accordingly. Most bedding plants are herbaceous annuals, and most of these are propagated by seed. Note the greenhouse of bedding plants in Figure 23-8.

Important practices are (1) germinating seed under optimum conditions of temperature, moisture, and oxygen supply, usually in steam-sterilized soil (a necessity for killing weed seeds and the organisms which produce damping off); (2) preparing the mixture for the container (in general, 1 part of finely shredded

Figure 23-7 A field of Easter lilies growing for the production of bulbs. *(Courtesy, F. S. Batson, Wiggins, Mississippi.)*

Figure 23-8 Production of bedding plants for spring sale. *(Courtesy, William H. Carlson, Michigan State University.)*

peat or well-rotted manure to 3 parts of soil); (3) transplanting the seedlings into the containers; (4) placing the containers on boards or gravel rather than on the soil (to restrict the growth of the root system and to produce a more satisfactory top growth); and (5) watering and fertilizing (to ensure a moderately rapid growth). At present, two types of containers are used: small, flat trays and peat pots. Both types have advantages and disadvantages. The first type requires fewer handling operations, but it does not permit grading for uniformity in the preparation for market. The second type permits grading for uniformity, and, since the pot and plant can be transplanted as a unit, there is little if any disturbance to the root system in the transplanting operation.

In large-scale production of plants, mechanization of soil- and container-handling operations markedly reduces the cost of production and increases profits. For example, conveyer belts are used to carry peat and soil to the shredder and to carry the prepared soil mixtures to the containers. Machines are used to fill the containers. Other machines place seeds at the proper spacing in the containers. An entire flat can be seeded as rapidly as it can be moved into position in the machine.

Coleus, balsam, geranium, pansy, petunia, marigold, salvia, snapdragon, verbena, vinca, tomato, and pepper are frequently grown as bedding plants. Excess height can be a problem in bedding-plant production, but height is controlled by recommended applications of B-Nine.

HOME FLORICULTURE

Principal factors concerned are (1) relative humidity of the home, (2) type of container, (3) composition of the soil, (4) application of commercial fertilizers, and (5) control of pests.

Relative Humidity

As stated in Chapter 4, the rate of water absorption should equal the rate of transpiration; and the principal factors influencing the rate of transpiration are the leaf area, temperature, light intensity, wind velocity, and relative humidity. Thus, with the environmental factors constant the amount of water lost per unit of time will be more or less proportional to the leaf area. In other words, plants of the same kind with a large number of leaves under the same environmental conditions will lose more water per unit time than plants with a small number of leaves. On the other hand, with the same kind of plant and with the leaf area constant, the amount of water lost will be determined largely by the environmental factor in the minimum. In the home, the amount of moisture in the air or the relative humidity is likely to be the limiting factor. This is particularly true of non-air-conditioned homes in the winter and of homes during the time the air conditioning equipment is in operation. Since air conditioners take water out of the air, they lower the relative humidity.

According to authorities the optimum relative humidity for human comfort is between 50 and 60 percent. Consequently, if the humidity is within or near this range, the use of pans of water to increase the relative humidity would seem to be impractical. Fortunately, under these conditions all that is necessary to avoid deficits of water within the tissues of plants is to maintain the soil in a moist condition. If, however, the air becomes exceedingly dry, the use of shallow pans of water would not only promote human comfort, but it would also lower the rate of transpiration.

Type of Container

As previously pointed out, the living tissues of the root system are constantly respiring, and in respiration free oxygen is taken in and carbon dioxide is given off. This free oxygen flows from the outside air into the pore spaces of the soil, and the carbon dioxide flows from the pore spaces to the outside air. If the pore space becomes saturated because of overwatering and/or lack of drainage, the supply of oxygen becomes limiting and the concentration of carbon dioxide assumes toxic proportions. As a result, the respiration and growth of the root system are impaired, and the ability of the root-hair zone to absorb water and essential raw materials is reduced. Thus, adequate drainage of the soil is essential at all times.

From the standpoint of drainage and aeration, plant containers may be divided into two groups: (1) porous, or clay pots, and (2) nonporous, e.g., glazed, ceramic, rubber, glass, china, and plastic pots. With the porous pot, since water escapes through the pores, watering at frequent intervals is necessary and, unless extreme care is taken, the soil is likely to become dry. However, adequate aeration is likely to exist at all times. With the nonporous pot, watering at relatively wide intervals is necessary and, unless extreme care is taken, inadequate aeration is likely to be a problem. With both types, provision for drainage of excess water is necessary. This is accomplished by placing coarse gravel, pebbles, or pieces of broken clay pot in the bottom and over the hole of the container. In this way, the drainage of excess water is facilitated and a continuous supply of oxygen is available for the respiration and growth of the root system.

Composition of the Soil

As previously pointed out, the root system of plants in pots and similar containers is restricted, the volume of soil in the pot is low, and frequent watering is necessary. Thus, soils for plants in pots should retain large quantities of available water and essential raw materials and permit the rapid diffusion of carbon dioxide and oxygen in the respiration of the root system and the soil organisms. For these reasons, the soil should contain adequate quantities of highly decomposed organic matter, such as well-rotted manure, peat or leaf mold, fine to medium-coarse sand (usually builder's sand is satisfactory), and loam, such as sandy loam, silt loam, and clay loam. The highly decomposed organic matter forms humus which retains water; the sand promotes aeration and drainage; and the fine clay particles of the loam combine with the humus particles in the formation of a clay-humus complex which serves as a storehouse for the adsorption of essential cations. Although different species vary in their soil requirements, in general, the following mixtures on a volume basis should be satisfactory: (1) 1 part highly decomposed organic matter, 1 part builder's sand, and 2 parts well-drained silt loam or clay loam or (2) 1 part highly decomposed organic matter and 3 parts well-drained sandy loam.

Application of Commercial Fertilizers

When a seedling plant is placed in a finish pot, the amount of essential raw materials in the soil is usually sufficient for the initiation of growth. As the plant develops its root and top system, the essential raw materials are rapidly exhausted. As a result, applications of commercial fertilizers are necessary.

In general, commercial fertilizers may be applied in the liquid or dry form. However, applications in the liquid form are usually more satisfactory, since the essential elements are immediately available, and, if high-analysis mixtures are used, most of the ingredients will dissolve readily and more or less completely in water. To avoid possible harmful effects, the liquid fertilizer should be applied when the soil is moist at the recommended concentration. If the soil is dry or if concentrations above the recommended dose are used, plasmolysis of the absorbing system is likely to take place, which is manifested by drooping or wilting of the leaves and by a corresponding reduction in growth and in general health of the plant.

Control of Pests

Principal pests are aphids, mites, mealybugs, and scales. As stated in Chapter 15, all these pests have sucking or rasping mouth parts and their mode of feeding is quite similar. In general, they pierce the epidermis of the leaves and young stems and suck the manufactured compounds from the internal tissues and reduce the chlorophyll content of the photosynthetic surface. In this way, the rate of photosynthesis is reduced.

Control measures are general and specific in nature. General methods consist of (1) carefully inspecting the plants before they are brought into the home and removing insect pests which may be present; (2) inspecting the plants at frequent intervals, particularly the underside of the leaves and young stems; and (3) supplying the plants with optimum moisture and essential elements. Specific measures vary with the nature of the pest. Spraying the tops with nicotine sulfate for aphids, syringing the leaves and stems with a fine mist of water for spider mites, and swabbing the bodies of mealybugs and scales with a ball of cotton soaked with rubbing alcohol are generally recommended.

QUESTIONS

1. State the three factors which have been responsible for the development of commercial floriculture.
2. What is the FTD?
3. You want to send a gift of flowers to a friend living in another region. How would you go about it?
4. Under what conditions do you consider flowers a luxury? Under what conditions do you consider them a necessity? Explain.
5. Show how the law of mass action applies to the prepackaging of cut flowers.
6. In your opinion which is more important in prolonging the life of cut flowers in a room—a low transpiration rate or a low respiration rate? Explain.
7. How does the production of cut flowers in greenhouses differ from the production of cut flowers in cloth houses?
8. Under conditions of high light intensity and high transpiration, pompon chrysanthemums and asters grown under cloth produce longer stems, larger leaves, and larger flowers than comparable plants grown outdoors. Explain.
9. In your opinion, in what region of the country could cloth houses be used for the longer period? Explain.
10. Locating cloth houses near large trees should be avoided. Give two reasons.
11. The exclusion of bees by cloth houses prolongs the life of cut flowers. Explain.
12. What are the differences and similarities in growing plants in a garden and growing plants in pots, hanging baskets, or similar containers?
13. In general, the use of steam-sterilized soil is required for the production of bedding plants. Explain.
14. The containers of plants grown for bedding purposes should be placed on bare boards or on gravel rather than on soil. Explain.
15. Give an advantage and a disadvantage in the use of trays and peat pots for the growing of plants for bedding purposes.
16. What is the difference in morphology of bulbs and corms?

17 The forcing of bulbs and corms is entirely a carbohydrate utilization process. Explain.

18 In the greenhouse production of flowering bulbs and corms, extensive root development is necessary before foliage and flower production. Explain.

19 During the rooting period of bulbs and corms, soils should be low in nitrates. Explain.

20 Lily bulbs 5 to 7 inches (13 to 18 centimeters) in circumference produce from three to five flowers per plant, and bulbs 8 to 10 inches (20 to 25 centimeters) in circumference produce from five to seven flowers per plant. Explain.

21 Under optimum conditions in the greenhouse, large gladiolus corms flower two weeks earlier than relatively small corms. Explain.

22 When field-produced gladiolus spikes are cut, at least four leaves should be left on the remaining stalk. Explain.

23 Given two lots of lily bulbs of equal size for planting in the field. Lot A was planted late and harvested early, and it produced from one to two flowers per plant; lot B was planted early and harvested late, and it produced from five to seven flowers per plant. Explain.

24 Outline the characteristics a plant should possess to be adaptable to the relatively adverse conditions of the home.

25 Some home makers have difficulty in maintaining plants in the home in a healthy condition, others have no difficulty. Explain.

SELECTED REFERENCES FOR FURTHER STUDY

Ball, V. 1972. *The ball red book*. 12th. ed. Chicago: Geo. J. Ball. A book similar to a farmer's almanac, filled with diversified information on flower production: the first half, entitled "General Greenhouse Subjects," deals with topics such as growing bedding plants, air pollution problems on bedding plants, soil sterilization, soilless mixes, greenhouse cooling, greenhouse plastics, and pot and flat varieties; the second half, entitled "Cultural Notes by Crops," lists dozens of flower varieties and discusses extensively how they should be grown.

Laurie, A., D. C. Kiplinger, and K. S. Nelson, 1968. *Commercial flower forcing*. 7th ed. New York: McGraw-Hill. Standard text and growers manual, which is a complete study of commercial flower production based on experimental evidence, with information on greenhouse construction, heating and cooling, soils, fertilizers, propagation, major and minor crops, bulbs, pot and bedding plants, foliage plants, wholesale marketing, production costs, and the specific requirements of major floral crops.

Nelson, K. S. 1966. *Flower and plant production in the greenhouse*. Danville, Ill.: The Interstate Printers and Publishers. Practical information on the floriculture industry; greenhouse structures; plant metabolism as related to sunlight, water, air temperature, soils, fertilizers, insect and disease control; and growing cut flowers and pot-plant crops under greenhouse conditions.

Chapter 24

Horticultural Nursery Industry

There is a way to do it better. Find it.

Thomas A. Edison

Economic principles and increasing demand for nursery stock are changing nurseries from seasonal to year-round businesses. According to the Horticultural Research Institute, the horticultural nursery industry consists of four major enterprises: wholesale firms (37 percent), landscape firms (32 percent), garden centers (20 percent), and mail-order firms (2 percent), with noncategorical (9 percent).

The wholesale nursery grows plants in bulk for sale to landscape nurseries, garden centers, fruit orchards, retailers, and institutions. Wholesale nurseries are production units whose objective is to produce marketable plants. Most propagate their own cuttings and grow the plants for the retail market; however, some produce only rooted cuttings, while others grow small plants to a marketable size. The wholesale nursery requires a large initial capital investment. The nursery needs to be located in the country to avoid high land prices and taxes, near a plentiful water source, and near a well-maintained highway. Also, labor usually is cheaper outside cities. Some of today's large nurseries were located, at first, close to cities. As a city grew, the land that a nursery owned increased in value, and the nursery made a profit by selling it. When the nursery moved, it was able to establish a modern operation with up-to-date equipment. A container nursery can be moved completely.

Some landscape nurseries grow plants, but their main concern is drawing plans. The landscape architect's plans are often used as a basis for a planting job. Most landscape-nursery personnel also sell plants, with homeowners as the major source of buyers. However, factories and office buildings demand landscaping services as well. In general, landscape nurseries are located near large population centers.

Garden centers sell plants, fertilizers, mulches, tools, and other supplies that residents use in their gardens and yards. Centers may grow their own stock or may buy it and care for it until it is sold. The facilities typically include a showroom and office building, a shade house, a greenhouse, and ample parking space as shown in Figure 24-1. Many garden centers offer landscape planning and installation services. Garden centers are also usually near large population centers.

Mail-order nurseries are much like specialized wholesale nurseries. Almost all their stock is field grown, as plants are shipped so as to be economically feasible. These nurseries print catalogues that are distributed to the public. Customers order and receive the stock by mail. These nurseries are usually located where the land is inexpensive and there are plentiful supplies of water.

Figure 24-1 A garden center, consisting of shaded display area, sales center and office, and centrally located parking space. *(Photo by Warren Uzzle, Raleigh, North Carolina)*

NURSERY PLANT PRODUCTION

Growing Plants in Containers

Container-grown plants have become the major means of production in the nursery industry. Knowledge of soil mixes, container type, potting operation, irrigation, commercial fertilization, weed control, and cold protection is necessary to produce plants successfully.

Soil Mixes Components for growing plants in containers vary from 100 percent Canadian or German sphagnum peat moss to 100 percent fine sand. Perlite or vermiculite may be used instead of all or part of the sand requirement. Rice hulls, shavings, sawdust, and bark can be used in place of peat. Leaf mold, manure, and other compost materials are not used in the soil mixture because these materials are unstable under steam or fumigation treatment. Leaf mold, manure, and other compost materials generally shrink 33 percent, whereas 1 cubic foot (2.7 decistere) of baled peat moss breaks into approximately 1.5 cubic feet (0.4 decistere) of loose material.

A special soil mix for each plant variety has many disadvantages. Different soil mixes and compost piles require the use of more soil piles, which cost more than fewer larger piles, more storage bins, and a larger mixing area. Small piles also make mechanization of soil mixing uneconomical. A large number of components used in the soil mix result in unpredictable plant responses, preventing the scheduled production of plants. John Innes Institute pioneered research for a reliable, uniform mix that could be used for a variety of plants. The basic component of this mix is composted organic material. This material, however, is not uniform in its chemical and physical properties, and the amount of composted material required consumes nursery space.

Researchers at Cornell University developed the "peat-lite" mixes. Unlike the John Innes mixes, these are soilless. The two main ingredients are milled sphagnum peat and perlite or vermiculite. Lime and fertilizer are added when mixing. Superphosphate and a 5-10-5 analysis are the other ingredients usually added.

The most recent scientific approach to soil mixes has been done at the University of California. Researchers have concentrated on developing a soil mix that has desirable chemical and physical properties, and, by the addition of chemicals at mixing time and by supplementary fertilization, they have created the desired mix without having to consider other variables. For this soil mix to give optimum results, a basic understanding of the system and its use and elements is needed. The proper preparation of mixes requires certain procedures. The peat to be used should be sprinkled one or two days before use. The ingredients should be kept in low level piles handy to the operation. Fertilizer should be broadcast over each pile. The ingredients should be shoveled from pile to pile until thoroughly mixed. After planting, a water breaker should be used on the hose to prevent sand from settling out and leaving a crust on top of the can. University of California soil mixes are designed for support, moisture supply, aeration, and nutrient supply. These mixes reduce the need for specific soils for different plants. Almost any soil will give support to a plant, although it may need to be supplemented by stakes and ties. Since the supply of soil moisture must be continuous for satisfactory plant growth, the University of California soil mixes are designed to retain moisture. High salinity has the effect of making water unavailable to the plant and should be guarded against. Closely tied to moisture supply is aeration. A soil that holds water too long retards aeration. For optimal plant growth, mineral nutrients should be continuously available to the plant. The soil mix must provide the 12 necessary elements, or the growing procedure must allow for the addition of the elements.

Sometimes the soil itself contains materials toxic to plant tissue. Manganese can reach toxic levels if the soil is treated with steam for sterilization purposes. Heat and chemical treatments can cause organic matter to break down and form toxic water-soluble organic matter that releases toxic levels of soluble salts. Choice of soil mixture, leaching and aging of the soil, and planting immediately after steam treatment are measures used to prevent soil toxicity. There are several different methods of soil pasteurization. It is referred to as *soil pasteurization* because the objective is to kill only the harmful organisms and weed seeds. The soil should be moist before it is pasteurized. It requires five times as much heat to raise the temperature of 1 pound (454 grams) of water 1°F (0.6°C) as it does to raise the temperature of 1 pound (454 grams) of soil 1°F (0.6°C); however, heat plus moisture is more effective than heat alone in killing pathogens and weeds.

There are several advantages of using steam for soil treatment. Steam improves the physical structure of the soil and is faster, easier, and cheaper in large operations and more effective than chemicals. Steam also can be used around people and plants. The major disadvantage is the high initial cost of equipment.

A chemical used for treating soil must be one that kills fungi, bacteria, insects, and weeds. It should be inexpensive, harmless to people and equipment, effective in and on the surface of the soil, and not leave chemical residue in the soil. Though there are no chemicals with all of these characteristics, three kinds of chemicals are available: (1) those for disease control, (2) those for insect and nematode control, and (3) those for weed control. Disease control is the most difficult. The greatest disadvantage of chemical treatment is the time required for complete aeration. Methyl bromide is one of the most widely used soil fumigants because it requires less time to treat the soil and less time for aeration. The chief disadvantage of methyl bromide is that it is extremely toxic to man and also has to be applied at temperatures of 60°F (15.9°C) or higher.

In Florida and other locations with organic soils, containers are sometimes filled with soil only. In some southern states and in northern Illinois, pine bark and sawdust are used in plant containers. Occasionally sand, soil, or peat is mixed into the media. Like sawdust, fresh wood residues create a problem as they decompose by binding nitrogen in the soil.

Figure 24-2 Converted concrete mixer used for preparing soil mix for container-grown plants.

The soil mixture in which the plant grows is a most important factor in successful cultivation. This mixture should be uniform in grade and chemical components. It should be stable under steam and chemical fumigation, otherwise toxic materials such as water-soluble organic matter, manganese, or soluble salts may develop. It should mix uniformly, be easy to aerate, and be resistant to leaching. It should be low in fertility and not require a large addition of gypsum and lime. In addition, the soil mix should be inexpensive, moisture retentive, light in weight, and not apt to shrink.

Commercial Fertilizing Essential elements may be applied in dry form or in a water solution to container-grown plants. When fertilizer is applied as dry material, it is incorporated into the soil mix and added to each container individually as it is needed. One method is to use spoons of the desired size to measure the fertilizer for each plant. A dispenser can be used to enhance uniformity and to speed up the operation. Also, tablets of fertilizer consisting of various components can be used. Liquid fertilizer can be applied by spraying or by adding it to the irrigation system. The fertilizer can be injected into the watering system of smaller nurseries with a simple pump that adds a certain unit of fertilizer per volume of water. In some of the larger nurseries a complex system is used. Huge tanks, prefilled with fertilizer, are brought to the nursery and attached to the watering system shown in Figure 24-3. Intricate electrical controls feed precise amounts of liquid fertilizer into the water at the appropriate time. These nurseries apply fertilizer with each irrigation and then leach the soil mix in the cans by watering once a week, thereby preventing salt buildup. The advantages of irrigation fertilization are obvious. Labor costs are reduced, and the plants are kept supplied with optimum essential-elements at all times. The possibility of a disasterous accident (weed killer in the irrigation water), the wasting of fertilizer, and the increased need for weed control in areas outside the containers are the chief disadvantages of this system.

Plants have different essential-element requirements at different times of the year. In preparation for cold weather, plants should be fertilized heavily with phosphorus and potassium and have nitrogen withheld. The soil in the containers should be scientifically tested several times a year to determine fertilizer requirements.

The trend in the nursery industry today is toward the use of long-lasting or slowly releasing fertilizers. A magnesium ammonium phosphate fertilizer is available that, with one application, provides essential raw materials for several years. Having to make only one application of fertilizer is the ideal as far as the concept of slowly releasing fertilizers is concerned.

Containers The use of containers for growing nursery stock is expanding rapidly. The first type was the used food can. These cans were cheap, strong, and available. However, they required considerable space for storage and rusted quickly. A second type of steel container, made specifically for growing plants, is painted and tapered. In general, cans of this type protect the root systems from being crushed, allow the roots to slide out easily during the transplanting operation, and permit stacking for storage. However, they conduct heat quite readily.

Figure 24-3 Prefilled tanks of liquid fertilizer and injector system for applying commercial fertilizers when irrigating container-grown plants.

As a result, the root systems are frequently damaged by extremely high and low temperatures. A third and more satisfactory type is made of stiff plastic. These containers stack easily, require little storage space, and are economical (especially in the smaller sizes). Plants can be easily removed from them. Since they conduct heat slowly, the roots are protected from sudden changes in temperature; however, they may allow the roots to be damaged in shipping. Polyethylene bags are being tried in California. For moderately large, fast-growing plants, treated continuous stave-type, seamless bushel baskets are, in general, satisfactory. The chief advantage is that they are inexpensive. However, they deteriorate relatively quickly and give little protection to the roots.

The ideal container has the following characteristics. It is strong, rustproof, stackable, light in weight, neat in appearance, inexpensive, provides insulation, and is not affected by temperature and chemical changes.

Potting Operation Almost every nursery has a different method for potting plants. The simplest method involves a worker on a stool behind a pile of mix, with a pile of cans on one side and a flat of liners on the other. This operation becomes more efficient when the worker has someone to supply the materials and to carry away the potted plants. A machine to mix the soil speeds up the operation even more.

Machines have been developed that fill the can with soil, punch a hole for the plant, put the plant in the can, and water the plant. (See Figure 24-4) Usually a front-end loader tractor places the soil in the hopper. An elevator carries the soil to the potting machine. The pots with the soil in them are carried by a conveyer to a machine that either punches or bores the proper size hole in the soil. The plants are placed in the hole, and the conveyer carried the potted plant to a truck or power cart. By using this type of operation, nine people can pot approximately 18,000 gallon (3.8-liter) cans per day.

Figure 24-4 A potting operation. Left: close-up of machine for filling cans with soil mix, punching a hole in the soil, and planting the plant. Right: contained plants departing via conveyor belt.

Another system of potting plants, which requires less equipment, is to place 300 gallon cans on a flat-bed trailer. Three front-end buckets of soil are dumped on top of the cans. The soil is leveled flush with the top of the containers by the use of a board. One worker makes a hole in the soil, two more workers plant the rooted cuttings, and another waters them. With this system a four-worker crew can pot approximately 5,000 plants per day.

Mechanized potting is essential in a large nursery with volume sales but would be impractical in a small nursery, which has no way to care for or to market large numbers of plants. An operation that is comfortable for the people doing the work, that results in high-quality plants, and that is reasonably fast and efficient is the goal.

Growing Beds Growing beds should be prepared by level grading and firm rolling. Then they should be sterilized with a soil fumigant and covered with 4-mil or thicker black plastic. Each piece of plastic should overlap the next one and not be stretched, since it may split when it contracts in cold weather. Once the bed is prepared, a center line should be established down the length of the bed. This keeps the rows straight and leaves a place to put irrigation pipes. When spacing the cans, a standard measure should be used to ensure uniformity. This measure can be constructed of two-by-fours with the measurement marked on them. Staggering the plants of one row between the plants of the preceding row allows more growing room and more sunlight for each plant. Spacing for various-sized cans is suggested as follows: for 1-gallon (3.8-liter) cans, 12-inch (30-centimeter) centers with 12 inches (30 centimeters) between rows; for 2-gallon (8-liter) cans, 18-inch (46-centimeter) centers with 18 inches (46 centimeters) between rows; for 3-gallon (11-liter) cans, 28-inch (71-centimeter) centers with 28 inches (71 centimeters) between rows; for bushel (35.2-liter) baskets, 48-inch (122-centimeter) centers, with 48 inches (122 centimeters) between rows. In newly canned plants where the plants do not reach the edge of the containers, can-to-can placement can be used with no harmful effects. In setting up growing beds, walkways should be placed between the beds. In beds of 1- or 2-gallon cans, walkways perpendicular to the centerline should be left every 6 feet (1.8 meters). For larger cans, the normal spacing should be enough so that it is possible to work among the plants without leaving the walkways.

Irrigation Systems Basically only two types of irrigation systems, the overhead and the "spaghetti-type", are workable for container-grown plants. The spaghetti-type, as shown in Figure 24-5, consists of a fairly large pipe that is connected to a water source, with small flexible tubing originating at the pipe and running directly to the container. The advantages of this system are lower water use, less fungus disease on the foliage, and fewer run-off problems. However, labor costs of setting up and maintaining this system are fairly high. The rotating, or overhead, type consists of (parallel construction) lightweight, movable aluminum pipe or underground plastic pipe. The pipelines are usually placed on the ground in an aisle in the center of the bed of plants.

Figure 24-5 The "spaghetti-type" irrigation system used for watering trees supported by stakes in containers.

Weed Control The most widely used weed control program is a multistep process. The soil mix is fumigated with steam or chemicals. The beds on which the containers are to be set are cleared of weeds and covered with black polyethylene, as shown in Figure 24-6. Treatment of soil, containers, and beds before planting helps control weeds in containers. Removing weeds that emerge

Figure 24-6 Beds of container-grown plants with plastic for weed control and pine trees for winter protection.

before they have a chance to develop seed reduces the development of new weeds. Other nearby areas where grasses or weeds are growing should be mowed or sprayed with herbicides. Fiberglass or rubber discs cut to fit around the plant stem and close to the sides of the can are a recent development. These discs allow fertilizer, water, and air to reach the roots in satisfactory amounts, while presenting a barrier to weed seedlings.

Cold Protection The method for low-temperature protection of container-grown nursery stock varies considerably in different areas. In warm-temperate subtropical climates, jamming the containers close together under a grove of pines is sufficient protection. (See Fig. 24-6) This practice insulates the root balls of all the plants except the ones near the perimeter of the block. A natural or synthetic windbreak will keep dry winter winds from "burning" the foliage. Winter damage and death are caused by the rate of water absorption failing to keep up with the rate of transpiration. Water in the container in the form of ice does not allow sufficient moisture intake in order to combat the loss of water by transpiration, which is a result of the cold air and its low relative humidity.

Unheated, polyethylene greenhouses that have been used for mist propagation areas during the summer, as shown in Figure 24-7, can be converted into completely enclosed structures for protection against cold. Plants are crowded together in these structures. At night the temperature drops and the water in the soil may be frozen. When sunlight enters through the plastic, its energy is trapped inside and the temperature rises. In this way, the water becomes available to the plant, and all processes can go on normally. At night when the

Figure 24-7 Plastic greenhouses used for mist propagation of plants and for winter protection of plants.

temperature falls outside, the temperature inside falls more slowly. The prevention of rapid temperature changes is advantageous to plant growth and development. Holding houses or "greenhouses" should be of sturdy construction for maximum protection. A steady draft could seriously damage the plants affected by it. Ample space for work, such as repotting and fertilizing, and a center aisle large enough for a wheelbarrow are essential. On a warm winter day, temperatures inside a polyethylene house can rise very high. The foliage of the plants in the house is jammed together. These conditions, combined with high humidity inside the house, can cause leaf damage. A thermostatically controlled ventilator fan can solve the problem of ventilation. Having doors that open at both ends of the house is sometimes sufficient. Maintaining an adequate moisture level is necessary to prevent cold damage. In early fall or late summer, low nitrogen, high phosphorus and potash fertilizer should be applied. This stops rapid growth and allows carbohydrates to build up in the plant. Plants with stored carbohydrates are less likely to be damaged by low temperatures than plants with no stored carbohydrates.

Sprinkler systems may also be used for cold protection. In general, ⅓ inch (0.8 centimeters) of water per hour is needed. When water freezes, heat is given off. Ice on the foliage also is a buffer against cold winds. Once this method of cold protection is begun, it should be continued until the temperature rises.

Growing plants in containers as a large scale operation is a relatively new idea. This method has many advantages. More plants per acre can be grown in containers than can be grown in the field. (See Figure 24-8) Thirty-five thousand plans in 1-gallon (3.8-liter) containers can be grown on 1 acre (0.4 hectares) of land with approximately one-third of this acre used for walkways. The chief advantage of container-grown plants is that they are more adaptable to the new production programs than are the field-grown plants. It is easier and more economical to apply water and fertilizer to container-grown plants. Plants grow-

Figure 24-8 A large wholesale nursery for production of container-grown plants.

ing in containers are sprayed, dusted, and pruned more easily than field-grown plants, and weeding is minimized. Another advantage is that when moved, container-grown plants have less root disturbance than field-grown plants; therefore, there is a higher percentage of marketable plants. Some growers report that 90 percent of their container-grown plants are marketable.

The greatest disadvantage in using containers is that there is a danger of the plants becoming pot-bound because the containers in which they are planted are not large enough for extensive root growth; therefore, the growth of the plants is stunted. If an extremely pot-bound plant is sold to a customer, it usually will not grow when it is planted. Another disadvantage is that many landscaping companies want to buy large trees and shrubs to use in their work. Large trees and shrubs have to be grown in very large containers which are expensive and hard to move. These containers do not fit into a standardized operation and require a nursery specially set up to handle them.

Growing Plants in the Field

Plants started from cuttings or from seeds may be grown in containers or they may be planted in a bed called a *nursery row*. When large enough to plant in the soil, they are lined out in a field.

Lining Out Lining out consists of transplanting the nursery stock from the propagation unit to the production unit. An individual plant is called a *liner*, and methods of lining out vary greatly with the kind of plant. In general, the plants are lifted, sorted, root-pruned if necessary, placed in transplanting shelters, and placed in transplanting boards if hand-transplanted and in the transplanting machine if machine-transplanted. Many kinds of transplanting machines have been developed.

Commercial Fertilizing The fertilizing of field-grown plants differs from that of container-grown plants in manner and rate of application. Fertilizer can be applied in the field by side dressings. For large trees that are far apart in the row, fertilizer may be applied by hand in a circle around the tree. This circle should match the drip line of the plant where the most absorption by the roots is taking place. The rate of application must be higher in the field than in containers because mobile essential ions move away from the plant, and essential ions may be carried away by leaching. Because of soil variation in different parts of the field and occasionally in the same place, soil tests are a valuable aid to the nurseryman. This is especially true in regard to soil acidity and liming requirements. Many essential ions become unavailable at the extremes of the pH scale.

Top Pruning The type of top pruning required depends upon the use of the plant. For example, the lower limbs usually are removed from trees that will be used for shade, as shown in Figure 24-9.

Figure 24-9 Lower limbs of field-grown trees removed in training landscape shade specimens.

Growers prune by pinching the liners, removing approximately one-half of the last year's growth in the spring before new growth starts. Pruning causes side shoots to be produced; therefore, a dense plant develops. Sometimes plants are pruned periodically during the growing season; however, plants should not be pruned in the fall because new growth will be tender and may be killed by cold weather. When plants are sold bare-root, one-third of the top is removed to reduce the amount of water lost in transpiration.

Root Pruning Root pruning consists in removing portions of or all the younger root sections. The object of this practice is to develop extensively branched root systems at the base of the tree. In this way recovery from transplanting is facilitated. This practice is particularly necessary for the transplanting of evergreens. Can you think of any reason for this? In general, root pruning is done in the fall, and many types of power-drawn cutting blades have been developed. One type of root-pruning machine is a large hydraulically powered knife in the shape of a semicircle. The knife enters the soil on one side of the plant and moves under the plant severing roots. It then emerges on the other side of the plant.

Weed Control Weed control is one of the biggest problems of nursery personnel. It is estimated that the hand weeding of an ornamental nursery costs approximately $3,000 per acre, whereas weed control by chemicals costs approximately $100. Generally, a combination of chemical and mechanical methods is used.

Most herbicides are available in granular, liquid, or wettable powder form. One of the most important factors in applying a herbicide is the degree of accuracy to which the equipment is calibrated. The sprayers for herbicide application are usually pulled behind a tractor. Their pump is either powered by the power-take-off of the tractor or by a gasoline engine mounted on the sprayer. The nozzles are set low to the ground and mounted on steel booms as shown in Figure 24-10. Hand-carried sprayers are available. These sprayers make use of a hand pump that exerts pressure on the contents. Generally, low pressures of 20 to 40 pounds per square inch (1.4 to 2.8 kilograms per square centimeter) are used in herbicide applications to minimize drift and also to reduce gallonage per acre. Tractor speed, constant pressure, and nozzle height are among the most critical aspects of herbicide application.

Irrigation Systems Field-grown nursery stock is watered by either a furrow-type system or by sprinklers. For information on these systems, see Chapter 13.

Plant Removal Plant removal consists of removing the plant with part of its root system from the soil. Thus, water absorption stops or is greatly reduced. However, transpiration continues to take place. As a result, plant removal is done in the fall, winter, or early spring when transpiration is at a minimum. Two rather distinct methods of plant removal are used: (1) lifting and moving without soil around the roots (bare-root) for deciduous plants and small evergreens and (2)

Figure 24-10 Herbicide sprayer with constant pressure and with nozzles set low to the ground.

balling and burlapping for large evergreens and other plants which require considerable care in transplanting. Lifting may be done by hand with a sharp spade or by machine, and balling and burlapping are usually done by digging a trench at the required distance from the base and around the root system of the tree and enclosing the mass of soil and roots in burlap. In this way, the root system within the ball of soil supplies the leaves with the necessary water.

Machines, which are powered hydraulically, have been developed for digging up evergreen and large deciduous plants in the field. These machines are self-contained or powered by a compressor or a tractor. One type, in Figure 24-11, has four large blades in a frame that opens, encircles the tree, and then closes. The blades, using water as a lubricant, work their way into the soil until they meet at the bottom. They are angled so as to form a ball with straight, sloping sides. When all the roots are cut, the blades raise the ball of roots out of the soil, and the ball may then be burlapped. If the plant is to be used immediately, it can be carried in the blades of the machine to the spot where it is to be planted. Before digging the plant, a hole is prepared using the same machine. When the tree is lowered into place and the blades removed, a perfect fit is assured.

Storage Temperatures from 31° to 35°F (−0.6 to 1.7°C) not only permit a low rate of respiration but assist woody plants out of the rest period later. High relative humidity permits a low rate of transpiration without excessive development

Figure 24-11 Hydraulically powered machine for digging and balling plants in the field.

of molds and fungi. Abundant ventilation supplies sufficient oxygen and carries away the carbon dioxide and heat of respiration.

Inside storage of nursery stock requires the use of specially constructed houses with work rooms, packing rooms, and storage rooms. The storage rooms contain a series of deep shelves with vertical partitions. Each nursery stock can be stored separately and maximum use of the storage house can be maintained. The use of storage houses permits grading, packing, and shipping during the winter, particularly in cold climates. Outdoor or heeling-in storage requires the use of well-drained sandy loam with adequately drained subsoil, so that plants do not become waterlogged in wet weather. Thus, nursery stock is heeled-in or removed at any time that the soil is not frozen. Usually, the storage field is located adjacent to an all-weather road and the packing shed.

Packing Packing consists of preparing and packing the nursery stock for shipment. The type of packaging depends on the variety, the size of the stock, and whether it is deciduous or evergreen. Evergreens are shipped individually because of the necessity of balling and burlapping. Deciduous stock is shipped bare-root with the roots packed in moist sphagnum moss, peat, or shingle tow, with the top covered with either dried stalks or rushes and with the entire plant wrapped in burlap or placed in a box lined with waterproof paper. Because parcel-post shipments are frequently mailed in airtight containers, shipments by express are usually preferred.

Many nursery supervisors choose to grow their stock in soil because they want to grow large plants such as trees. If the soil conditions at a nursery are adequate, large plants are grown more easily in the field than in containers, usually with less irrigation. Another advantage is that roots are protected from low temperatures by the soil. However, more space is required, the types of plants that can be grown are limited, the growing medium is more unpredictable, labor costs of plant removal are high, and fertilizer and pesticide requirements are higher than for plants growing in containers.

Turfgrass Production

Turfgrasses have many functions. They markedly reduce erosion; they reduce glare and cool the immediate surroundings; and they serve as an outdoor carpet. Turfgrasses may be placed in two groups: (1) warm-season and (2) cool-season. Some of the most common are presented in Table 24-1. What kind of lawn plants are used in your community?

Vegetative Propagation Turfgrass is produced vegetatively by either sod, plugs, stolons, or sprigs. The sod establishes new growth by producing new roots. Therefore, the sod is cut thin (approximately ½ inch, or 1.3 centimeters), because if it is cut thick (2 inches, or 5.1 centimeters, or more), it may fail to

Table 24-1 Warm-season and Cool-season Lawn Plants

Kind of plant	Leaf		Plant height, in.	Sod‡	Ability to maintain itself	How propagated
	Width*	Color†				
Warm-season Plants						
Bermuda	n.	d.g	2–3	d.	Low	Sod, sprigs, seed
Fine-leaved Bermuda	n.	d.g.	2–3	d.	Low	Sod, sprigs
St. Augustine	m.b.	l.g.	3–4	m.d.	High	Sod, sprigs
Centipede	b.	l.g.	3–4	d.	High	Sod, sprigs
Carpet	b.	l.g.	2–3	d.	High	Seed
Zoysia	n.	d.g.	3–4	d.	High	Sod
Cool-season Plants						
Kentucky blue	n.	l.g.	3–4	m.d.	High	Seed
Italian rye	n.	g.	6–8	m.d.	High	Seed

*n., narrow; m.b., moderately broad; b., broad.
†l.g., light green; g., green; d.g., dark green.
‡d., dense; m.d., moderately dense.

produce adventitious roots and remain just a layer of sod. Plugs are sod or stolons that have been separated into single pieces of grass and rooted. Plugs are used on a limited basis. Stolons are grown in sandy soils so that the roots can be shaken clean. A 1-foot (30-centimeter) row of the more vigorous grasses planted to be harvested as stolons will grow to cover an area of 4 to 7 square feet (0.4 to 0.6 square meters) in one growing season and can be used to plant approximately 500 square feet (46 square meters). The pieces of stolon are 2 to 3 inches (5.0 to 7.6 centimeters) long with 1 to 3 buds per stolon. If the stolons are placed in a pile, the pieces will produce heat that could possibly cause damage unless the stolons are cooled by ice or cold-water spray. If the temperatures are high, a fungicidal dip is used. Dry stolons are stored at 31°F (−0.6°C) for several weeks but should be soaked when removed from storage. One square yard (0.8 square meters) of sod produces 1,000 to 6,000 sprigs. The sprigs should be 2 to 4 inches (5.1 to 10.2 centimeters) long. Peat or loam should be used for most sod production because sandy soils will not keep the roots intact during the cutting, rolling, and transporting of the turf. These soils provide moisture for the grass while it is being transported and established.

Seedbeds It takes an entire growing season to grow a crop of turfgrass; therefore, in preparing the seedbed there is not an opportunity to repeat the tillage to correct mistakes. The seedbed is tilled deeply to at least 18 inches (46 centimeters) to increase noncapillary porosity, to break up impervious layers, to incorporate organic matter in the soil, and to stimulate nitrification. The seedbed is irrigated with at least 3 inches (7.6 centimeters) of water to settle the soil before seeding. Fertilization, weed control, and limitation of crusting are important factors in the establishment of turf. To control weeds without cultivation, a contact herbicide is used. The seedbed is covered with ¼ inch (0.6 centimeters) of vermiculite or 1/16 to ⅛ inch (0.2 to 0.3 centimeters) of sand, peat, or straw. If the seeds are sown deeper, fewer seeds germinate, but those that do come up will be more vigorous.

Commercial Fertilizing Recommended practices for cool-season grasses are (1) to apply a fertilizer with a ratio of 4:1:2 after establishment (2) to apply phosphorus and potash once in the spring based on results of soil testing, and (3) to apply nitrogen two to four times a year but not to apply it between June 1 and September 1 because of the dormancy of the grass. The warm-season grasses are fertilized with a 4:1:2 fertilizer. Phosphorus and potash are applied in early spring and late summer. Centipede and carpet grass are low fertility grasses and require only one application of phosphorus and potash in early spring. Nitrogen is applied every four to eight weeks during the summer, depending on the species.

Sod Removal Machines are used for harvesting sod. A vibrating blade moves ¾ inch (1.9 centimeters) beneath the soil, cutting the sod in 1-foot (30 centimeter) strips and cutting the strips into shorter lengths. The rectangles of sod are placed on pallets or into trucks. A 3-inch (7.6-centimeter) ribbon of sod is left between each strip to repropagate the field. The field should be disked and rolled after harvesting, and additional soil should be added.

MARKETING

Marketing involves presenting goods and services to the public in an attractive manner so that the public will purchase the product. No matter how good the quality of the plants, a profit cannot be made unless the plants are sold. The many details involved in selling plants should be considered. The stock of every grower should be inspected by a state inspector for disease and insects. Before plants are shipped to another state, a dealer's license and an inspection certificate are required. In some areas, a retail merchandiser's business license for sale to the public is necessary.

Wholesale Nursery

A wholesale nursery should not sell plants to the public. Public sale requires too much time and too many records. Furthermore, many retailers will not buy from a wholesale nursery that sells to the public. A retail merchandiser's business license may also be required if public sales are made. Wholesale nurseries deal with retail nurseries and retail outlets. Their clientele know what plants they desire. Most companies employ salespeople who handle around 30 percent of the sales. Approximately 25 percent is contracted by mail and 20 percent by telephone. The remainder is attributed to standing orders and other methods. The broadleaf evergreen is the major seller. Deciduous trees, fruit and nut trees, deciduous shrubs, and narrow-leaf evergreens follow next in volume. The most important aspects of wholesale marketing are good plant quality and dependability. The ability or failure to keep a client's business may depend on whether or not the wholesaler delivers the right order at the right time.

Landscape Services

Landscaping used to be considered a luxury, but today most people consider it a necessity; therefore, the landscape construction business is growing rapidly. A landscape construction company should not be combined with a wholesale nursery unless the two are operated by separate personnel; however, many landscaping companies keep a few trees and large shrubs at the office so that the customers can select the larger plants to be used in the landscape themselves. Landscape nurseries offer planting services and usually offer design assistance as well. The landscape nursery should have a display planting for the public to see. The planting should be attractive but not so elaborate as to make upkeep a hindrance. Another good advertising technique is to donate services to a worthy project that will attract press coverage and be visible to the public. If the landscape-nursery personnel are to do landscape design, they should have some training in the area. They should begin with small jobs such as home grounds. As their talents and reputations grow, they may begin to do landscaping of large-scale projects. They should keep in touch with what the landscape architects in the area are doing, as they are usually aware of design trends throughout the whole country. Since landscape-nursery personnel plant the stock they sell, they must decide whether or not to guarantee it. It has been the norm for landscape nurseries to give guarantees and to add a percentage to the bill to cover losses that are certain to occur at times.

Garden Centers

Garden centers sell plants and gardening accessories directly to the customer. Many homeowners seek gardening advice from the garden-center manager. Therefore, the manager should be as knowledgeable as possible about the stock being sold. Attractive display is of upmost importance in a garden center. The tools and chemical products on the shelves should be labeled. It is also important to label the plants in the display beds and keep these beds free of weeds and trash.

The type of clientele shopping at the garden center should be taken into consideration. Weekday shopping is usually done by a large percentage of women, but on weekends there is usually an equal number of men and women. Also, stores near newer housing projects usually have a young married clientele. A wise manager will note the trends in the area and supply the garden center accordingly. People seem to "shop around" for nursery supplies. When they find a place they like, however, they patronize it more than they do other shops. Plant quality, dependability, line of products handled, sales personnel, parking, professional assistance, neatness, location, and price are factors, ranked in order, influencing why a particular garden center is preferred. It is encouraging that customers rank plant quality first and price last. It is very important to control the inventory of a garden center. If the center has too much stock, the cost of the insurance and the amount of working capital will be too high. A profit and loss statement should be obtained each month. A well-developed garden center should contain a display area, a sales area, a parking area and drives, and a service section. These areas should be arranged in proper proportion and balance to provide an attractive landscape.

Display Area The display area contains plants arranged to show the effective use of landscape materials. It may show foundation plantings, pleasing combinations of trees and shrubs, or various kinds of gardens, such as the informal, formal, rose, azalea, camellia, or bulb. The display area greatly encourages impulse buying, an important feature of the garden center's business.

Sales Area The materials being purchased are presented in the sales area. This area contains reception and display room for materials such as seeds, bulbs, spray materials, catalogue, and books, the landscape or drafting room, storage and wrapping area, toilet facilities, greenhouses for house plants, lathhouses, vents or terraces for potted plants, and pergolas for vines.

Parking Area and Drives The parking area and drives provide ample space for customers' cars and free access to all parts of the sales and display areas. In this way the loading of bought materials is facilitated.

Service Area The service area contains facilities for the maintenance of display and sales areas. Principal features are workrooms for potting plants and other garden operations; storage bins for soil, peat moss, manure, and sand; storage space for pots, labels, commercial fertilizers, pest-control materials, and wrapping materials; garage and parking space for employees; and heeling-in space. Note the arrangement of various areas in Figure 24-12.

Marketing Requirements Marketing requirements for retail nurseries are similar to garden centers. Attractive display beds, high-quality plants, adequate

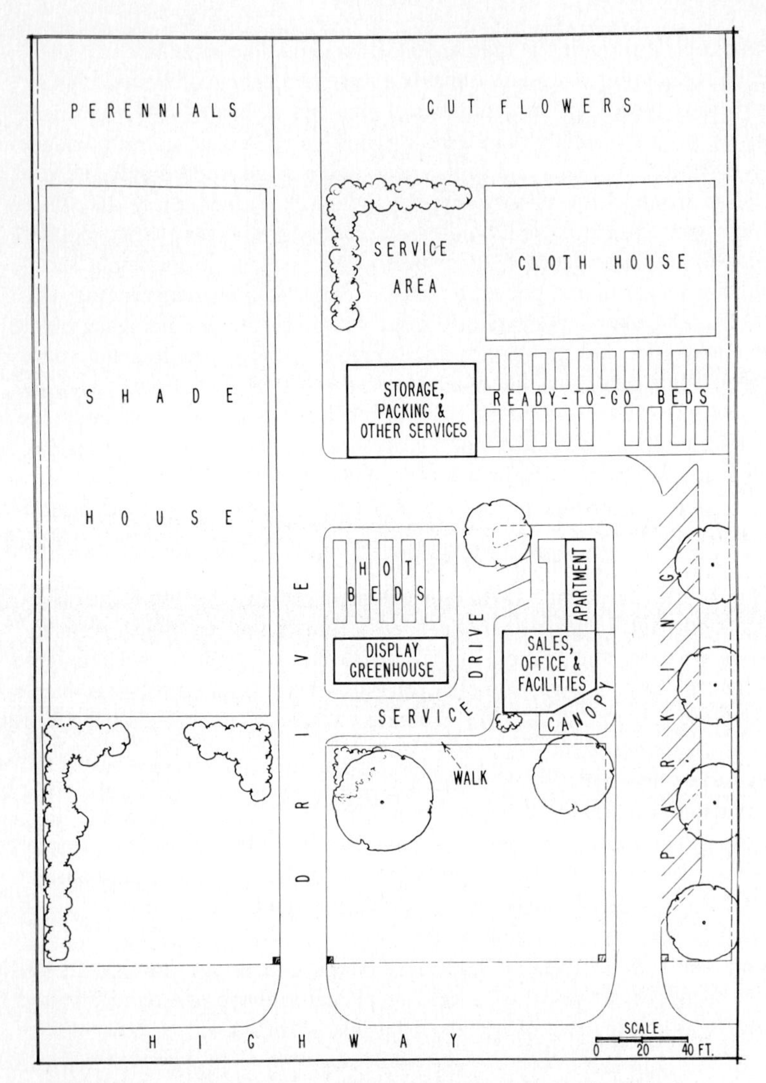

Figure 24-12 Arrangement of a sales unit in southern Mississippi. *(Courtesy, F. S. Batson, Wiggins, Mississippi)*

parking, and dependability are important. Since a retail nursery deals more exclusively with plants, the nursery manager is expected to know more about plants and thereby can give professional assistance that most garden-center managers are not qualified to provide. In one study, most of the people interviewed chose a nursery because of its reputation, while only a few chose a nursery because of its advertising.

QUESTIONS

1 State the objectives of the propagation unit, the production unit, and the sales unit.
2 Outline the cycle of production of container-grown nursery plants.
3 Applications of nitrogenous fertilizers, if made in late summer, prolong vegetative growth, delay maturity of the tissues, and make plants susceptible to winter injury. Explain.
4 What are the four major enterprises of the nursery industry?
5 A warm, rainy fall followed by a sudden, killing frost usually results in marked injury to nursery stock. Explain.
6 A warm, sunny fall is most favorable for conditioning nursery plants for the winter. Explain.
7 In general, top pruning should be avoided in late summer. Explain.
8 Show how root pruning favors carbohydrate accumulation.
9 What is meant by *balling and burlapping*?
10 Explain the different procedures of digging up field-grown nursery plants.
11 Discuss soil mixes for producing container-grown nursery plants.
12 Outline chronologically the cycle of production for field-grown plants.
13 The roots of wrapped nursery stock require air. Explain.
14 What does a well-developed sales unit contain?
15 The sales area of nurseries should provide an attractive landscape atmosphere. Explain.
16 What are some advantages of growing plants in cans?

SELECTED REFERENCES FOR FURTHER STUDY

Baker, K. F. 1957. The U. C. system for producing healthy container-grown plants. *Univ. of Calif. Div. of Agri. Sci. Man.* 23. A book concerned with cost reduction through uniformity of soil-medium operational requirements such as potting, watering, and fertilizing, which stresses sanitation as a basic necessity for healthy container-grown plants—clean soil, clean stock, and clean environment—includes discussions on helpful and harmful soil pathogens, and takes into consideration all aspects of using the U. C. system for a controlled medium.

Patterson, J. M. 1969. Container growing. *Amer. Nurseryman*. A book full of the facts, pitfalls, and satisfaction in starting your own nursery business written from Mr. Patterson's personal experience in growing container plants in Georgia, with discussions on finances, growing operation, labor, marketing, and bookkeeping, in an attempt to help the reader anticipate problems in nursery production.

Pinney, J. J. 1971. Beginning in the nursery business. *Amer. Nurseryman.* The nursery business divided into the classifications of garden center, landscape nursery, mail-order nursery, and agency nursery with a chapter pertaining to each one, as well as information on finance, record-keeping, and container-grown plants.

Chapter 25

Horticulture and the Landscape

Modern man with his great knowledge has it within his power to create on this earth a paradise beyond his fondest dreaming.

John Ormsbee Simonds

The people of the world are slowly realizing the damage that has been done to our environment. Hopefully, this resurgence of interest in nature and our environment is not just a passing fad. The field of landscape design will be of increasing importance in any effort to preserve and protect our world. In studying landscape design it is essential to maintain an open mind. Ideas from the past and the present should be observed and analyzed. However, the designer should not concentrate on what is ''modern'' or ''traditional'' but consider what is the most useful for the intended purpose. Each individual landscape should contribute to the development of a useful and beautiful environment by using the raw materials of land, plants, water, and space. In achieving this goal, the skills of engineering, architecture, horticulture, ecology, geology, and the social sciences are employed. Thus the development of a successful landscape design requires both science and art.

Figure 25-1 The creation of a series of waterfalls from an existing creek and a few stones showing clearly the rewards of working with, instead of against, nature.

The landscape is ever changing. It is alive because it consists of living plants–trees, shrubs, and flowers. Plants grow; flowers bloom and fade; and rain, snow, and sunshine change the landscape's appearance. Also, modes of expression change from time to time, either as a matter of fashion, for economic reasons, or for reasons arising from the mores of a particular era or place.

BASIC PRINCIPLES OF LANDSCAPE DESIGN

Landscape spaces are organized through the use of the basic principles of unity, balance, accent, focalization, scale, proportion, harmony, and rhythm. To develop a functional and beautiful landscape, the ground forms, structures, and plants are organized into a pleasing composition of spaces that satisfy these principles.

By grouping, arranging, or placing the different parts of the design so that they appear as a single unit, a sense of oneness is achieved. The design should present a pleasant picture from several angles. Unity also may be expressed as simplicity. Unity is achieved in a design by the use of plants similar in texture, form, and color; by noticeable repetition and transition from one grouping to another; by enclosure, which sets the scheme apart; or by developing a relation between the lines that create the pattern. The principle of unity is violated when walks, buildings, and other areas are not logically related to the overall plan. Unity is

Figure 25-2 A flat terrain and symmetrical house emphasized by a symmetrical landscape plan. (*Photo by Warren Uzzle, Raleigh, North Carolina*)

not obtained when many plants and flower beds are placed on the front lawn in competition with the house or when many different ideas or accent plants are used in the same area.

Balance means symmetrical or asymmetrical equilibrium. In symmetrical balance, there is equal balance on each side of any imaginary axis by exact duplication of plant material in line, form, and color, as in Figure 25-2. Asymmetrical, or occult, balance is the dissimilar placement of unlike objects or masses on either side of any unstressed axis to create visual equilibrium, as in Figure 25-3.

Figure 25-3 Balance meaning motion. The viewer's vision moves from level to level, but there is harmony in repetition of plant and building materials.

Accent provides emphasis. Without accent a design may be dull, static, or uninteresting. Accent may be obtained by specimen plants, change of line, use of water, lighting, variation in forms, proper use of accessories, contrasting colors, or contrasting textures.

Focalization is the climactic, or dominant, point in the design. The various parts of the design lead toward this final focal point, which attracts and holds the attention of the viewer. Note the fountain in Figure 25-4. In a formal design, the focal point is often a terminal feature at the end of the axis, such as a statue, birdbath, sundial, arbor pool, or plant composition. In some cases of formal design, the central feature of focalization is located at the crossing of two axes. This is called *Central Motive Scheme*. In an informal design, the various parts of the composition normally lead the eye to a final focal point of climax, such as attractive groupings of garden furniture or plant material.

Scale denotes the relative size of objects and plants. This takes into consideration the plant volume as well as the size of both foliage and blossom. Many large-leafed plants would be out of scale in a small garden, as would large and heavy-looking flowers. In a landscape composition, the scale of the objects is established by the introduction of anything that shows a person's height, such as steps, seats, doors, and hand rails.

Proportion is the pleasing and proper relationship of one part of a design to another part and to the whole. If any part seems large or ungainly in comparison with the rest, it will not give complete satisfaction. The interrelation of the size of one part or object to another should also be considered in designing a space.

Harmony means fitting together the various parts of a composition without a clash. A harmonious relationship among all parts of the design is the objective.

Figure 25-4 Arching limbs and brick paths pointing the way to a restful fountain in the sunlight. (*Photo by Warren Uzzle, Raleigh, North Carolina*)

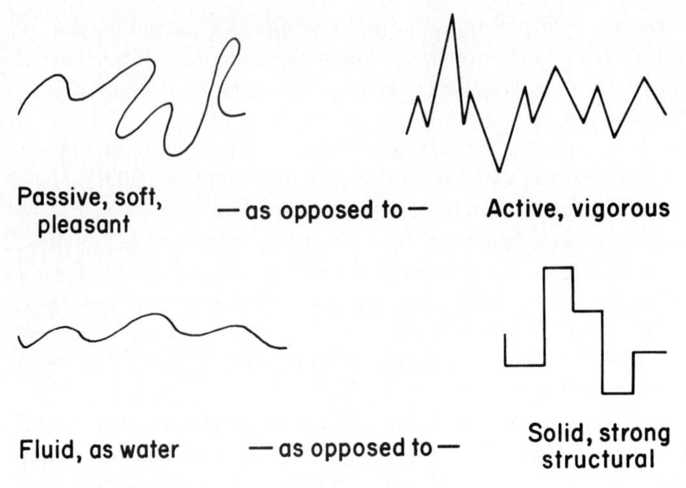

Figure 25-5 Abstract line qualities.

Rhythm is a repetition of elements that creates the feeling of motion as the eye is directed through the design. Rhythm is usually created by repetition and transition. It may be obtained by repeating groups or drifts of plants, by a row of trees, or by steps. When the eye jumps from form to form and color to color without a gradual repetition of pattern and color, there is no design. Learning to feel the rhythmic beat of color and form in design is a stimulating experience.

Just as some colors can affect the emotions—red is exciting, green is soothing—so can lines stimulate or relax. Long, flowing curves are considered passive, soft, and pleasant; whereas, jagged, pointed lines tend to be active and vigorous. Straight lines and square corners imply strong, solid, structural qualities, as opposed to curved, fluid lines. (See Fig. 25-5) These abstract qualities of line and color evoke moods and emotions through landscapes.

COMPONENTS OF LANDSCAPE DESIGN

Plant Materials

> Never underestimate the value of a handsome tree. Protect it, build your house and garden compositions around it, for it offers you shade, shadow, pattern against the sky, protection over your house, a ceiling over your terrace. It can also provide an enviable example of living sculpture.—*Thomas Church*

Plant materials are used to serve definite functions or solve specific problems. It is a mistake to scatter shrubs over the yard and try to show off each plant. This detracts from the overall design and may also create a mowing problem. The various plant materials—shrubs, trees, ground covers, and grass—are used in different ways in the design for accent, softening, screening, surfacing, shade, framing, and background. Shrubs are used as accent and specimen plants, as foundation plantings, and as borders and hedges. A shrub used as a specimen plant should be a perfect example of that plant's special characteristics. It is usually planted alone, possibly against a fence or in an easily viewed spot. It can be incorporated into a border, but it should have a different height and may vary from the other plants in texture, form, and color. Repeating the accent plant at several places in the landscape gives unity to the design.

Accent shrub plantings can be used to attract or focus attention on the front entrance of the house, and both shrubs and trees can be used for this purpose in the family living area. Normally these are the only two areas where accent plantings should be used. Do not destroy a design by using upright evergreens or other accent plants all along the front of the house or at the corners. In order to accent, plants should possess one or more distinctive characteristics. They should be outstanding in form, texture, size, or color or have a combination of these qualities. Often an accent plant or group of plants is most effective when used with other plants that emphasize the characteristics of the accent plant. For example, a plant with a coarse texture used in association with plants having a fine texture would make a good combination. The same principle works with form, color, or size.

Shrubs for softening, in foundation plantings, should be slow growers that do not require heavy pruning. Using few shrubs in the most effective places is generally better than using a large number per unit area. Rounded and spreading

Figure 25-6 A multitrunked crape myrtle supplying shade and color in the summer and sculpture in the winter.

Figure 25-7 Rocks, cotoneaster, santolina, and pebbled pavement creating the textural contrasts that make landscapes interesting.

forms usually serve better as foundation plants than stiff, upright forms. The plants should soften the meeting of the lines of the house with the lines of the ground. Plants should be placed in masses at the corners of a building. One plant will not usually do the job. Use a small group of shrubs spaced so that they will mass or grow together upon reaching maturity. Medium- and small-sized shrubs work well for one-story houses. Two-story houses may require large- or medium-sized shrub plantings to do an adequate job. The best shrubs for softening a landscape are irregular and inconspicuous in form but not unusual in foliage color and texture.

Shrubs are often used in border planting arrangements to separate areas, to screen unpleasant views, to serve as wind barriers, and to give privacy or provide an enclosure. When planning a shrub border or mass planting, use several shrubs of one type before changing to another species. For example, locate 10 or 12 camellias in a group or row; then continue the border with 8 or 10 red photinias. If further extension of the border is needed, a third choice of shrub may be used or more camellias may be planted. This approach will result in unity and harmony in the design and a well-planned border. To prevent monotony, accent plants may be introduced into the shrub border. The choice of shrubs to serve as screening is nearly limitless. The best shrubs for screening purposes are large in size, irregular or oval in form, and standard in color and texture. Those with an ultimate height of 6 to 12 feet (1.8 to 3.7 meters) that are dense and low-branching are best.

Considering the number of years required to grow trees, they should be valued and given special consideration in the landscape. Wooded lots present special problems to the builder, and many valuable trees are destroyed through careless grading practices. Even a few inches of topsoil removed from or added around the base of a tree may result in its death. Designers should take special care to include and take advantage of any existing trees on the lot, as illustrated in Figure 25-8. Trees provide protection from undesirable winter and summer winds, and they screen unsightly views. For winter wind protection, evergreen trees are best. Deciduous-tree plantings, dense in twig and branch mass, also afford excellent wind protection.

Nearly all small trees grow larger than even the largest of shrubs. Therefore, they can be used most effectively where extra tall or massive screening is needed. Small trees may be used alone or combined with shrubs in a screen or border planting. The variation in height that trees provide in a border planting is an interesting and pleasant change. Summer shade from trees is an asset to both indoor and outdoor living areas. To shade a roof or wall of a one-story house, plant medium- to large-sized trees as close as 15 to 20 feet (4.6 to 6.1 meters) from the side or 12 to 15 feet (3.7 to 4.6 meters) from the corner of the building. The lower branches should be removed as the tree grows and the canopy allowed to reach over the roof. Trees should be planted for shade on the lawn and patio, but the vegetable garden or flower beds should not be shaded. For framing purposes, trees are generally located in front of and out from the corners of the house. The best trees for framing are medium to large in size and round, oval, irregular, or pyramidal in form. Background trees are located behind the house and are important along with the framing trees in providing a total setting. Background trees should become large enough to visually break the roofline of the house when viewed from the front.

Figure 25-8 A wooded lot that makes it easy to relax and enjoy nature's landscaping, with the house absorbed into the landscape. (*Photo by Warren Uzzle, Raleigh, North Carolina*)

There are many different types of grass that will make an attractive and relatively durable surface. A lawn can be walked on and is pleasant to sit and lie on. It does not reflect heat from the ground to cause discomfort. It keeps itself clean and provides a surface for play. The color of grass and the lines that can be developed by the lawn area fit well with almost any landscape development. There are many varieties of grass, each suited for a particular area and situation. Depending on the type, grasses are either sodded, seeded, or sprigged and should be planted only after careful preparation of the soil. All grasses will need to be fertilized, weeded, watered, and mowed, so maintenance must be considered when planning a lawn. The people who will care for the lawn should estimate how much time they are willing to devote to it.

The term *ground cover* applies to low-growing, individual plants set very close together so that the soil cannot be seen. Although they are difficult to establish in sunny areas due to the invasion of weeds, they are very effective in shade. They should be used in areas that will not receive much traffic. Once established, they are relatively maintenance-free, useful in reducing erosion, and have a pleasing appearance. If ground covers are to be viewed close up, they should have a fine texture and dense growth. If they are to be viewed from a distance, their growth patterns can be looser.

Many beautiful and practical gardens contain no annual or perennial flowers, depending for their color on various flowering trees and shrubs. However, flowers and bulbs may be used as a supplement to the woody plants in a landscape design. They require a great deal of maintenance and are appropriate only in the yards of gardening enthusiasts, where they will receive the best care. Flowers are not designed to stand alone; they should be grouped in beds or in borders. Where space permits, pleasing effects can be achieved by allowing flowers to naturalize. Flowers need a dark backdrop, such as a fence or a grouping of shrubs. Flowers of varied and beautiful colors can be used, but they should be toned down with the use of browns, greens, and whites. Where cut flowers are desired, separate beds are used. These beds can be visible if well maintained or may be hidden in some part of the outdoor living area. In formal gardens, flower beds are an essential part of the development; these beds should be bordered and neatly arranged. Many flowers can be effectively grown and displayed in attractive containers; they can then be kept in an inconspicuous part of the garden until they have reached their peak of beauty, at which time they can be grouped around a patio or entryway. This will eliminate the need for frequent transplanting of flowers and yet maintain a colorful display where it will be most appreciated.

The home vegetable garden can be made to provide an adequate supply of vegetables for the family during the growing season and for canning and storage for winter use. In many cases, vegetables are not otherwise available because of the distance to markets or lack of purchasing power. Homegrown vegetables are higher in quality because they are consumed or preserved sooner after they are harvested. To a certain extent, vegetables can be substituted for more costly food. Thus, the importance of the home vegetable garden is apparent.

If properly cared for and managed, a small plot of land is excellently productive. For example, garden authorities state that a properly managed ½-acre (0.2-hectare) garden will produce enough vegetables for the average farm family the year round. Because garden crops differ in their requirements and behavior, careful planning is essential. As previously stated, some mature in a relatively short time and make their best growth in the spring or fall. Others thrive best in the summer. Some require the entire growing season. Some are tall; others are relatively short.

The vegetable garden is an integral part of the landscape plan and should be planned and considered as such. Two rather distinct types of management are used: (1) as a distinct unit and (2) in combination with the flower garden and/or home orchard. Helpful rules in planning and operating are: (1) plant in rows—100 feet (30 meters) of any crop is usually sufficient; (2) place the perennials—asparagus, rhubarb, and small fruits—on one side of the garden to avoid interference with the tillage operations of the annual crops; (3) plant crops together that have similar methods of culture; (4) rotate the crops; (5) use the wheel hoe wherever possible; (6) supply water during times of drought; and (7) keep the garden free from weeds.

An important garden practice, succession cropping, is simply a short rotation that conserves space and labor, e.g., early maturing crops, such as spinach, lettuce, endive, green onion, or mustard, followed by bean, pea or tomato, pepper, and eggplant; cabbage, cauliflower, carrot, beet, pea, and kohlrabi followed by snap bean, cowpea, or corn; beans followed by late cabbage, cauliflower, or sweet corn; sweet corn followed by bean, beet, lettuce, turnip, carrot, mustard, or spinach.

In general, the soil of the garden should be deep, friable, high in organic matter, well drained, and slightly acid. Liberal supplies of barnyard manure, combined with deep plowing and thorough preparation of the seedbed, are essential in successful home or farm gardening.

Like the vegetable garden, the home orchard should be considered an integral part of the landscape plan. On relatively large properties, the trees are grown together as a unit and are grown primarily for their fruit. However, on small properties the trees may be grown for their landscape value as well as for their fruit. For example, peach trees are both useful and ornamental. They display flowers in the spring, produce fruit in the summer, and provide an attractive color in the fall, whereas flowering ornamentals, such as the redbud, flowering peach, or dogwood, display flowers and provide an attractive color only. Thus, fruit trees may be utilized for their landscape value as well as for their fruit.

Important factors in the selection of appropriate kinds and varieties of fruits are (1) resistance or susceptibility to pests and (2) adaptation to the local environment. As is well known, certain fruits are susceptible to many insects and diseases and others are relatively free. In general, fruits that are susceptible require the application of dusts and/or sprays, which must be applied at the right time and in proper dosage. This practice is relatively expensive and time-consuming and frequently requires special equipment. Thus, when facilities, time, or interest are limited, greater satisfaction is likely to be obtained with fruits that are relatively free from pests. Note the relative freedom from pests of the crops listed in Table 25-1.

An important factor of the environment is temperature. As discussed in Chapter 5, horticultural crops may be divided into two groups: (1) cool-season and (2) warm-season. Thus, for the reasons given in Chapter 5 and with other factors favorable, cool-season fruits thrive best in the northern half or in the elevated sections of the southern half of the United States, and warm-season fruits thrive best in the southern half of the country. In other words, all the fruits listed in Table 25-1 will seldom be grown in any one home orchard. For example, in California the English walnut takes the place of the pecan. In the northern states, the apple, peach, cherry, and black grape take the place of the pecan, muscadine grape, and fig, and along the Gulf Coast and in many parts of Florida, citrus fruits take the place of the apple and peach.

Structural Materials

There are three general principles that are basic to the use of materials for any aesthetic purpose. First, every material has certain inherent characteristics that tend to produce structural forms. When the properties of materials are not suited to the constructed forms, the structural forms produced will, in turn, have a definite disrupting effect on the landscape design. If skillfully assembled, they will make an organically unified design. Second, materials have importance and character only in relation to other materials and to a specific situation. They cannot be judged by themselves; they cannot have importance in a vacuum. An effect is created by the combination of the specific way in which each material is handled, the quantitative and qualitative relations that are established between that material and other materials around it, and the nature of its

Table 25-1 Suggested Fruits for a Home Orchard

Fruit	Trees or plants	Number of varieties recommended	Planting distance, ft	Relative freedom from pests
Pecan	4	3	60 × 60	Free
Apple (on dwarf stock)	8	7	20 × 20	Not free
Peach	12	7	20 × 20	Not free
Plum	6	3	20 × 20	Not free
Cherry	3	3	20 × 20	Not free
Muscadine grape	4	2	10 × 30	Free
Other grapes	8	3	10 × 10	Not free
Brambles	40	2	5 × 5	Not free
Strawberry	400	2	3 × 1	Not free
Fig	4	2	15 × 15	Free

effect upon the space from which it is viewed. Third, a material is used not for its own sake but primarily to organize space for people to use. The more obvious functions out-of-doors, such as ground cover, erosion control, screening, shade, and color, are really all portions of the overall objective of space control. Landscape construction should extend the house or building into the outside area. For this reason, it is best to repeat the materials, such as wood, brick, and stone, used in the building construction in the landscape features. In addition, since landscape construction materials should harmonize with the building materials in terms of texture, color, and scale, the proper choice of materials is necessary.

Of prime consideration in the planning of a landscape design is the surface that will be applied to the earth. In some instances, the earth can be left in its natural state, as in a wooded area. In most cases, however, some type of natural or artificial surface is used. The expense, durability, and ease of maintenance vary considerably with the different materials used, so it is important that these surfaces be selected carefully. Coarse pine bark and chips from brush shredders such as those used by power line clearance crews are readily available and fairly inexpensive. They make a satisfactory surface for walks in private areas or wherever there is not much traffic. Their colors are natural and tend to tone down the color scheme. However, such materials are not very durable and must be replaced frequently. Gravel and rounded river rocks are more durable. They can be applied directly to the ground or laid over plastic. These materials can be used for walks where there is moderately heavy traffic. When using gravel or rocks as surfacing material, it is best to enclose the area with wooden or metal borders at ground level to keep the rocks from being scattered on the lawn. All these materials can be used in planting beds as a mulch to prevent the growth of weeds and to preserve moisture.

Asphalt blacktop has some interesting characteristics for use as a surfacing material. If laid properly, it is quite durable. It also adds contrast by creating dark areas. However, asphalt does have limitations. A large area covered with asphalt tends to be dull. It becomes soft when heated and can be marred by lawn furniture. It will also crack and sink in places if not applied properly. Bricks are used extensively for walks and patios. They provide a very durable surface and are easy to lay. They can be laid on a sand base with sand used to fill the joints, in the temperate climates of the world. However, in colder climates the bricks should be set with mortar. Bricks are available in different shapes, sizes, colors, and finishes. They can be arranged to form a wide variety of patterns. Bricks can produce a warm visual effect in the landscape. Concrete is inexpensive, readily available, and adaptable to almost any situation. It is very durable and provides a smooth, even surface for a variety of recreational uses. Concrete can be poured in many different designs and colors. Its surface can be modified by staining, brushing, and tapping in pebbles. The use of brick or wood dividers breaks the monotony of a large expanse of concrete. Precast concrete slabs, bricks, flagstone, and patio blocks can all be laid in sand, providing not only ease of installation but the added benefit of permitting a terrace to be built around a tree without depriving it of needed oxygen and essential elements. This type of terrace can also be rearranged as needs change, as in the case of converting a children's play yard to a planting bed or patio.

Wooden decks are increasing in popularity, mainly because of their usefulness as problem solvers. (See Fig. 25-9) They originated because of the need to provide a level outdoor living area on a hillside lot but are also suited to gradually sloping yards. In many cases, extensive grading around valuable trees is necessary to create a level spot for a terrace; a deck could be built around these trees. Decks built on the same level as a floor inside the house create a feeling of spaciousness, making the deck appear to be a part of the inside space.

The primary function of steps is to provide a safe and comfortable transition from one level to another, as shown in Figure 25-10. In constructing steps in the landscape, the use of a combination of materials is suggested. Wooden risers with soil or grass treads are natural and attractive but can present maintenance problems. Risers of railroad ties are rustic. Brick treads and risers offer many variations in design. Bricks used in steps should be set with mortar. Concrete and bricks used together present an attractive step design, and brick walks and wood steps provide good contrast. Precast concrete slabs and concrete blocks are often used in conjunction with other materials to create attractive outdoor steps.

Figure 25-9 Patio and wooden deck with stairs that tend to combine the two floor levels and become a part of each other for total indoor and outdoor living. *(Courtesy, Robert E. Marvin and Associates, Landscape Architects, Walterboro, South Carolina)*

Figure 25-10 Wooden risers that seem to fit naturally into the contours of the site.

Structural materials can also be used advantageously in the landscape as vertical dividers and ceilings. Tall fences and walls create privacy and screen unsightly areas. Low fences direct traffic. Retaining walls terrace a slope or create a raised planting bed. (See Fig. 25-11) A sunscreen built of wood provides a partial roof over a patio; the addition of glass or plastic will make it rainproof. All of these materials should be chosen to harmonize with the house or building—a stone wall would be inappropriate next to a brick foundation; a rustic fence would be out of place with a formal home. Walls are generally of two types—dry, retaining walls and free-standing, mortar-type walls. Dry walls can be built of wood or stone and are not set with mortar but have soil packets between the structural material used. Wood used in a dry wall should be prepared with a preservative; stones should be long enough to anchor in the bank. Dry walls should always lean into the bank. Concrete blocks can be used in the garden for walls and fit best in an informal design. Bricks form a more elegant wall. If a wall is lower than 2 feet (61 centimeters), an 8-inch (20.3-centimeter) width is sufficient, while 12-inch (30-centimeter) walls are recommended for heights over 2 feet (61 centimeters). Bricks can be laid to form an open-lattice effect. Although stone walls are not easy to build, when well built, they are very attractive and fit in many types of landscape designs.

Boards, plywood, split rails, 2-inch-square stakes, and panels made of a variety of materials can be used alone or in combinations to create fences. A board fence provides maximum privacy, but it takes a lot of lumber to build one. Care must be taken to avoid the feeling of imprisonment that can be caused by a high board fence. Setting the boards at an angle to the plane of the fence within

Figure 25-11 Visual effect of a retainer wall softened by the Shore Juniper—note the drainage holes in the wall from the raised planting bed. (*Photo by Warren Uzzle, Raleigh, North Carolina*)

the frame of horizontal supports produces a louvered fence. This type of fence gives some privacy and allows breezes to pass through. Louvered fences let more light into the garden, giving a less somber effect than a solid fence. Using 1- or 2-inch-square stakes in front of a solid fence of board or plywood produces an interesting effect. The stakes are nailed to the upper and lower horizontal supports and are set an equal distance apart. This is called a *board-on-board fence*. Split rail fences are used for decoration and as traffic directors. They are well suited to ranch style houses and natural landscapes. Basket-weave fences using redwood or other materials give complete privacy. Interesting shadow patterns are produced by a fence of this type.

RESIDENTIAL LANDSCAPING

A house should be more than just a structure equipped for eating and sleeping, garnished with a few shrubs for decoration. No matter how large or small, it is a place where the habits, hobbies, pleasures, and friendships of the entire family can find expression. Before beginning the construction of the house, the unique characteristics of the lot should be considered in relation to the structure. The contours of the lot, the existing trees, and the views, both good and bad, will determine to a great extent the most desirable placement of the house. Major living areas can be located to take advantage of a natural view, a prize shade tree, or an unusual rock formation. Unfortunately, the landscape designer is often called in to rework an existing, overgrown, or inefficient garden or to landscape an already existing structure.

The landscape designer needs to assess not only the site but also the particular needs of the family. Site analysis should begin with a detailed topographical map, showing contours and other natural features, existing trees, street right-of-way, and the location of good views to develop and bad views to screen. The designer should plan to visit the site several times to learn its landscape character. John Ormsbee Simonds, in *Landscape Architecture*, quotes a Japanese designer's own method of site analysis:

> If designing say, a residence, I go each day to the piece of land on which it is to be constructed. Sometimes for long hours with a mat and tea. Sometimes in the quiet of the evening when the shadows are long. Sometimes in the busy part of the day. . . .
>
> And so I come to understand this bit of land, its moods, its limitations, its possibilities. Only now can I take my ink and brush in hand and start to draw my plans. But strangely, in my mind the structure by now is fully planned, planned unconsciously, but complete in every detail. It has taken its form and character from the site and the passing street and the fragment of rock and the wafting breeze and the arching sun and the sound of the falls and the distant view.

The designer will then want to interview the family to learn not only their interests and recreational needs but also the amount of time they are willing to spend maintaining their property. Most families will not be able to afford a complete landscape plan all at one time. Often the installation of a major item, such as a patio or swimming pool, will be postponed for several years. The designer will, nevertheless, want to prepare the plan with all of the present and future needs included. A temporary surface, such as grass or pine needles, could be used in the location of a future pool or patio.

The landscape plan is put on paper—not only to facilitate the design process, since errors are easier to erase than to dig up and transplant, but also to convey the designers' plans to the various contractors. The lot, its significant natural features, and the actual location of the house should be plotted on graph paper using a scale of 1 inch = 10 feet (2.5 centimeters = 3.0 meters) or 1 inch = 8 feet (2.5 centimeters = 2.4 meters). Using tracing paper over this graph, rough outlines of the various use areas can be sketched, reworking this general outline until each area—the public areas, the service area, and the private living area—is in proper relationship to the house, the street, and the unique characteristics of the lot. The basic design should then be worked out, with the major areas of construction —walks, drives, patio, and the areas in need of screening with fences or planting—drawn in. (See Fig. 25-12) The final phase of the landscape plan (which most people think of as "landscaping") is the placement of trees, shrubs and ground covers, keeping in mind not an individual plant, but a general category of size, shape, and texture. Thus a preliminary plan might specify an "evergreen shrub, 6 to 12′ (1.8 to 3.7 m) for shady area." The specific variety of plant can then be chosen from within this general classification.

People want the landscape to be both useful and beautiful. Thus, everything is planned for the convenience of the homeowner and also combined into a pleasing design. A workable, beautiful garden and yard should be the aim. This cannot be

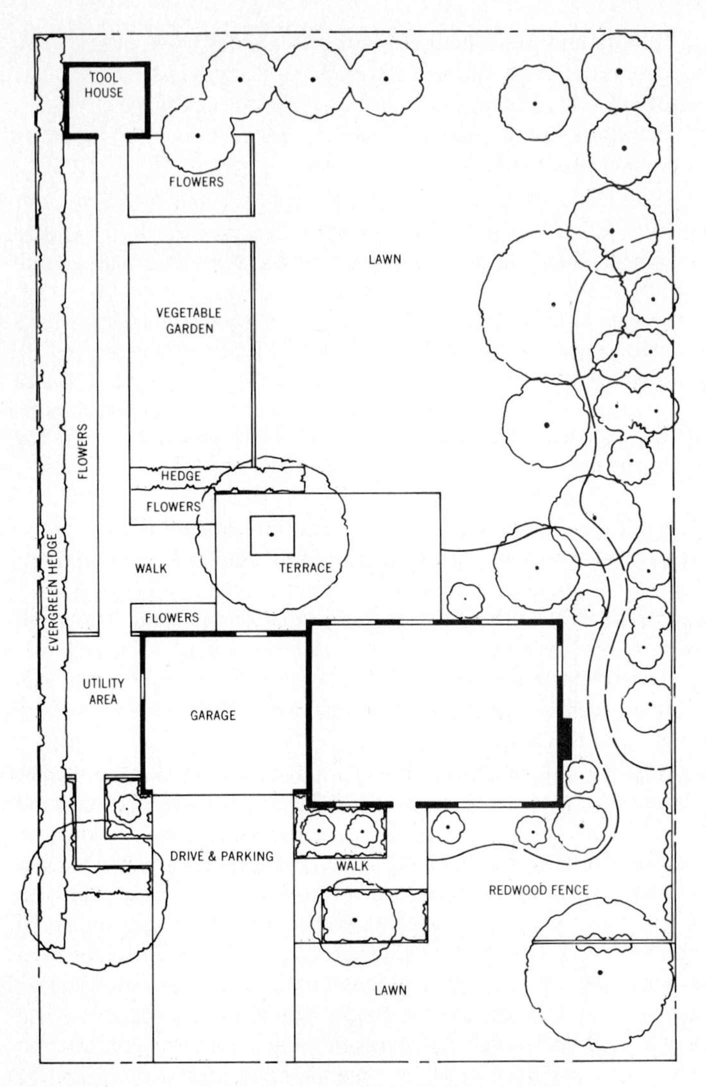

Figure 25-12 Basic plan for a residential landscape design.

accomplished without careful planning. In general, the residential landscape is divided into three spaces: the front, or public space, the service space, and the private outdoor living space. These spaces may be compared to the rooms of a house, each designed for a specific purpose, but related to and dependent upon the others. A landscape space, or volume, is composed of three parts: the ground, and whatever organic or structural surface is applied to the ground; the verticals, which include the outer walls of the building, the trunks of trees, and any fences or planted screens; and the ceiling, which may be the open sky, the leaves and branches of trees, or a roof or trellis. The verticals in the landscape determine, to a great extent, the emotional quality of the space. A vast space without verticals tends to frighten and humble an individual. The introduction of a tree into this space helps the person relate to the space. An enclosure, such as a fence or a row of plants, serves to relax and reassure. In the opposite extreme, too many verticals, as in a maze, excite and confuse. Although the sky is the obvious ceiling to the outside space, it becomes an element in the design when viewed through a canopy of leaves or a latticed roof.

The Public Space

The portion of the lot between the house and the street is called the *public space*. While nearly everyone agrees on the purpose of the private living space, there is a general disagreement as to the function and thus the form of the public space. Tradition commits us to a large, uncluttered front lawn, perhaps a center walkway to the street, foundation plantings, and a few mature shade trees. However, the necessity for offstreet parking, buffers against street noise and dust, development of a view and privacy for the living areas in the front of the house, and the desire for a distinctive and individualized ''outdoor entrance hall'' have led many landscape designers to alter the traditional form of the public area. Traditionally, emphasis has been on creating a showpiece, to present an elegant picture to the passersby. Currently, more and more public areas are being developed for the benefit and use of the homeowners and to provide an inviting and convenient entrance for guests. However it is developed, the public space should be attractive, of durable construction, and designed for minimum maintenance. The guest entrance should be designed for the least confusion and the most pleasure. In designing the public space, plan first for the location of the basic parts—the driveway system, walkways, and entryway. The size and shape of the driveway system will be determined by the number of family vehicles, location of garage or carport, the need for off-street guest parking with convenient access to the front entrance, and the need for a turnaround space to eliminate the necessity for backing out into a street. A drive designed for one car should be at least 10 feet (3.0 meters) wide, while one serving two cars should be at least 18 feet (5.5 meters) wide. The drive and turnaround should be close to the property line, and in most cases, lead straight from the street to the house. On a large lot, a semicircular drive can help emphasize the entrance to the house. Off-street parking areas for guests can be incorporated into the driveway system but should be easily accessible to the front entrance. For side by side parking, a width of 10 feet (3.0 meters) per car is required. Although the necessary size for turnarounds will vary with the car, a radius of 18 feet (5.5 meters) is suggested as the minimum. Drives may be surfaced with a relatively inexpensive material such as gravel; this can be very satisfactory, especially if the drive is edged with wood or bricks to prevent the gravel from scattering. Gravel would not be appropriate, however, in a climate where frequent snow shoveling is necessary. A more permanent and low-maintenance surface can be obtained at a higher initial cost with the use of asphalt, concrete, or brick.

Walks should be placed only where they are needed; this can usually be determined by observation of traffic patterns since walkers will tend to take the shortest, most level route. The walks should be kept on a scale with the house and be constructed of harmonious materials. The main walk should be between 4 and 6 feet (1.2 and 1.8 meters) wide to permit two people to walk side by side to the entrance. This walkway should be surfaced with a permanent material, such as concrete, asphalt, brick, or stone, to permit ease in walking and should include steps if the grade is uncomfortably steep. Steps outside should be designed differently from those used inside the house. Out-of-doors, people generally walk faster with normal strides and tend to be less careful than when they are in the confined space of a building. Therefore, risers on outside steps should not be as steep as those inside a building and the treads should be deeper. If steps are necessary on long steep banks, it is better to break the long climb with a landing after seven or eight steps than to have a continuous tiring climb. Garden steps should be well lighted by low lights with shields to reduce glare if used at night. Steps should be in harmony with walks, paths, and other landscape features. Secondary walks, those going from an entrance to something other than the drive, need not be as wide, nor is it necessary to construct them of sturdy materials. For instance, flagstone or bark can be used for secondary walks. When the walk runs parallel to the house, at least 6 feet (1.8 meters) should be left between the walk and the house for plantings.

Perhaps one of the most abused ideas in design is that of foundation planting. Years ago, houses were built high off the ground with several feet of foundation or lattice-work that had to be covered. Here, the foundation planting had its beginning. The two basic principles that should be kept in mind are harmony and unity. Two to four types of plants should be chosen for the foundation planting and used exclusively. This creates repetition and builds unity. Harmony is maintained between the house, the plants, and the landscape. The corners of the house and the sides of the doorway are where the lines of the house and those of the ground acutely meet. Therefore, these spots will almost always require plants. Corner groupings of three to five plants should be higher than those near the doorway. The groupings on either side of the doorway should be low-growing or dwarf plants whenever possible. The use of conical shaped evergreens in the foundation planting should usually be avoided, as these plants tend to emphasize rather than soften the vertical lines of the house. A ground cover can be used to unify the corner and doorway plantings. In symmetrically designed houses, the plantings need to be identical on both sides; in asymmetrical designs, the groupings on each side of the entrance should be of the same type plant, but the grouping on the long end should be larger. This acts to "shift the weight" back toward the center of the house and achieves balance. With very tall houses, trying to soften the corners with shrubs results in loss of scale; therefore, by grouping slow-growing shrubs at the corners and planting groupings of ornamental trees in the yard, the corners will remain in scale. Wing plantings are used to make a house appear longer, direct eye movement to the house, and separate the public and private areas. In wing plantings, smaller shrubs should be near the house with the shrub height increasing as it approaches the property line. Foundation plantings should emphasize good architectural features and mask poor ones. When the foundation is high or unsightly, some type of construction or planting is needed to minimize the distance between the ground and the first level of the house. Series of broad terraces or decks are preferable to a steep flight of steps leading to the entrance. In this instance, a retaining wall backed with fill could be placed several feet out from the foundation and planted with low shrubs and ground covers. Where the floor level of the house is nearly even with the ground and the architectural lines of the house are acceptable, elaborate foundation plantings are usually unnecessary. The distance from the house to the street, along with the amount of traffic on the street, determines the possible need for some type of screening.

The presence of existing trees and the shade they create indicate the desirability of using ground covers rather than grass. In cases where a lawn is appropriate, it should not be cluttered with flower beds, birdbaths, or other bric-a-brac. Ideally, it would remain green all year round; in some climates, however, this is impossible. The public space is to be used and seen. A very simple design that is well cared for is far better than an elaborate design that is allowed to become shabby. Be careful not to design a public area that will require more care than the homeowner is willing to give.

The Service Space

Garbage cans, clothesline, dog run, and woodpile, along with storage for gardening equipment, boats, and trailers, are included in the service space. This space is located where it is screened from the street, from the neighbor's view, and from the outdoor living area and is convenient to the kitchen and outside entrance. If a woodpile is kept in the service area, it should be convenient to the room containing the fireplace. A storage place for garden implements or anything that needs to be kept handy but out of sight belongs in the service area. Screening can be achieved with walls, fences, plants, or a combination of these. One idea is to use a fence to screen the area from the neighbors and to use plants to block the view from the private and outdoor living space. The service space should be kept as small as possible and yet large enough to serve its purposes.

The Private Outdoor Living Space

An indication of good landscaping is not only a house that invites you in but also a garden that invites you out. (See Fig. 25-13) The private space is designed as an

Figure 25-13 Private deck and intimate garden that invites you out of the house, with the wood deck constructed around the existing dogwood tree. *(Courtesy, Robert E. Marvin and Associates, Landscape Architects, Walterboro, South Carolina)*

extension of the indoor living space; a place where the family can, in relative seclusion, pursue their various hobbies, entertain their friends, enjoy a quiet dinner, or simply sit and contemplate the view. A thorough analysis of the site reveals such things as the necessity for screening an objectional view or enlarging a pleasant one, the need for fencing to ensure privacy, or the possibility of locating the deck under or around an existing shade tree. Naturally these requirements differ with each situation. The recreational needs of the family and the ages of the children determine the design of a large portion of the living area. Families with small children need a play yard conveniently located so that their activities are easily supervised. In addition to the usual swings and sandbox, a yard for the children could include a circular hard-surfaced track for riding toys. The ground under swings and play equipment can be surfaced in a soft material such as crushed bark or sand, and the play area should be located in a shady area. (See Fig. 25-14) This section of the private space is planned for change as the youngsters grow. Older children enjoy a large lawn for outdoor games such as volleyball, badminton, and football, or a hard-surfaced area with a basketball hoop. If a swimming pool is to be included in the plan, or considered as a future addition, many factors will have to be taken into account, such as the need for a fairly large, level area adjacent to the living area, kitchen, and bathroom, and the need for protective fencing, which demands a high initial expense and involves considerable maintenance. For some families, a swimming pool is a worthwhile investment and will determine the nature of a large portion of their outdoor living. (See Fig. 25-15) Families who are interested in gardening

Figure 25-14 Canvas tent providing shelter from the sun while giving a circuslike atmosphere to a play area.

Figure 25-15 Swimming pool designed to look like a garden pool reflecting the shape of the entire garden, with pool, brick walls, and patio designed to be subservient to the overall landscape design. *(Courtesy, Robert E. Marvin and Associates, Landscape Architects, Walterboro, South Carolina)*

beyond the usual maintenance of lawn and shrubs may want to set aside a sunny spot for vegetable and cut-flower gardens, collections of different kinds of plants, or herb gardens. If the vegetable garden is well kept screening is not required.

The outdoor living space should be an extension of the indoor living area. To achieve this, it must be easy to see and travel between the two. An important part of any outdoor living area is a space for entertaining. This can be a terrace, a patio, a wooden deck, or simply an open space of lawn. The indoor living area should open directly onto this area. Convenient access to the kitchen also facilitates outdoor cooking and eating. Shade should be provided for this area so that sitting outside in the summer will be pleasant. A portion of this area may be covered so that all outdoor activity will not have to be moved inside in case of rain.

Although the temperature of a terrace can be regulated by trees, screens, or roofs, the designer will certainly want to take into consideration the direction a terrace will face. A southern exposure will have sun all day in every season. This may be desirable in a cool climate, but in a hot climate it will be necessary to provide shade for this south terrace with a large tree or a latticed roof. A patio oriented toward the west will be pleasant in the mornings, but the late afternoon sun will make it uncomfortably hot and bright. Some type of screen will be necessary to make this terrace usable. A terrace facing the east will be most desirable in hot climates. It will receive the benefit of morning sun, with welcome shade in the heat of the afternoon. In cooler weather, however, the east-oriented terrace may never warm up enough to be comfortable. The north-facing patio will be the coolest, since portions of it may never receive any direct sun.

Patios and terraces will receive much traffic. Therefore, they should be composed of durable materials such as concrete, flagstone, or brick. If the slope of the area is too great for a terrace or patio, a wooden deck may be built. If properly constructed, a wooden deck contributes beauty to the area. Waterproof electrical outlets and lights installed around these areas will add to the enjoyment of the area and prolong the time during which it can be used. Direct lighting with floodlights causes an unpleasant glare. Downlighting, however, involves placing the lights high, about 20 feet (6.1 meters) above ground and shielding them if necessary. Uplighting involves reflecting the light from a source on the ground against a structure or a plant grouping. Interesting shadows result.

A large lot in the country will not ordinarily require enclosure for privacy in the living area. In an area where the houses are close together, however, enclosure is a necessity. Fences provide enclosures using the least amount of space. The wide variety of fencing materials available should fit most landscaping requirements. Open fences give a sense of enclosure and privacy without creating a closed-in feeling. Solid fences provide a greater sense of privacy. Simple fencing materials are better because they do not detract attention from the landscape. Fencing material should harmonize in texture and color with the house itself. Different complementary fencing materials can be used in the same plan with more attractive and/or expensive materials used around the patio and simpler materials used for the rest of the enclosure. It is best not to change fencing materials at a corner because this attracts attention to that spot. Short panels of fencing set at right angles to the side fence tend to break up the area and can lend a pleasing effect. These panels may be used as a backdrop for specimen plants. All fences should be of the same height in a single design. The posts for fences are usually 8 feet (2.4 meters) apart and are set in concrete below the frostline. Walls or masonry may be used for enclosure but are more expensive and require more skill to construct. Where space allows, plants serve as ideal screening material. Plants give a relaxed feeling to the area. The screen should be composed primarily of one type of plant to achieve unity. Specimen trees incorporated into the screen give a nice appearance and can emphasize the importance of an area.

Although in some parts of the world, people rarely use their outdoor area during the winter months, this does not reduce the need for creating a view from the living areas. Therefore, the designer should suggest plants that have good qualities all year. Pleasing bark and branching effects of deciduous plants, fall color, a good selection of evergreens, and colorful fruiting effects of plants are some of the elements that must be considered when planning a year-round outdoor living area. (See Fig. 25-16) The outdoor living space also can improve life inside by improving the climate immediately surrounding the house. Noise, dust, and harsh sunlight can be alleviated by careful planning. Urban dwellers are constantly pelted with the sounds of planes and freeway traffic. Yet, something as simple as a tree or shrub offers a solution. A row of trees or shrubs in the backyard of a private home will serve as a sound barrier against these noises.

Figure 25-16 Bark textures of birch and pine harmonized with azaleas, liriope, and water to provide an intriguing garden trail.

The following example for developing a lot for a family's use and enjoyment may be helpful in understanding the landscaping of a residence. Consider a four-member family of the average middle-income bracket. Economic forces will play an important part. Taxes are high, building is expensive, and lots are small. These factors have dictated a home that will not be large enough to satisfy all the needs and hobbies of this family. The result is a scattering of the family in pursuit of their hobbies and friends. The children have disappeared on their bicycles and the grown-ups in their cars. This is a serious condition and at first glance seems hopeless. Careful landscape design can develop this small home on this small lot in such a way that the family's hobbies and space needs are met so that the family can return home for its hobbies, inspirations, and satisfactions. This will require an acute awareness and complete use of the site. The garden will not be designed primarily for plants but primarily for the family. Why own a quarter of an acre lot and live cooped up and separated from nature in 2,000 square feet (185.8 square meters) of walled-in house? Why not replace the old idea with a new concept that divides spaces on the entire piece of property for this family's use and pleasure? The cost of building will not be increased, but additional space for living will be created. Why not change some traditionally solid walls to glass and gain privacy by building fences on the property line? Why not angle the glass walls to the proper exposure so that there is sun in the rooms during the winter months and shade in the summer months? Why not place these glass walls in proper location to deciduous trees, or plant trees if necessary, so that they give additional shade in summer but drop their leaves and let the sun flood the room in winter? Why not extend floors through these glass walls, thereby increasing the visual space inside and the living space for family and friends outside? Why not drop walls back and extend walls in other directions to save trees and give a feeling of being closer to nature? Why not go even farther and make glass walls larger, and let nature be the theme and the key to color, lines, forms, and textures? This house would cost no more, yet its possibilities are as great as the imagination, its limitations only those of influences of the past. (See Fig. 25-17)

LANDSCAPING PUBLIC BUILDINGS, SCHOOLS, AND INDUSTRIAL AREAS

> Understanding how people use and value the spatial environment is the key to planning sites that fit human purposes.—*Kevin Lynch*

Building contractors apparently feel it is easier to build on a leveled site and replace the landscaping later than to try to build around existing trees and contours. There is a growing tendency among people, however, to insist that all possible care be taken to work within the natural limits of a site. The shoppers and salespeople at a shopping center, the workers in an industry, and the students at a college are all soothed and refreshed by a well-planned landscape. Business

Figure 25-17 The inside environment extended to the greater outside environment with only an expanse of glass separating the two. *(Courtesy, Robert E. Marvin and Associates, Landscape Architects, Walterboro, South Carolina)*

people are discovering that customers gravitate toward a drive-in or service station under the welcome shade of a tree; apartments with well-designed, functional landscapes attract and keep tenants; workers are more efficient if their lunch break includes a stroll through a leafy park. There is a growing trend toward converting the main business streets of a city into a parklike mall, limiting automobile traffic to adjacent streets, and leaving the area free for casual strolling and window shopping. Thus, the relation of nature to human happiness and health is slowly being realized. The actual formulation of a landscape design of a building to be used by the public is not so different from that of a private home; it is simply based on a much larger scale. The designer considers the three areas—public, service, and private—in much the same way.

In the public space, access to the building is provided through driveways and walkways. These must be of very durable materials, such as concrete or asphalt, but should complement the structure. The parking area is, in most cases, of prime consideration. In general, ideal parking lots are designed to be pleasant as well as functional, with dividing islands of greenery, trees for shade, and contours for variety. The walkways and entrances to the structure are designed in a manner similar to that of the private home—with an attempt to create an inviting and pleasant approach to the building, avoiding unpleasant confusion along the way. Most public buildings need some sort of service entrance for delivery of supplies, trash pickup, and outside storage. This area should be screened with fencing or plant material. A very necessary, but often neglected, portion of the landscape of public buildings is the equivalent of the "private space." This should be an area sheltered from the noise and confusion of the streets, filled with a variety of plant material and places to sit and relax. The plant materials used around a public building should be extremely durable and suited to the location. Fragile plants that need pampering are doomed in such a situation. As with a private home, the type of plant material and the amount of grass used will depend to a great extent on the availability and enthusiasm of the maintenance staff. Planting areas should be carefully and heavily mulched to minimize the necessity for hand weeding. Lawn areas should be edged around trees and planting beds to eliminate the need for hand trimming. If there are no existing shade trees around the building, new ones should be planted. In addition to their aesthetic value, trees will help to control the climate within the building by deflecting winds and shielding summer sun.

QUESTIONS

1. What material of the landscape is most effective in softening noises? Explain.
2. How can accent be obtained in a landscape design?
3. What are the cultural ecological factors that should be considered in selecting plants for a landscape composition?
4. What are the three areas of the home grounds? What is the purpose of each of these areas?
5. Why is it almost impossible to locate landscape elements such as drives, walls, and patios before a design is prepared?
6. What methods are used to achieve unity in a landscape design?
7. Enumerate the advantages of a home vegetable garden.
8. Make a simple plan for a home vegetable garden showing the proper arrangement for perennials and long-season and short-season crops.
9. Outline the steps in properly preparing a vegetable garden previous to planting.
10. Make a list of vegetables that can be grown in your locality.
11. Contrast the peach and the redbud for use in landscaping small properties.
12. Show how fruit trees can be used to screen out undesirable views and to provide a background for the house.
13. In general, the average homeowner is more successful with bramble fruits, figs, and strawberries then with peaches, apples, and pears. Explain.
14. Explain the relationship of economics, aesthetics, glass, and garden to space.
15. What is meant by incorporating livability into the design of the outdoor living area?

SELECTED REFERENCES FOR FURTHER STUDY

Church, Thomas. 1969. *Your private world*. San Francisco: Chronicle Books. A discussion of the landscaping of specific city, suburban, and country gardens beautifully illustrated with photographs and based on the author's belief that each landscape should be an extension of the personalities and needs of the residents it surrounds, incorporating the author's landscape designs, which are as varied as the individuals he has served, and his garden designs, which are intimate and serene.

Lynch, Kevin. 1971. *Site planning*. Cambridge, Mass.: The M.I.T. Press. How to plan, layout, and landscape a site for housing, shopping centers, industrial parks, or open spaces by using practical approaches that take into consideration mobility, existing structures, management and control of the site, and cost, as presented by the author with a special sensitivity to human activity, both social and psychological, and to how humans interact within and respond to the landscape site.

Simonds, John Ormsbee. 1961. *Landscape architecture*. New York: McGraw-Hill. A book conveying the action in natural surroundings and the attempt to capture these feelings in landscaping and site planning by leading the reader through the process of forming a design for a site, with drawings and photographs illustrating landscape design.

Chapter 26

Scope of and Careers in Horticulture

Horticulture is . . . Food for body and soul
a bunch of hot peppers ready for drying,
a field of pumpkins, and bright marigolds,
the apple a day that keeps the doctor away,
grape jelly and jam all store-sacked for buying.
Provitamin A and vitamins B, C., and E,
a poinsettia for Mom's Christmas, a potted mum for me,
wide-spreading elms and tall, sturdy oaks,
lilacs and petunias for old-fashioned folks,
cold apple cider, sparkling wine and champagne.
A head of lettuce for a fresh green salad,
a corsage for Grandma on her birthdays,
cool grass under bare feet,
manicured golf greens and mowed lawns smelling sweet,
orchards and gardens and beautiful highways.
People working for a better world.[1]

[1]Adapted from *HortSci. 6, June 1971, by Mrs. Betty Darden, Mississippi State University.*

THE NATURE OF HORTICULTURE

The term *horticulture* is derived from two Latin words, *hortus,* meaning a garden, and *cultura,* meaning cultivation. In ancient times, the gardens of large estates were surrounded by high walls or similar structures, and the crops usually cultivated within these areas were tree fruits such as apple, pear, peach, date, pomegranate, or fig, and/or many vegetable, flower, and ornamental plants. As a result, the term *horticulture* in the original sense referred to the cultivation of crops within a protected enclosure. Quite frequently this area was called a *garden.* Thus, *horticulture* referred to the culture of crops grown in gardens.

At present, fruits, vegetables, flowers, and ornamentals are grown not only within the home grounds, but also in large quantities on a commercial scale. In other words, many horticulture crops throughout the world are grown primarily as business enterprises. Thus, horticulture has at least two aspects: the amateur and the professional. In addition, with the establishment of agriculture experiment stations as an integral part of colleges and universities, combined with the use of the scientific method in the solution of problems, a comparatively large amount of scientific information on the behavior of horticultural plants and products has become available. This information, or literature, constitutes the science of horticulture, and the application of this scientific information constitutes the technology of horticulture. Thus, *present-day horticulture* may be defined as *the science and technology involved in the production, processing, and merchandising of fruits, vegetables, flowers, and ornamentals.*

The Broad Scope of Horticulture

Horticulture is a wide field and includes a great variety and diversity of crops. In general, these crops are classified as follows: (1) crops grown for their food and/or vitamin and mineral content, for example, apple and sweet potato, (2) crops grown for their beverage properties, for example, tea and coffee, (3) crops grown to beautify the environment, for example, the grasses of a well-kept lawn, or to express human sentiments, for example, cut roses, and (4) a miscellaneous group, including crops grown for perfume or spices. Each crop has its own science and technology. Consequently, horticulture may mean one thing to one person and another thing to another. To the teacher, research worker, extension specialist, or landscape gardener, it is an applied science and a profession; to the commercial producer, the wholesaler or the retailer of horticulture products, or to the manufacturer or seller of horticulture supplies, it is a business; and to the home owner, it is a fascinating hobby.

The Impact of Horticulture

In general, the population of any given country consists of units—one or more individuals living under one roof. These units may consist of one person, for example a young or old bachelor or widower, or a young or old spinster or widow; two or more unmarried men, or two or more unmarried women; or more frequently, the individual family—the father, the mother and the children. Irrespective of the number and nature of each unit, all individuals within a unit require: (1) a continuous and abundant supply of high-quality food at reasonable cost, and (2) an environment which elicits the positive side of human behavior. With reference to the first requirement, horticulture provides many essential foods and vitamins, substances which contribute to a well-balanced diet. Examples are fruit and vegetable crop juices and fresh and preserved fruits for breakfast; fruit and vegetable crop juices, pot-herbs, or greens, baked or French fried potatoes, baked or scalloped sweet potatoes, salads, and many kinds of pies for lunch and/or dinner; and potato chips, nuts, and fresh fruit for snacks. In fact, an examination of the banquet menu presented in Figure 26-1 or a trip through a modern supermarket presents convincing evidence of the importance of the many kinds of fruits and vegetables which are always available to the consumer. With reference to the second requirement, horticulture is the sole supplier of the many kinds of ornamental trees, ornamental shrubs, and herbaceous flowering plants which are used to beautify the environment and to help stimulate the expression of the positive side of human behaviour. Here again, a modern ornamental plant nursery or garden center, a community of well-landscaped homes, a landscaped park or picnic area, or a landscaped expressway or parkway presents convincing evidence that the aesthetic phase of horticulture is an integral part of man's environment. In fact, ever since man became a tiller of the soil, he has always liked to have plants in close association with him. This explains why man seems to be happier in the presence of green plants than in the absence of them, and why when he leaves the country for the city, he takes some of the country with him.

Tomato Juice
Roast Sirloin of Beef au Jus
Idaho Baked Potato
Blue Lake Snap Beans
Tossed Green Salad
Hot Rolls and Butter
Coffee
Hot Apple Pie

Figure 26-1 The menu of a typical banquet at the student union of Mississippi State University. *(Courtesy, G. L. Haralson and R. H. Goodwin, Mississippi State University.)*

CAREER OPPORTUNITIES IN HORTICULTURE

In my Father's house are many mansions.—*John 14:2*

Since horticulture is a wide field, career opportunities are many and varied. In general, they are classified as opportunities in (1) the production of horticulture crops, (2) the buying and selling of horticulture products and the manufacture and/or merchandising of materials and supplies essential to the production of horticulture products, (3) the processing of horticulture products, (4) the landscaping of public and private properties, and (5) the research, teaching, and extension phases of horticulture.

Career opportunities in production include owning and/or managing orchards, vegetable farms, commercial greenhouses, commercial flower gardens, and ornamental crop nurseries. In general, these opportunities require a working knowledge of the effect of the environment and practices which affect the photosynthesis, respiration and water absorption, and transpiration rates, and the unique characteristics of the crop concerned. For example, successful orcharding requires a knowledge of varieties and pollination requirements and the effect of fertilizers, irrigation, spraying, and pruning practices on the market productivity of the plants; successful greenhouse production requires a working knowledge of varieties and the effect of the water supply, temperature, light and essential-element supply on plant growth and development.

Career opportunities in buying and selling include buying and selling horticulture products for wholesale distributors, hotels, hospitals, the armed services, and chain stores or supermarkets; selling fresh or processed fruits and vegetables, cut flowers, potted plants, or nursery stock on a wholesale or retail basis; and selling fertilizers, growth regulators, pesticides, and equipment used in the many horticulture industries. Buying horticulture produce requires a knowledge of varieties, the physiology of maturity, or degree of ripeness, market classes or grades, type of defects, and the effect of the environment on the storage life of the product concerned. The management of produce in retail stores requires training in packaging, in displaying the product to the best advantage, and in anticipating the needs of the consumer.

Career opportunities in processing include owning or managing canning, freezing, dehydrating, or pickling plants, serving as a food technologist in the processing of fruits and vegetables, and doing research work in the prepackaging and merchandising of flowers, fruits, and vegetables. For example, the potato chip industry requires a thorough knowledge of the effect of frying and packaging techniques on the quality of the chips; and the pickling industry requires a knowledge of varieties and the effect of the fermentation process on the quality of the pickled product.

Career opportunities in landscape design and related fields include owning and operating a landscape design company or a tree surgery and maintenance company; serving as a superintendent of parks, estates, corporate communities, and recreational areas, as a turf consultant for golf courses, playing fields, and recreational areas, or as a "plant doctor," either in a private capacity, or for a city or a group of adjacent communities. Success in landscape design requires ability in freehand and other types of drawing, training in color and design, and a working knowledge of landscape properties and the physiology and morphology of trees, shrubs, and other plants used in landscape plantings.

Career opportunities in research, teaching, and extension include investigating horticulture problems in both public and private institutions. Examples of research include the U.S. Department of Agriculture, colleges and universities throughout the world, seed companies which develop new varieties, and manufacturers of materials and equipment used by growers. Teaching opportunities include teaching horticulture and related subjects in high schools, in technical institutes, and in colleges and universities throughout the world. Extension opportunities include extending the results of research in horticulture to all the people. Other opportunities include serving as a writer or editor of a horticulture or garden magazine or working for the horticulture section of agricultural magazines or the garden section of weekly or daily magazines or newspapers. Each capacity provides an opportunity to extend the results of research with horticulture crops to all the people.

HORTICULTURE AND THE INDIVIDUAL STUDENT

As pointed out previously, the field of horticulture is very wide. Thus, it has an appeal to the potential commercial producer, businessman, landscape designer,

Figure 26-2 A laboratory class in commercial floriculture. *(Courtesy, C. O. Box, Mississippi State University.)*

Figure 26-3 A distinguished teacher lecturing to his class on commercial floriculture. *(Courtesy, C. C. Singletary, Mississippi State University.)*

investigator, teacher, and extension specialist. All of these positions have characteristics in common and characteristics which make them distinct from each other. Three important common characteristics are (1) a profound and abiding interest in crop plants, (2) a willingness to work constructively, first at low-level jobs, if necessary, and (3) an active interest and a positive attitude toward people. The first leads to a working knowledge of crop-plant behavior and an appreciation of the fundamental processes of plant growth and development and the effect of the environment and practices on these processes. The second provides the experience necessary to hold high-level positions and leads to a full appreciation of all phases of the horticulture industry. The third enables individuals to acquire knowledge about how to get along with people. For these reasons, the first two years of study in college are practically the same for all majors in horticulture. In general, the student is required to take courses in basic science, usually basic botany and basic chemistry, in basic mathematics, in English, usually composition and speech, and in the social sciences and the humanities, usually political science, history, economics, and sociology.

HORTICULTURE CURRICULA AS A WELL-ROUNDED EDUCATION

Do horticulturists acquire a well-rounded education? Does the course of study help a student to attain success in other fields? Are the students aware of the opportunities for leadership in the local community, and are they willing to help solve current problems? A partial answer to these important questions is based on the fact that not all graduates in horticulture remain in the field. In fact, a recent survey shows that 25 to 30 percent of the graduates enter other fields: medicine, dentistry, law, various businesses not directly related to horticulture or agriculture, and military science. What are the reasons for success in other fields? A study of the horticultural curriculum of most institutions shows a well-selected core of studies in the English language, including composition, speech, and business correspondence, and in the social sciences and humanities, including history, government, philosophy, religion, and psychology. Further, and as previously pointed out, students may elect a relatively large number of courses. In this way, students, particularly in their junior and senior years, are given the opportunity to make decisions with respect to their life's work. This is in keeping with the belief that the purpose of a college education is to provide opportunities for the student to learn to think for himself and to stand on his own feet.

QUESTIONS

1 Contrast horticulture in ancient times and horticulture today.
2 Since the establishment of agriculture and horticulture experiment stations throughout the world, the literature of the science of horticulture has increased rapidly. Explain.
3 Present-day horticulture may mean one thing to one person and another thing to another. Explain.
4 Student A plans a career in the management of a contiguous group of commercial apple orchards, and student B plans a career in the the genetics and physiology of the apple. Name three courses which both students should take and three courses which each student should take relevant to the particular field of interest. Give reasons for each.
5 Student A plans a career as field manager of a large vegetable-crop canning company, and student B plans a career as manager of a vegetable growers' cooperative marketing association. Name three courses which both students should take and three courses which each student should take relevant to the particular field of interest. Give reasons for each.
6 Given two joint owners of a large ornamental nursery. One is primarily interested in the growth and production of the plants, and the other is primarily interested in the marketing of the finished product. Outline a course of study for each of the two partners. Give reasons for your selections.
7 Career opportunities in the processing and consumer packaging of fruit and vegetable crop products are increasing. Explain.
8 A profound and abiding interest in crop plants is an essential requirements for success in all phases of horticulture. Explain.

SELECTED REFERENCES FOR FURTHER STUDY

Brochure on career opportunities in horticulture available at your local landgrant college or university.

Dowdell, D., and J. Dowdell. 1969. *Careers in horticultural sciences.* New York: Julian Messner. A discussion of career opportunities in the wide field of horticulture.

Tukey, H. B. 1965. The image of horticulture. *Proc. Conf. Undergrad. Ed. in Hort. Sci. of the Amer. Soc. Hort. Sci.* Mt. Vernon, Va. A forthright discussion of the various aspects of horticulture—as a science, an art, a business, a profession, and as a hobby.

Index

Page references followed by *t* indicate tables.